ORIGINS AND EVOLUTION
BOEING B-47 STRATOJET & B-52 STRATOFORTRESS

SCOTT LOWTHER

TEMPEST BOOKS

Published in Great Britain, 2021
by Tempest Books
an imprint of Mortons Books Ltd.
Media Centre
Morton Way
Horncastle LN9 6JR
www.mortonsbooks.co.uk

Line art copyright © Scott Lowther, 2021
3D art copyright © Rob Parthoens, 2021

All rights reserved. No part of this publication may be reproduced or transmitted in any form or by any means, electronic or mechanical including photocopying, recording, or any information retrieval system without prior permission in writing from the publisher.

ISBN: 978-1-911658-76-4

The right of Scott Lowther to be identified as the author of this work has been asserted in accordance with the Copyright, Designs and Patents Act 1988.

Typeset by: Burda Druck India Pvt. Ltd.
Printed and bound by: Gutenberg Press, Malta

Cover artwork by: Rob Parthoens

Contents

Introduction	005
Chapter 1 – B-47 Evolution	007
Chapter 2 – B-47 Development and Projects	043
Chapter 3 – B-52 Evolution	100
Chapter 4 – B-52 Development	165
Chapter 5 – B-52 Projects	188
Chapter 6 – B-52 with Nuclear Propulsion	212
Chapter 7 – B-52 Miscellaneous Designs	225
Chapter 8 – B-52 as a Carrier	236
Chapter 9 – B-52 Re-Engining Studies	314
Chapter 10 – B-47 and B-52 Derived Transports	323
General data tables	370
References	376
Index	382

Introduction

As hardly needs repeating, the Boeing B-52 Stratofortress is an icon of American air and military power. Developed just after the Second World War, it has been in service for more than 60 years – and there is every reason to believe that some of them may still be lumbering through the skies a full century after the breed first flew. This could be because the design is so supremely magnificent that it cannot be readily replaced… or because the American military procurement system is so mired in bureaucracy and expense that keeping aged aircraft flying well past their retirement date has simply proven cheaper than getting something new. There is validity to both views.

The Boeing B-47 Stratojet paved the way, embodying advanced technologies and becoming the archetype of the sleek swept wing jet aircraft. Yet the B-52, perhaps curiously, began life not as an enlarged clone of the B-47 but as a much less forward-thinking aircraft. In its earliest incarnations, the B-52 was straight-winged and dragged through the air with propellers. That B-52 would doubtless have had a far shorter lifespan than the plane that was actually produced.

The development of the B-52 went on for a number of years however, years in which aeronautical knowledge was advancing rapidly. What was learned on the B-47 was transferred to the B-52. And newer learning still was added, leading to the design that Boeing eventually sent skyward.

A handful of the steps taken by Boeing's designers to produce the B-47 and the B-52 have appeared in various publications over the years, but the actual evolution of both aircraft, even before the initial prototypes were made, was a process involving dozens, even hundreds, of design changes small, large, and utterly total.

This book attempts to flesh out those evolutionary charts a little further, though it is by no means complete. As well as the evolution from first concepts to prototype, the various major variants that were built are included, as are a number of unbuilt projects for derivative designs. Unsurprisingly, there are more of these available for the B-52 than for the B-47; the B-47 was a fantastic aircraft, but it was a first generation jet and was quickly rendered obsolete. The B-52 has had many decades to rack up studies.

Some truly remarkable derivatives proposed have been proposed for the B-52, including nuclear propulsion, liquid hydrogen fuel and even supersonic variants. In service, the B-52 has proven itself to be a capable carrier of other aircraft, having served as a launch pad for everything from small remotely piloted gliders, to the hypersonic X-15 rocket plane to the orbital Pegasus.

Many more designs for research and operational aircraft were put forward to use the B-52 as a carrier aircraft, and a number of them are included here.

As was done with *Lockheed SR-71 Blackbird: Origins and Evolution*, the designs have all been recreated here as line drawings, useful for model builders. A source grade is included for each diagram in the data tables at the back to indicate the clarity or otherwise of the vintage diagrams these new images were based upon.

Many of were based on diagrams found in the Boeing Historical Archives; others in reports, papers and seemingly random blueprints found in a multitude of widely scattered sources.

Scott Lowther
aerospaceprojectsreview.com

B-47 Evolution

Two of Boeing's most famous aircraft of the Second World War are the B-17 and B-29 bombers. While the B-17 is certainly an aircraft worthy of merit, the B-29 was something else entirely. With a pressurized fuselage and an analog computer control system for remotely controlled machine gun turrets, the B-29 was the very definition of 'futuristic'. Add in the fact that it was the first aircraft armed with nuclear weapons and the B-29 seems nearly science fiction come to life.

As advanced as the B-29 was at the time however, the propulsion system was entirely conventional: four radial piston engines driving four propellers. When the B-29 was originally designed in 1939, though, there were no practical alternatives. Turbojet engines were in their infancy and rockets had laughably poor fuel efficiency. But as the war progressed, jet engine technology leaped ahead. The British Power Jets WU centrifugal flow turbojet was first run in 1937 and further developed into the Power Jets W.1.

The W.1 was the first British turbojet to power a flying aircraft, the Gloster E.28/39. It was not, however, the first turbojet aircraft to fly – that accolade would go to the German Heinkel He 178. The Heinkel first flew on August 27, 1939, while the Gloster first took to the skies on May 15, 1941. At that time the United States had no official turbojet engine or jet propelled aircraft programmes in active development. The Lockheed L-1000 axial flow turbojet was promising but was at a preliminary level of development and had little official backing.

In April 1941, Major General Hap Arnold of the US Army Air Forces visited Britain and witnessed flights of the Gloster E.28/39. The technology impressed him enough that the following month he officially requested that the British share the technology. General Electric was selected to produce a version of the W.2B/23 engine (a more powerful engine than the W.1, refined and put into production as the Rolls-Royce Welland I to power the Gloster Meteor fighter) as the J31. This was successful and the engine went on to power early American jet aircraft such as the Bell P-59 Airacomet and the Ryan Fireball. This early engine was too small to power a bomber-sized aircraft.

General Electric began development of a new engine entirely of its own design in 1943. The TG-180 was to be an axial flow turbojet, an all-around better form of turbojet than the centrifugal flow J31 and its successor the J33. Thrust was substantially higher than that of the J31… high enough that it could be reasonably envisioned as a bomber engine. As the TG-180 design was finalized, it received a military designation as the J35 and its details were provided to aeronautical companies so they could design aircraft powered by it.

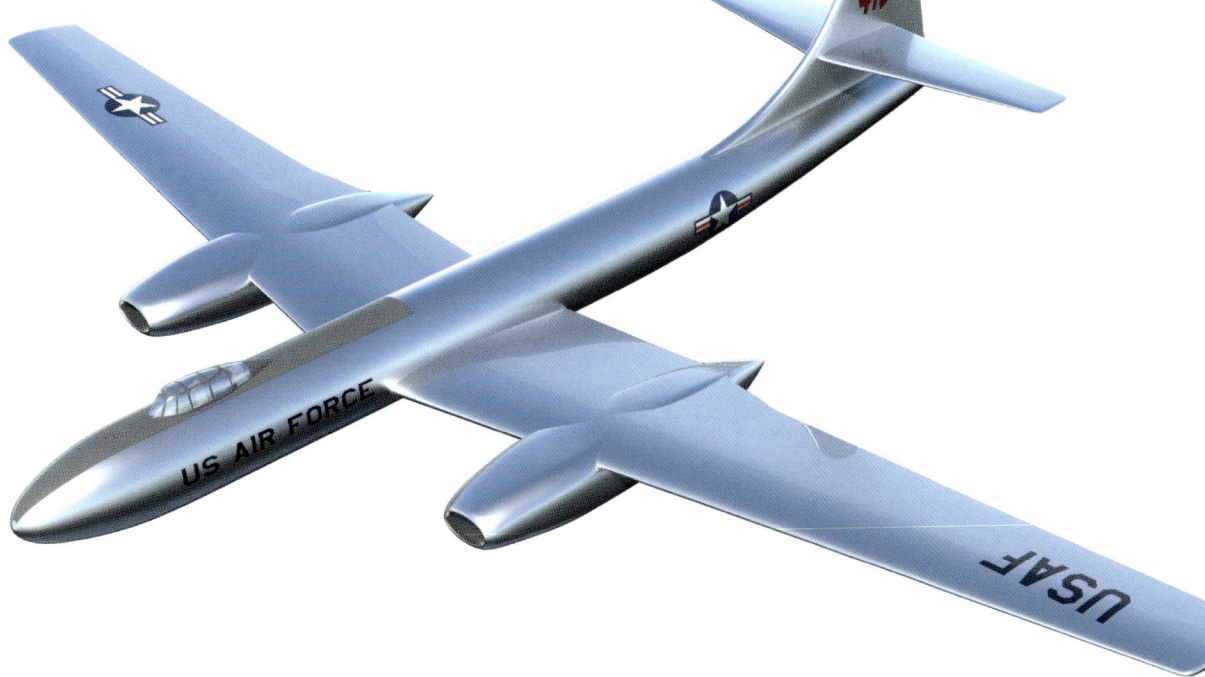

Boeing Model 413
SCALE 1/240

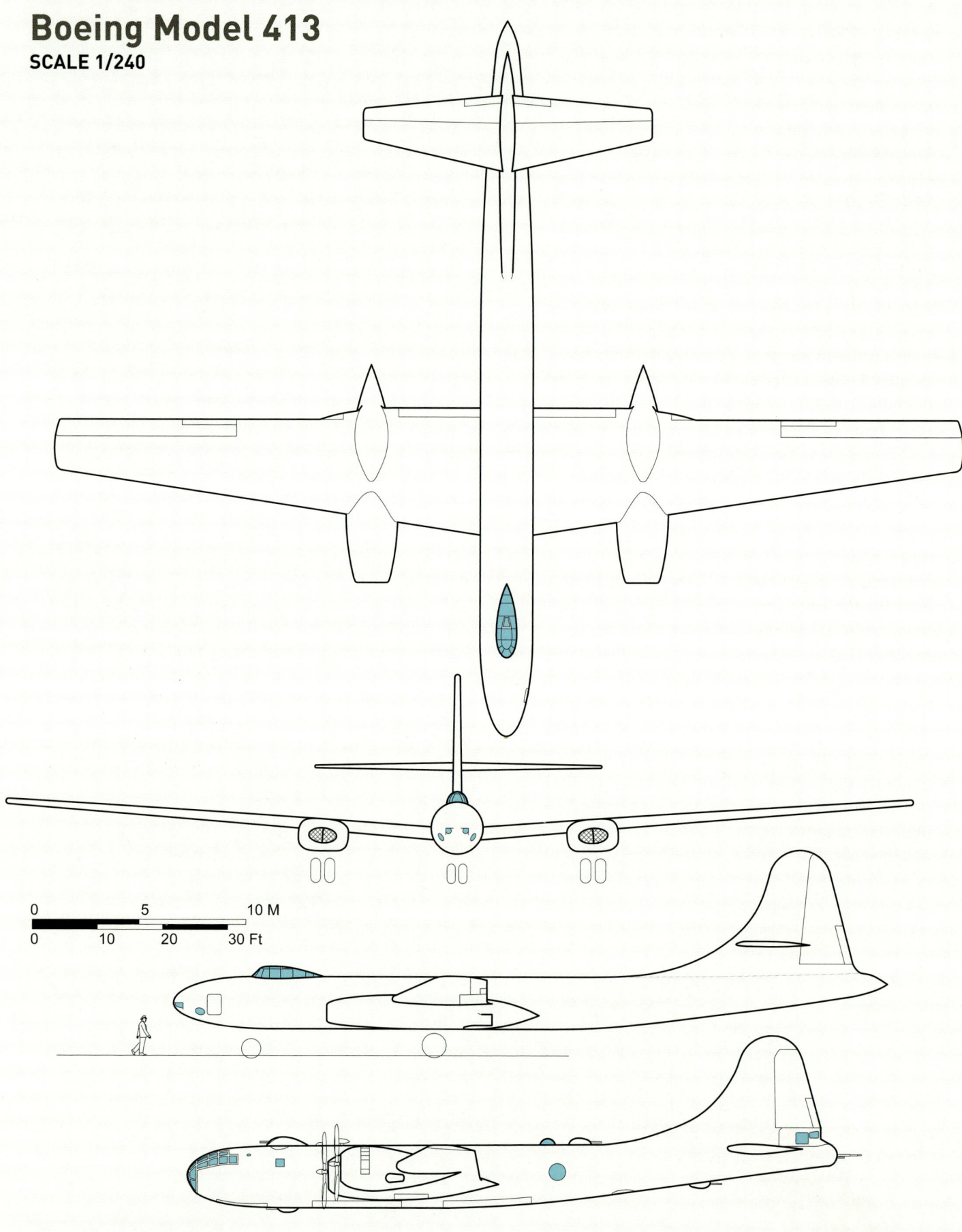

Boeing Model 413

Boeing jumped on the TG-180. The earliest known Boeing aircraft project to make use of this new engine was the Model 413, designed in response to an October 1943 RFP for a jet-powered photo reconnaissance plane. The Model 413, assigned in November 1943 and finalized in early 1944, was more or less a scaled-down B-29 configuration with one dual-jet pod under each straight high aspect ratio wing. The fuselage was somewhat smaller in diameter than that of the B-29 (8ft rather than 9ft 6in), so the pilot – there was no dedicated co-pilot station, rather a radio operator/gunner who could take the pilot's place – was given a raised canopy. It does not appear that the Model 413 was armed with either defensive or offensive weapons. But in order to perform its reconnaissance role it was equipped with a small bomb bay containing six M46 photoflash bombs. These devices contained 25lb of flash powder which would burn for 0.20 seconds, producing a peak light intensity of 500,000,000 candlepower. This was sufficient to light up a ground target at night for photography.

Performance details such as speed and range are not currently available. However, it can be surmised by the lack of a co-pilot that flight durations were reasonably short; doubtless speed was relatively high for the time and range was relatively short. A wind tunnel model of this configuration was tested, sans canopy.

Boeing Models 422, 424, 425, 426

By the time Model 413 was submitted in January 1944, the USAAF's requirement for fast photo reconnaissance planes had already gone to the Hughes F-11 and the Republic F-12. But the Model 413 showed promise and with the prompting of staff at Wright Field the idea was continued with an eye towards becoming a true bomber. Four preliminary aircraft designs were produced in March 1944. Available data on these designs is minimal, though the informal request from the USAAF for a jet bomber called for a top speed of 500mph, a range of 2,500 to 3,500 miles and a ceiling of 35,000 to 40,000ft with an offensive payload of 16,500lb bombs. Four designs were produced at one time, Models 422, 424, 425 and 426. Model 423 was an unrelated low cost transport design.

Model 422 used the basic design of Model 413 but with the four turbojets replaced with two turboprops. It's not clear which turboprop engines these were, but they were meant to power contra-rotating six (total) bladed propellers. This would have resulted in a lower top speed than that of Model 413, but increased fuel efficiency and likely increased payload and gross weight. This allowed for a meaningful bomb load to be carried within two fuselage bomb bays. Additionally, Model 422 was to be fitted with defensive guns in the nose and tail, unlike the apparently unarmed Model 413. The tail turret was much like that of the B-29, while the nose turret included a small raised bubble canopy for the gunner ahead of the pilot's canopy. The aircraft had a tricycle landing gear arrangement with the main landing gear folding up into the fairings aft of the turboprops.

Model 424 was configured much like Model 413 but reduced in size, the fuselage dropping to 7ft in diameter and with shorter fuselage length and wingspan (each about 82% of Model 413's dimensions). It retained the engines and nacelles of Model 413. The nose was substantially recontoured, the underside uplifted to give the bombardier a sizable flat panel of armored glass to peer through. The diagrams seem to show that the pilot's canopy was a single blown piece with an escape hatch cut into the top.

Model 425 was Model 422 modified with Model 413's engine nacelles. In effect it was just Model 413 turned into a bomber. And

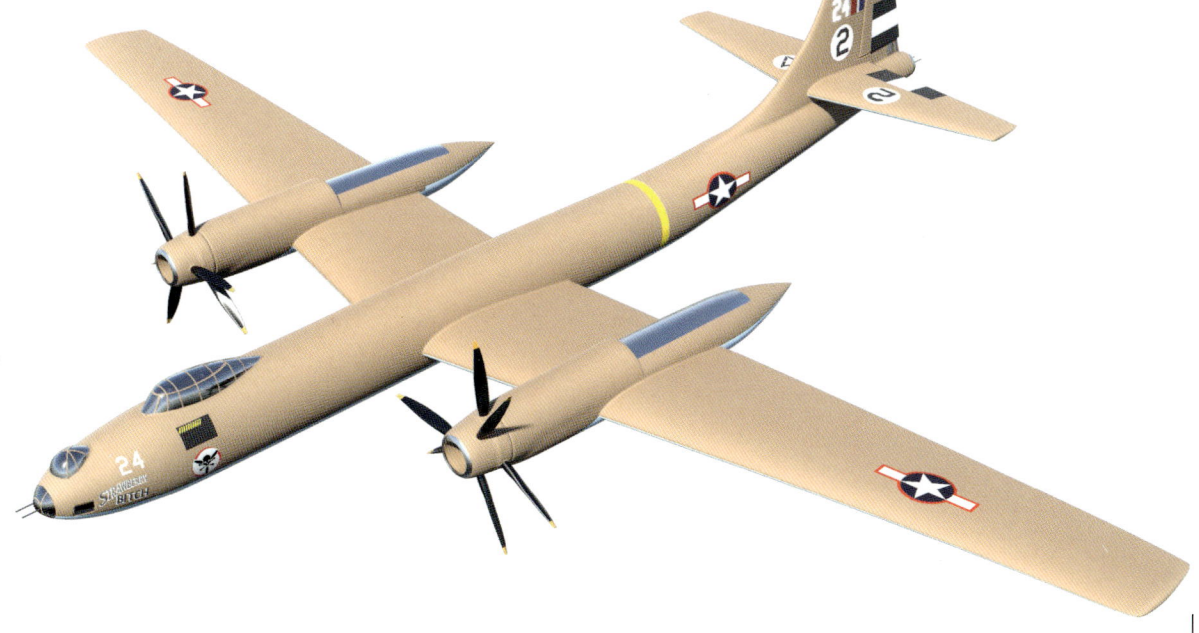

Boeing Model 422
SCALE 1/250

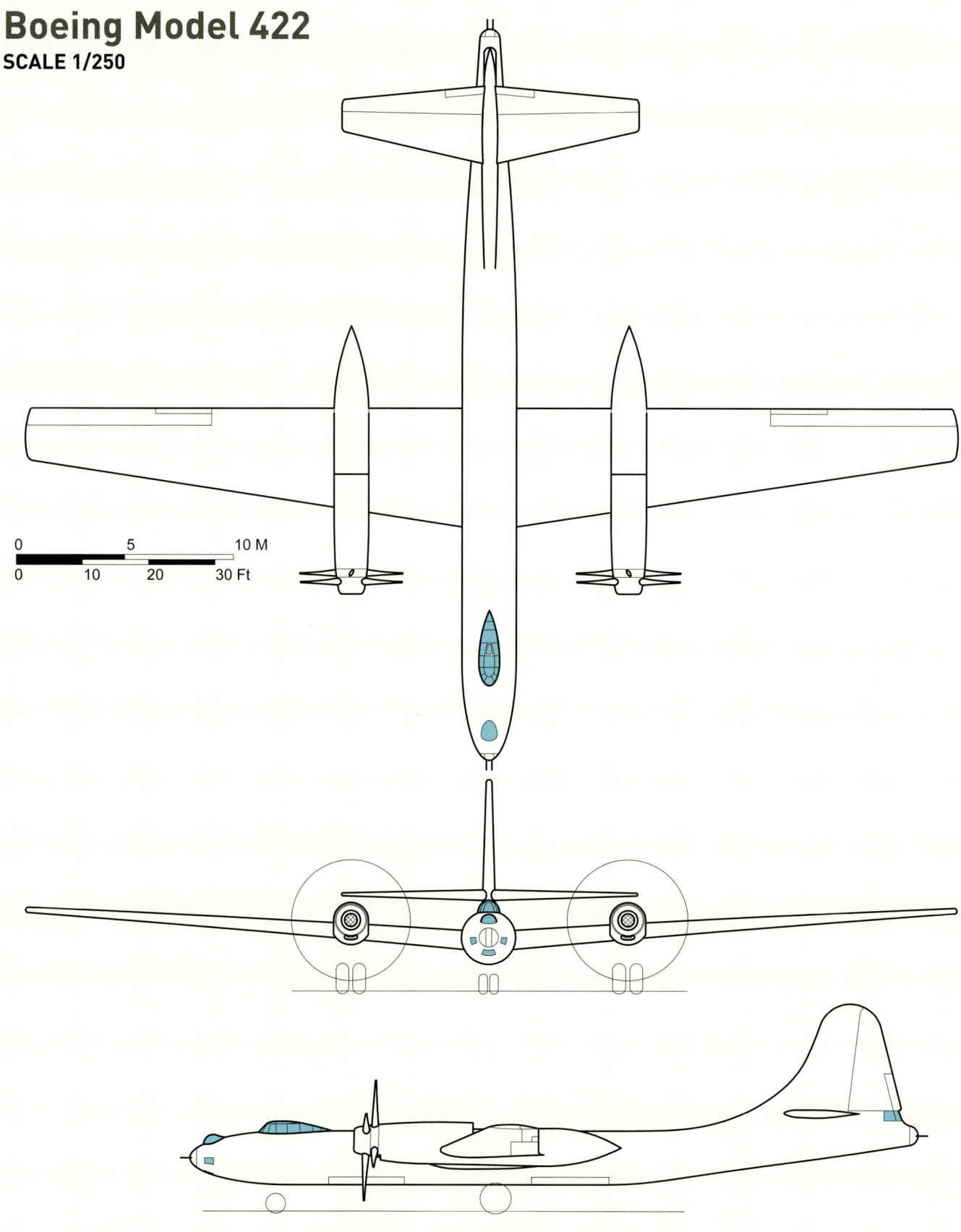

Boeing Model 424
SCALE 1/220

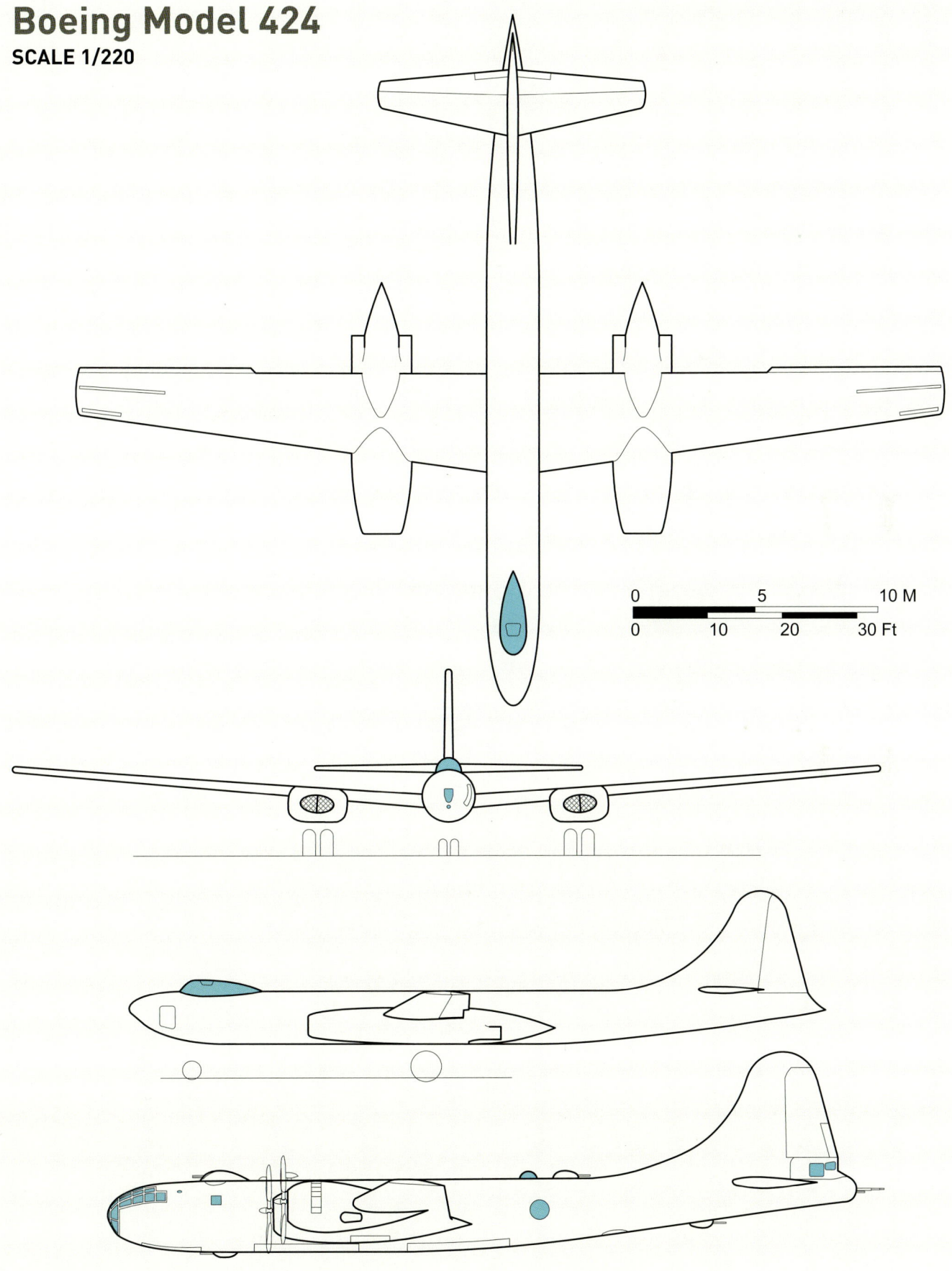

Boeing Model 425
SCALE 1/250

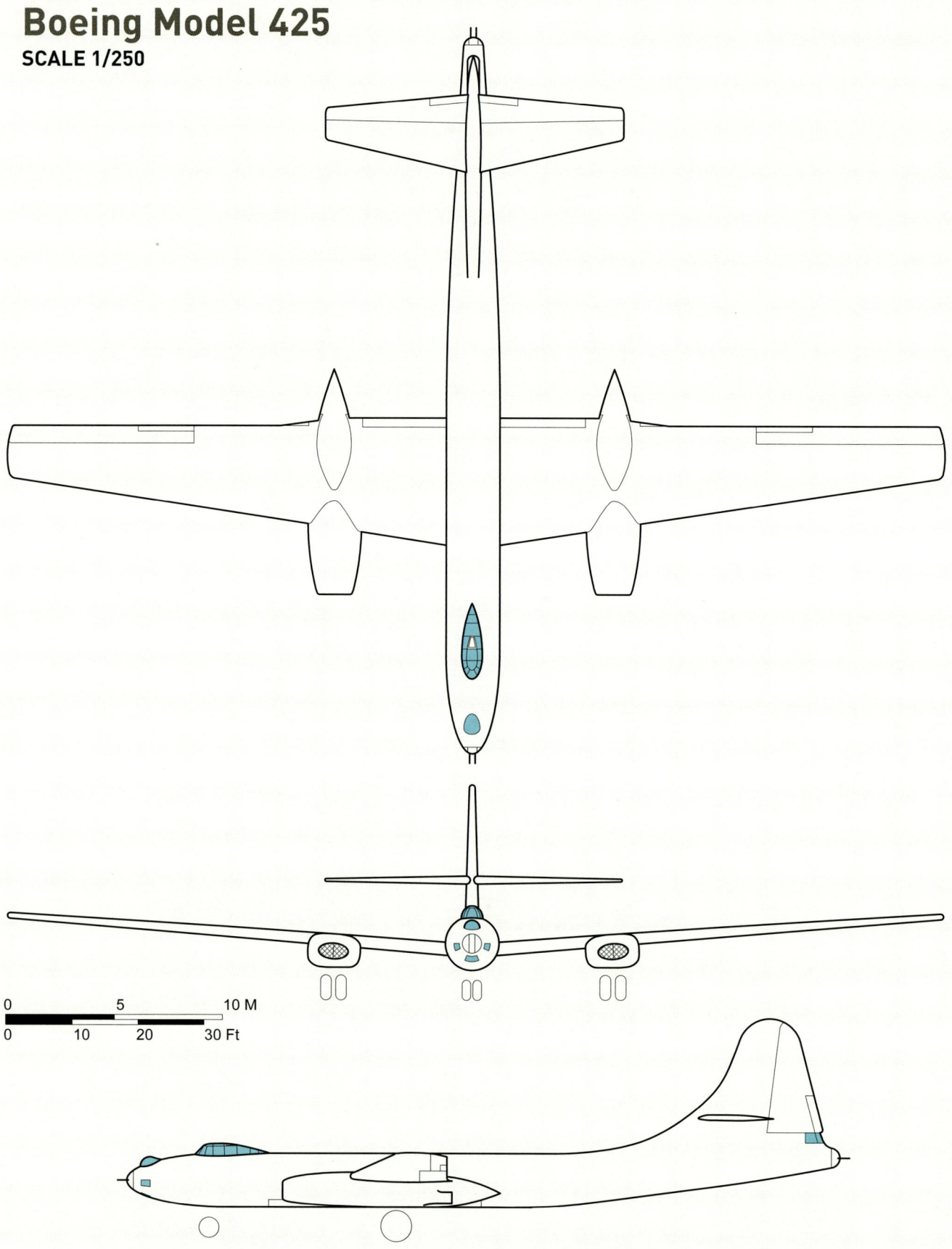

Boeing Model 426
SCALE 1/220

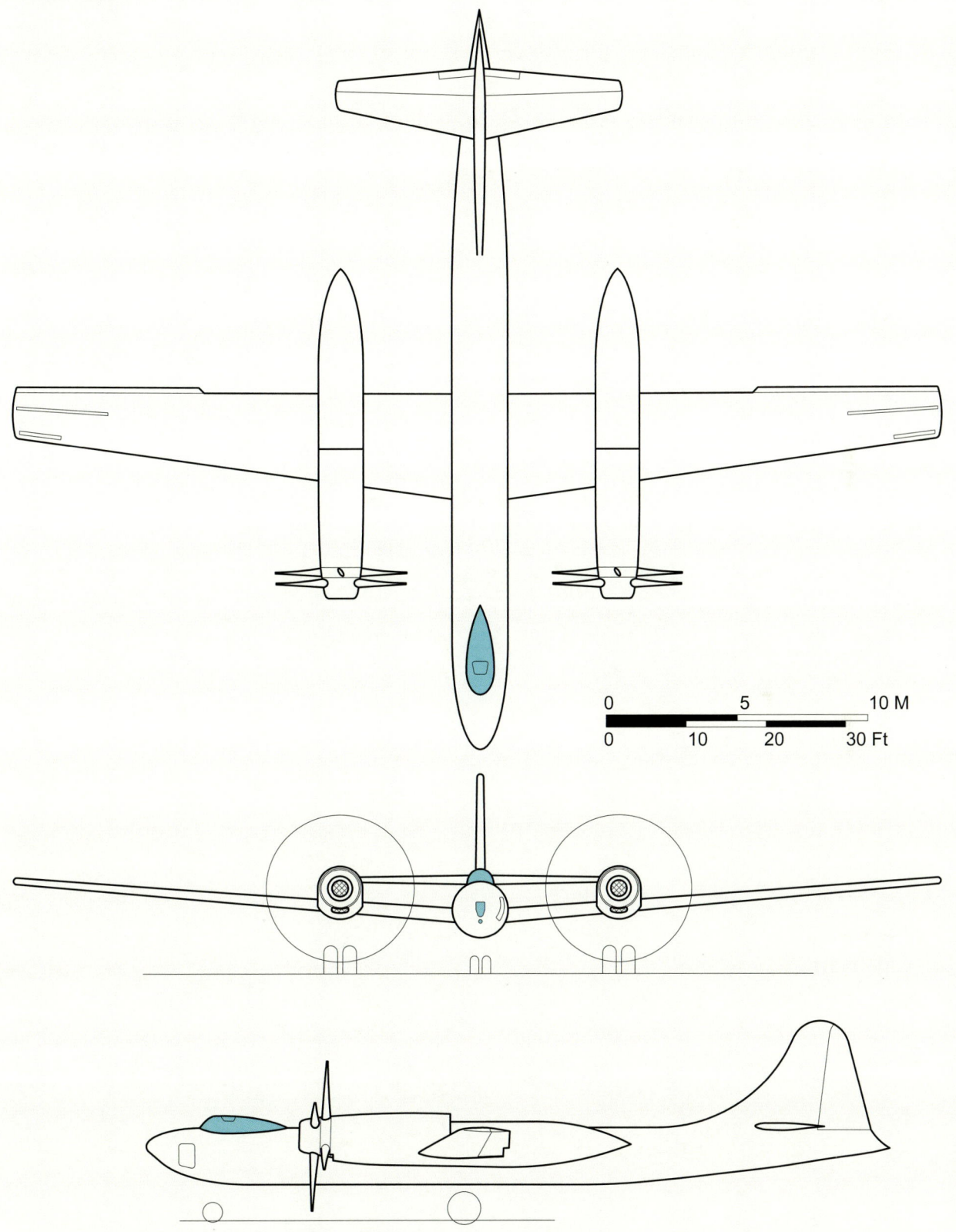

Boeing Model 432
SCALE 1/220

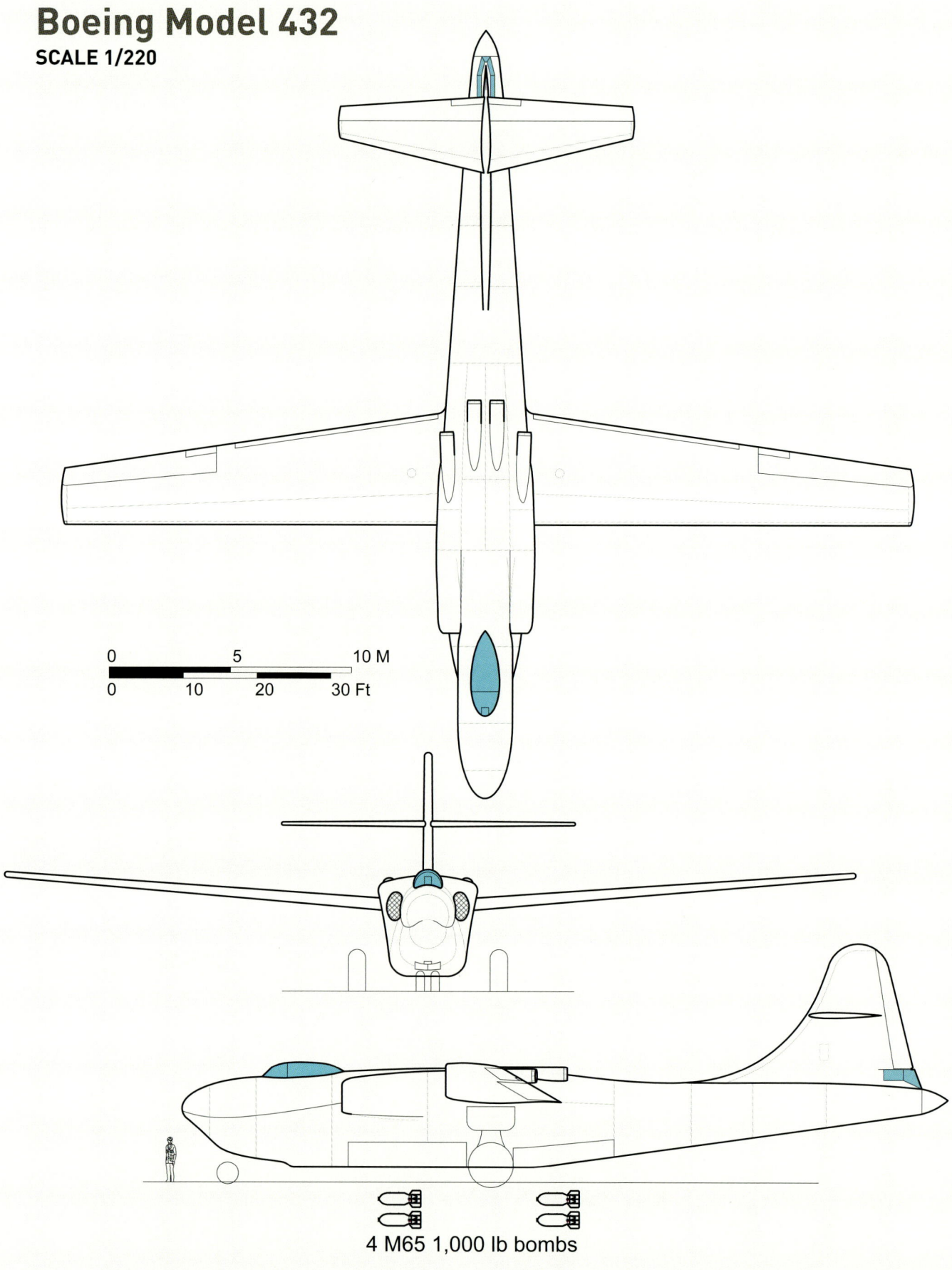

4 M65 1,000 lb bombs

Model 426 was Model 424 configuration and replaced the jet engine nacelles with the turboprops from the Model 422.

These early designs were not everything the Boeing engineers could have wished. They were essentially just the B-29 shape with jet or turboprop engines stuck on it – without taking best advantage of the new possibilities now opened up. Wholly new engines suggest wholly new applications, including new ways to integrate them into the airframe. Wind tunnel testing showed that the dual nacelle merged directly into the wing produced more drag than expected. So the Boeing designers went back to the drawing board with a clean slate.

Boeing Model 432

Rather than having the turbojets hanging from the wings, in September 1944 the Boeing designers of the Model 432 decided to try integrating them into the fuselage. This left the wings utterly clean, undisturbed by engine nacelles, inlets, exhaust and landing gear. The four TG-180 turbojets were lined up side-by-side on the upper surface of the fuselage, fed by a pair of shoulder inlets. The engines would exhaust along the upper surface of the fuselage, doubtless raising issues with heat and shear. The fuselage had a rather bulbous nose, a distinctly non-circular cross-section and a number of odd contours. The tail was very much in the family of the B-29. The result was an aircraft with a remarkably clean and aerodynamic wing, but a fuselage that looked like a mutant pig. It's a strange bit of engineering dissonance, where a wonderfully aerodynamic wing was married to a fuselage that seemed opposed to clean airflow.

Model 432 was to be fitted with a tail turret with two .50 machine guns with 500 rounds each. The design bomb load was 8,432lb, but by offloading fuel the bomb load could be substantially increased to 1 x 10,000lb bomb, or 4 x 4,000lb bombs, 6 x 2,000lb bombs, 18 x 1,000lb bombs or 24 x 500lb bombs.

As odd as Model 432 seems, in December of 1944 Boeing submitted it to the USAAF. The design was approved and in January 1945 was given the military designation XB-47. Funds were approved to begin development and work on the B-47 was under way.

Boeing Model 446

Studies around Model 446 are little known. This was a derivative of Model 432 with six turbojets rather than four; wind tunnel tests were carried out to study alternate locations for the fifth and sixth turbojets. Currently only photos of wind tunnel models are known, showing relatively large nacelles added to the wingtips, to the rear of the fuselage and another configuration with the same nacelles stuck on the ends of the horizontal stabilizers. What advantage there would be in engines in that position is unknown; perhaps it was thought that it would be useful from a weight balance point of view. Additionally engines at the ends of the horizontal stabilizers would be closer to the aircraft centreline than wingtip engines. At this early stage in the development of jet aircraft, some design eccentricities can be forgiven. The need for two additional engines must have been clear as six engines became standard on XB-47 designs moving forward.

Things change

Within a few months, captured German data on high speed aerodynamics started to become available to the designers of the XB-47. In April of 1945 American forces liberated the German town of Völkenrode, a seemingly sleepy little burg valuable for just one thing: the Hermann Göring Research Institute. There German aerodynamicists had carried out studies of high

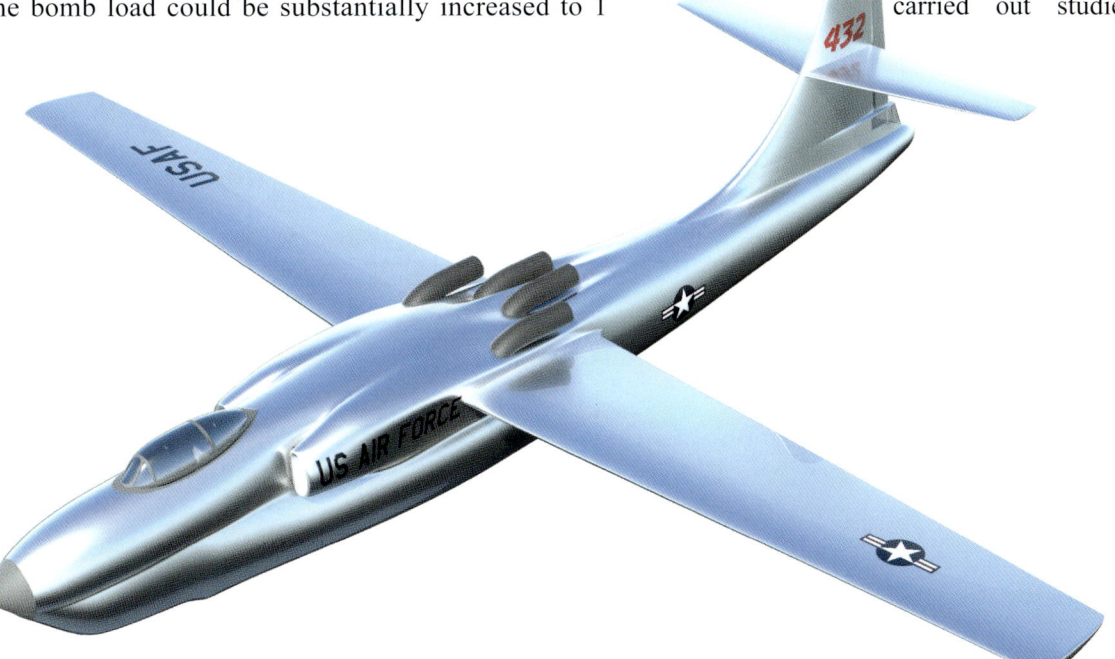

Boeing Model 446
Wind Tunnel Model
SCALE 1/250

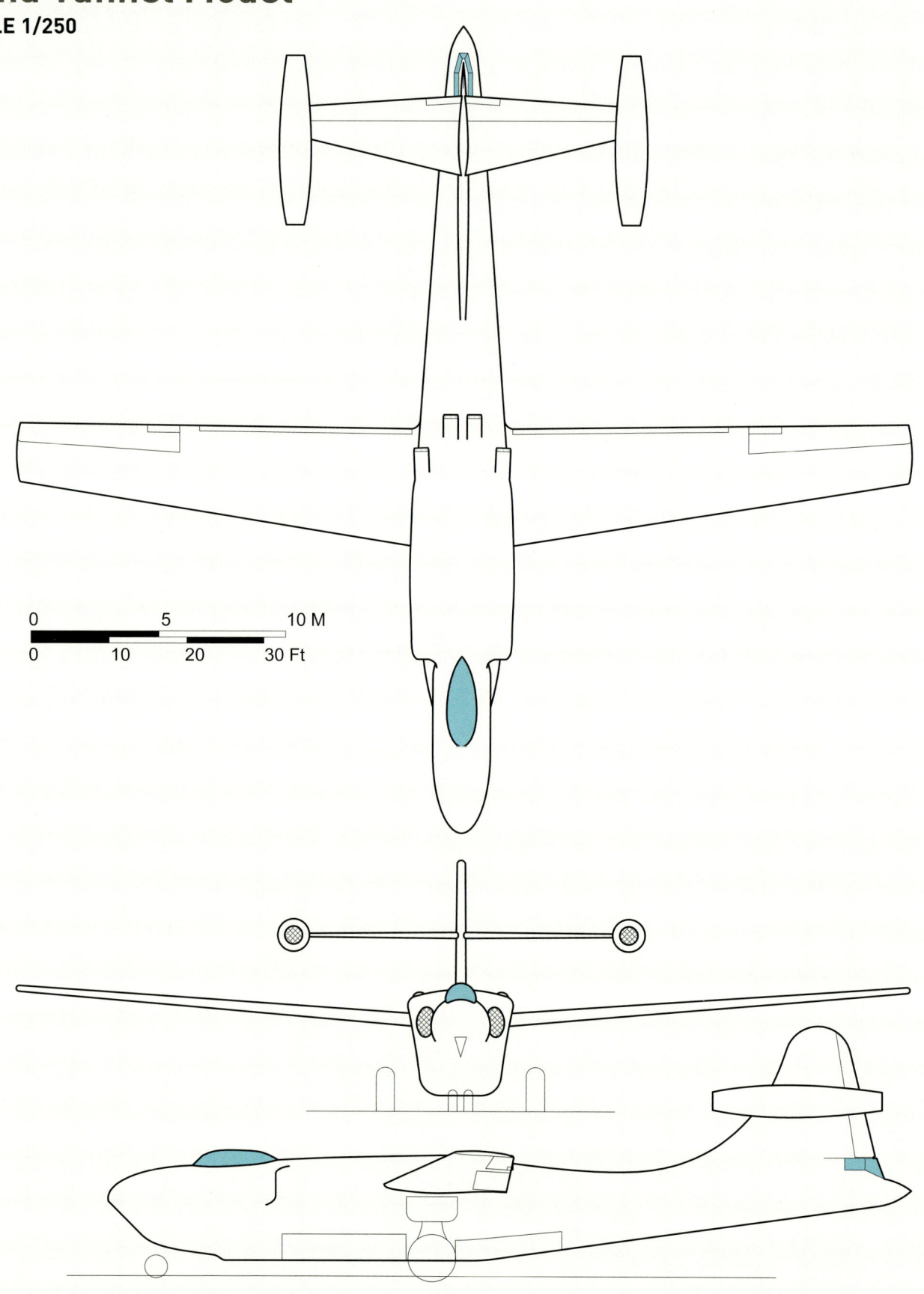

Boeing Model 446
Wind Tunnel Model
SCALE 1/250

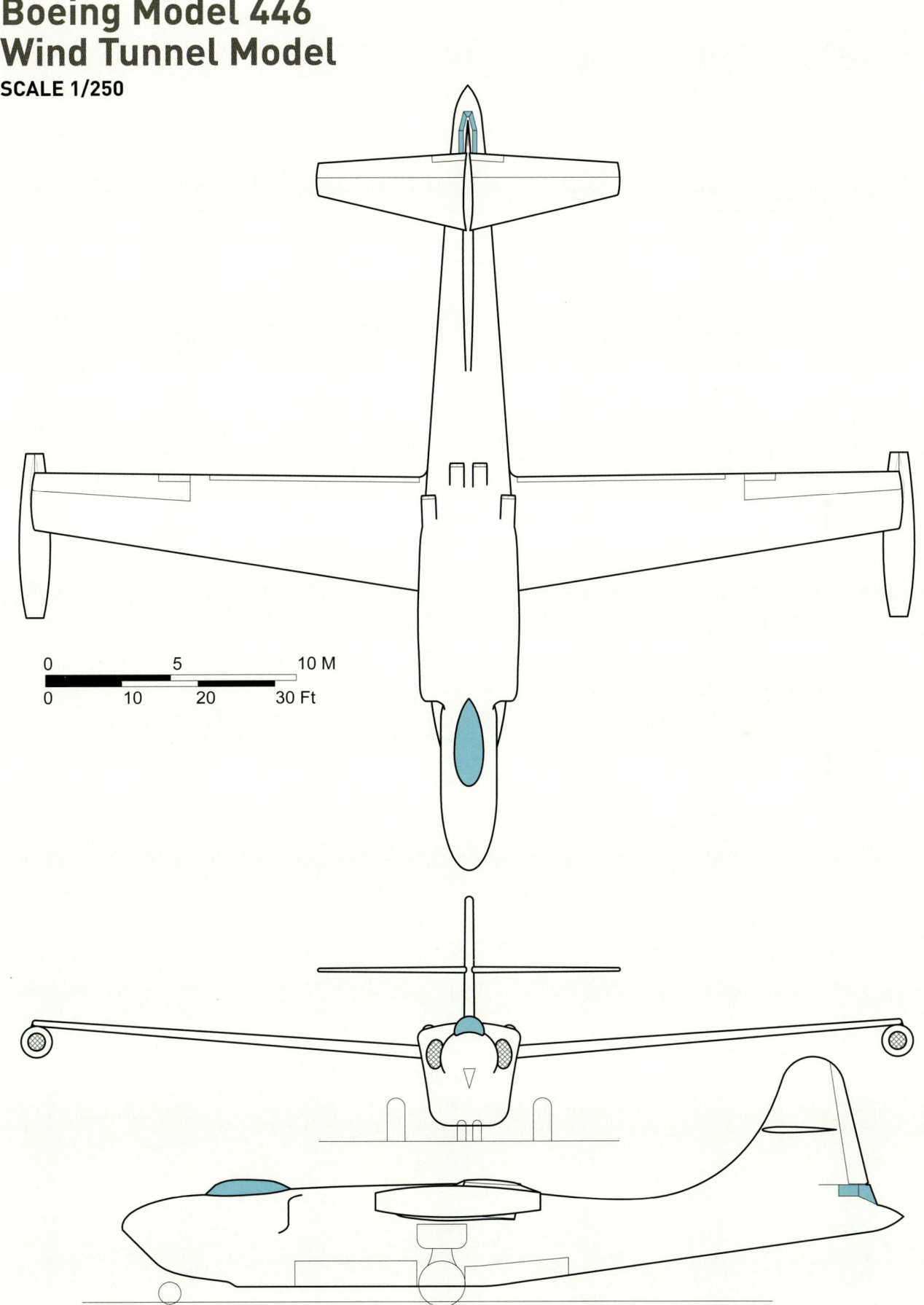

Boeing Model 446
Wind Tunnel Model
SCALE 1/250

0 5 10 M
0 10 20 30 Ft

speed aerodynamics and their data fell into American hands. At the time, those Americans included an aerodynamicist from Boeing.

German aerodynamicists had expended considerable time, effort and resources in studying models of wings of varying sweep and German aeronautical designers had begun to incorporate that information into their new designs. There are innumerable books, articles and websites describing – to variable degrees of accuracy – the designs of 'wunderwaffen' that Germany was working on in order to dominate the skies, but it is undeniable that swept wings and their high-speed advantages were well understood.

Many designs were produced to take advantage of that. One that has become well known and which was doubtless made available to Boeing was a Focke-Wulf design from August 1944 for a single-seat bomber. This aircraft, which has become known post-war as the 'Focke-Wulf 1000 x 1000 x 1000', was designed to carry a 1,000kg payload 1000km at 1000km/h. It had a sleek fuselage with complex curvature, but more importantly, distinctly swept wings. Swept-back wings were, as it turned out, the shape of the future.

A normal airfoil is relatively flat on the underside and rounded or humped on the upper side. Consequently, as air flows over the airfoil cross-section shape of a wing, it briefly speeds up. His increase in speed is correlated with a decrease in pressure. The difference between the higher pressure on the underside of the wing and the lower pressure on the upper side is what produces lift. This is all fine and good but as an aircraft approaches the speed of sound a slight increase in airspeed over the top of the wing can mean that the air is moving at or faster than the speed of sound, relative to the wing. When this happens, drag can increase substantially as well as other unfortunate flow instabilities.

A way to delay this was discovered in the mid-1930s by German aerodynamicists Albert Betz and Adolph Busemann, professors of aeronautics at the University of Göttingen. When a wing is swept back, the development of a supersonic shock wave is delayed, so that a swept wing at transonic speeds has notably lower drag than a straight unswept wing. When this fact was revealed by Betz and Busemann in a 1935 conference, few took much note as aircraft of the time were far from attaining transonic, never mind supersonic, speeds. But by the start of the Second World War it was a different story, and by the last year of the war the Germans were fielding aircraft that were dangerously close to transonic and seriously studying supersonic aircraft. The advantages of swept wings were thus of great importance and were highly studied. The Focke-Wulf bomber took full advantage of this development.

The Focke-Wulf design had not only swept wings but also swept tail surfaces. Additionally, the fuselage had a sudden reduction in width where the wings attached to it. This would have resulted in a form of 'area ruling' which also has definite aerodynamic advantages at high speed. It had two conventional turbojets just inboard of mid-span, hanging underneath the wings. The turbojets are notable for being angled outwards; this would have increased drag and slightly reduced effective thrust, but in the event that one engine failed – a not uncommon event with first generation jet aircraft – the remaining engine would have a better thrust vector and could thus propel the aircraft more effectively. The landing gear was unusual in that it had tricycle arrangement with the main gear folding up into the wing roots and the nose gear folding up into the nose of the fuselage... but also a fourth wheel folding up into the central fuselage. This fourth wheel was slightly larger than the wing wheels and was presumably meant to take up the weight of the single SB 1000 bomb.

This aircraft was designed at only a preliminary level, but its appearance and features were remarkably prescient. Numerous aspects of it would show up in the final B-47 design. Whether or not these details were intentionally copied is unclear, but it is unlikely. Boeing certainly made use of the data coming in from Germany, but the company – along with the National Advisory Committee for Aeronautics, other American aviation companies, and various universities and scientific institutions in America – also carried out innumerable wind tunnel tests of their own. The German findings were not simply taken as divine revelation, but were proven and understood. The fact of aft-swept wing drag reduction was sufficiently important that the XB-47 went back to the drawing board in order to take advantage of it.

Swept Wing Model 432

A wind tunnel model of Model 432 was modified with swept wings. Based on photos, the wing used was probably an existing 432 wing, repositioned and with modified wingtips. The dihedral of the Model 432 wing was turned into a slight anhedral and the tail surfaces appear to have been unchanged. This model used the standard four engines of Model 432, rather than the six of Model 446. This may have been only a preliminary study to see what aerodynamic benefits could be had by changing an established design, rather than a serious engineering study.

Model 448

Model 448 of September 1945 was an outgrowth of Models 432 and 446. Aerodynamically and structurally refined from the earlier concept, Model 448 was built along lines previously established but appears

Focke-Wulf 1000x1000x1000 A
SCALE 1/90

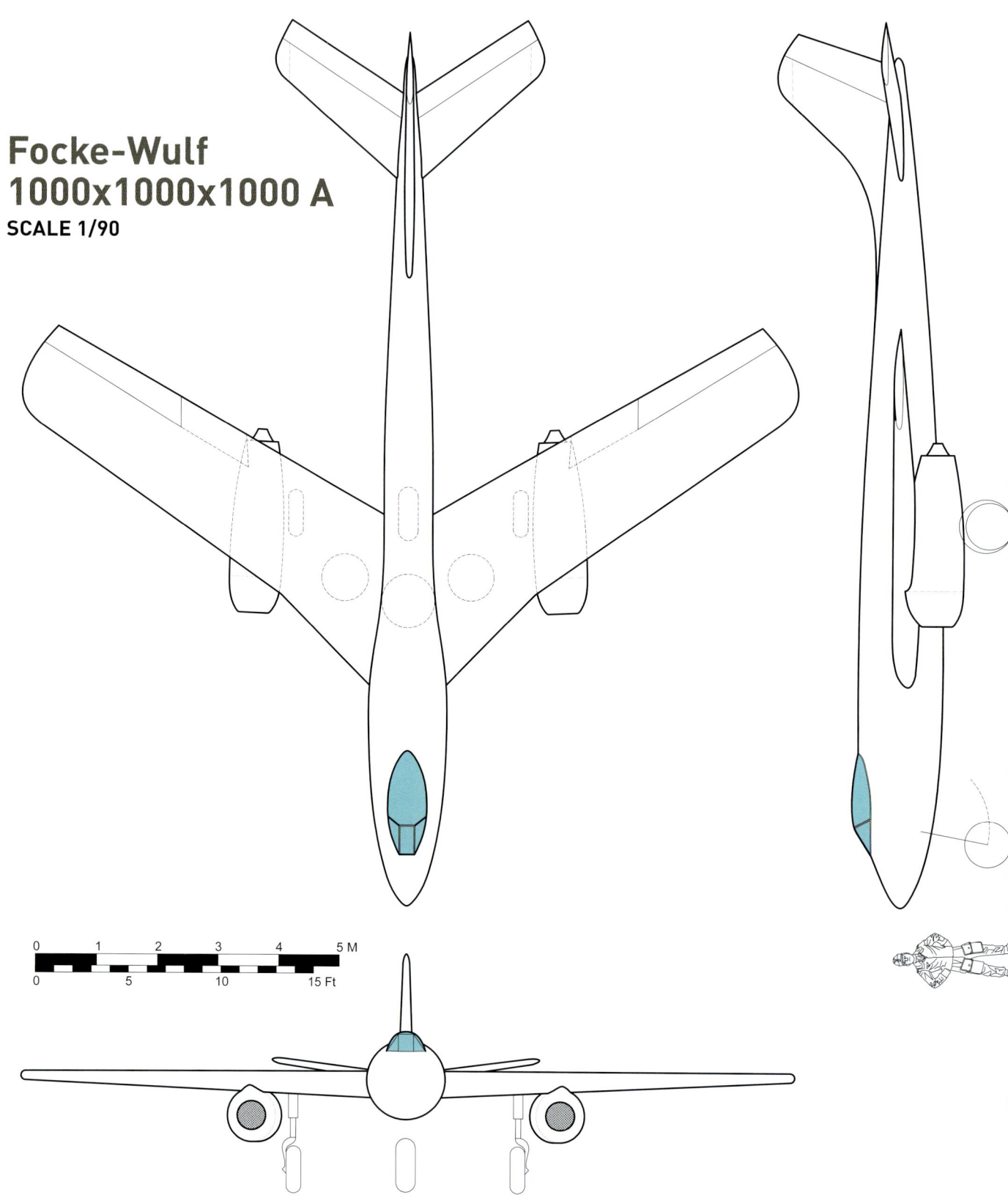

substantially more advanced and futuristic. Four of the 4,000lb thrust TG-180 turbojet engines were fuselage mounted just aft of the cockpit, in a position similar to those on the Model 432 (but substantially further forward) and two were mounted in the tail cone, similar to engine positions tested on Model 446.

The forward engines were fed by a common inlet in the nose, forming what looked like the gaping mouth of a whale shark. The inlet bifurcated, the division creating a passageway between the cockpit and the bombardier's position. The crew compartment was pressurized. The rear engines were faired into the tail,

Boeing Model 432/448
Hybrid Wind Tunnel Model
SCALE 1/220

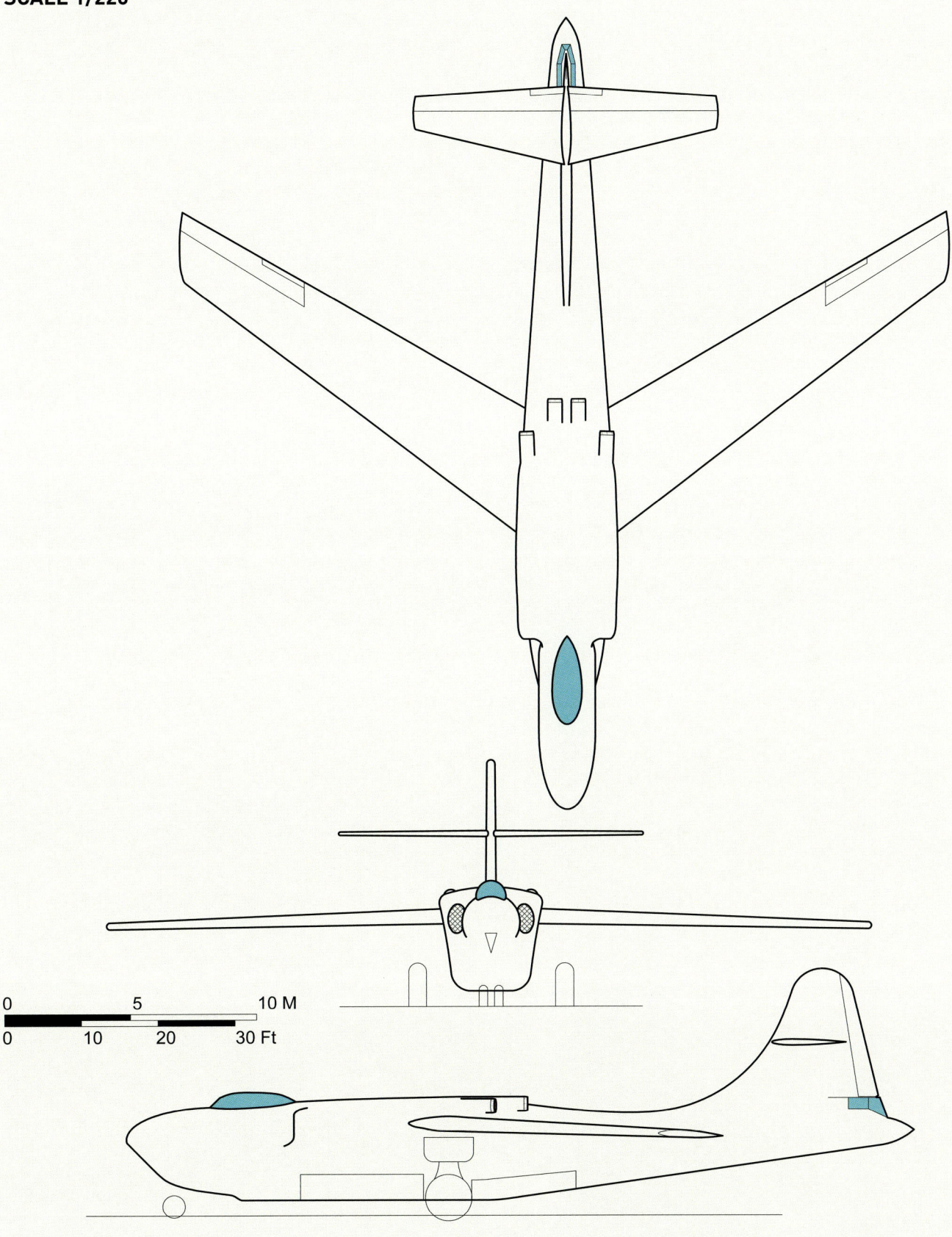

with the flush inlets being long and somewhat complex. This left the wings remarkably clean and swept aft, as had been studied with the modified Model 432 model. The vertical stabilizer was still remarkably reminiscent of that of the B-29, but for the first time the horizontal stabilizers were also aft-swept.

Model 448 used a conventional tricycle landing gear arrangement, all retracting into the fuselage. To aid in navigation and accurate bombing, a radar unit was located under the nose, just forward of the nose gear.

Defensive armament for Model 448 included two .50 calibre machine guns in the tail (contained within a jettisonable tailcone and equipped with radar scanner; the gunner controlled the guns remotely from the forward fuselage) with 600 rounds each. Offensive armament could be 16 x 500lb bombs, 8 x 1,000lb bombs or a single 22,000lb 'Tallboy' bomb.

Putting the engines on the top of the forward fuselage produces an immediately obvious problem: the jet exhaust blasting along the surface of the fuselage. Not only does this lead to the risk of structural damage through blast and heat, it also somewhat reduces engine performance. This odd design feature was inherited from the previous design, Model 432. The Army Air Forces were understandably skeptical of the concept. At this point in the development of turbojets, technology was advancing rapidly and new engine designs were being constantly proposed. If Model 448 had been built and new engines were introduced, integrating them into the airframe could have been problematic. And putting the engines atop the fuselage would have complicated maintenance, something early jets needed a lot of. Perhaps worse, if an engine were damaged – from bullets, shrapnel or simple mechanical failure – the engine would project flames and debris into the fuselage itself. Boeing had to go back to the drawing board. The fuselage-mounted engines were relocated to underwing pods in the Model 450… a big step towards the final B-47 configuration.

Model 450

The next design was a major step forward. The Model 450-1-1 (still officially designated the XB-47), dating from October of 1945, took the swept wing of Model 448 and married it to an entirely new fuselage. It had the same six engines as Model 448, but relocated them away from the fuselage. Each wing had a dual-engine pod suspended beneath with a pylon and another engine stuck to the wingtip. With this design, the engine exhausts came nowhere near to impinging upon the aircraft structure. A catastrophically failed engine would also pose a much lesser risk to the aircraft, though the loss of a wingtip engine would create a definite thrust imbalance.

This engine arrangement had a structural benefit. The Model 448 wing was clean, with the only weight hanging from it being the fuselage. The wing was thus a single cantilevered beam, all of the weight concentrated at the wing root. But Model 450-1-1 suspended two engines and the associated nacelle and pylon at about the one-third span point, and another engine at the wingtip; this helped to spread the load across the span of the wing and modestly reduced the strain on the wing root. This advantage was especially important given that the wing of the model 450-1-1 was not only swept 37.5° but, compared to similarly sized but earlier bombers, very thin at 12%. The tail surfaces – ALL of the tail surfaces – were at last aft-swept as well.

Model 450-1-1 is clearly recognizable as the B-47. But it had a number of important differences from the aircraft that was eventually built, many as a result of the April 1946 inspection of the full-scale plywood mockup. It was generally well received by Army Air Forces personnel, but a few suggestions were made. In particular, the tricycle landing gear that Boeing had designed for Model 450-1-1 was complex,

Boeing Model 448
SCALE 1/150

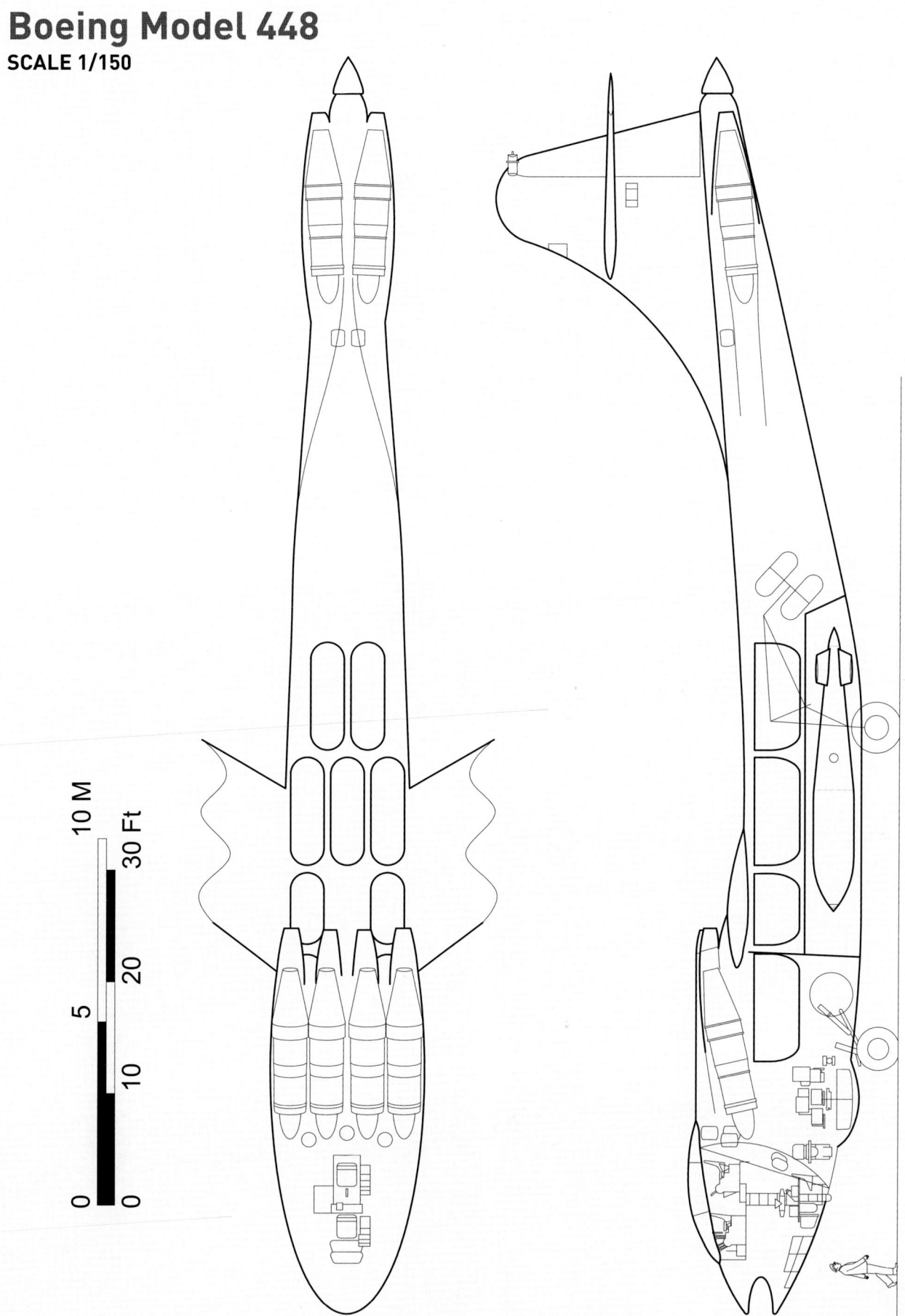

Boeing Model 448
SCALE 1/220

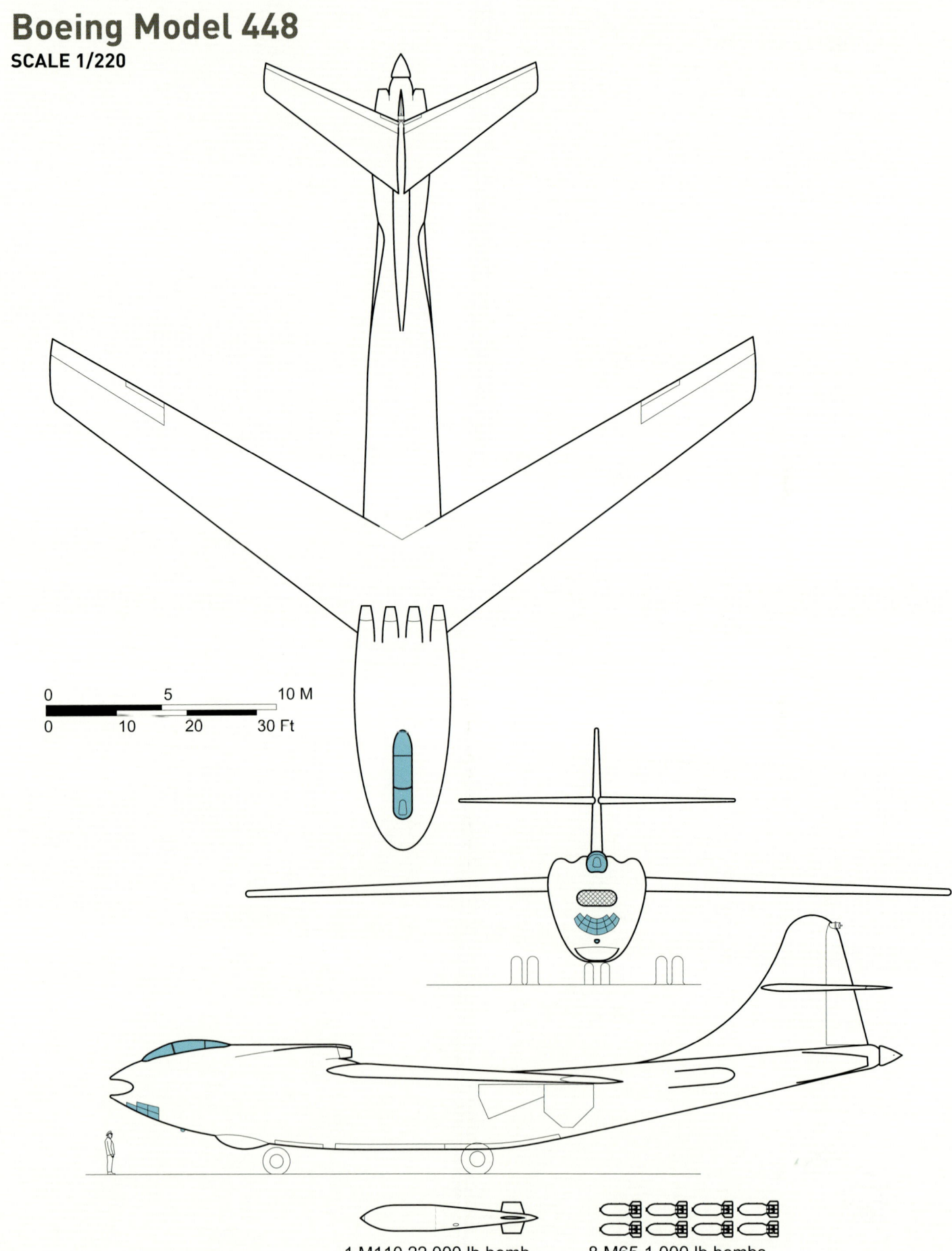

1 M110 22,000 lb bomb

8 M65 1,000 lb bombs

heavy and consumed substantial volume within the fuselage. In fact, it required so much volume that the fuselage could not contain the gear; blisters on the exterior surface showed where the main wheels projected through what would have been the undisturbed outer mold line. The fuselage was also of constant cross-section for a considerable length; this made construction easier, but was not aerodynamically optimal.

A single long bomb bay could contain the multiple smaller bombs – or one giant 'Tallboy' bomb – that the XB-47 carried. But instead of monolithic doors as long as the bomb bays, the XB-47 was equipped with multiple shorter-length doors that would open as needed for the weapons being deployed. For defence it had a remotely operated tail turret with two .50 calibre machine guns. The gunner was given a periscope above the forward fuselage (just at the rear of the raised cockpit canopy) and one directly below, projecting through the fuselage underside. A radar scanner was planned for eventual inclusion within the tail cone to aid in aiming.

Further examination showed that increased wing area was needed, so the span was increased by 16ft. This slightly increased range, but also substantially reduced the takeoff run and raised service ceiling. As the wings were extended, the wingtip nacelles were moved to below the wings and somewhat inboard. These changes produced the Model 450-2-2. Apart from the wing/engine changes and some additional windows on the nose, Model 450-2-2 appears little different from Model 450-1-1.

The next design iteration, unsurprisingly Model 450-3-3, incorporated the suggested changes and recent updates. The tricycle landing gear was dispensed with. Knowingly or not, the Boeing designers took another feature from the wartime Focke-Wulf design: the Model 450-3-3 had a 'bicycle' landing gear arrangement. In Boeing's case, the new aircraft had a pair of conventional nosewheels, a pair of main wheels stowed in the fuselage slightly aft of the midpoint, and quite small 'outrigger' gear which would deploy from the dual-engine nacelles. The outriggers were meant to provide stability on the ground rather than take up substantially landing loads.

The engine nacelles themselves changed. The length of the nacelle was reduced; the single elliptical inlet that fed both engines was replaced with two independent inlets. A fairing was added underneath the new inlet to stow the outrigger gear.

With the main landing gear simplified, the fuselage was refined. The change in contour was not extreme, but it did make the fuselage look more elegant. At last, the XB-47 had moved almost entirely away from its pre-jet technology forebears. It was a modern, optimized jet-propelled aircraft. Compared to the piston engine bombers of the Second World War, it was gloriously futuristic. And as it turned out, the Boeing design was also gloriously futuristic compared to its direct competitors.

When Boeing's designers began looking at turbojet propelled bombers, they were hardly alone. Secret as the technology started out, the entire American aircraft industry soon wanted in.

Douglas XB-43

America's first jet bomber is somewhat perversely also one of its least known. The Douglas Aircraft Company, in a private study, produced a design for an unconventional piston engine bomber with engines buried in its fuselage, driving contra-rotating pusher props at the rear of the fuselage. To keep the props

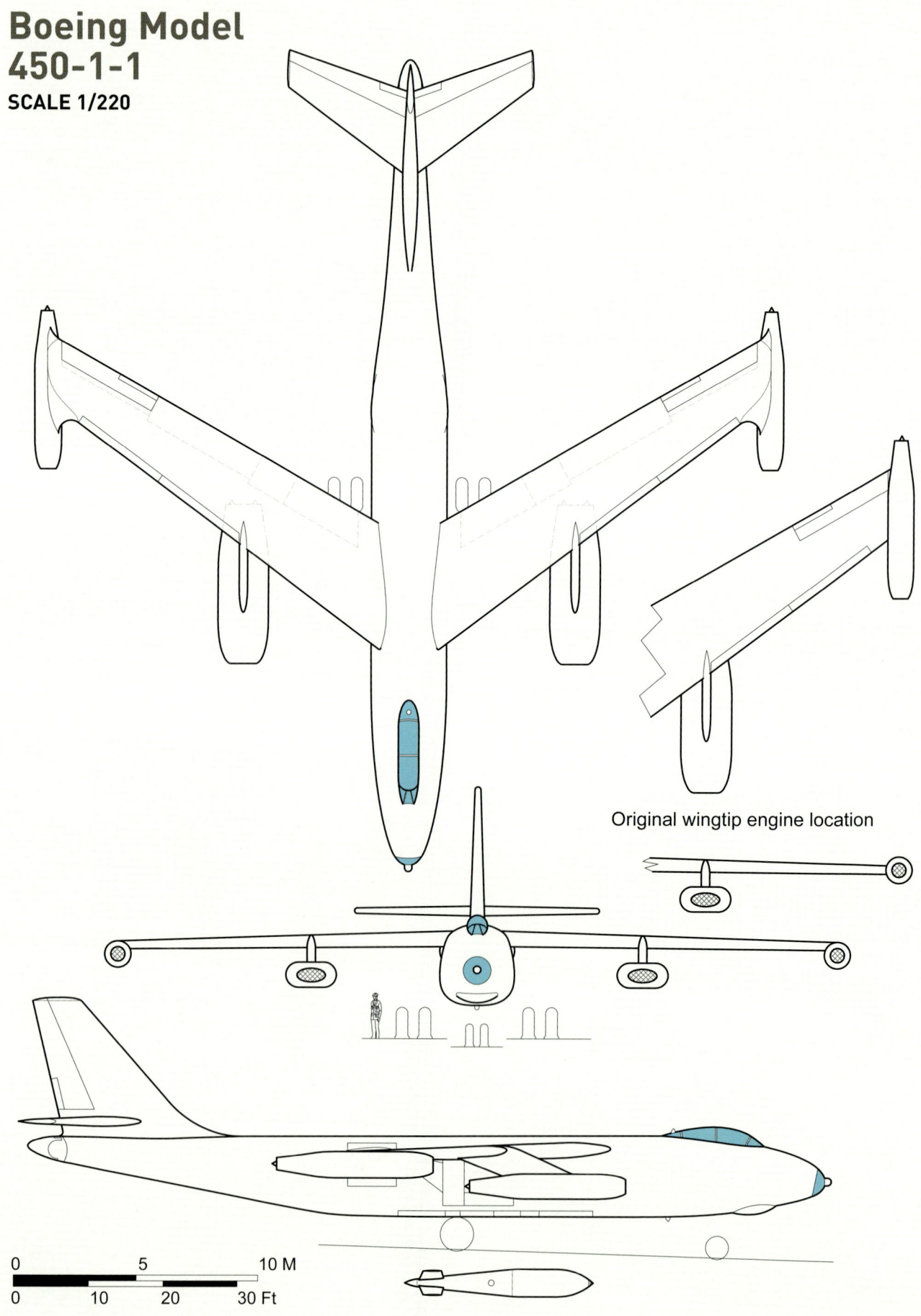

Boeing Model 450-1-1
SCALE 1/220

Original wingtip engine location

Boeing Model 450-2-2
SCALE 1/220

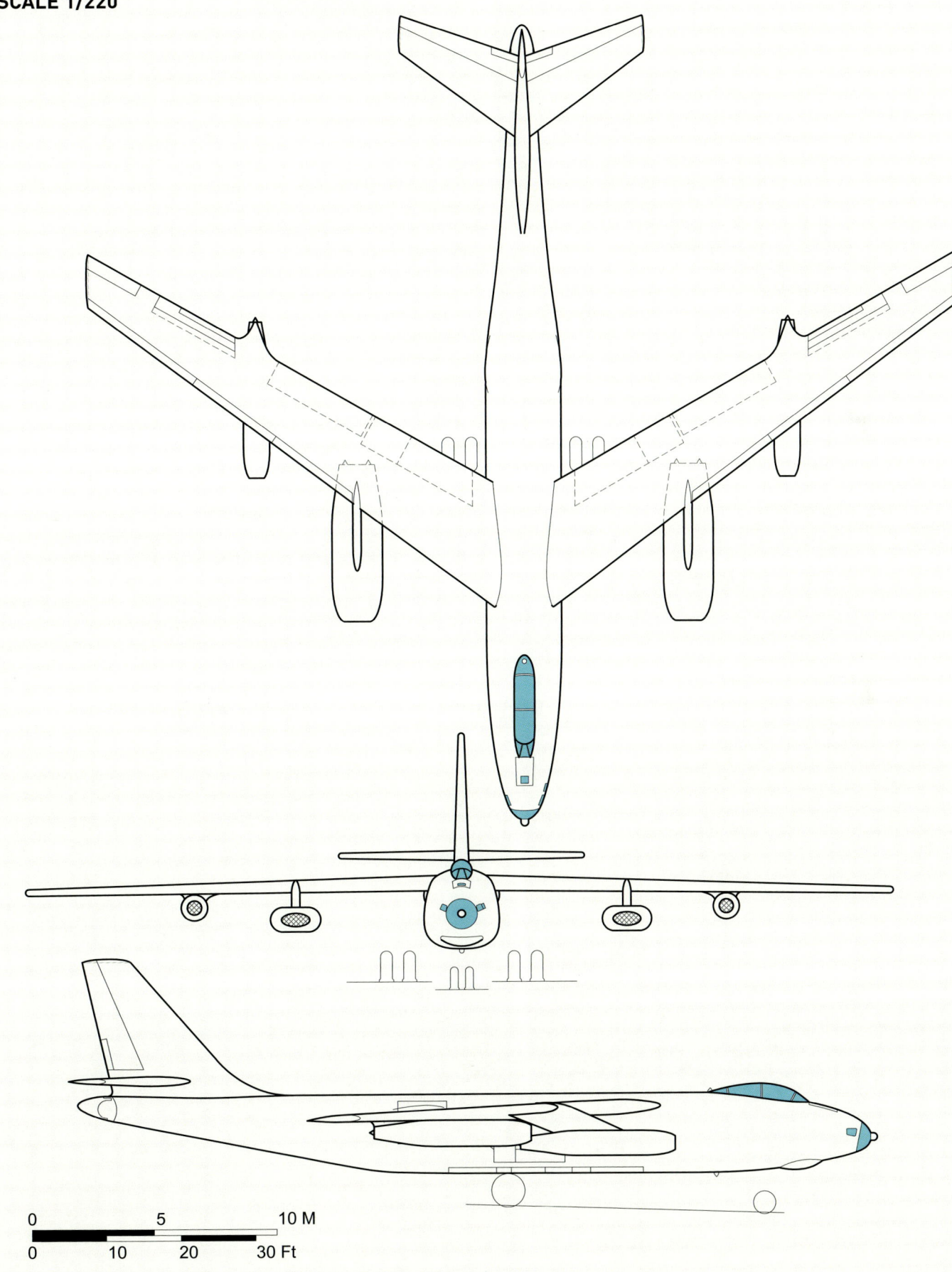

from striking the ground on takeoff and landing, the aircraft was fitted with a ventral fin as well as a normal dorsal one.

The resulting design was extremely aerodynamically clean and, when presented to the Army in May of 1943 in an unsolicited proposal, was well received and promptly received a contract to build two flying examples and one static test article of the XB-42 (initially 'XA-42') 'Mixmaster'. It was a remarkable aircraft, capable of great speed and surprising performance; it was the size of a medium bomber but it could carry 4 x 2,000lb bombs around 5,000 miles. The first prototype began flight tests in May 1944, the second in August of that year. The two aircraft were visually distinguishable by the cockpit canopies: the first prototype had separate small 'bug eye' canopies of the kind that Douglas was at the time rather enamored with, using them not only in several proposed concepts but also on the first of the C-74 Globemaster cargo planes. As with the C-74, the separate canopies of the first XB-42 prototype proved low-drag, but also made communication between the pilots difficult. So the second prototype was equipped with a standard single canopy.

Testing showed the XB-42 to be an impressive aircraft, but the end of the war loomed. Coupled with a few technical issues, this slowed adoption of the type, and as the war wrapped up the Army decided that piston engines were the wave of the past. Jet engines were the future. But the XB-42 nevertheless tried to get in on that future. The second prototype crashed in mid-December 1945 but the first prototype was given a second lease on life by the addition of two new engines: Westinghouse 19XB-2A axial flow turbojets, one under each wing. These boosted top speed from 410 to 488mph during testing in 1947. This was not enough to merit further development though.

The idea to add two jet engines to the B-42 dated back to October 1943… a response to the same Army Air Forces RFP that resulted in the Boeing Model 413. And along with a B-42 with additional small turbojets, Douglas proposed a more radical modification: a B-42 with the piston engines replaced outright with turbojets. As a result of the Douglas response to the RFP, the Army agreed and the B-42 static test article was sent back to the shop to be converted to full jet power. The piston engines removed, the tail structure modified, the ventral fin deleted (no longer needed to prevent props striking the runway), the dorsal vertical stabilizer enlarged and shoulder inlets and ducts added, the development of the XB-43 'Jetmaster' proceeded quickly. But the jet engines – the same General Electric TG-180s that were to power the Boeing Model 413 – were slow to develop.

The XB-43 finally flew in the middle of May 1946, well after the war and just as jet technology was beginning to dominate. A second B-43 – the YB-43 – was built, flying a year later. The XB-43 ended its flying career in 1951, the YB-43 in December 1953. At one point Douglas had high hopes of a production contract, but another jet aircraft design ended those dreams.

Douglas planned several types of operational B-43: a ground-attack version that had 16 .50 calibre machine guns and 36 unguided rockets under the wings, and a bomber version with 6,000lb of bombs in an internal bay. Both versions would have had a remote control turret in the tail with twin 0.50 calibre machine guns. Drop tanks as well as RATO bottles were planned to be carried under the wings as necessary. As of April of 1945, the plan was for the operational B-43 to have a canopy that somewhat melded the separate bug-eyes of the first XB-42 and the conventional canopy of the second XB-42; this unusual arrangement is shown in the diagram. In all likelihood, an operational B-43 probably would have had the conventional canopy of the XB-43.

B-47 Evolution

Douglas B-43
SCALE 1/144

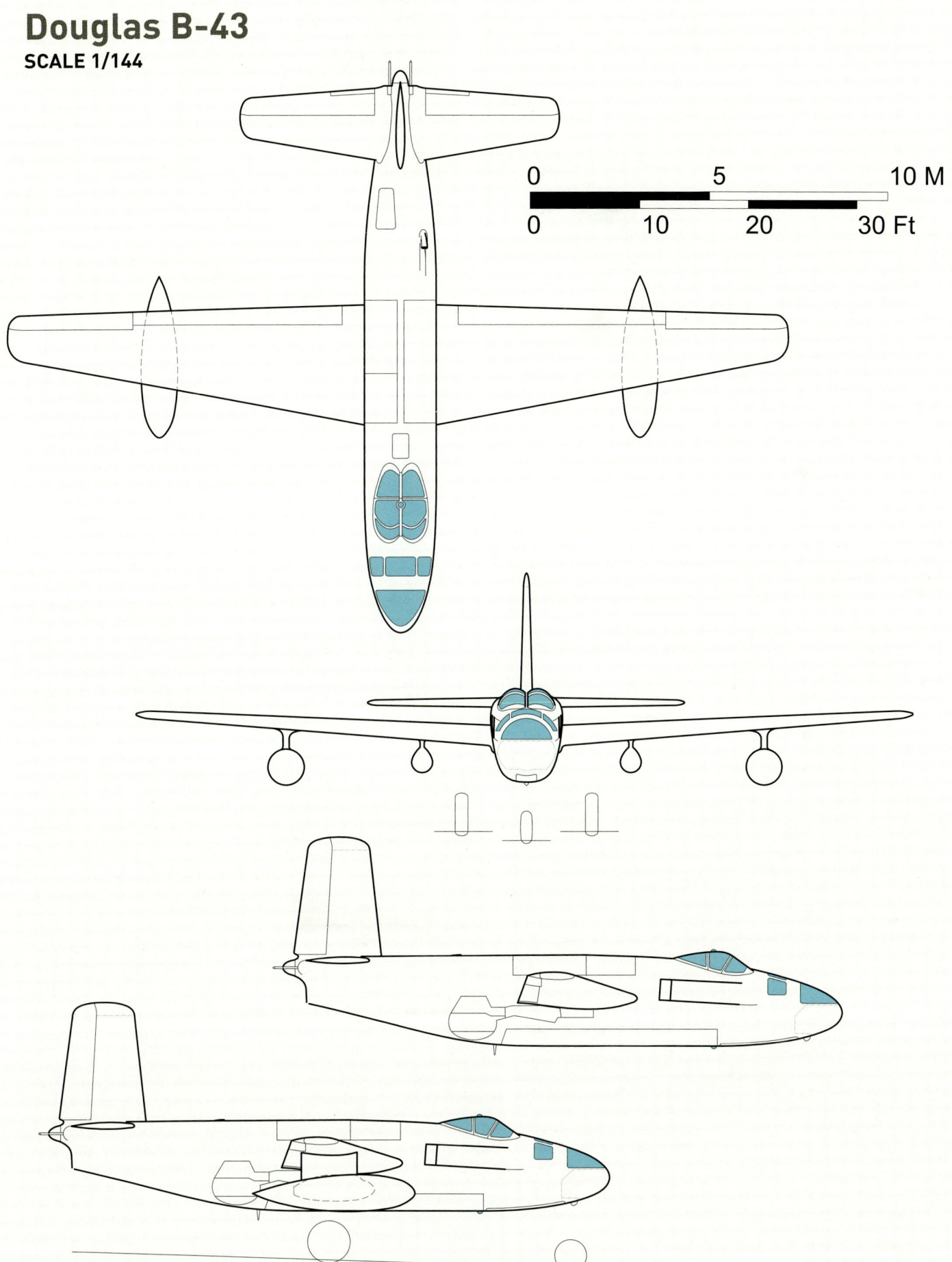

29

North American B-45

Responding to the same RFP for a jet bomber as Boeing, North American came up with a design that was broadly conservative and conventional, a solid straight-winged design with four TG-180 turbojets. The engines were in dual engine pods stuck directly to the undersides of the wings. Within their nacelles, the engines were located fairly far forward, just under the wing leading edge; long ducts took the exhaust to just aft of the trailing edge of the wing. The horizontal stabilizers were angled upwards with a dihedral of 12° in large part to get them out of the way of the jet exhaust.

Landing gear was of conventional tricycle type, the main wheels being of large diameter but relatively thin in order to fit within the wings, inboard of the engines. All in all the design, granted the designation B-45 'Tornado', looked very much like a late Second World War American bomber, but with jet engines. It was a safe bet, and North American was awarded a contract to construct three prototypes. The first XB-45 flew in March, 1947, demonstrating the successful nature of the design.

As with the company's own P-51, the B-45 had laminar flow straight wings for low drag at relatively high speeds. In this case, high speed was initially a little over 500mph... a hundred miles per hour slower than the B-47, but faster than the Mustang. It was a definite advance over prior bombers in its class and it was safe and practical enough that the US Army Air Forces – or the United States Air Force, after September, 1947 – put it into production and service. The B-45 saw service as a bomber in the Korean War, and was the first jet bomber shot down by a jet fighter. Designed to carry Second World War-era conventional bombs, it was modified to carry early nuclear weapons. To this end, on November 5, 1951, a B-45 dropped a 31 kiloton Mark 7 nuclear bomb over the Nevada Test Site during test 'Easy' of Operation Buster-Jangle, the first drop of a live nuclear weapon from a jet powered bomber. Modified into the RB-45C reconnaissance vehicle, the Tornado overflew China and the Soviet Union. The B-45C variant was the first jet bomber equipped for inflight refuelling, which greatly extended its range.

The North American B-45 had something of a twin: the Soviet Ilyushin Il-28. A jet bomber that first flew a year after the B-45, it had a remarkably similar configuration, with the biggest visual differences being the distinctly aft-swept stabilizing surfaces on the Il-28 and the Soviet plane having two large engines rather than four small ones. The biggest difference, though: the production run of the B-45 created 143 aircraft, while the Soviets stamped out more than 6,000 of theirs, the Chinese making a further 300. The B-45 ended its service days in the late 1950s (though a few test aircraft continued to fly into the 1970s), but the North Koreans continue to operate the Il-28.

Consolidated XB-46

Boeing's entry was seen as high risk but high reward, while the North American design was seen as safe and conservative. Consolidated-Vultee and Martin, however, produced designs that were somewhat riskier than the North American concept but not as potentially promising as the Boeing design. Thus North American got early production contracts, Boeing was given time... and Consolidated-Vultee and Martin were each given contracts for demonstrator aircraft.

Consolidated-Vultee was at the time hardly unfamiliar with bombers. Its B-24 Liberator had been built in vast numbers; the under development B-36 would be one of the largest aircraft to ever fly and would serve admirably in the post-war years. This gave the firm confidence that its design (Model 109, submitted in December 1944) would be not only a successful aircraft, but would win the contest. The initial contract called for three demonstrator aircraft, which Consolidated Vultee began work on following a successful mockup review in February 1945. But the contract was cut to just a single aircraft in late 1945; partly this was due to a general slowdown in military development contract at the end of the war, but it was also due to a planned diversion of the XB-46 funds to help fund the development of another Convair (as Consolidated Vultee was increasingly referred to) project, the forward-swept tailless design XA-44. The XA-44 would gain the designation XB-53 in June of 1946... and would then be cancelled in December of that year.

Consolidated Vultee's entry was remarkably similar to the original Boeing Model 413 in configuration. But for all their comparable attributes, the Consolidated design was unarguably one of the most graceful and handsome aircraft ever to grace the skies. As with the Model 413 and even the B-45, the layout of the XB-46 was conventional: a tubular fuselage with straight shoulder-mounted wings, a dual-engine nacelle directly below each wing, a conventional tail unit. But Convair made every effort to make the aircraft as sleek and aerodynamic as possible. The fuselage was long and gradually tapering at both ends; the wings were clean and relatively high aspect ratio.

The result was an aircraft with a remarkable resemblance to a high performance sailplane; reportedly, if the aircraft lost power at 40,000ft it could glide 200 miles. For flight at low speed the wings sported Fowler flaps that extended across 90% of the span. The engine nacelles were rounded pods

North American XB-45
SCALE 1/150

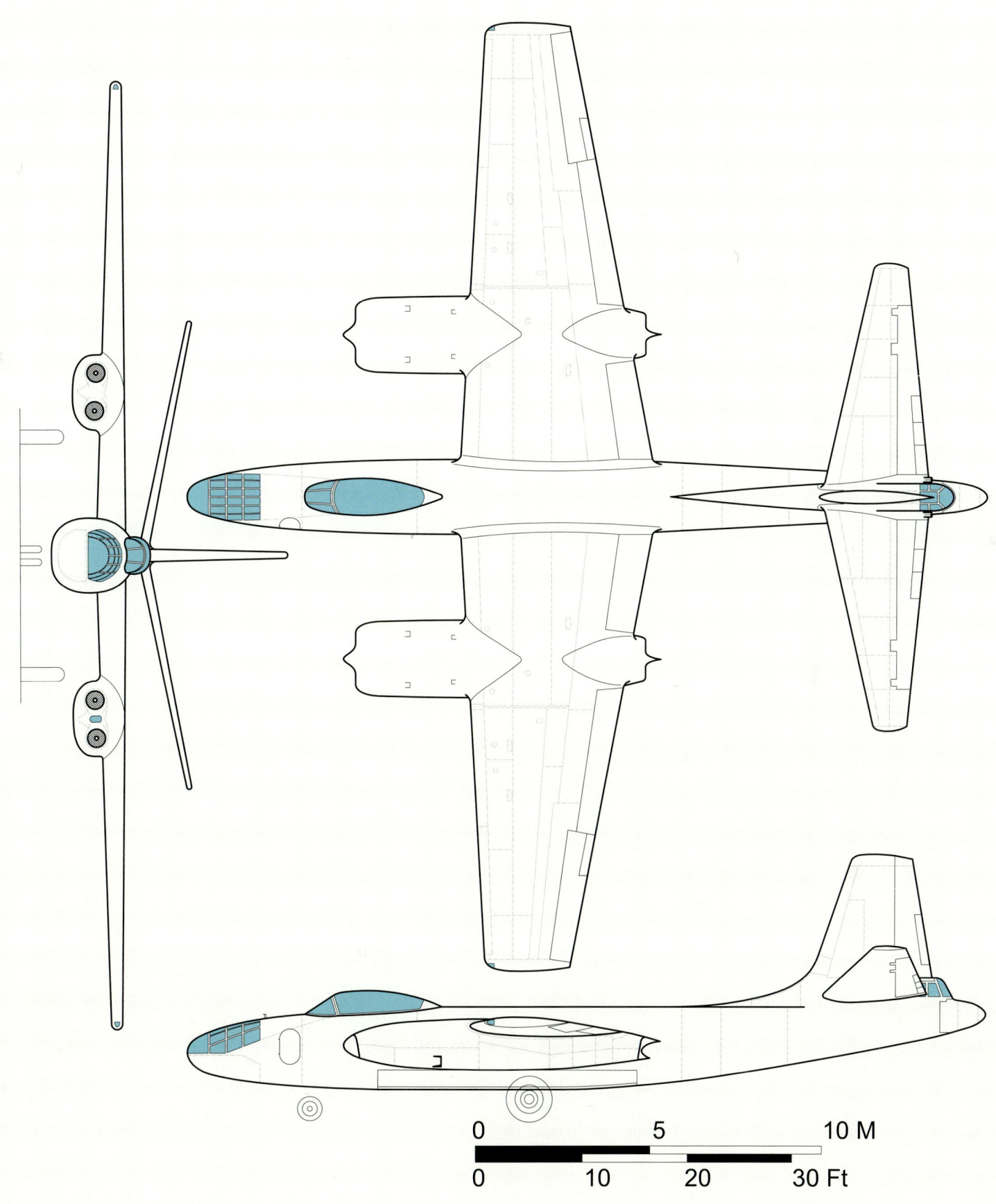

Consolidated Vultee XB-46
SCALE 1/200

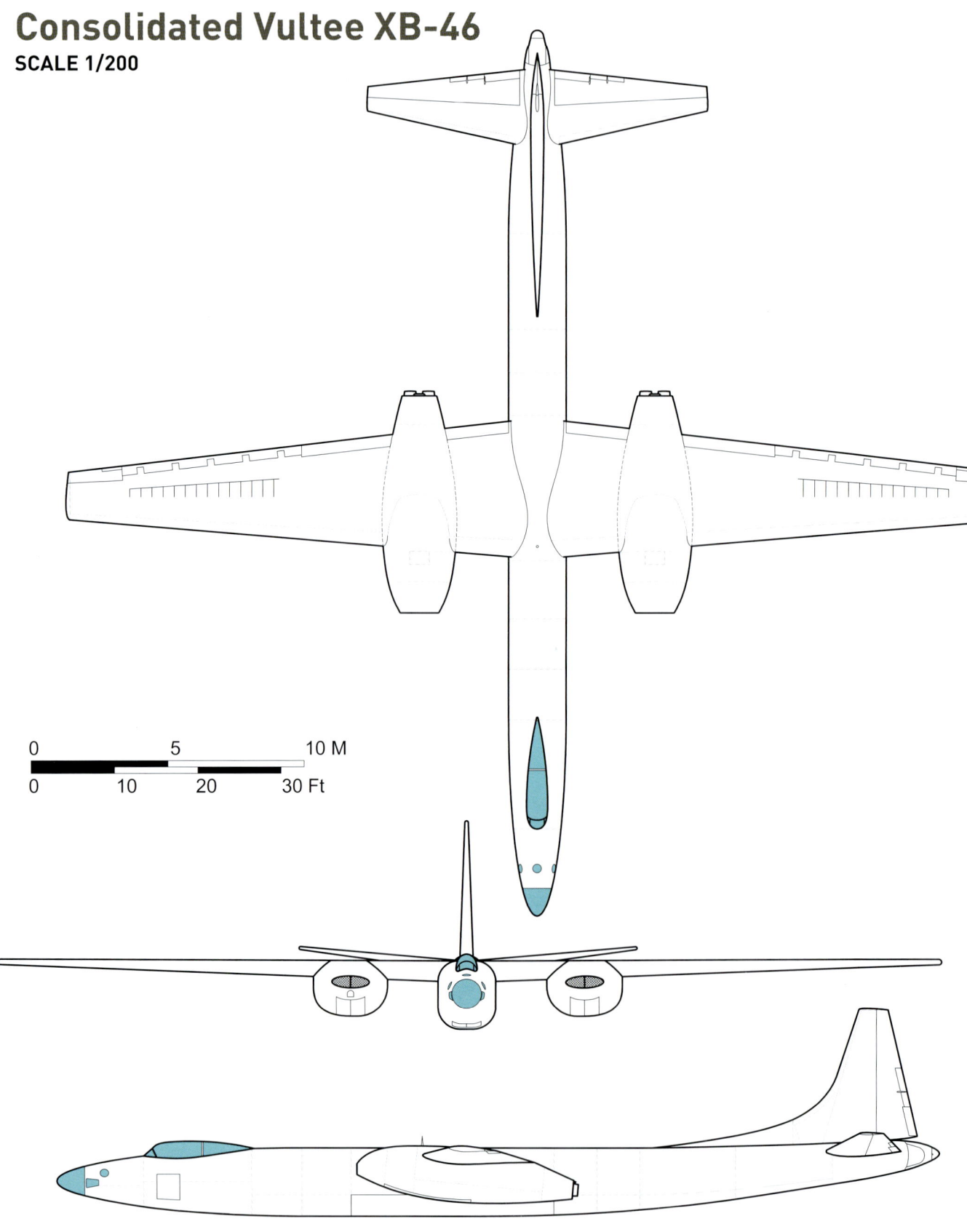

containing two engines each, fed from a single oval inlet; the inlet duct bifurcated, splitting around the main landing gear bay. The engines exhausted from separate nozzles at the tapered rear of the nacelles, just aft of the trailing edge of the wing.

The XB-46 was designed to be all-pneumatic, with no hydraulic or electric actuators for mechanical systems. The engines that Convair wanted for the plane, General Electric J47s, were not available for the prototype and Allison J35s had to be used instead, resulting in a substantial reduction in maximum thrust. The operational B-46 would have had a remotely operated tail turret equipped with two .50 calibre machine guns. There were also proposals to mount four fixed 20mm cannon or a twin-.50 turret in the nose. What exactly a sizable bomber like the B-46 would do with four cannon fixed forward in the nose is anyone's guess; it seems unlikely that it would be used for dogfighting and it seems a little large for low altitude ground-attack.

The XB-46 was a large aircraft with a span of 113ft and a length of 105ft 9in. It did not have the B-29's 141ft wingspan but it exceeded the B-29's length of 99ft. It was substantially lighter than the B-29, both in terms of gross weight and payload, and had a shorter range. However, it could go nearly 200mph faster than the B-29.

The single XB-46 flew for the first time in April, 1947. It flew well and was liked by its pilots, though it did have a number of problems with oscillation and a list of mechanical and electrical issues. But even though on the whole the test flights were promising, it was up against the B-47, and the XB-46 flew for the last time in 1950 before being scrapped.

Martin XB-48

The Glenn L. Martin Company submitted its Model 223 proposal for preliminary engineering and mockup construction on December 9, 1944; one week later the Material Division in Washington, D.C. authorized wind tunnel models and testing. A cost-plus contract for further testing was approved in April 1945. In June 1945, Martin submitted a proposal to actually build the Model 223, dubbed the XB-48, and the Army Air Forces approved the construction of two flying prototypes. The XB-48 was seen from the outset as a fallback option in case the Boeing design failed.

Where the XB-46 was sleek and gorgeous, the direct competitor from Martin, the XB-48 was… well, the opposite. The two XB-48 aircraft were stout beasts with six engines; laid out with the same basic plan as the B-45 and XB-46. But the fuselage of the XB-48 was as graceful and elegant as a grain silo, with a long straight section with a flat underside and fairly blunt nose. Rather than a harbinger of the jet age, the XB-48 looks almost as if it would have been at home next to an Avro Lancaster.

The XB-48 had what its designers called a 'horizontal bomb bay', rather than a conventional 'vertical' bomb bay. By this they meant that where a conventional bomb bay filled the fuselage from top to bottom, the XB-48's only filled roughly the lower half of the fuselage. The wing spars and several fuel tanks took up the volume above the bay. It might not have had much depth, but the bay was quite long… more than long enough to accommodate the Grand Slam, as well as a wide variety of other bombs from 250 to 4,000lb. The bomb bay doors ran the full length and retracted within the outer mold line of the aircraft, reducing drag.

The landing gear was, like the B-47's, of 'bicycle' arrangement, with one two-wheel bogey just ahead of the bomb bay and another just aft of the bay. Small outrigger gear retracted into the outboard side of the engine nacelles.

The one truly unique feature of the XB-48 was its engine nacelle design. Like the Boeing B-47, it had three turbojet engines per wing, but it arranged them in a way that no other aircraft ever has, before or since. The three engines were arranged side-by-side in a single nacelle, but with sizable rectangular cross-section ducts between them. The theory was that this would reduce drag compared to a single unbroken nacelle. The NACA was consulted in 1945 and initially concluded that a single 'egg' with three engines would be more aerodynamic. The issue was to be resolved with a series of wind tunnel tests… but these were cancelled when it was realized that the actual aircraft would be completed before the wind tunnel tests could be carried out.

The first flight of an XB-48 was in June 1947. The two aircraft flew well, but slowly compared to the B-47. The unusual nacelle configuration produced more drag than hoped; at high speeds the ducts produced a high airspeed venturi effect. Martin could see the writing on the wall for the design – the B-48 was simply never going to be as fast as the B-47. So in January 1949 the company proposed to convert the second XB-48 into the Model 247-1. The jet engines and nacelle would be removed and replaced with four XT-40 turboprops. The resulting aircraft would be slower – which the XB-48 already was, compared to the competition – but it would be more fuel efficient. The Air Force, however, turned this idea down. Flight tests of the two existing aircraft ended in mid-1951 with both airframes being scrapped.

While Boeing's B-47 proved itself, it did so in an era of rapidly advancing technology and constantly shifting priorities. The 1940s had seen a constant stream of new aircraft, and the rise of jet technology promised more of the same. So it comes as no surprise that in 1948 the US Air Force requested proposals for a jet bomber that was aimed at being something of a replacement for the B-47. The B-47, after all, had begun life in 1944, a full three years earlier… and that was an eternity. The previous three years had seen aircraft progress from propellers to jet engines to rockets, from 400mph to breaching the sound barrier. Thus even though the B-47 had not yet had its first flight, it was clearly time to look at what would come next.

Project MX-948 called for studies on the next generation of medium jet bombers. Information on this project and the proposals that followed along is sparse and in some cases uncertain.

Martin XB-48
SCALE 1/175

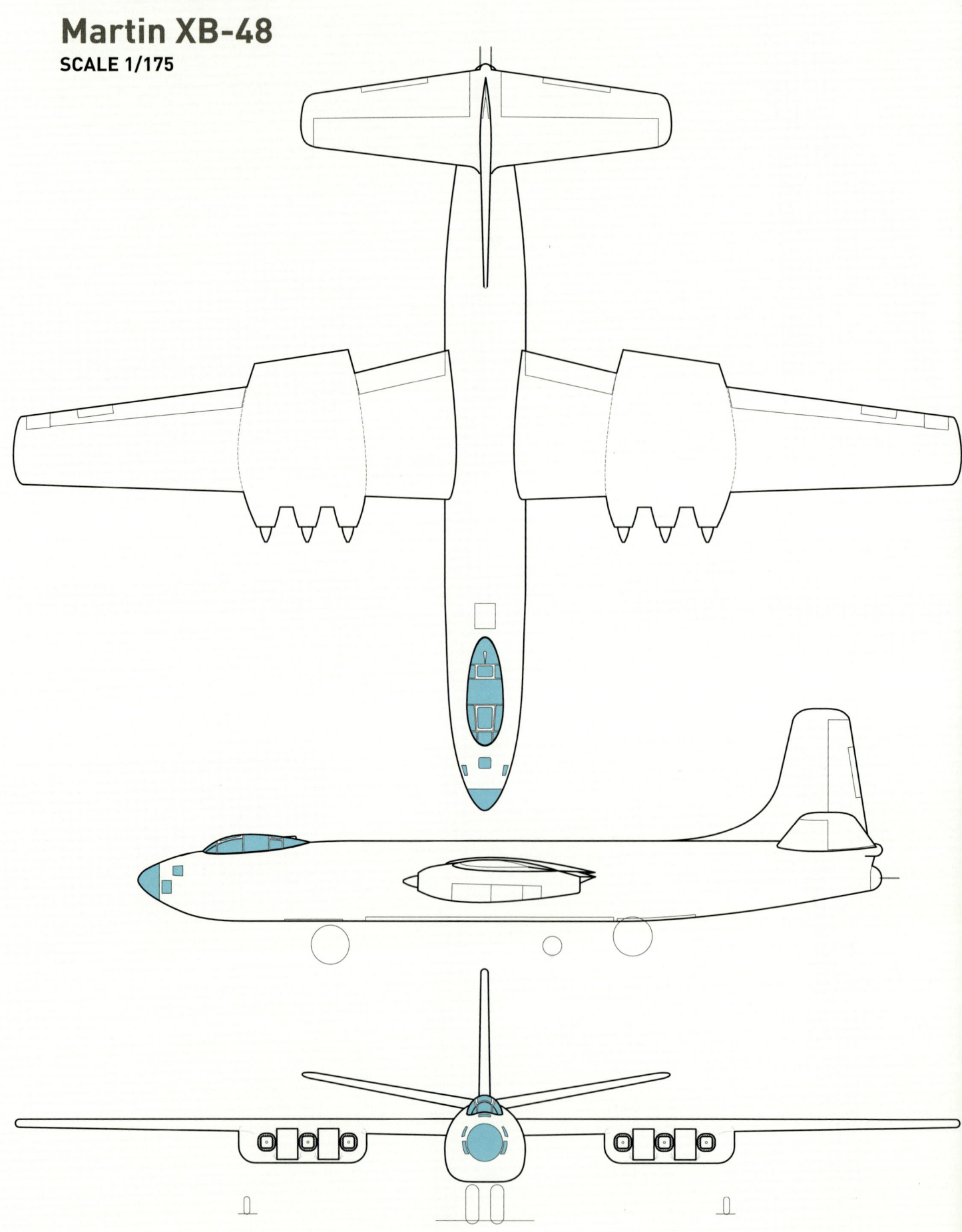

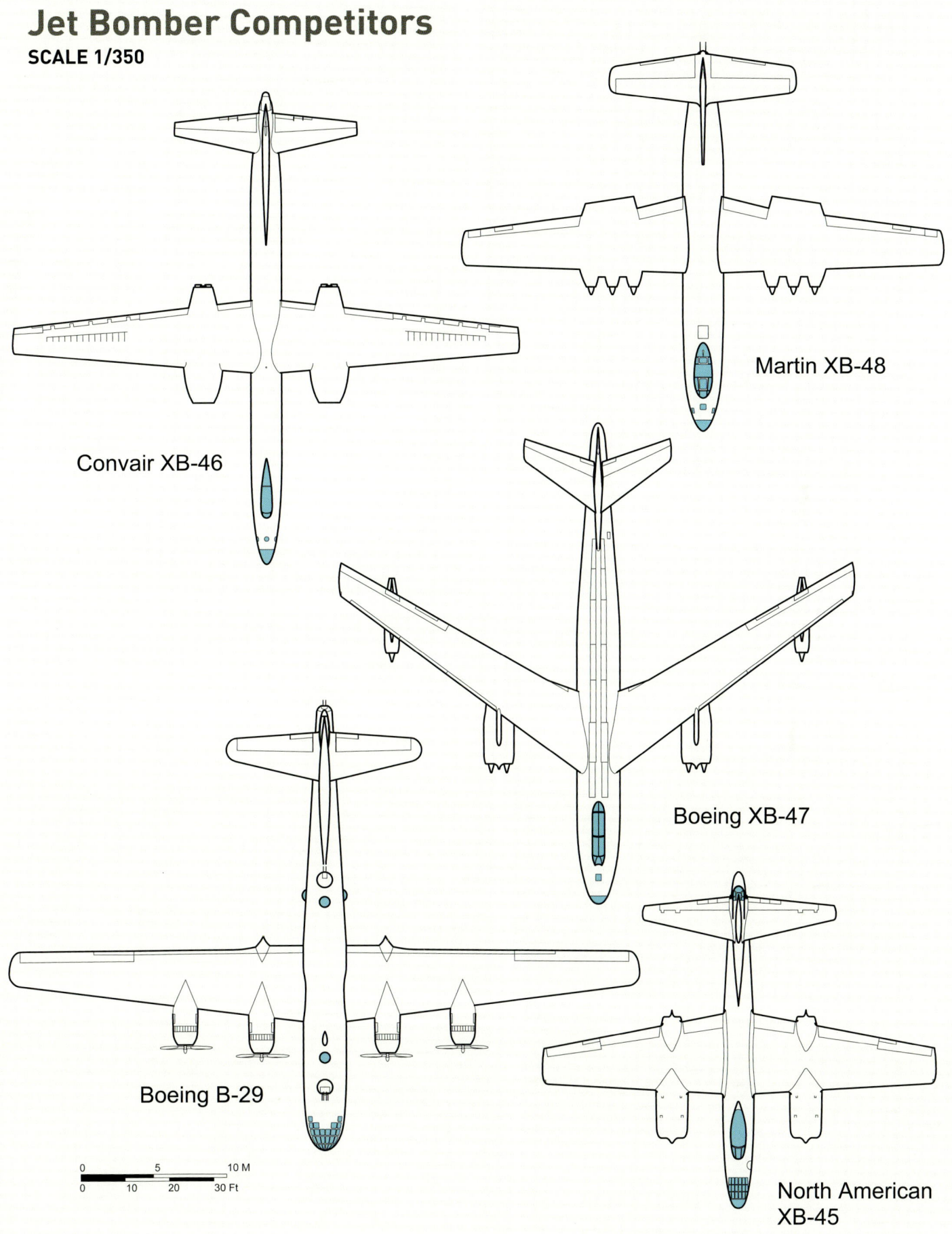

Lockheed L-173

Lockheed had not participated in the 1944 jet bomber contest, but the company tried to correct this with the L-173. Some number of designs were studied, including jet propelled, turboprop powered and mixed systems, but the only design known with any certainty is L-173-15, which seems to be the final configuration. As with the B-47, the L-173 was a six-engined jet bomber with swept wings, but that about ends the similarities. It featured a circular cross-section tubular fuselage with a crew of eight up front and two gunners in a separate unconnected pressurized section in the rear fuselage (this seems unlikely and unusual… a crew of ten is a shocking number for a medium bomber meant to replace the three-seat B-47, and while an available inboard profile depicts eight crew in the front, no crew positions in the rear fuselage seem to be visible). For defense the aircraft was equipped with a retractable spherical turret under the forward fuselage with two 20mm cannon, a turret on the upper rear fuselage just ahead of the vertical stabilizer with two 20mm cannon and a tail turret with four 20mm cannon. The abundance of defensive guns appears to indicate that Lockheed thought the L-173 would operate in a space swarming with enemy fighters able to exceed its altitude and speed performance.

The engine arrangement for the L-173 was a bit unusual. Two were located in a nacelle just inboard of half span… but unlike the B-47 nacelle, the engines were stacked vertically, with the upper engine staggered slightly forward of the lower engine. The reason for this is not clear; doubtless it would have made maintenance access to the engines, especially the upper engine, rather interesting. A single engine nacelle was mounted below each wingtip.

The wings were aft-swept, but less so than on the Boeing B-47, with 20° sweep at the quarter-chord line. Also as with the B-47, landing gear was of the bicycle type with outriggers dropping from the engine nacelles.

Northrop N-31

In January of 1948, several designs for the Northrop N-31 series flying wing bombers were unveiled. It is uncertain whether these designs were initiated for MX-948, but the timing seems about right and the aircraft seem to fit the role. These aircraft were advancements upon the existing B-35 and B-49 flying wing bomber designs, using somewhat different planforms. They featured more highly swept wings of somewhat shorter span and included central fuselages with noses that projected well ahead of the wing leading edge. The fuselage provided the volume necessary to carry a single Grand Slam bomb.

Two designs were initially put forward, N-31 and N-31A. Data on them is sketchy. N-31 used an all-jet propulsion system while N-31A used turboprops. Both designs had a range of 2,781 miles but where N-31 could reach a top speed of 520mph, N-31A was slightly slower at 510 mph.

Revised versions of N-31 were revealed in May 1950. Specific aircraft designations are unavailable for them, the information coming from a promotional brochure which does not provide nomenclature. Two different engines were considered, with either two or four engines per plane.

A crew of five was contemplated for these designs, with two crewmembers in a tandem cockpit similar to that of the Boeing B-47… and indeed for almost all medium bomber designs of the era. Two other seats were located in the leading edges of the wing roots and were provided with large windows for forward visibility. A tail stinger was provided with a remote gun turret containing two or four machineguns (probably .50 calibre). Inflight refuelling was planned for long range bomb runs; speed and manoeuvrability were expected to be such that fighter interception would be extremely difficult.

The primary version was equipped with two Turbodyne V (also known as the XT-37) turboprop engines, each driving a six-bladed counter-rotating propeller. The Turbodyne was a large turboprop engine developed in-house at Northrop; each engine could put out 10,400hp. The alternate version was equipped with four Allison XT 40 turboprops, providing a total of 30,000hp. The propeller arrangement was divided into four six-bladed counter-rotating props. Otherwise the design was essentially identical to the Turbodyne V variant. Performance was lower than that of the Turbodyne V version.

Republic AP-42

The Republic response to MX-948 is believed to be the AP-42 concept, for which minimal data is available. In fact, all that seems to now exist for it are a few photos of a display model. Consequently, anything that can be said about it must be taken to be provisional or speculative. Hopefully someday the proposals or other design diagrams and data will come to light.

The model depicts a swept-wing aircraft with a spindle-shaped fuselage and a narrow-chord V-tail. The most unusual feature was the arrangement of the engines; at first glance it appears to have two nacelles with two engines in each… but it actually has four engines in each nacelle. Two engines were stacked vertically and fed from an elliptical inlet at the front, and exhausted through two nozzles on the outboard surface of the nacelles, about two thirds of the way back. Two more vertically-stacked engines were located at the extreme rear of the nacelles; they were fed from inlets on the inboard side of the nacelles, just forward of the outboard exhausts, and exhausted from the tail of the nacelles.

Lockheed L-173-15
SCALE 1/288

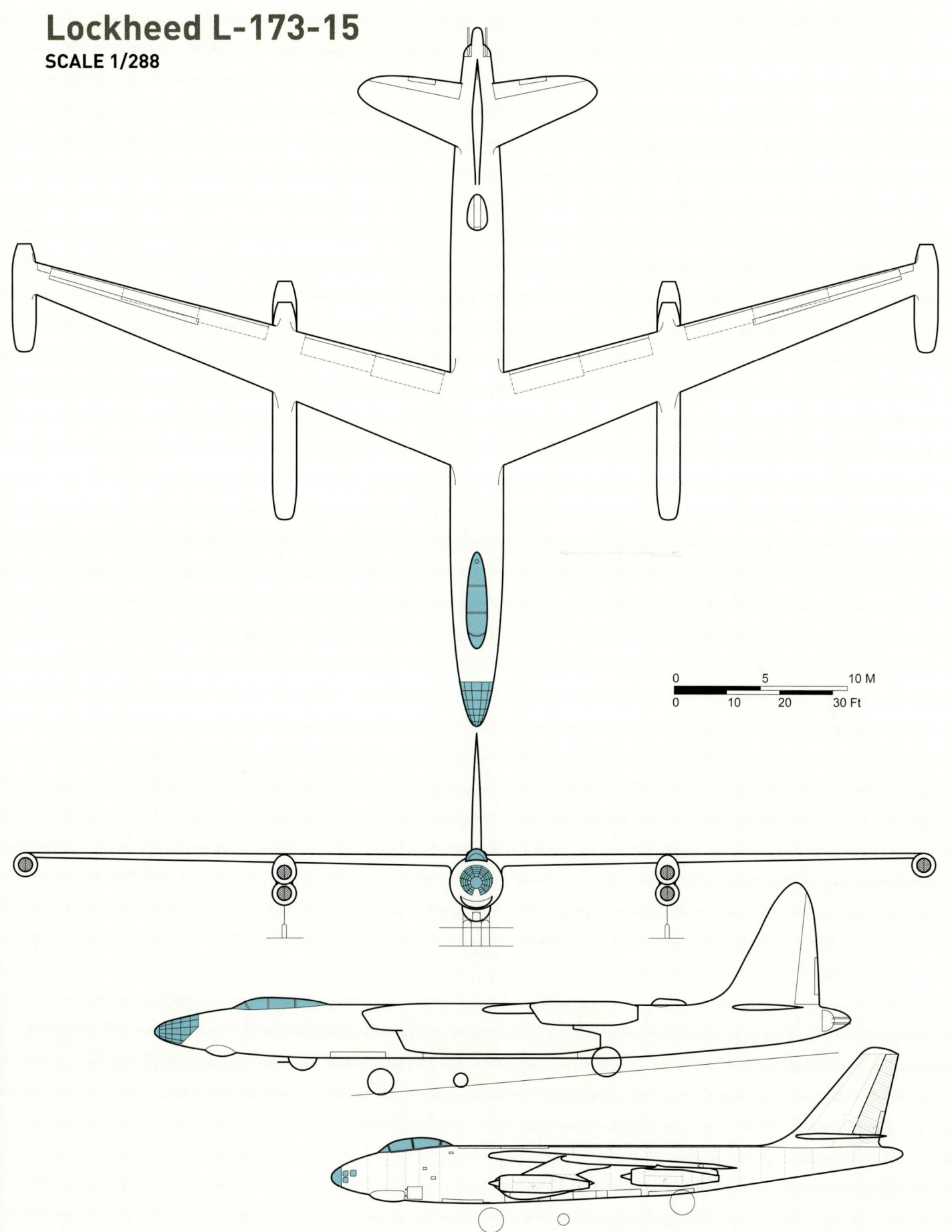

Northrop N-31, Turbodyne Engine
SCALE 1/300

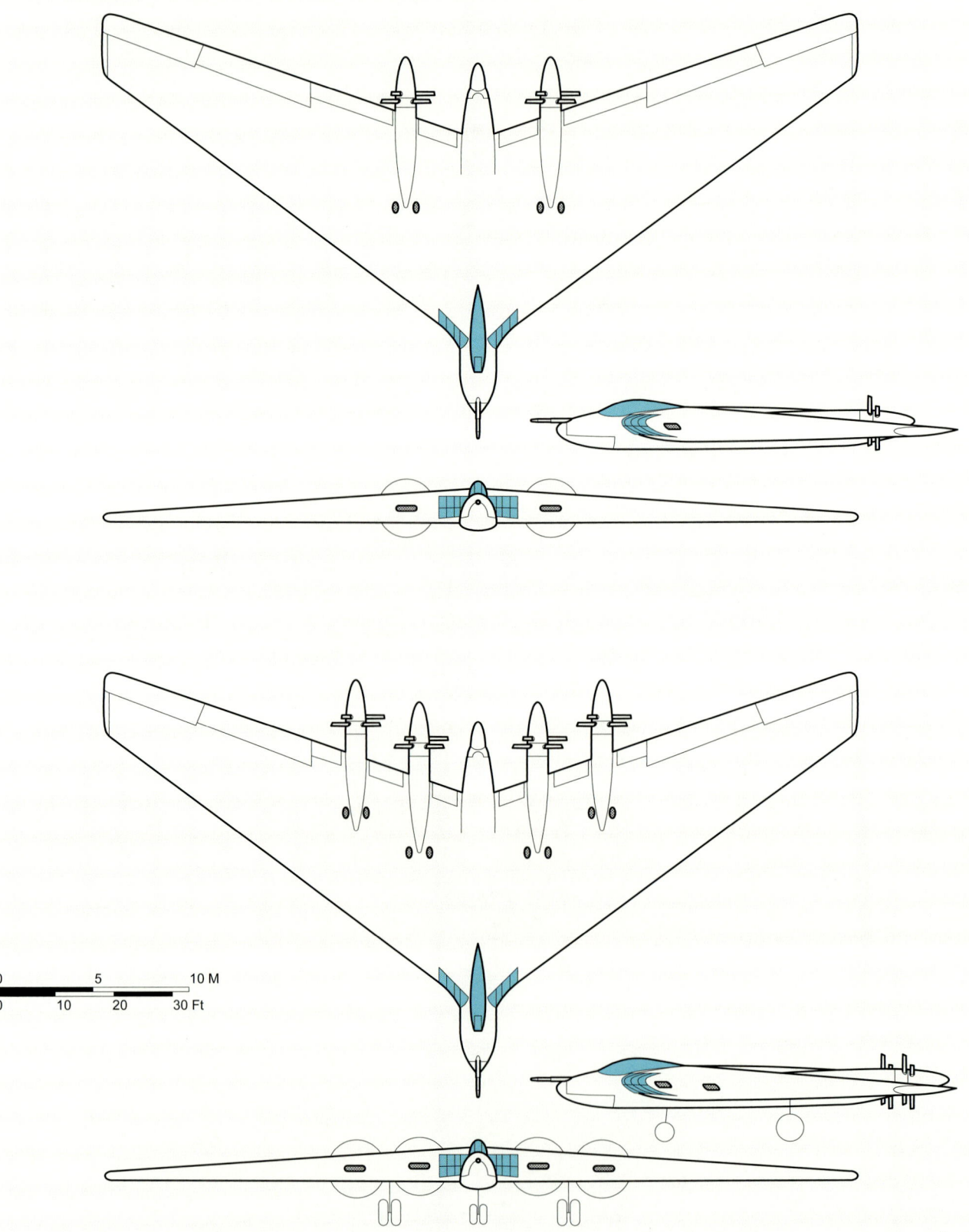

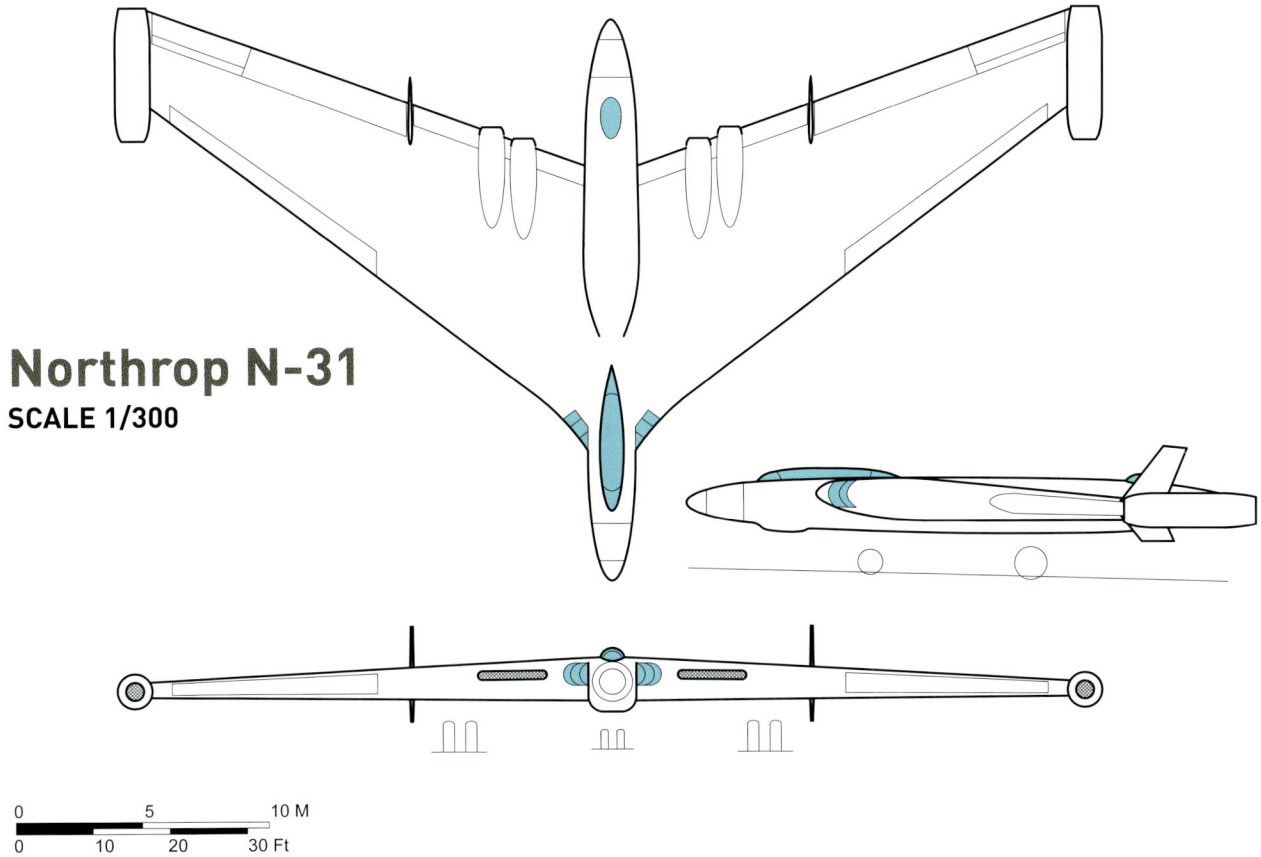

Northrop N-31
SCALE 1/300

The sole indication of the size of the aircraft is a crude human figure next to the model in the photos. Dimensions for the aircraft were assessed from that figure when this author created diagrams; due to uncertainty both in the accuracy of the little figure and in assessing the figure, dimensions given in the data table should be taken with a grain of salt.

Douglas Model 1126

Douglas' proposal for MX-948 was the conventional-looking Model 1126 from August 1948. This aircraft looked like nothing quite so much as the Boeing Model 450-1-1 from 1946, with twin-engine nacelles pylon-suspended beneath swept wings, and single engine nacelles on the wingtips. At least two major variants of the Model 1126 are known, differing largely in the wing. One had a span of 102.2ft, the other 140ft. The fuselages were also a bit different in length; the smaller design was 119.9ft, the longer, 128ft 2in. The diagram included here is of the smaller design.

Compared to the Lockheed, Northrop and Republic designs, the Douglas vehicle seems almost shockingly conventional. But Douglas was not to be outdone. In order to get the maximum possible range, the smaller version would be fuelled in mid flight. But in-flight refuelling had not yet been as perfected and standardized as it has since become, and Douglas' suggestion was unique to say the least. Instead of meeting up with a tanker aircraft and taking on fuel through a probe or a hose, the 1126 would rendezvous with a B-36 bomber carrying a relatively gigantic 'drop tank' above the fuselage. The 1126 would – almost magically, it would seem – latch onto the tank with cables. The B-36 would detach the tank and the 1126 would reel it in, latch on and tap into it. The idea seems outlandishly unlikely to work in actual practice, especially in wartime; but this was an era when parasite fighters were being seriously proposed to be carried in bomb bays or attached to wingtips. The larger version was probably not intended to have to undergo such in-flight machinations in order to attain its maximum range.

The Model 1126 had seven crew in the forward fuselage, and a tailgunner in a rear pressurized section. He operated an unusual system composed of two separate turrets, each with two 20mm cannon, one turret on either side. Gunners in the forward section operated retractable turreted cannon located above and below the fuselage just behind the cockpit.

The fuselage was distinctly non-circular in cross section, having a flat underside akin to that of the Martin XB-48. Performance and weight data is currently unavailable for the Model 1126, but cutaway artwork depicts it with a Grand Slam-like bomb, but notably smaller in diameter. It's unclear if this was meant to represent an unusual conventional bomb (possibly a deep penetrator of ground or armour) or if it was a stand-in for a nuclear bomb.

Republic AP-42
SCALE 1/300

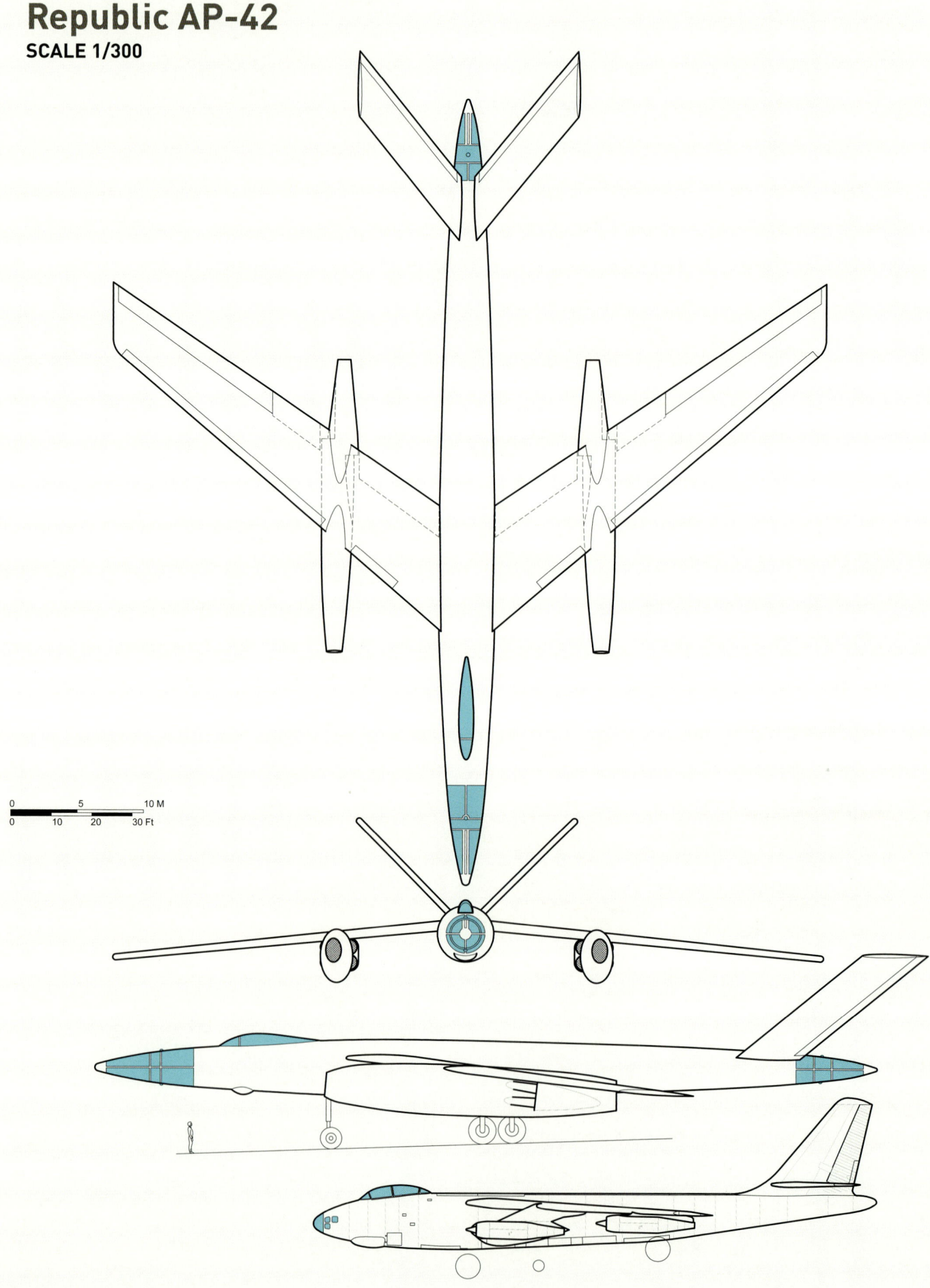

40

Douglas Model 1126
SCALE 1/220

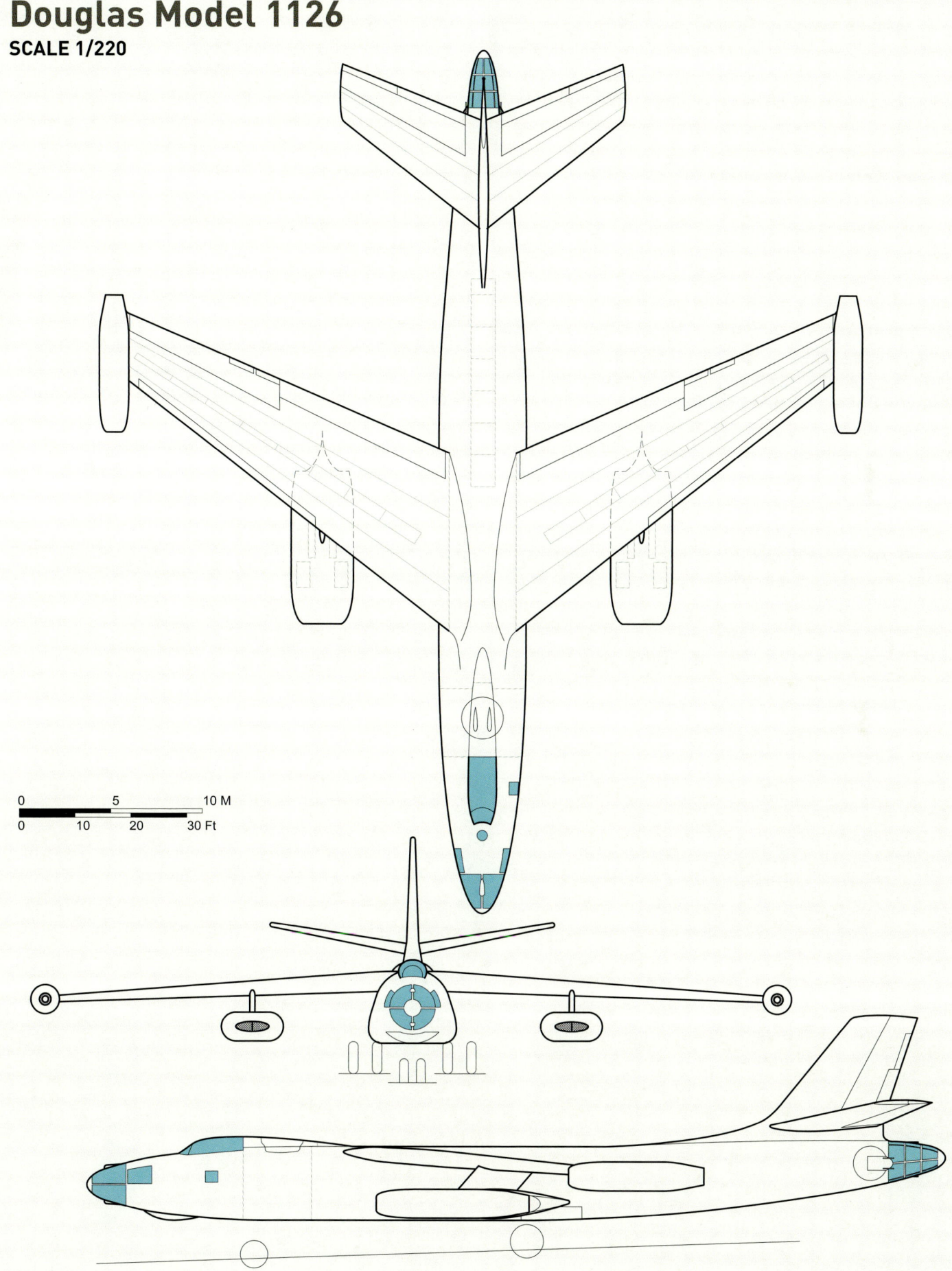

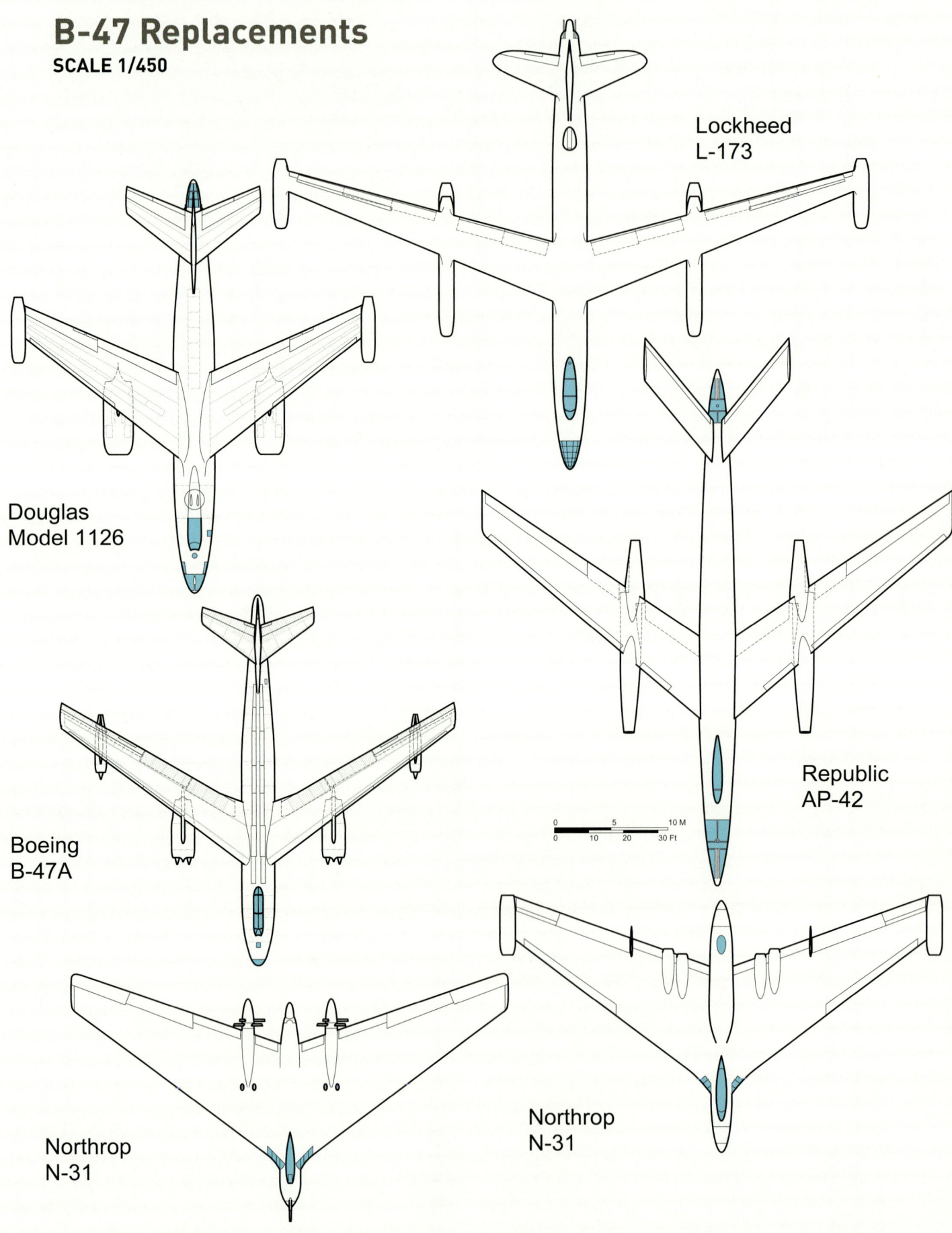

CHAPTER 2: B-47 Development and Projects

The fixed-price contract that turned the Boeing Model 450-3-3 dream into the XB-47 reality was signed on July 10, 1947. This called for the construction of two prototype aircraft for flight testing. But work on the two XB-47s had actually begun more than a year earlier when the US Army Air Forces had given Boeing the go ahead to build them in June 1946. Construction began soon thereafter at the Boeing plant in Seattle. They were not built with production-style jogs and manufacturing, but instead were essentially hand-made. Nevertheless, in what today would be considered blistering speed the first of them was rolled out on September 12, 1947 to a combination of astonishment and apathy. The design looked amazing but it seems many people at the time thought its chances of success were poor.

Boeing XB-47 (Model 450-3-3)

The XB-47s as built were indeed futuristic looking. The public was largely unfamiliar with swept wings; in fact, just a month later the Bell X-1 piloted by Chuck Yeager would break the sound barrier for the first time… using straight wings.

But while the XB-47 did not have the X-1's now-antiquated unswept wings, it did have something in common with the X-1 – rocket propulsion. The early TG-180 engines were marginal at best for heavyweight takeoff. So to assist, the XB-47 had fittings for rocket-assisted take-off units, colloquially often known as 'RATO bottles' built into the port and starboard sides of the rear fuselage.

Each side could hold nine solid Aerojet rockets (with a mix of potassium perchlorate and asphalt propellants), each producing about 1,000lb of thrust for a total of 18,000lb. This did not quite double the 22,500lb of thrust that the jet engines could produce, but for the dozen seconds that the rockets operated they could hurl the aircraft up off the runway, and if the pilot wanted to abort a takeoff, he had to wait until the rockets burned out. The solid rockets were considered an interim solution to the problem of marginal takeoff thrust; with the inert weight of the steel bottles and the vast clouds of impenetrable smoke, what Boeing really wanted was a liquid rocket system. A system like that would weigh less and perform better, but it was not yet ready. So solid rockets were the next best solution.

The first prototype – which took to the air for the first time on December 17, 1947 – was fitted with TG-180 B1 turbojet engines (AKA J35-GE-7), while the second prototype had TG-180 C1 engines (J47-GE-9). Both engine types

43

Boeing Model 450-3-3
XB-47
SCALE 1/220

Focke-Wulf mit 2 HeS 109-11 to scale

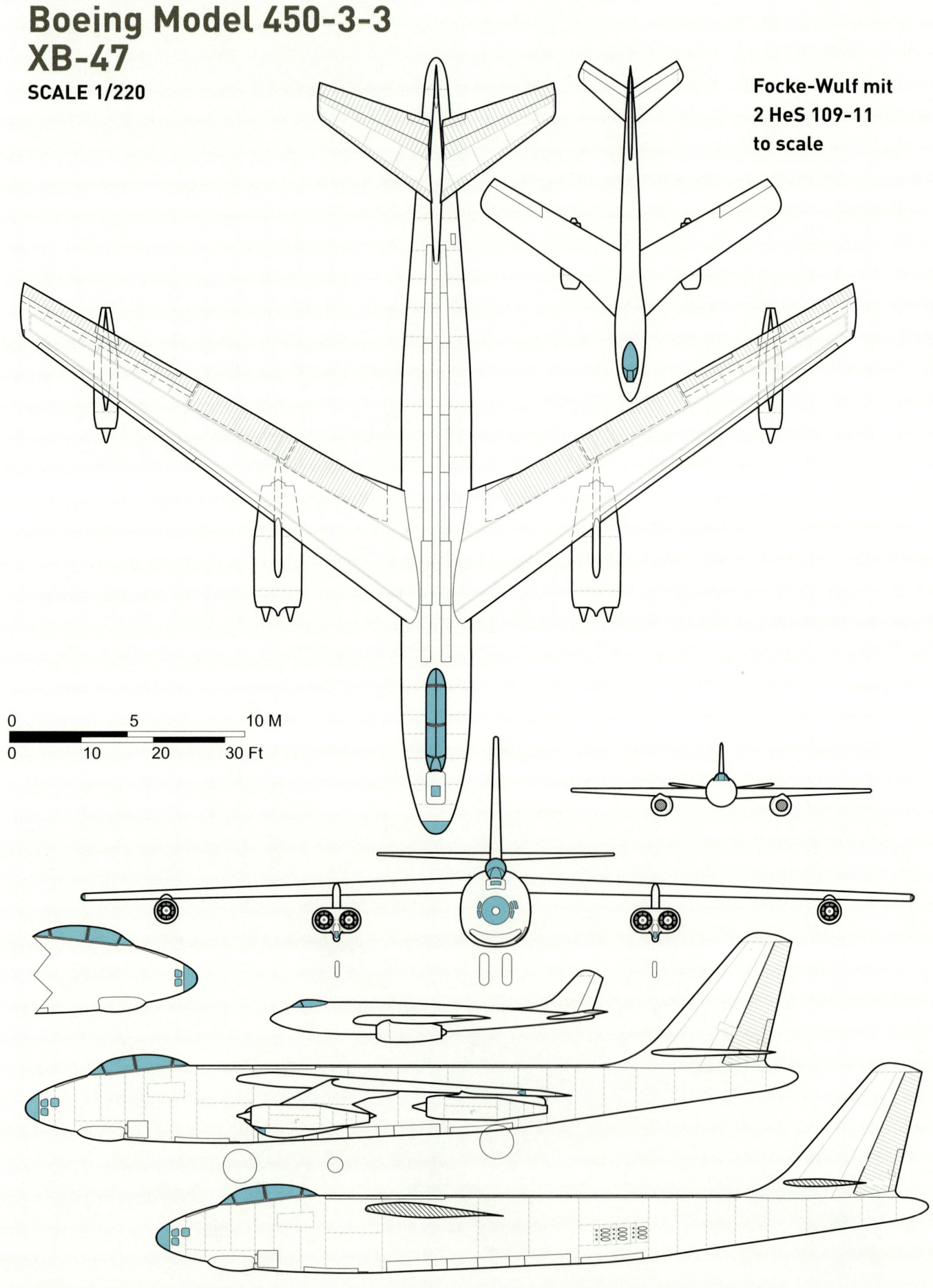

produced 3,750lb of thrust each. They might have been marginal at best for takeoff, but for cruising flight at the desired speed and altitude these engines were just fine.

The configuration of the XB-47 was largely that of the production B-47s that would follow, and the original design turned out to be an excellent one. One of the few major changes involved the first prototype's leading edge wing slats, positioned outboard of the engine pylons. Intended to help the B-47 during low speed landings, testing showed that it didn't need them and they were never again included.

The XB-47 was not equipped with weapons, either offensive or defensive. The three crew each had an ejection seat. The canopy would be blown off to allow the pilot and co-pilot to eject safely while a hatch in the top of the nose, ahead of the pilot's canopy, would blow off to provide a safe exit for the navigator. The main entrance was a hatch located on the lower port side of the forward fuselage.

Boeing B-47A (Model 450-10-9)

Ten B-47A aircraft were built (in Wichita, rather than Seattle), with first rollout in March 1950 and first flight in June. They were essentially identical in configuration to the XB-47 with little to visually distinguish them. They were used largely for training and testing, with functional bombsights, navigation equipment and autopilots installed. Only the seventh of the ten B-47As had a functional tail turret, an Emerson A-2 with two .50 calibre machine guns. The rest had an aerodynamic fairing instead. Problems with Emerson – technical and managerial – meant that defensive armament was simply not available for use on the B-47A. Propulsion was provided by six General Electric J47-GE-11 turbojets (5,200lb of static thrust each) and 18 Aerojet 15KS-1000 solid rockets for takeoff.

The first B-47A was delivered to the NACA Langley Aeronautical Laboratory in July 1952 for flight testing. The B-47 was so far into the unknown in terms of design and performance that the NACA wanted to study the aeroelastic effects of high speed on the structure – essentially they want to watch the plane bend and warp at high speed. To this end they added a long pitot tube to the nose with angle of attack and angle of yaw sensors, as well a many more sensors to watch how the plane reacted. Above the fuselage, behind the cockpit, the NACA added an 'optigraph', a camera system that recorded and measured changes in the position of spots on the wings

Boeing B-47B (Model 450-67-27)

The B-47B was the first operational version of the B-47 with 399 built. The first one flew in April 1951 and at first glance the most obvious change from the B-47A was to the nose: gone was the symmetrical glazed nosecap. Now fitted was an asymmetrical metal nose with a protruding Y-4 horizontal periscopic bombsight to one side. A small hemispherical glass dome with a retractable metal cover gave the bombardier sitting in the nose a view of the target area below. Coupled with the APS-23 ground-viewing radar, the bombardier had a good shot at spotting, and with luck, hitting the target.

The end of the vertical stabilizer was squared-off, no longer rounded as it had been on the XB-47 and B-47A. The arrangement of the RATO bottles was also slightly adjusted; each side still had three rows of three columns, but the middle row was mover forward slightly.

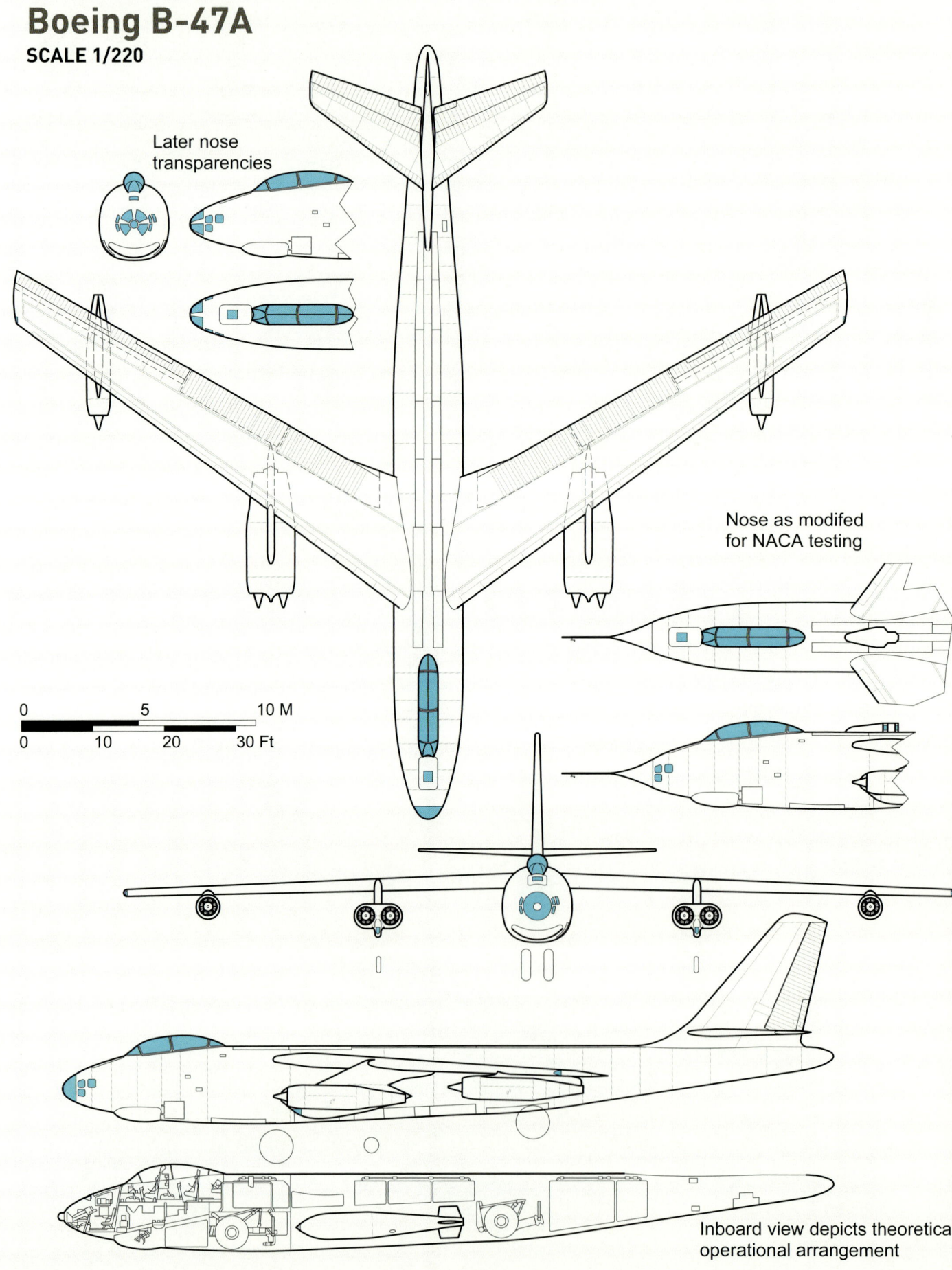

Boeing B-47B
SCALE 1/220

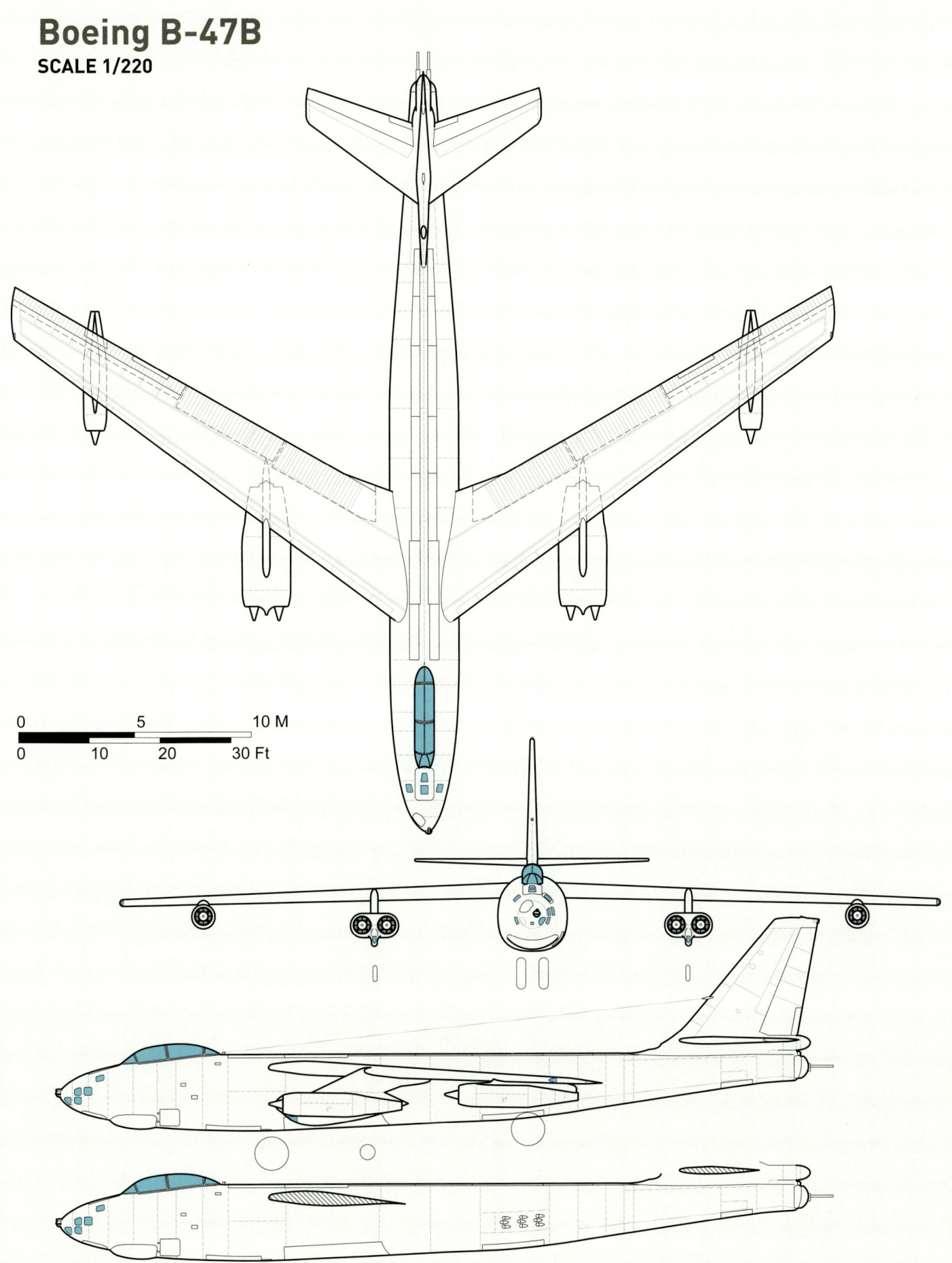

The B-47B was at last given both offensive and defensive weapons. The bomb bay was produced in two sizes – the long bay as introduced on the XB-47 and built on the B-47A, and a shorter bay. When originally designed, the bomb bay had needed to carry the largest 'blockbuster' bombs, both for themselves, and to serve as a proxy for the largely unknown world of atomic ordnance. But as the 1950s dawned, thermonuclear weapons were becoming understood. Megaton-class bombs were hefty chunks of canned sunshine to be sure, but nowhere near as massive as the Tallboy. Consequently the bomb bay was shrunk, just big enough to carry a single sizable thermonuclear weapon. The rest of the space was filled with a new fuel tank, boosting range.

The problems with Emerson meant that early B-47Bs were still delivered with no defensive armament, just an aerodynamic fairing at the tail. A number of the Emerson A-2 turrets were available, though they were not what was really wanted; still, two .50 calibre machine guns were better than nothing so with some modifications and improvements these were converted into B-4 turrets and 282 of them were installed on B-47Bs, starting in late 1952. Installing the B-4 turret resulted in a relatively long 'tail', and when it aimed left or right it worked as a small but effective rudder, playing hell with yaw control at high speed.

The General Electric A-5 turret with two 20mm cannon solved this problem. It was an outgrowth of the tail turret used on the Convair B-36; a spherical 'ball' turret that did not work like a rudder. Beginning in early 1955, the A-5 was introduced onto new-build B-47s and was retrofitted onto existing airframes.

One change was particularly unpopular with the crew: in order to save weight, the ejection seats were removed. In the event of a catastrophe in flight, the crew would be obliged to bail out the old fashioned way. A spoiler was added to the crew door on the lower left-hand side of the forward fuselage in the hopes that it would give them a fighting chance of gaining some distance from the fuselage before the airstream took hold of them.

An important improvement was the addition of an inflight refueling receptacle to the right side of the nose. A door would open to provide access, allowing the aircraft to take on fuel and vastly increase its range.

The B-47B was used to test the liquid rocket system as an alternative to the RATO bottles for takeoff. For propellants, the Aerojet YLR45-AJ-1 engines – two on each side of the plane, deployed outwards 20° for takeoff and stored within the fuselage afterwards – used white fuming nitric acid oxidizer and JP-4 fuel. This resulted in long, bright flames but no appreciable smoke. This was both good and bad: the lack of smoke made things a lot easier for any aircraft in line to take off behind the B-47, but the radiant heat from the bright exhaust torches did the skin of the rear fuselage no favours.

Tests carried out from 1952 through 1954 showed that the added thrust (5,000lb per engine) certainly reduced the takeoff roll, but added complexity, cost and even weight to the aircraft. Given the risks of fire and explosion from the propellants, especially the corrosive WFNA oxidizer which had a tendency to blow up if not treated with great care around organic contaminants, the liquid rocket system was abandoned.

The first 298 B-47Bs were fitted with General Electric J47-GE-23 engines, capable of 5,800lb of static thrust each. Later B-47Bs were given General Electric J47-GE-25 engines which produced 5,970lb of static thrust.

Boeing B-47C

No 'C' model of the B-47 was ever built, but that was not for lack of trying. The idea of replacing the six rather disappointing turbojets that equipped the early B-47s with four more powerful engines was kicked around for a while and gained some traction. Initially the aircraft appeared to be little more than a re-engined B-47, but it evolved into something else… something that pointed the way towards the future.

Boeing Model 450-16-17

In early 1949, Boeing produced Models 450-11-17 and 450-16-17. These were essentially the same design but with different engines: the former had four General Electric TE-TG-190-C11 engines (aka J47), while the latter had four of the more powerful Westinghouse J40-WE-6 turbojets. The latter, with substantially greater thrust (7,500lb to 5,610lb, military rating), seems to have received more study. However, while the J40 promised greater thrust, speed, altitude and range than the J47, the J47 had the benefit of actually working. The J40 turned out to be a protracted development nightmare, ultimately failing to produce a reliable production jet engine, and helping to usher Westinghouse out of the turbojet business.

The Model 450-16-17 had the round-ended vertical stabilizer of the B-47A and the asymmetrical nose of the B-47A. The engines were arranged as normal for the B-47, though the inboard pylons only had a single engine. The engine was offset to the outboard side of the pylon; this was done so that the outrigger stabilizing gear could fold up into the pylon itself rather than into an additional fairing built onto the engine nacelle. The diagrams show that the fuselage had the long bomb bay initially designed for the B-47 aircraft.

Boeing Model 450-16-17
SCALE 1/220

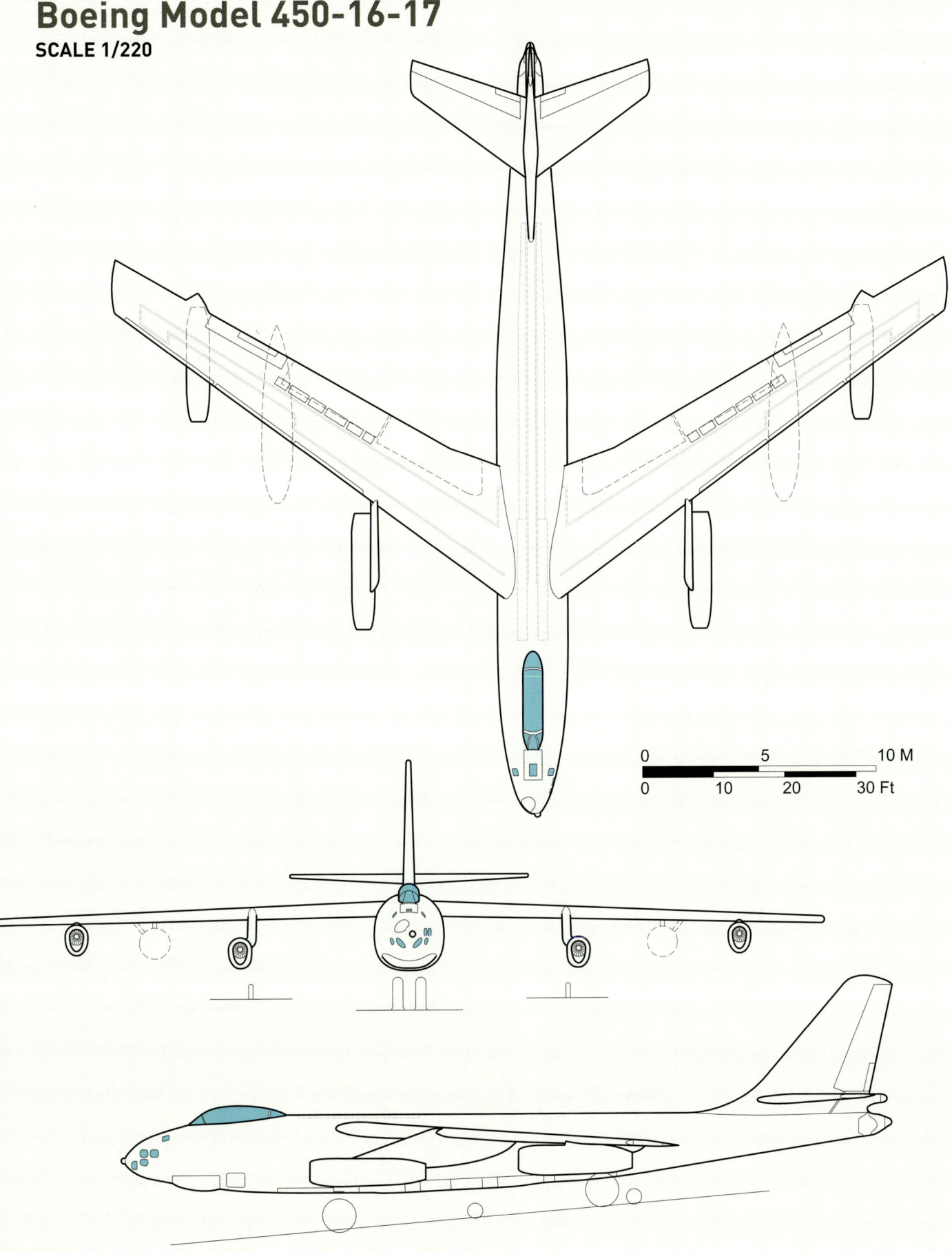

Boeing YB-56
SCALE 1/220

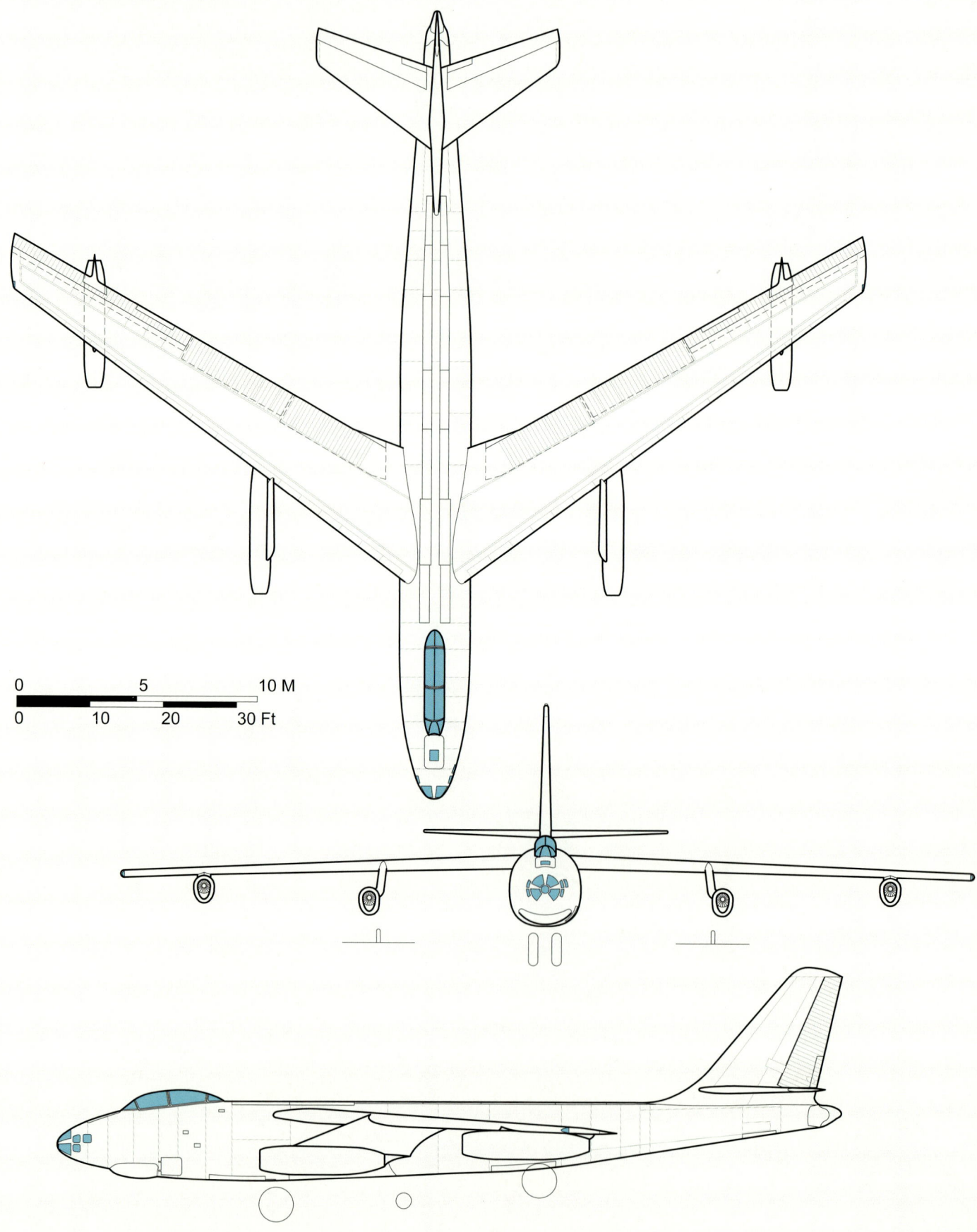

Boeing YB-56 (Model 450-24-26)

The same basic design was revisited in January 1950 and dubbed B-47C. The J40 engines were this time replaced with Allison J35-A-23 engines. These promised yet more thrust (9,700lb with afterburners; 8,200lb normal rating), and so initially looked very promising. As of early March 1950, first flight was projected for April 1951. Also at this time, the configuration change was felt to be important enough to merit a new aircraft designation – YB-56.

The YB-56 was to be essentially a B-47B with four engines arranged much as they were on the Model 450-16-17. In May 1950, contracts were drawn up to carry out the detailed design work as well as mockup construction of the YB-56. A B-47B was diverted from the assembly line for modification to the new configuration.

On August 14, 1950, the Air Force contracted Boeing to build 266 B-47Bs, 175 B-47Cs (the 'B-56' designation lasted only a few months before the Air Force changed its mind and went back to 'B-47C') and 35 RB-47Cs, definitely indicating interest in the design. However, the next day the contract was changed to 294 B-47Bs, 21 B-47Cs and 86 RB-47Cs. And then on December 27 the contract was changed again to 445 B-47Bs, 19 RB-47Cs and zero B-47Cs. Once again the contract was changed on April 12, 1951, to 471 B-47Bs, 48 RB-47Bs and 19 RB-47Cs. This was the end of the B-56 version of the B-47C, as it was cancelled in December 1952.

Boeing YB-47C/B-47X (Model 450-155-33)

In mid-March 1952, Boeing began work on a different version of the B-47C. Using the same engine layout as the prior iteration of the B-47C, it was intended from the beginning to use a different engine. The J35 engines had teething difficulties, reaching a maximum thrust not of 9,700lb but of 7,400lb. Consequently, other engines were planned. The initial thought was to use the Allison J71-A-5, capable of around 10,000lb of thrust. The J71 was itself undergoing development troubles though and was removed from consideration; it was replaced with the Pratt & Whitney J57. But the J57 was scheduled for use on the B-52, and by this point the B-52 was beginning to look like the next big thing. J57s were prioritized to the B-52 and could not be spared for the B-47C project.

The chord of the wings was broadened somewhat on the inboard section, slightly increasing wing area. But more obvious than the engines and the wing was a change in the fuselage. SAC commanding General Curtis LeMay, it seems, disliked the tandem seating of the crew in the B-47 and wanted a design with side-by-side seating. Thus the new B-47C, sometimes confusingly referred to as the B-47X, had an entirely new forward fuselage and cockpit. This looked utterly unlike the B-47 but it DID look like something that would come a bit later: the B-52. Just as he had disliked tandem crew seating in the B-47, he didn't like it any better in the XB-52 (or the YB-52). So when he

Boeing Model 450-155-33 YB-47C
SCALE 1/220

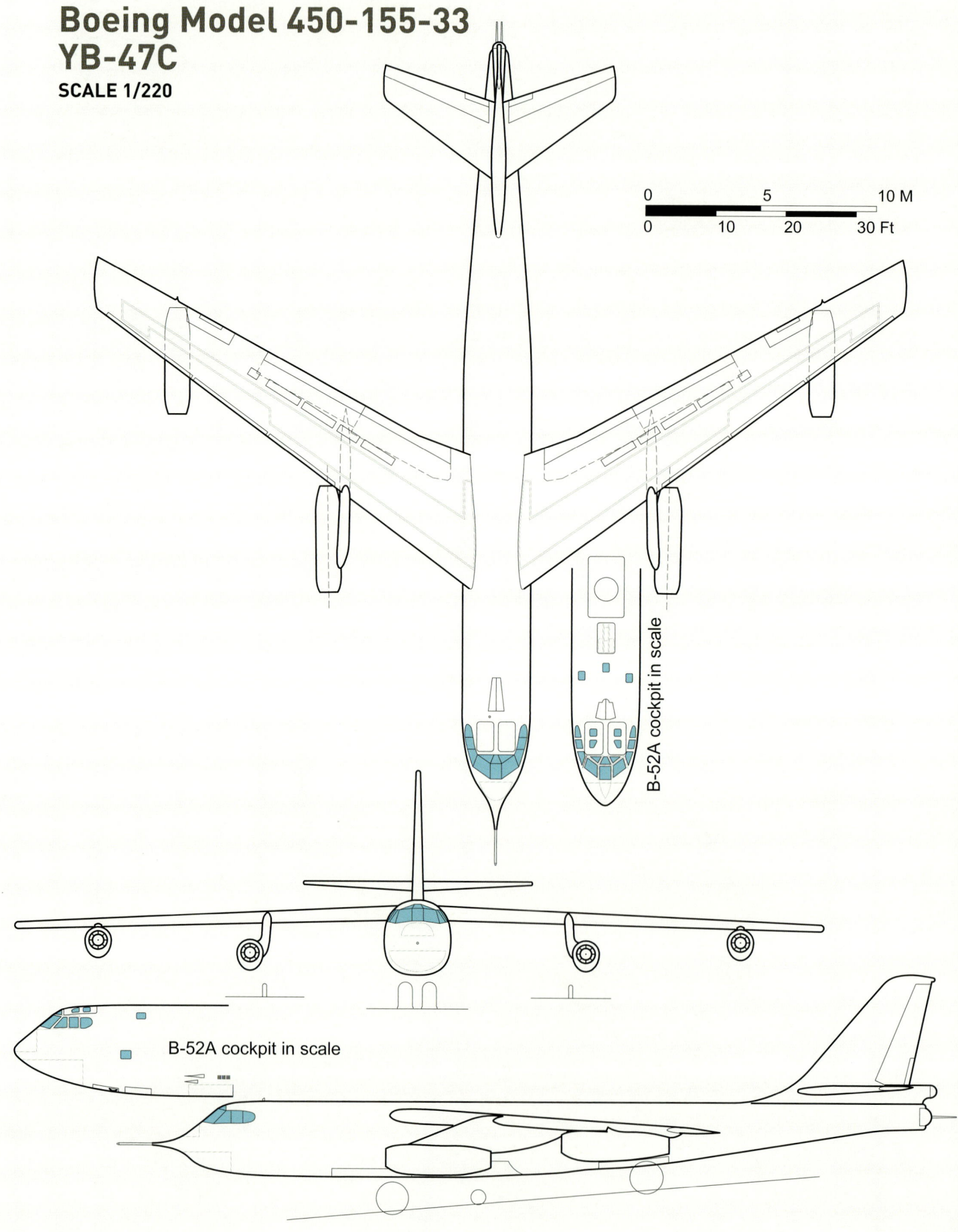

ordered Boeing to redesign the B-52's cockpit, they already had a layout ready to incorporate.

The B-47C forward fuselage as designed at this stage was a dead ringer for that of the B-52, just slightly smaller in depth and with a test flight pitot probe designed onto the nose. The crew of four (pilot, co-pilot, navigator-gunner and bombardier-assistant navigator) all sat in the forward fuselage and all were given ejection seats. The pilots sat up top, side-by-side behind the wide canopy; the navigators sat below and behind the radar, again side-by-side.

With this cockpit arrangement and the new, more capable J57 engines, Boeing designers had come up with an aircraft that should have impressed. LeMay liked the new cockpit but disliked the fact that this put the aircraft in competition with the B-52. As a result, the project was killed off.

Boeing B-47X/RASCAL

The B-47X was described as being intended to carry the GAM-63 'RASCAL' standoff missile. Described further in the DB-47E section, this was a relatively large liquid-fuelled missile that hung off the side of the aircraft in a rather ponderous fashion. No diagrams are currently available depicting this, so the reconstruction here is pieced together from the B-47C and DB-47E diagrams. It is provisional, but probably reasonably accurate. No weight or performance data is available. Given the considerable difficulties that the GAM-63 had, it is unlikely under any conceivable set of circumstances that the B-47X would have been built specifically to carry the missile.

Boeing Model 450-166-38

Dating from either the end of 1953 or early January of 1954, Model 450-166-38 was perhaps the final design

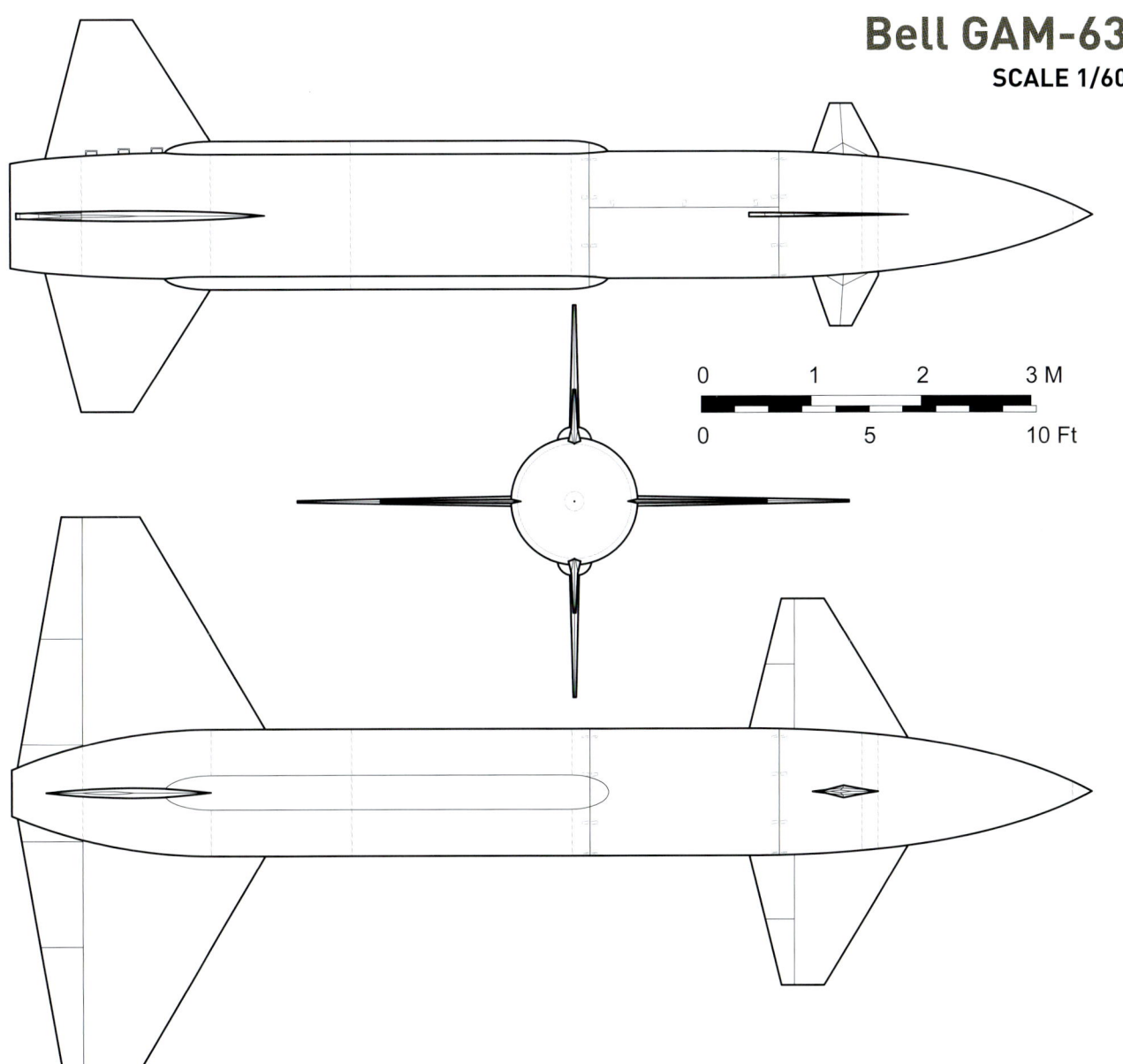

Boeing B-47X
SCALE 1/250

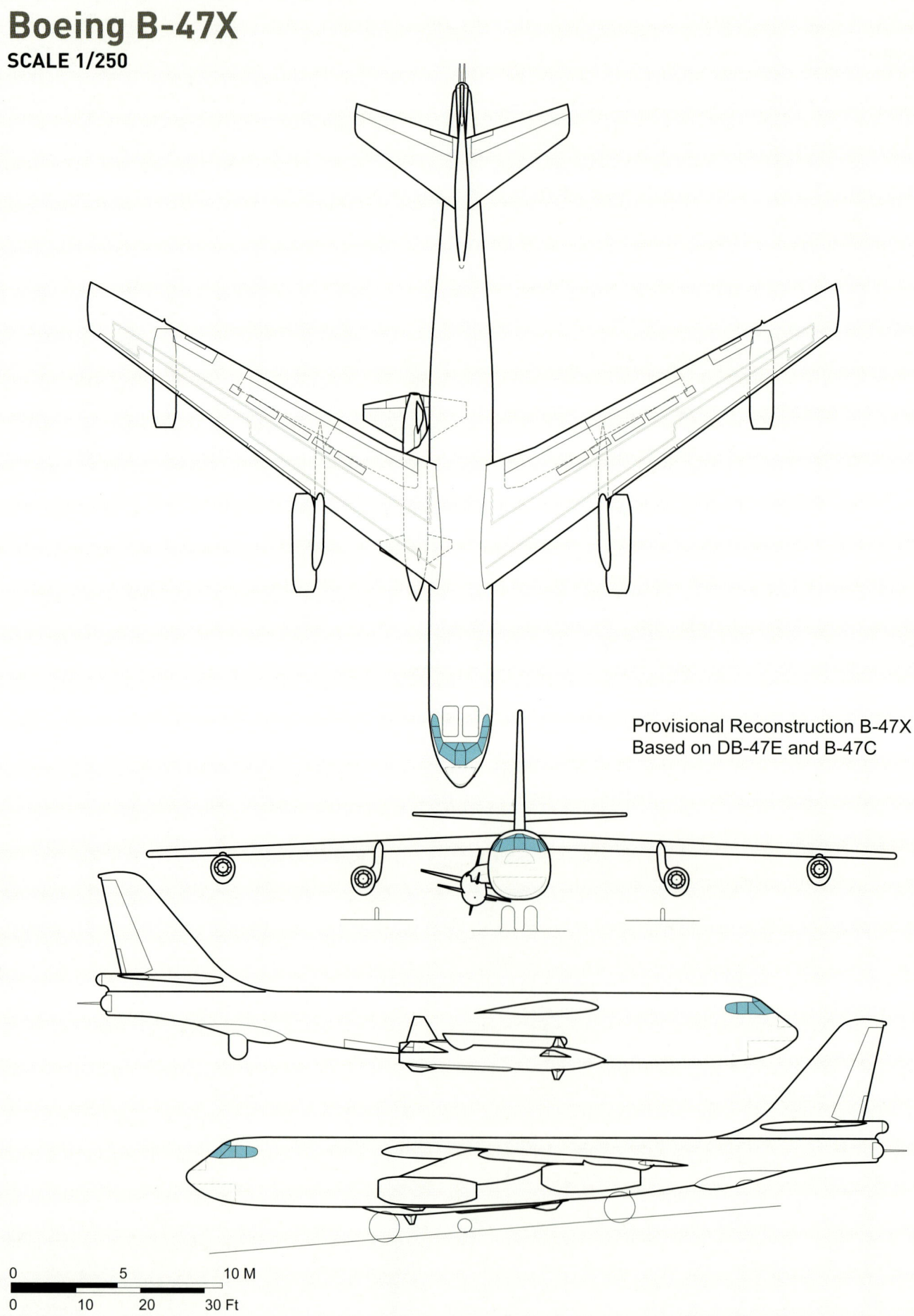

Provisional Reconstruction B-47X
Based on DB-47E and B-47C

B-47 Development and Projects

to try to incorporate four engines in the 'B-47C' pattern. The fuselage was essentially that of the B-47E; the four J57 engines were laid out as with the B-47C designs. The major change was a substantial increase in area of both the wings (33.5%) and the horizontal stabilizers (45%). The better engines and increase in wing area permitted a higher maximum ceiling, about a mile above the standard B-47E, as well as a combat

Boeing Model 450-166-38
SCALE 1/220

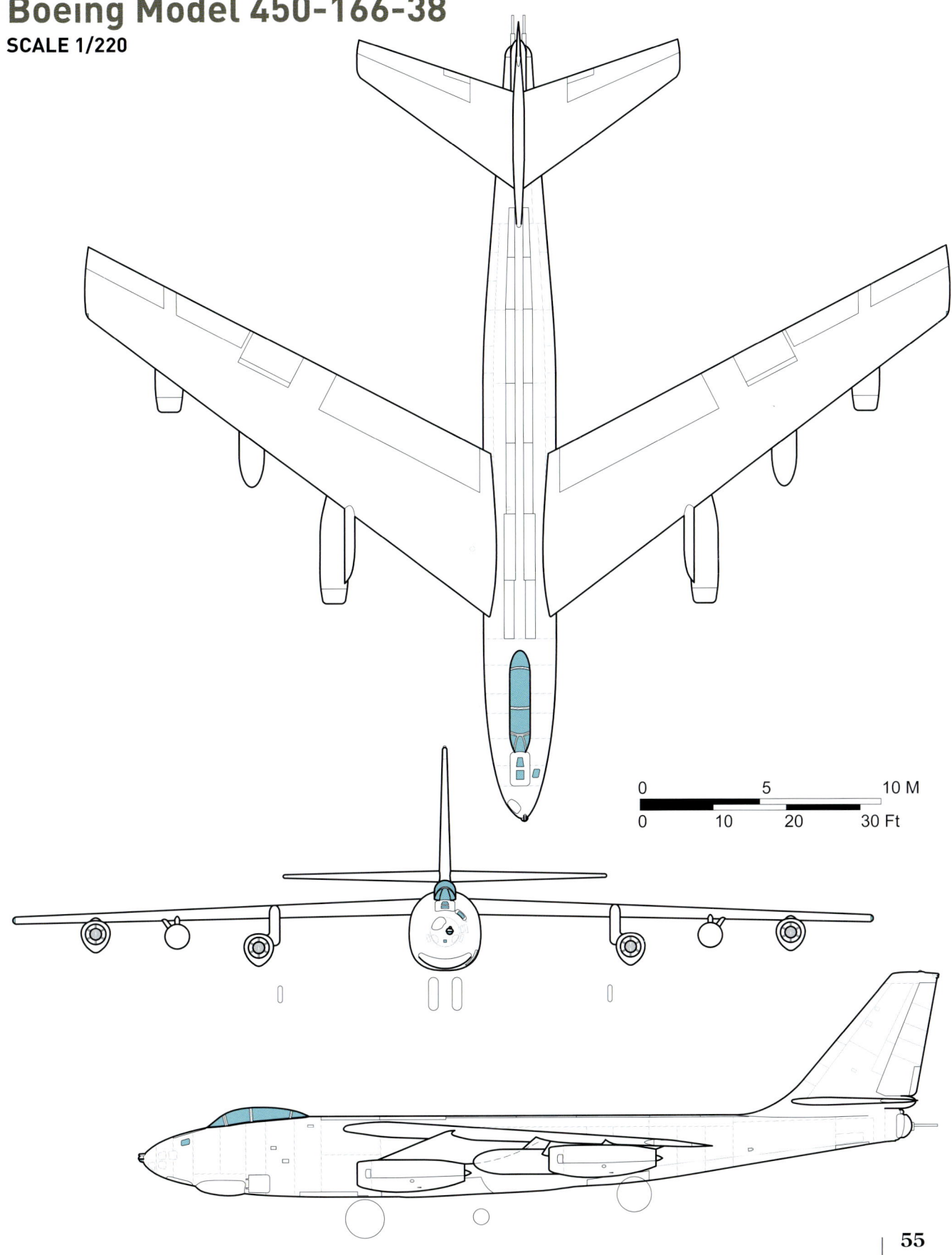

55

radius increase of more than 1,100 nautical miles. Maximum speed was also substantially improved. However, this all came at the cost of increased dry weight and greatly increased maximum takeoff weight; this necessitated a structural strengthening of the landing gear.

For offense it could have either the short or long bomb bay, with a maximum load of 10,000lb; alternatively it could carry a RASCAL missile externally. It retained the twin 20mm cannon in the tail with 700 rounds each.

Boeing B-47D

Turbojets gave the B-47 unprecedented speed for a bomber, far outpacing the best that its piston engined cousins could ever hope to achieve. However, early turbojets were fuel hogs, burning through propellant at astonishing – and range-discouraging – rates. But there was a compromise: the turboprop. A straight turbojet generates thrust by compressing and heating air by burning it with fuel, expelling the combustion products as a high velocity jet. But if that energy was used instead to provide the mechanical power to turn a propeller, the mass of air that would be pushed aft to generate thrust would be greatly increased. The average velocity of the air would be lower, but the mass flow rate higher. The end result is a notable increase in thrust and fuel economy, at the expense of a lowered top speed.

Another promising feature of turboprops over piston engines was the fact that the components of the turboprop rotate and don't reciprocate. The turbines simply spin, rather than banging back and forth like the pistons and valves in the earlier engines. This was expected to result in lower stresses, even with the high temperatures, producing engines that could operate much longer and be more reliable. After all, the installation of the massive radial engines of the Convair B-36 was designed so that flight mechanics could access them in flight and perform some basic servicing. And turboprop engines did eventually surpass the piston engine in the area of reliability.

Boeing Model 450-9-3

Turboprop derivatives of the B-47 were under study before the B-47 entered service. The first known of these concepts was the Model 450-9-3 (dating from May 1948 at the latest), very much an XB-47 with the six turbojets replaced with four Allison Model 500 turboprops. Each turboprop engine was housed in a separate nacelle, suspended beneath the wings under short pylons (notably shorter than the pylons used to hang the B-47's inboard dual-jet nacelle – important for moving the propeller arc upwards and away from the runway).

The Allison Model 500 was given an official government designation as the T40. It was not actually a single powerplant but instead two Allison T38 turboprop engines attached side by side and connected through a gearbox to drive contra-rotating propellers. The T40 was used and proposed on a range of aircraft (it was the powerplant of choice on aircraft as diverse as the Douglas A2D Skyshark, Convair R3Y Tradewind, North American XA2J Super Savage and the VTOL tailsitters the Convair XFY and Lockheed XFV). But the development of the T40 was problematic and the gearbox was perpetually troublesome; no aircraft fitted with that complex engine ever entered mass production or saw real service. While the T40 eventually failed, the T38 engines it was based on were further developed into the T56 (and then the Rolls-Royce T406), which has powered the C-130 Hercules for more than half a century.

The Boeing Model 450-9-3 was, depending on how you look at it, either a four-engine or eight-engine design. Either way, the nacelles were quite small, especially compared to the sort of space that would have been taken up with piston engines. The six-bladed propellers (three turning in each direction) would have been 15ft in diameter. Wing, fuselage and tail surfaces would have been essentially the same as the XB-47. Top speed at altitude would have been 510mph – substantially slower than the jet-powered B-47, but much faster than any piston engined bomber.

Boeing Model 450-30-10, Model 450-31-10

Models 450-30-10 and 450-31-10 from February 1950 were part of a series of medium bomber potential growth studies ranging from Model 450-25-10 through 450-64-10. The studies included both turbojet and turboprop variations on the B-47 theme, mixing and matching then-known powerplants in a wide range of configurations. Several are included here, though this is not an exhaustive list. These Boeing designs followed the publication of a 1949 RAND Corporation study which concluded that turboprop-equipped bombers were the best way to successfully penetrate deep into Soviet airspace. As often happens, a single report gained prominence for a moment and caused not only a stir of interest but a change in spending priorities.

Models 450-30-10 and 450-31-10 fitted a standard B-47B configuration with four T40 turboprops. The 450-30-10 tried something new with the location of the engines; instead of hanging below the wings in podded nacelles, here the Allison T40-A-8s were mounted above and ahead of the wings in long cylindrical nacelles that extended well past the

Boeing Model 450-9-3
SCALE 1/220

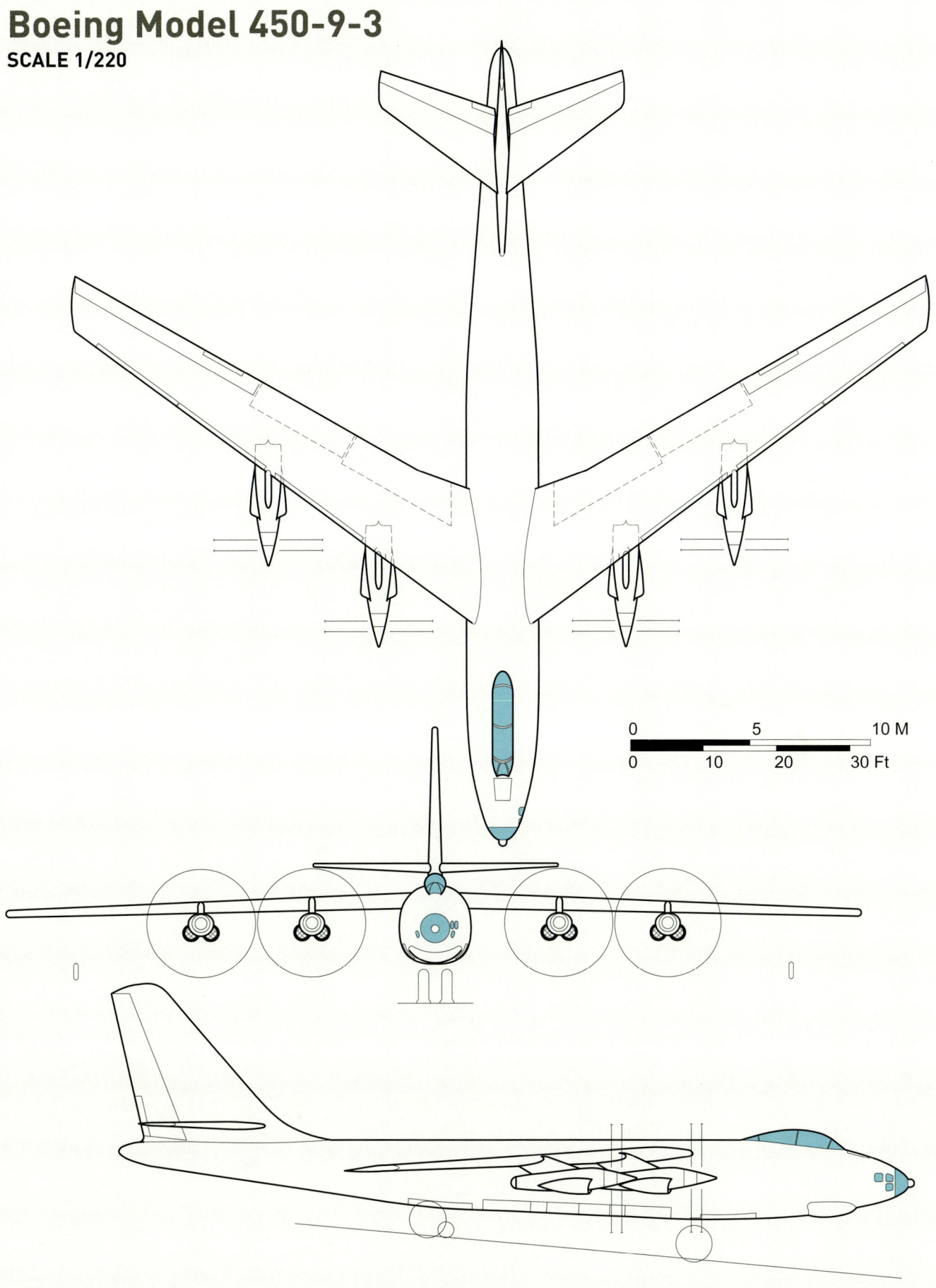

Boeing Model 450-30-10
SCALE 1/220

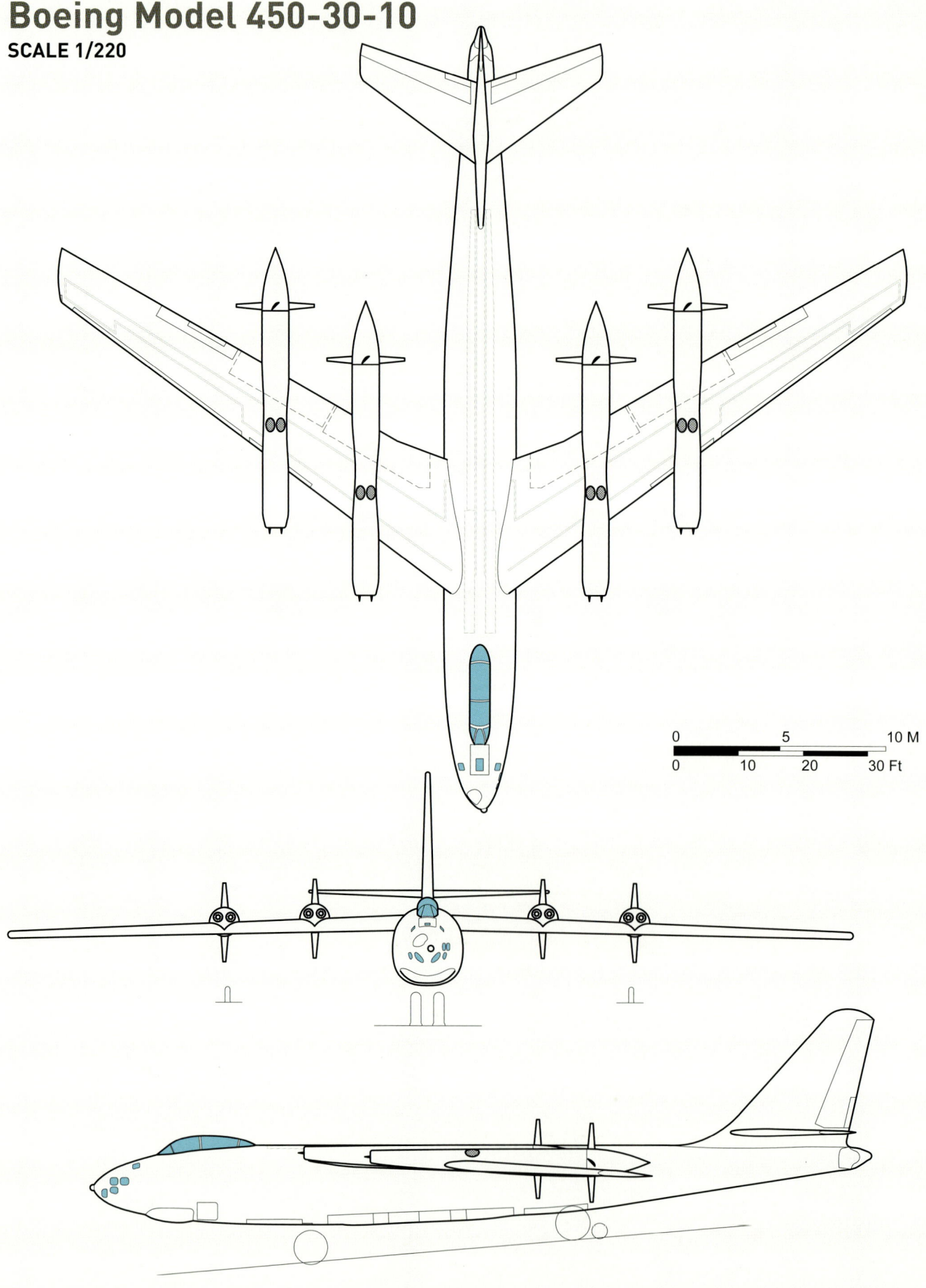

Boeing Model 450-31-10
SCALE 1/220

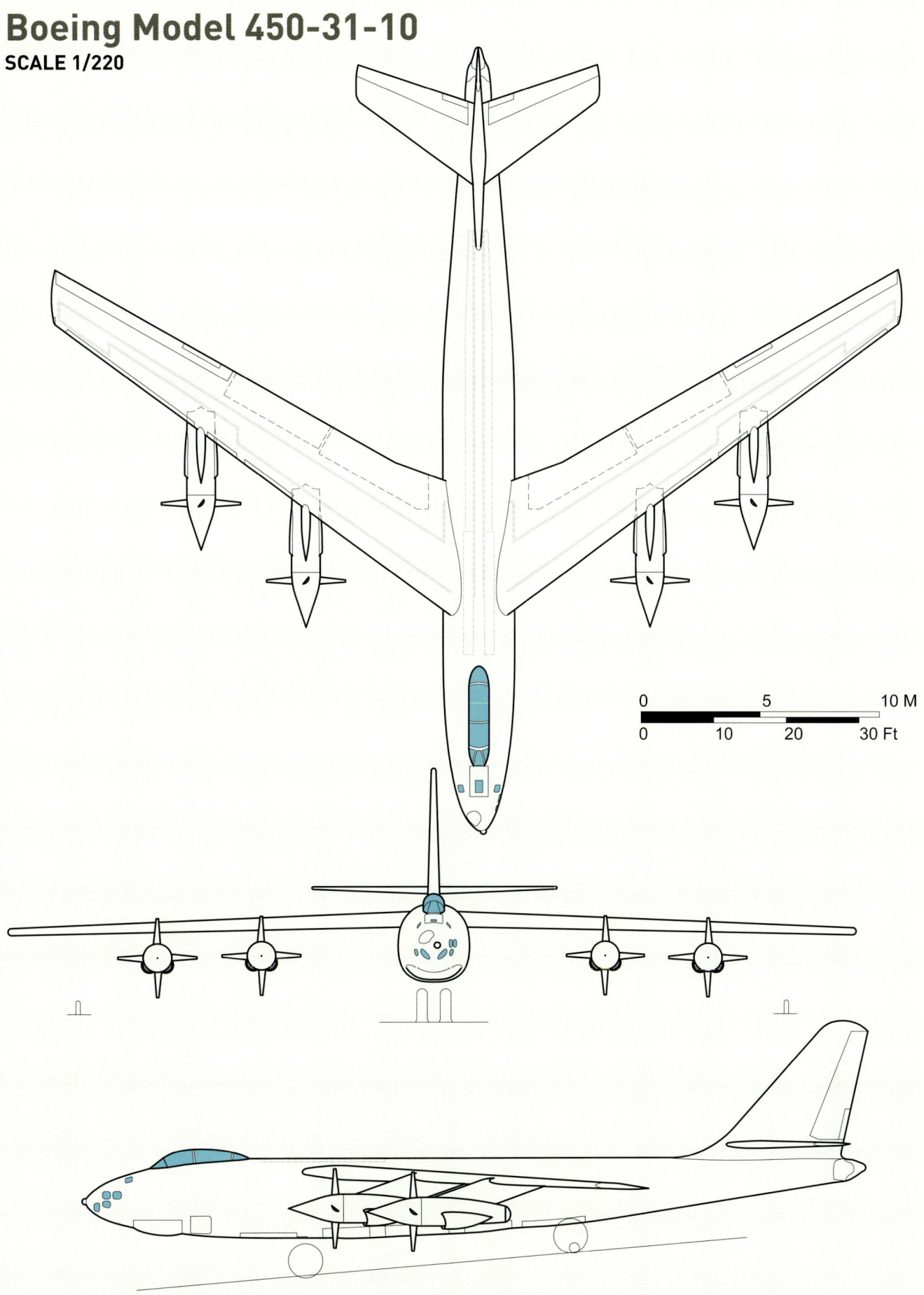

trailing edge. The exhaust vented through two nozzles at the top of the nacelles just aft of the leading edges of the wings. Access to the engines for maintenance and replacement should have been fairly convenient given their location ahead of the wings. The Model 450-31-10 used a similar podded nacelle layout to that of the 450-9-3, looking much the same except for one difference. That difference, which the Model 450-30-10 shared, was a lack of contra-rotating propellers. Instead, both designs featured pointed and seemingly somewhat bulbous propeller hubs with four surprisingly small diameter (11ft) propellers. What these propellers lacked in number and diameter they made up for in speed – the tip speeds were meant to easily exceed the speed of sound.

It should be noted that this engine was mated to supersonic propellers on the XGF-84H 'Thunderscreech'. The name was well chosen; with a single T40 and a three-bladed supersonic propeller it achieve dubious fame as being possibly the loudest aircraft ever built. While sitting motionless on the ground with the engine idling, shockwaves could be seen radiating away from the aircraft for a distance of some hundreds of feet. Nausea, headaches and seizures were reported among ground crew unfortunate enough to be anywhere near it. The idea of an aircraft fitted with four of these engines boggles the imagination; not only would it have turned the brains of ground crew into silly putty, low altitude overflights of foreign territory would likely have been met with charges of crimes against humanity. It seems fairly certain that the structure and skin of the B-47 would have suffered badly under the assault of sonic booms constantly raining down onto it from the tips of the propellers.

Boeing Model 450-32-10 and Model 450-33-10

Boeing's speculations about turboprops with supersonic propellers did not end there. Contemporary to the 450-30-10 and 450-31-10 (February 1950) were the 450-32-10 and 450-33-10. These differed in having Pratt & Whitney powerplants rather than Allison, and these were rather different powerplants. The 450-32-10 used a basic prop version of the successful J57 turbojet. Rather than being a regular turboprop, this featured a mechanically unconnected 'free turbine' sat in the turbojet's wake and was spun up by the exhaust gases. This free turbine then span the actual propeller. Unsurprisingly, this meant the propeller had to be at the rear of the nacelle.

Once again, the engines themselves were mounted well ahead of the leading edge with twin exhausts on the upper side, just behind the leading edge. A long drive shaft connected to the pusher propellers behind the wing trailing edge. The 450-33-10 used a similar engine, the JT3A, which was the company designation for the J57. In this case, it appears that the engine was to be modified into a conventional turboprop. The J57 was so modified a few years later, producing the XT57 turboprop which was to power the unbuilt Douglas C-132 cargo plane. The XT57 turned out to be the most powerful turboprop ever built in the US. Whether that's exactly what the turboprop for the 450-33-10 would have been is uncertain, but what is certain is that the 450-32-10 and the 450-33-10 would both have had 15ft diameter four-bladed supersonic propellers.

The powerplants of the -32-10 and -33-10 might seem less capable than those of the -30-10 and -31-10, given that there were only half as many J57/

Boeing Model 450-32-10
SCALE 1/220

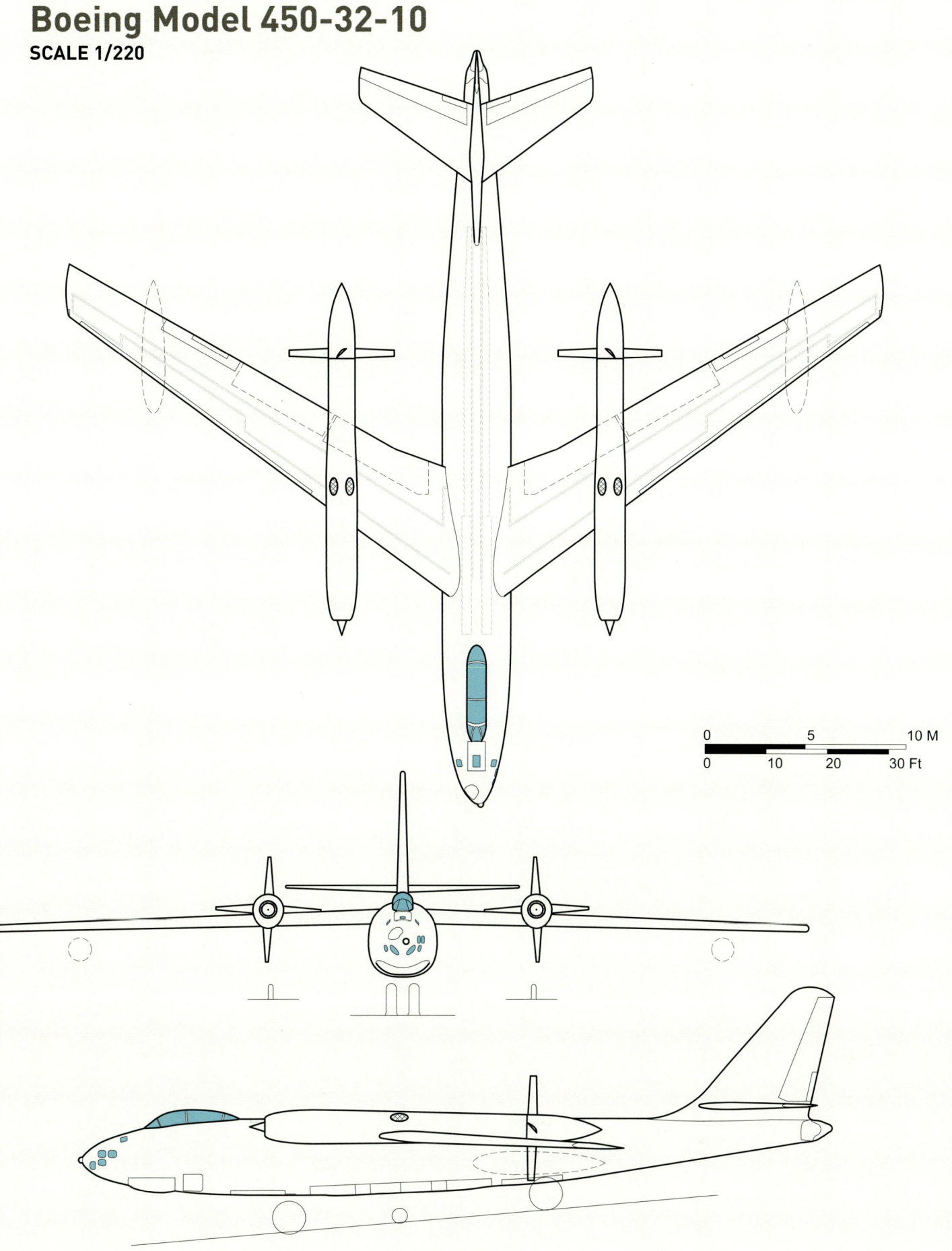

Boeing Model 450-33-10
SCALE 1/220

XT57 turboprops (two per plane) than there were XT40 turboprops (four per plane). But the XT40s only produced around 5,500hp each to the XT57's 15,000hp per engine. With half as many engines, the latter two aircraft would have had a third more power – but would have poured that power into what appear to have been virtually identical propellers. Attempting to extract adequate thrust from those props doubtless would have been an interesting and indeed deafening experience.

Boeing Model 450-59-10

Model 450-59-10 of May 1950 returned to the early notion of four podded Allison XT40 turboprops beneath the wings. This design in fact looks to be a warmed-over Model 450-9-3, except with a B-47B rather than XB-47 structure. Otherwise there is not much to distinguish them.

Boeing Model 450-60-10

The next design in the series, Model 450-60-10, also of May 1950, tried yet another propulsion configuration: four Pratt & Whitney PT-2E turboprops. These would be redesignated T34 and the Douglas C-133 Cargomaster was their sole major user. They were roughly as powerful (5,700hp for the initial versions) as the T40, but without the complex linking of two separate turbojets. The result was a somewhat more conventional looking turboprop engine nacelle with the open-nosed prop spinner serving as the engine inlet and the exhaust exiting the nacelle at the rear.

Boeing XB-47D (Model 450-162-28)

The goal of a turboprop-equipped B-47 finally came to pass with the Model 450-162-28. Design work, which entailed the modification of the B-47B airframe to replace the inboard turbojet engine pylons with turboprops, began in 1950 and was completed in February 1951, with a contract issued shortly after in April. A mockup was completed by January 1952. Two incomplete B-47Bs were moved from the assembly line to a separate area for modification and two Curtiss-Wright YT49-W-1 turboprop engines – derivatives of the J65 turbojet, itself a licence-made Rolls-Royce Sapphire – were fitted into all-new nacelles and equipped with 15ft diameter four bladed adjustable props. The undersides of the cylindrical nacelles were outfitted with sizable fairings into which the outrigger landing gear would stow. The outrigger gear now took on special importance: not just keeping the aircraft more or less level, but keeping the props from striking the ground. The outboard J47 turbojets remained, unmodified.

The first flight was projected for early 1953, but engine problems slowed progress and the first of the two B-47Ds would not lift off until August 26, 1955. The second flew on February 15, 1956, well behind schedule. The end result of all this was an aircraft that was very nearly as fast as the B-47B, but with much greater range. By that metric it was a successful demonstration of the utility of turboprops for high-speed long-range bombers. However, the two airframes only flew for 50 hours in total and during that time had required either an engine or propeller overhaul every 1.8 hours on average.

Additionally the B-47D exhibited some unfortunate flight characteristics such as poor engine out performance and porpoising while attempting to refuel from a KC-97. These difficulties, coupled with the rapid improvements in turbojets and in-flight refuelling techniques, technologies and methodologies, led to the idea and the airframes being scrapped. The final B-47D flight was in June 1956.

Boeing B-47E (Model 450-157-35)

The B-47E was the definitive version of the B-47; with 1,581 B-47E and RB-47E aircraft built, it was far and away the most-produced version of the plane. First flying in January 1953, early B-47Es were visually hard to distinguish from the B-47B. But soon the B-47E began to lose not only the windows around the nose but also the 18 RATO bottle ports on the rear fuselage. Even with General Electric J47-GE-25A engines with water injection and a maximum thrust of 7,200lb (which it could produce for only a brief period), the B-47E would struggle to lift off from shorter runways. It still needed the RATO assist, but the bottles were removed from the interior of the aircraft. Instead they were increased in number to 33 and attached to a 'collar' that fitted under the rear fuselage. The rockets would be fired in staggered order to spread out the thrust and after burnout – when the aircraft would hopefully be well airborne – the collar would be dropped, taking all its weight with it.

The ejection seats that had been taken away from the B-47B were returned for the B-47E, a change crews doubtless appreciated. The pilot and co-pilot would eject upwards, while the bombardier/navigator would eject downwards. Downwards ejection is obviously not the preferred method, particularly at low altitude or on the ground. But for most flight emergencies it was preferable to the much more difficult and time consuming process of bailing out through the side hatch.

It had been originally intended that the B-47 would fly high, evading interception through a combination of high speed and altitude beyond what Soviet interceptors could quickly achieve. But it soon became apparent that the B-47 did not fly either high or fast enough. If it flew fast and low, though, it

Boeing Model 450-59-10
SCALE 1/220

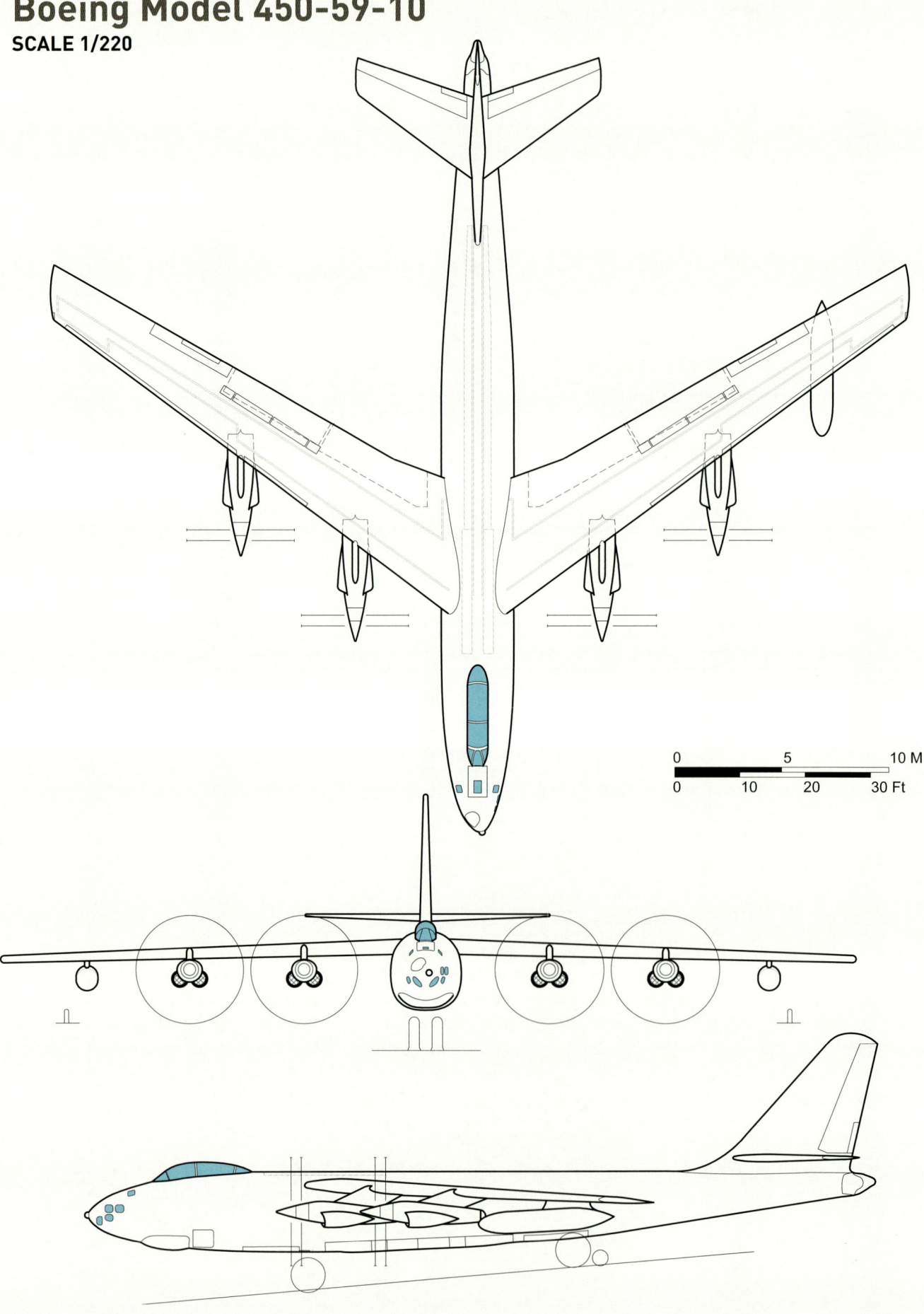

Boeing Model 450-60-10
SCALE 1/220

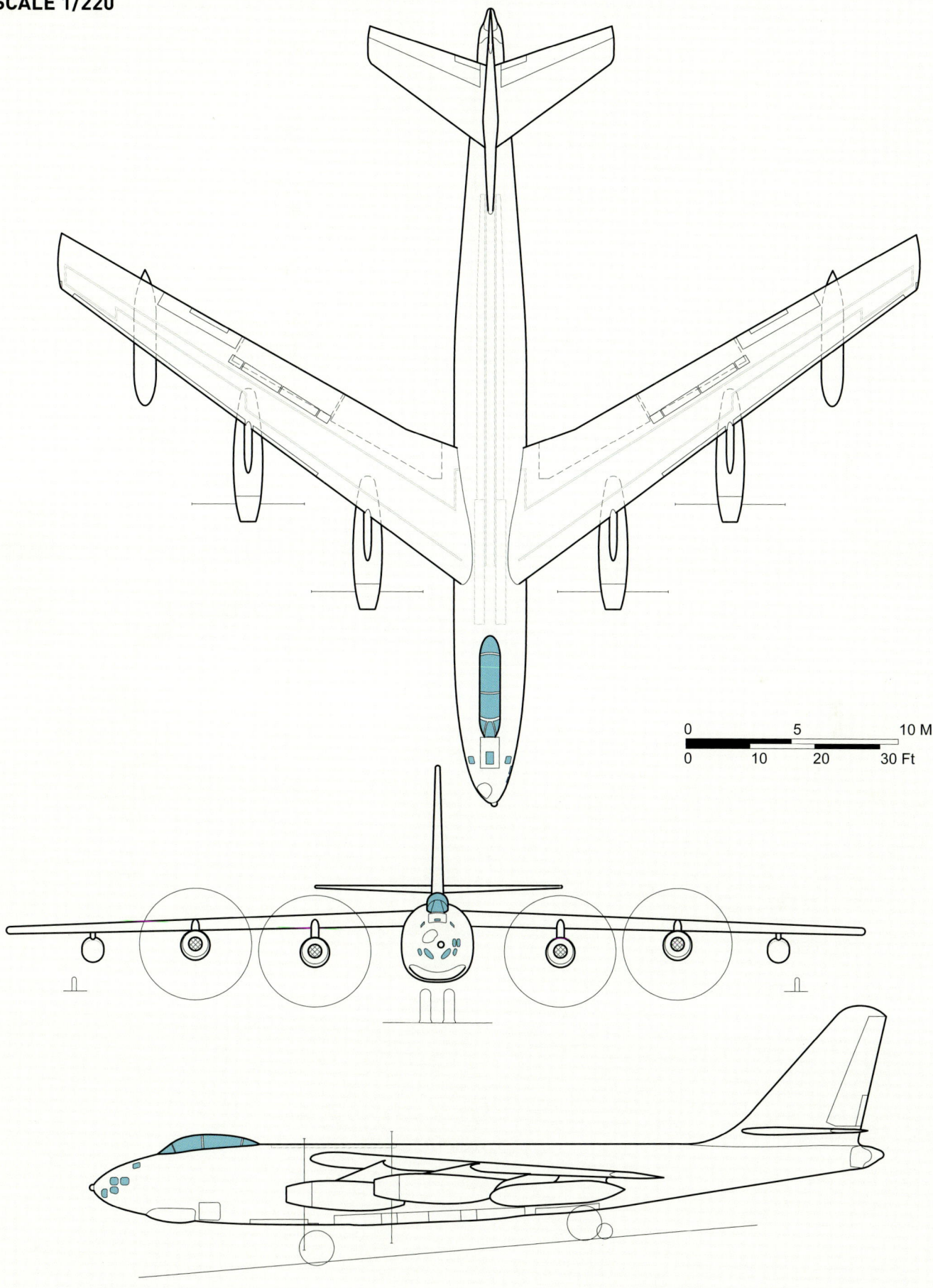

Boeing B-47D
Model 450-162-28
SCALE 1/220

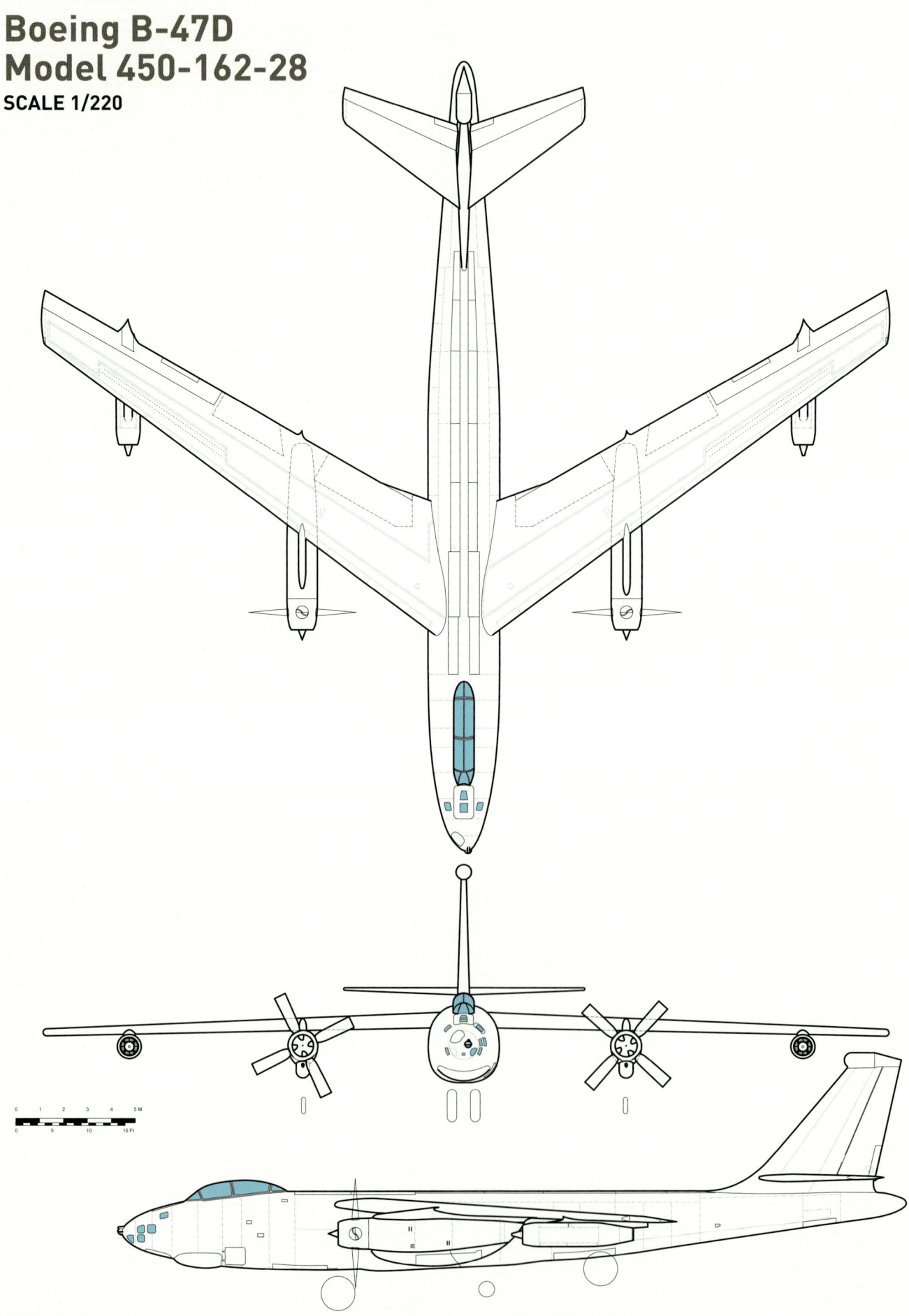

Boeing B-47E
SCALE 1/220

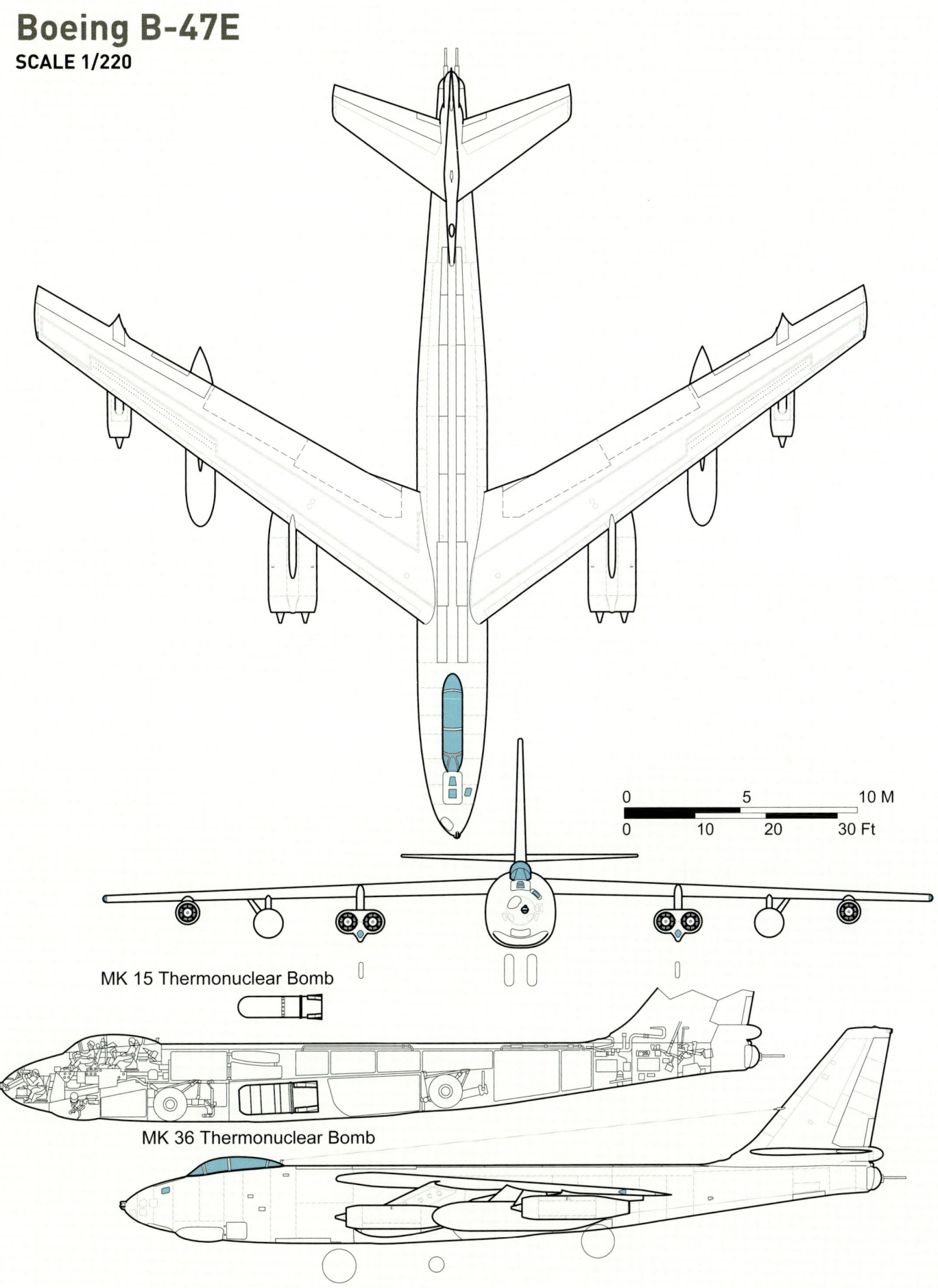

MK 15 Thermonuclear Bomb

MK 36 Thermonuclear Bomb

could evade detection by Soviet radar systems until it was too late. The problem with low level nuclear bombardment is that the bomber is not very far away from the bomb when it goes off. With the United States Air Force taking a generally negative view of mandating suicidal one-way missions, in 1955 another approach was worked out: the 'low altitude bombardment system' (LABS) where the B-47 flew low and fast, then suddenly pitched up and did a half roll. In this way, the aircraft was very quickly heading back the way it had come. But what made the process work was that while it was pitching up, it released the bomb at a specific point in the arc.

The bomb was then tossed high into the air to follow a ballistic arc; this gave the bomber the time it needed to get out of Dodge. Testing showed that bomb accuracy was good enough, given that the bombs were megaton yield. The LABS system was tested by B-47Es, with considerable study made of how the aircraft structure responded to the stresses and strains imparted by the challenging manoeuvres.

The Air Force wanted the B-47E in large numbers, and was determined to get them. Boeing-Wichita, however, was not able to produce them as quickly as the Air Force would have liked, so Boeing's main competitors, Douglas and Lockheed, were brought on board to build B-47Es under licence. Boeing eventually produced 931 B-47Es and RB-47Es; Douglas made 274 in Tulsa, Oklahoma; Lockheed made 386 in Marietta, Georgia.

The most numerous of the major variants of the B-47E was the RB-47E strategic reconnaissance aircraft. First flying in July 1953, this reconnaissance version was given a wide range of modifications, both permanent and modular. The nose was extended by almost 3ft to accommodate the specialist equipment needed by the camera operator, replacing the bombardier. The offset inflight refuelling receptacle was move to a position directly in front of the pilot's canopy. Curiously, the built-in 18-bottle RATO system used on the XB-47, B-47A and B-47B returned on the RB-47E. The bomb bay was filled with photographic equipment, with up to 11 cameras carried. It was a capable machine, but its service life was brief… the first operational B-47 of any kind to be retired was an RB-47E, sent to storage in October 1957.

In the end a total of 255 RB-47Es were built, though the last 15 were finished as RB-45K aircraft specifically designed for weather reconnaissance.

Boeing RB-47H (Model 450-171-51)

Electronic countermeasures and intelligence were the specialties of the EB-47E and RB-47H. The EB-47E was converted from the standard B-47E; several variants of the EB-47E were produced, but the main change was the addition of a pressurized capsule within the bomb bay. Two electronic warfare specialists occupied ejection seats of dubious effectiveness and operated the jamming systems that the aircraft was equipped with, raising the crew complement to five. The EB-47E would fly at high altitude along with other B-47s as they penetrated enemy airspace, using chaff and multiple active radar jamming transmitters to blind enemy sensors. The jammer aircraft themselves would of course be important targets. As Soviet surface-to-air missile technology improved, the survivability of high-altitude subsonic bombers soon fell into serious doubt and the mission seems to have ended in 1964. Another version, the EB-47E 'Tell Two', was festooned with external antennae for the specific goal of intercepting telemetry data being emitted from the Soviet launch site Kapustin Yar and the Tyuratam Launch Complex. During missions one of the three EB-47E (TT) craft would fly at high altitude over either the Black Sea or northern Iran to intercept Soviet transmissions.

The RB-47H was an important departure. Thirty-five of these electronic countermeasures and reconnaissance planes were built by Boeing-Wichita, the first flying in June 1955. The RB-47H is most readily distinguished from the rest of the B-47s thanks to a somewhat blunt black nose. This had the appearance of a radome, but was in fact a fibreglass fairing protecting a bank of 18 antennae for the AN/APD-4 electronics intelligence system. This automatically scanned through a range of radio frequencies, seeking out signals of interest. Once a signal was located, three onboard electronic warfare officers could examine it. They occupied a pressurized capsule in the bomb bay, a development of the capsule employed by the EB-47E. Numerous other receivers were located on the outer surface of the aircraft, most notably around the underside of the rear fuselage. These were all passive receivers. Yet more receivers were carried in pods hung under the wings or projecting from the sides of the fuselage.

The RB-47H had two feats that it could boast of. Firstly, it directly overflew Soviet territory many times in the quest to obtain electronic intelligence. And secondly, the RB-47H was the last variant of the B-47 to remain in Air Force service. The role of bomber had been supplanted by the mighty B-52 and Soviet surface-to-air missiles made overflights of subsonic and relatively slow-flying aircraft such as the B-47 unsafe. So on December 29, 1967, an RB-47H departed for Davis-Monthan to enter storage. The day of the B-47 was over, just a few days over 20 years after the first XB-47 had taken to the skies. The B-47's replacement, the Boeing B-52, stands a good chance of staying in service for nearly a century.

Derivatives

As with many successful aircraft, numerous modified versions of the aircraft were created beyond the normal variations of production runs. These include numerous and varied reconnaissance and electronic warfare versions, which are largely outside of the scope of this book due to their being fairly well known. But some other models were a bit more unusual and interesting.

Boeing DB-47E/RASCAL

World War II saw the development and deployment of the first practical – or nearly so – air launched guided rocket weapons, as well as the first liquid propellant rockets. It was probably inevitable that shortly after the war a large long-range surface attack missile would be proposed. In April 1946 the Air Force approved the MX-776 programme, the goal of which was a large liquid-fuelled missile to be carried by a bomber such as the B-36. Initially designated B-63, it was redesignated as the GAM (Guided Air Missile)-63 RASCAL, a somewhat tortured acronym for RAdar SCAnning Link. The missile was meant to be guided in order for precise attacks, but the technology was not yet there for the missile to guide itself so a complex system was planned.

Prior to launch the carrier aircraft would use onboard navigation systems to precisely locate itself so that it knew where it – and thus the missile – was in relation to the target. After launch, the missile would used an inertial tracking system to roughly locate itself. Final targeting was the unique aspect of the design, the missile being fitted with its own terrain scanning radar. As it approached the target, it would relay an image of the terrain back to the launch aircraft. The missile operator would compare this against known radar maps of the target area and relay course corrections back to the missile. The goal was a circular error probability of 500ft.

RASCALs were test-launched from B-50 and B-36 bombers, but in 1951 it was determined that the B-47 would be the operational carrier aircraft. The RASCAL was a very large weapon to be carried by the B-47; at 4ft in diameter and 32ft long there was no possibility of stuffing it into the bomb bay. So it was to be hung off the right side of the fuselage below the wing. Initially B-47B aircraft were planned, with required modifications turning them into DB-47Bs ('D' for 'Director'), but the B-47E was later found to be a better candidate. Thus both DB-47B and DB-47E aircraft were modified and flew with the GAM-63. Some 30 DB-47Bs were to be converted and six DB-47Es (two being YDB-47Es and two being JDB-47Es).

The performance of the RASCAL was, by later standards, rather anemic. It was not the sort of hypersonic ballistic missile that later decades came to recognize; instead it was essentially a rocket powered airplane. At launch it weighed 18,000lb with thrust provided by three Bell XLR67-BA-1 liquid propellant rocket engines. As each engine only produced 4,000lb of thrust, at ignition the thrust to weight of the missile was well under one: if it was sat on its tail and fired vertically, it would simply remain there while it burned off propellant. It was held aloft thanks to the lift produced by its substantial wings. After launch – normally at around 35,000ft – the RASCAL would climb to around 60,000ft. Maximum speed would be about Mach 2.95 with a range of only 75 to 90 nautical miles, the last 20 of which would be in a shallow dive. In the event that the missile could not be correctly launched, it could be dropped as a gravity bomb.

The purpose of the RASCAL was to give the crew some standoff distance from the target, rather than to extend the performance of the B-47. A range of only 90 miles, at the cost of added weight and substantial extra drag, actually reduced the B-47's reach. Worse, numerous test launches, and more numerous attempted but failed test launches, showed that the RASCAL was abysmal in reliability and dismal in accuracy. Much of this had to do with the stingy way in which the programme was funded, a result in no small part of Curtis LeMay's antipathy towards it. RASCAL was mercifully cancelled in late 1958, with the DB-47s being converted back to conventional bombers.

Bold Orion

Better by far than RASCAL was 'Bold Orion'. It was in some ways the same idea – a large missile hung off the side of the B-47, in this case, one of the YDB-47Es (a JDB-47E was also used). But instead of a liquid rocket airplane, the Bold Orion was a solid fuel rocket – a true ballistic missile.

A Martin programme, Bold Orion began in 1957 as Weapons System 199B. It was intended to be a technology demonstrator rather than an operational vehicle; Strategic Air Command had hopes for a larger missile to be carried by the B-52, but the technology was too new to leap right into an operational system. So existing components were used where possible. Initially it was a single stage vehicle using a Thiokol TX-20 'Sergeant', but range was not as hoped so a second stage in the form of an Allegany Ballistics Laboratory 'Altair' was added. With two stages, the missile would attain a range in excess of 1,100 nautical miles.

The test flights of Bold Orion were a mix of success and failure, with six rated as successes, six as failures. The final launch, on October 13, 1959, was a bit different: after a climb to 35,000ft, the DB-47E launched its Bold Orion on a nearly vertical trajectory,

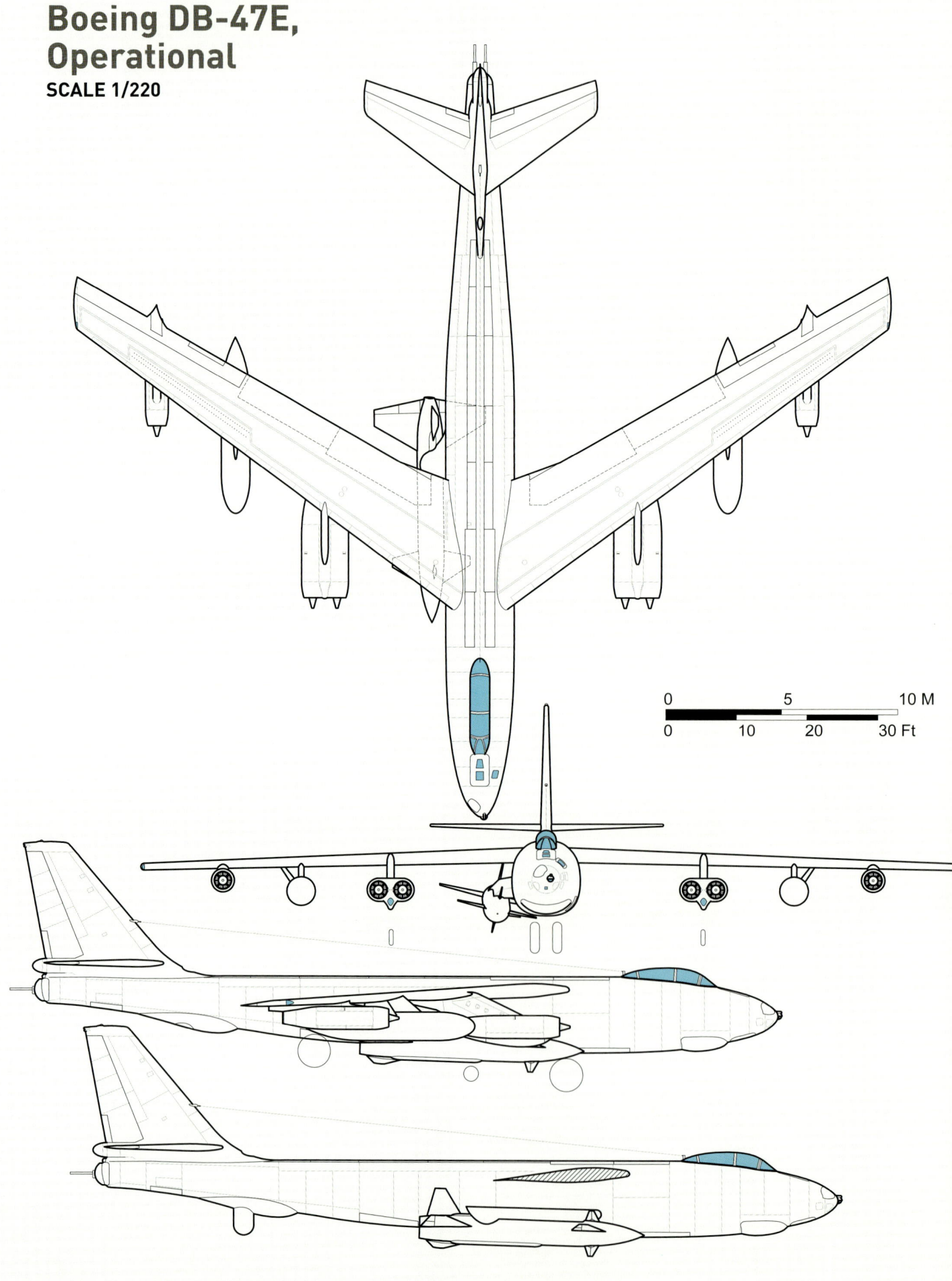

sending it around 160 nautical miles up. It targeted the dead Explorer VI satellite, orbiting at 156 nautical miles and missing it by four miles. As an anti-satellite test this was deemed a success, since an operational anti-satellite system would have been equipped with a thermonuclear device which would have fried the satellite's electronics at that distance.

That is a point to ponder: the B-47 emerged from the depths of the Second World War, a war that included soldiers armed with swords, cavalry charges and fabric covered biplanes. And eventually it demonstrated the ability to shoot hydrogen bombs into outer space.

Bold Orion
SCALE 1/60

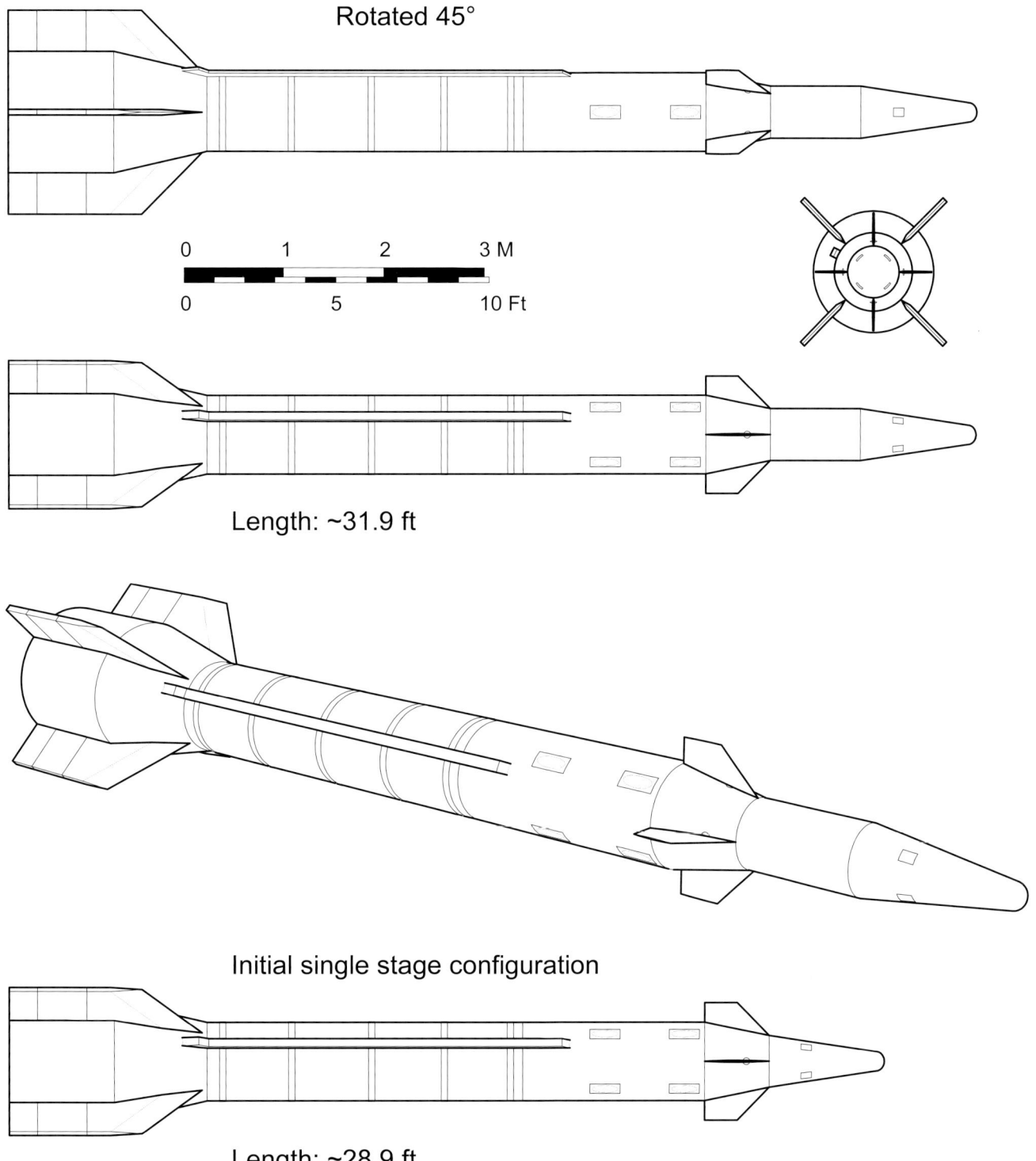

Rotated 45°

Length: ~31.9 ft

Initial single stage configuration

Length: ~28.9 ft

Boeing DB-47E+ Bold Orion
SCALE 1/220

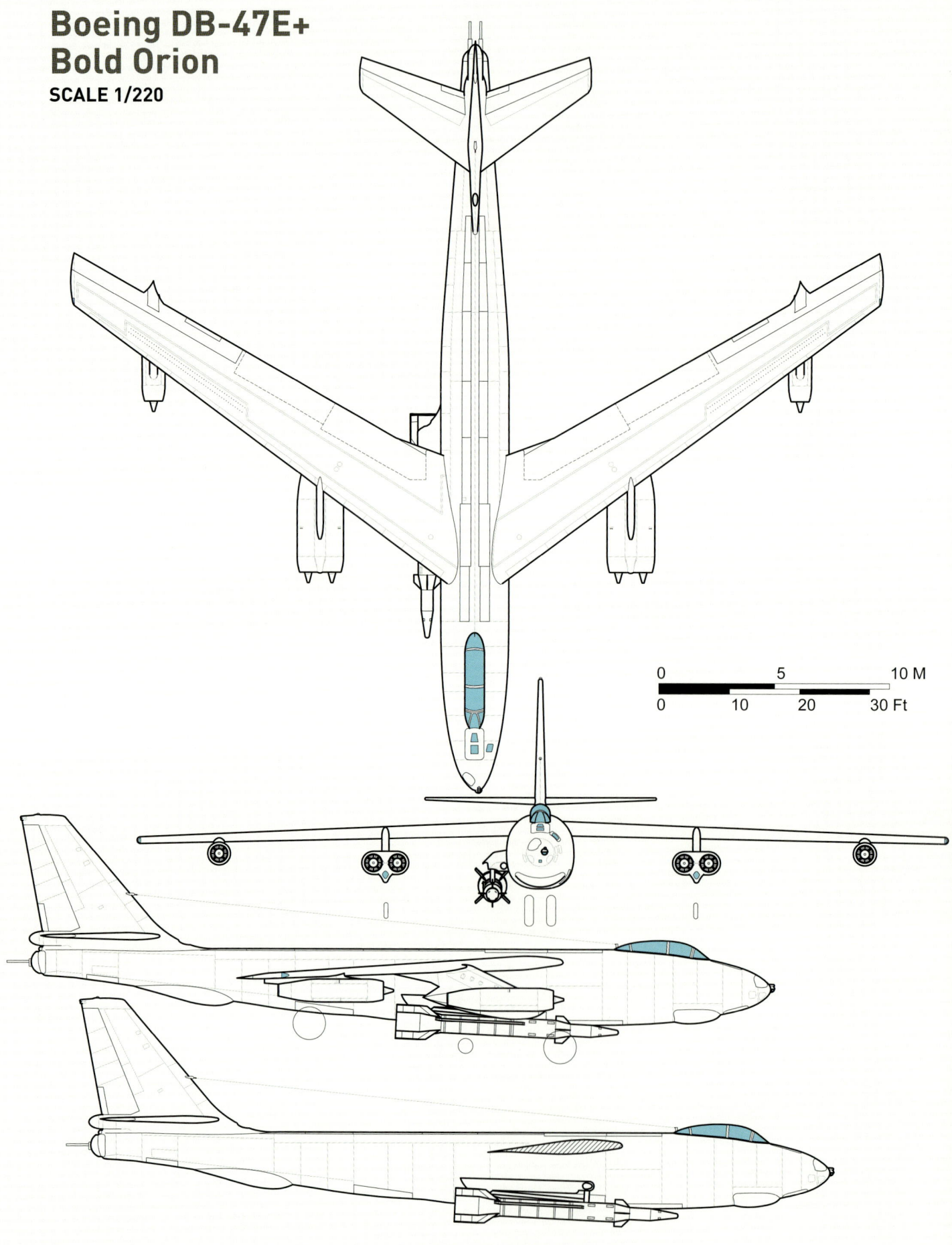

Crossbow

While the RASCAL and Bold Orion programmes were aimed at nuclear-armed strategic weapons, another programme tried to give the B-47 a tactical edge. The Radioplane GAM-67 'Crossbow' was a small subsonic turbojet powered unmanned aircraft; straight-winged with a cylindrical fuselage and an underslung inlet, it looked like a primitive cruise missile – a description which was not far off the mark. Originally designated the B-67, the Crossbow was intended to home in on Soviet radar transmitters, diving onto the target at nearly the speed of sound and setting off an unspecified conventional explosive.

Reportedly small nuclear warheads were considered for installation at one stage, a seemingly excessive approach to taking out radar installations. Two DB-47Es (unrelated to the DB-47Es used for RASCAL) were modified as carrier planes; each had four pylons added (one between the inboard and outboard engines on each wing, and one remarkably large pylon between the inboard engines and the fuselage). Additionally, the bomb bays were filled with a manned capsule.

Unfortunately, performance of the Crossbow fell below expectations since the little airplanes were slower than intended and their range was too short… shorter than the radars they were meant to take out, rendering their utility minimal since the B-47 carrier aircraft would be detected and interceptors vectored in well before the Crossbows could take out the radar. The first successful flights were in mid 1956, but the programme was cancelled in June 1957.

Canadair CL-52

Perhaps the unlikeliest B-47 to take to the skies was the Canadair CL-52, a one-off engine testbed. Structurally it was unusual, but what truly set it apart was the fact that it was the sole example of a B-47 operated by a foreign nation.

In the mid-1950s, the Avro Canada corporation was serious about getting into the advanced high-speed interceptor business. To this end, it was hard at work on the 'Arrow', which would prove to be an impressive Mach 2 aircraft when it first flew in 1958. To achieve that performance, it needed an equally advanced and powerful turbojet engine, the Orenda Iroquois.

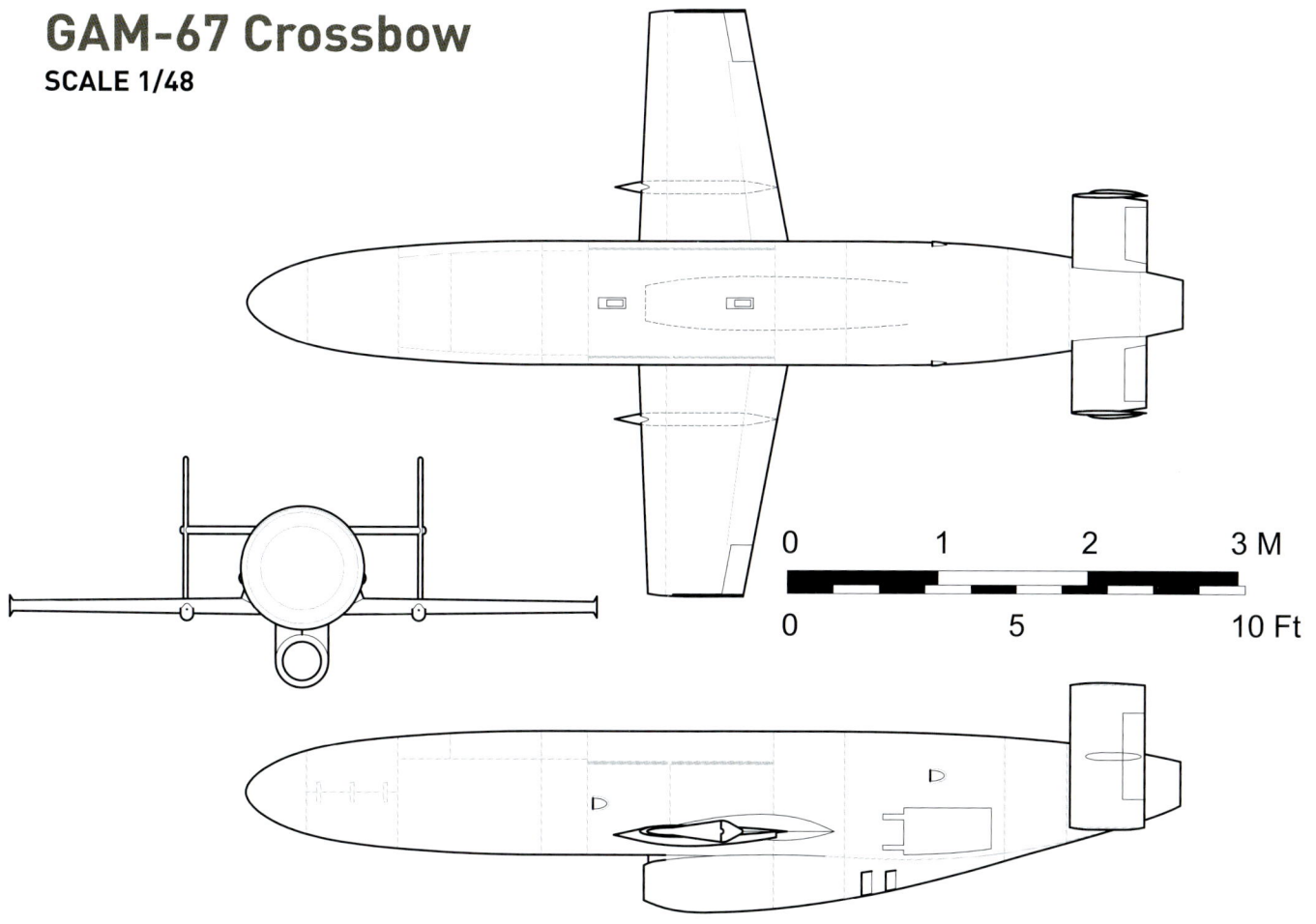

GAM-67 Crossbow
SCALE 1/48

Boeing B-47E
SCALE 1/220

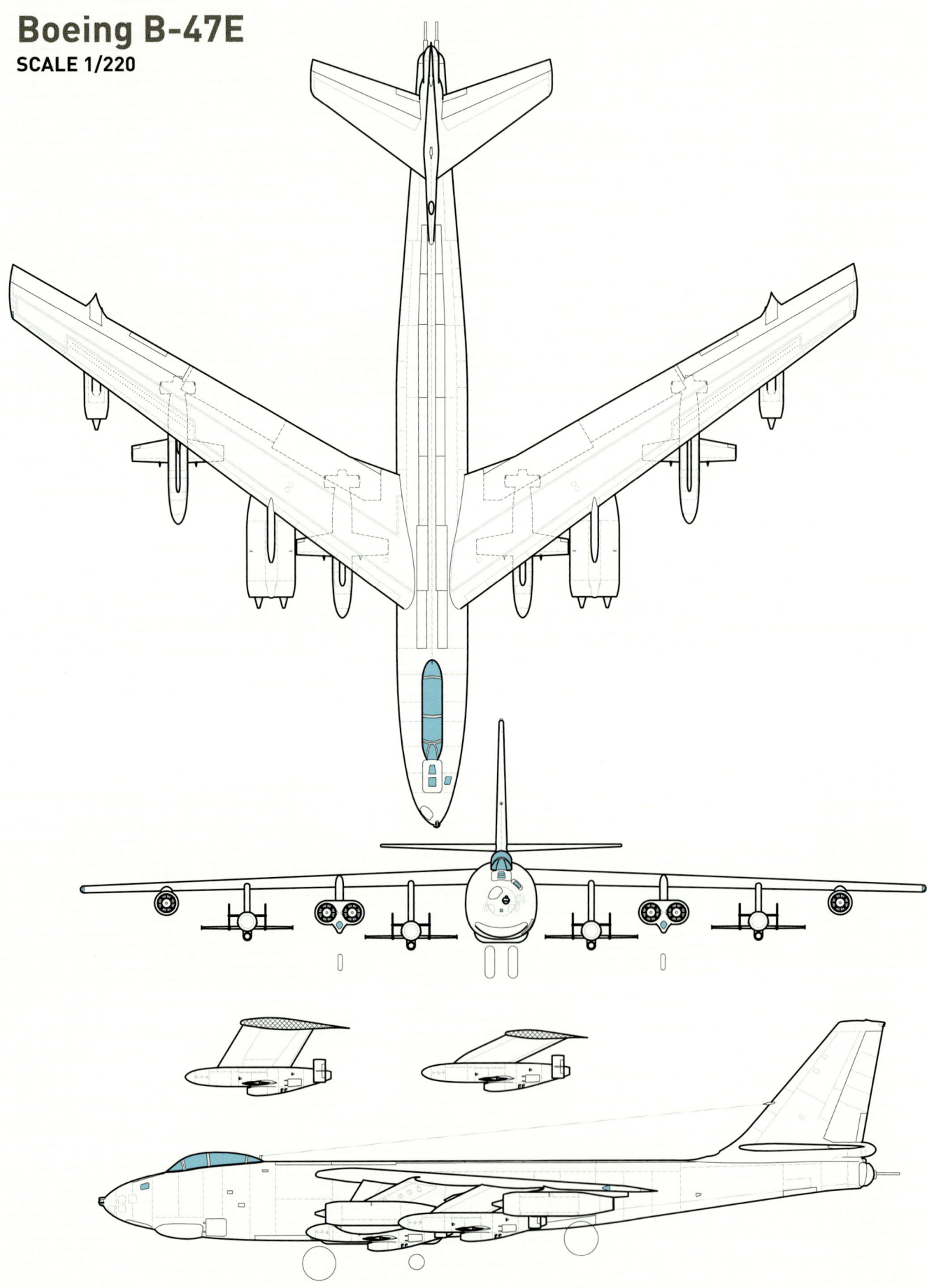

Canadair CL-52
SCALE 1/220

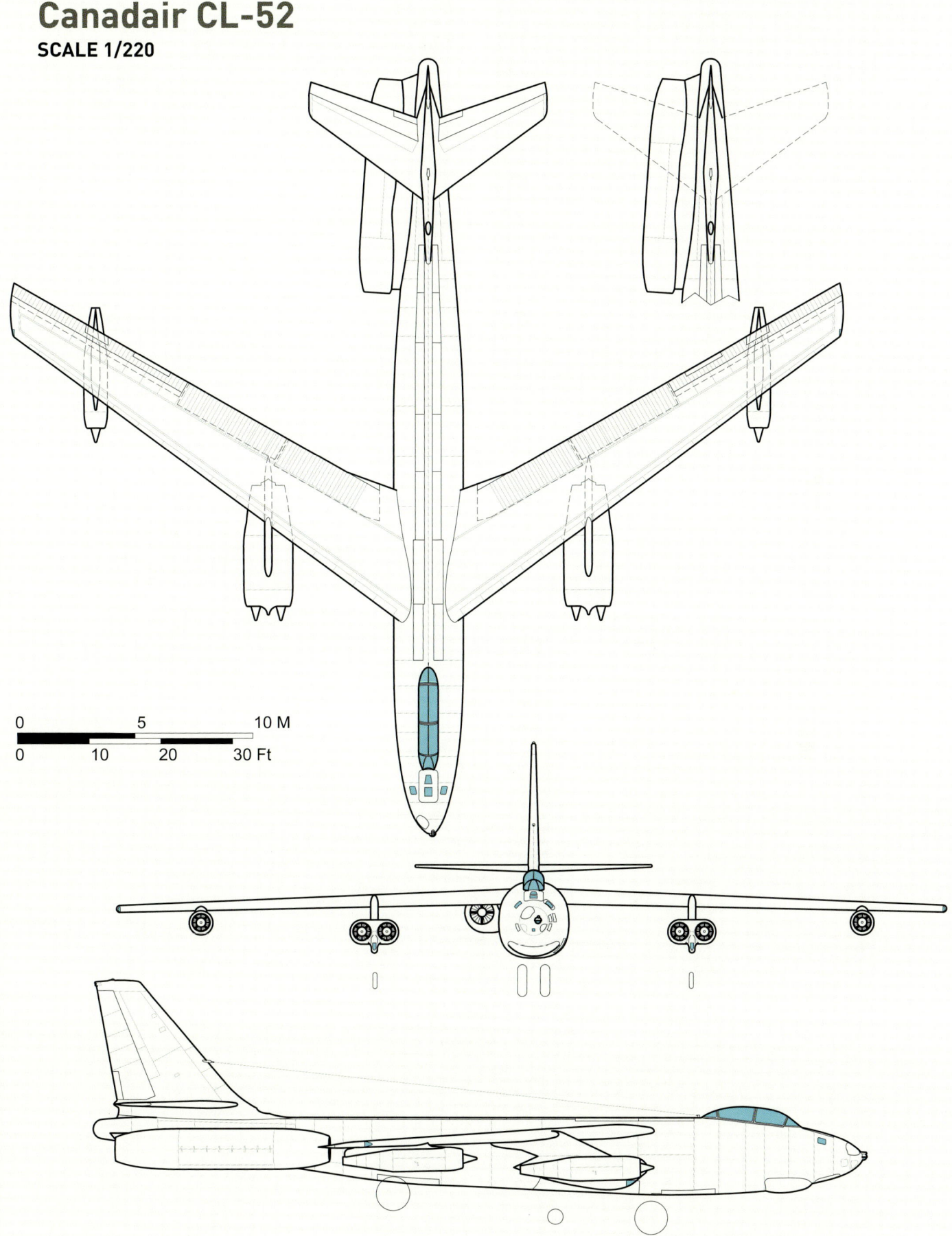

Avro looked at a number of aircraft to serve as a testbed for it – the right candidate would need to be large, fast, and readily modifiable. In the end the company determined that the B-47 would be best for the role and the USAF agreed to lend TB-47B number 51-2051 to the Royal Canadian Air Force in February 1956. Work began immediately to modify the airframe. A single Iroquois in a nacelle was attached to the right rear fuselage; the added weight so far aft meaning that 10,000lb of lead shot ballast had to be installed in the aircraft's nose. Fuel lines, instrumentation, structural enhancements and armour plating were added to the aircraft to accommodate the engine, which was toed-out 5° to partially make up for its asymmetric position. The modification work was done by Canadair, Ltd, under contract to Orenda Engines. Canadair designated the modified aircraft CL-52 and began flying on November 13, 1957. The asymmetrical thrust proved troublesome but manageable; at high power the three J47 turbojets on the starboard side were reduced to idle power while the outboard port engine ran at full thrust. On one flight, all of the J47s were shut down and the aircraft flew perfectly well simply using the Iroquois. Given that the engine was capable of 19,350lb of dry thrust and 25,600lb with afterburner, that is not too surprising.

The Avro Arrow, and the Iroquois along with it, was cancelled in February 1959. This decision had been debated furiously ever since, with the general consensus being that it was short-sighted (usually using colourful metaphors for emphasis). In the end, the CL-52 only racked up 31 hours of flight time with the engine running; only once did it attempt a full power (with afterburner) run – shedding several turbine blades and damaging the rear fuselage and horizontal stabilizer. The aircraft was returned to the USAF and scrapped at Davis-Monthan by mid-August 1959.

BRASS RING and WEARY WILLIE II

During the Second World War, Project Aphrodite took B-17 and B-24 bombers at the end of their service lives and modified them into remotely-piloted flying bombs. One last mission would see the planes filled with many more tons of high explosives than they would normally carry before taking off under the control of a pilot who would bail out shortly afterwards. The plane would then be radio-controlled by another pilot in another aircraft. Early television cameras allowed the remote pilot to direct the plane to its target then fly it directly into it. This proved to be an effective use of bombers that were meant for the scrap heap, though the programme was not without the occasional catastrophe.

As the era of atomic warfare loomed, the idea was revived. Project BRASS RING arose in 1950 as a way to allow the B-47 to carry a heavy nuclear weapon deep into the Soviet Union and also allow the crew to survive the trip home. The idea was that a single B-47 director aircraft could control one or more unmanned B-47s heavily laden with nuclear weapons. The unmanned aircraft would be flown as remotely piloted cruise missiles; the director aircraft would not be armed but instead filled with fuel for the return trip. One B-47A was converted to a DB-47A 'Director' aircraft and two B-47Bs were converted into MB-47B flying bombs.

After some effort, and a switch from North American to Sperry Gyroscope for the control equipment, the idea was made to work and by mid-1952 both the DB-47A and the MB-47Bs were flying. However, projected costs were excessive and with advances in retarded bomb design and the acquisition of forward bases for the B-47, the need for unmanned B-47s faded. BRASS RING was cancelled in April 1953.

By 1966, though, the idea had resurfaced. Now, as with Aphrodite, the issue was what to do with obsolete bombers. The notion of converting old B-47s into flying bombs was studied and the result was Project WEARY WILLIE II. This proposed the use of unmanned QB-47s controlled remotely by back-seat pilots in DF-100F fighters flying anywhere from 15 to 50 miles behind. A payload of 35,000lb of explosive slurry (presumably something like fuel oil mixed with ammonium nitrate) would be carried, not only in the bomb bay but within some of the aircraft's fuel cells. WEARY WILLIE II QB-47s were proposed for use against North Vietnamese targets such as bridges and even with the expenditure of entire bombers, the math worked. The loss of a single retired bomber was cheap compared to a bombing campaign carried out by wave after wave of F-105s, some of which were expected to be shot down. But even though it was feasible, the project was not advanced to actual practice.

Studies upon studies

Before the B-47 entered service, Boeing was already looking at ways to improve the design, to increase range or speed or payload. This was only to be expected – in the fast paced world of late 1940s aeronautics, if you weren't moving forward, you were being passed. And not just by your nearest competitor. At the time there were many companies that would have relished the prospect of replacing Boeing in the limelight, and Boeing knew it.

When it came to speed, the B-47 was a fast aircraft for its day but it was not supersonic. Boeing realized it was only a matter of time before a supersonic bomber was needed and therefore proposed several high speed and supersonic B-47 derivatives, all with recognizable B-47 fuselages and canopies… but with altogether new wings and engine arrangements.

Boeing Model 450-65-10

With the Model 450-65-10 series, Boeing studied a high speed – likely supersonic – version of the B-47. To achieve this, the long-span high aspect ratio wing of the B-47 was removed and replaced with a highly swept wing of shorter span, the configuration taken from the contemporary Model 484 supersonic bomber studies. The six podded engines were replaced with

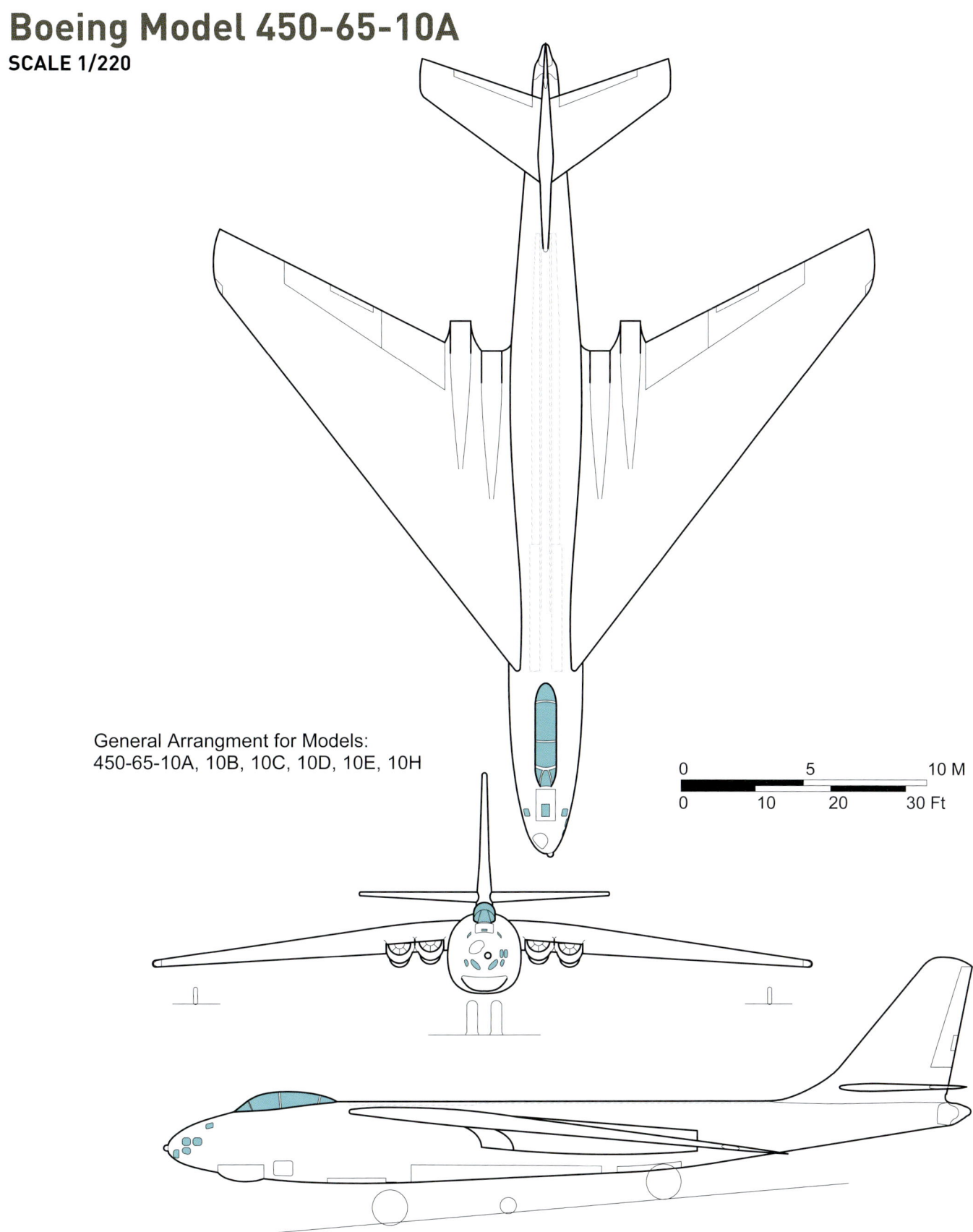

Boeing Model 450-65-10A
SCALE 1/220

General Arrangment for Models:
450-65-10A, 10B, 10C, 10D, 10E, 10H

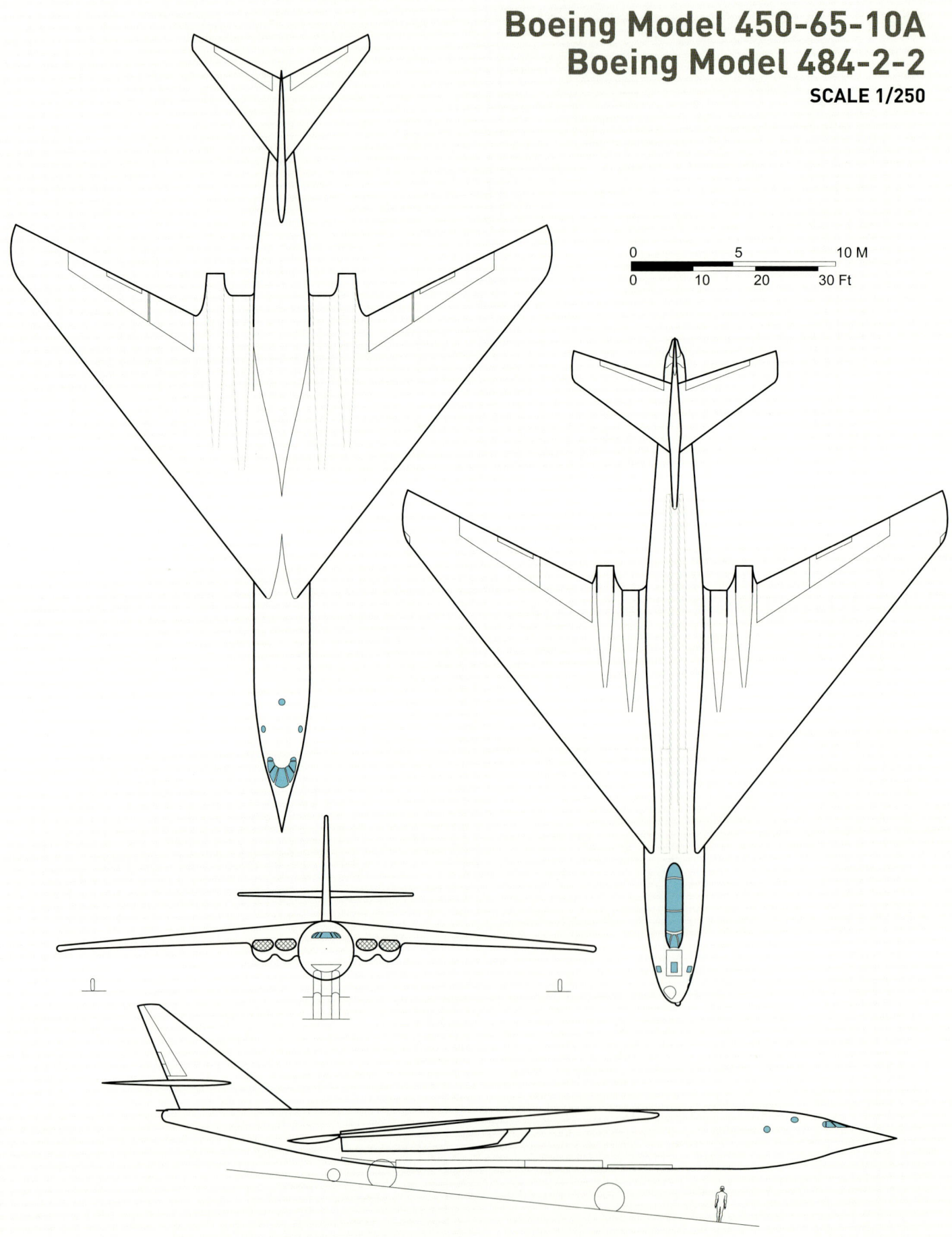

Boeing Model 450-65-10C
SCALE 1/220

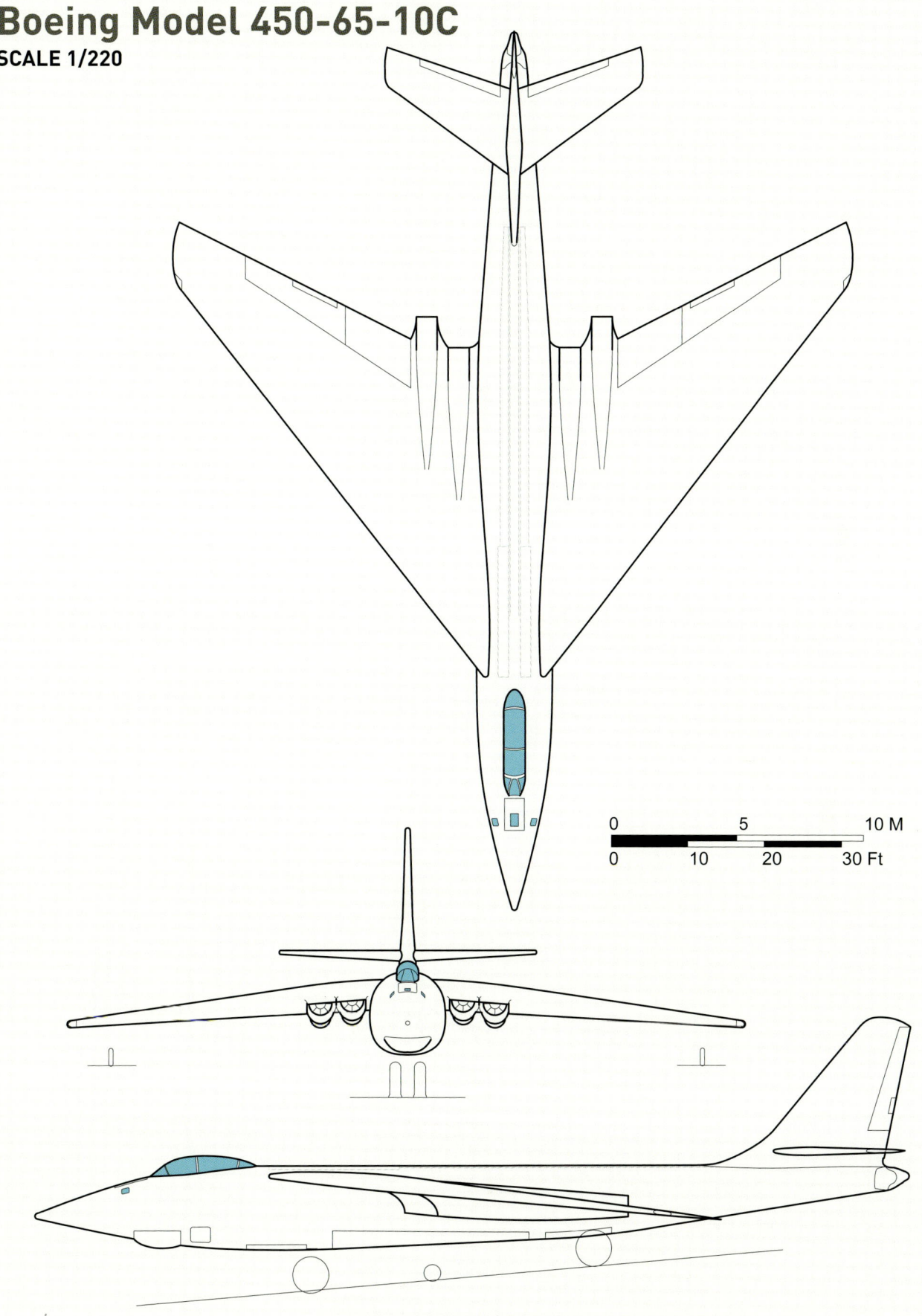

Boeing Model 450-65-10F
SCALE 1/220

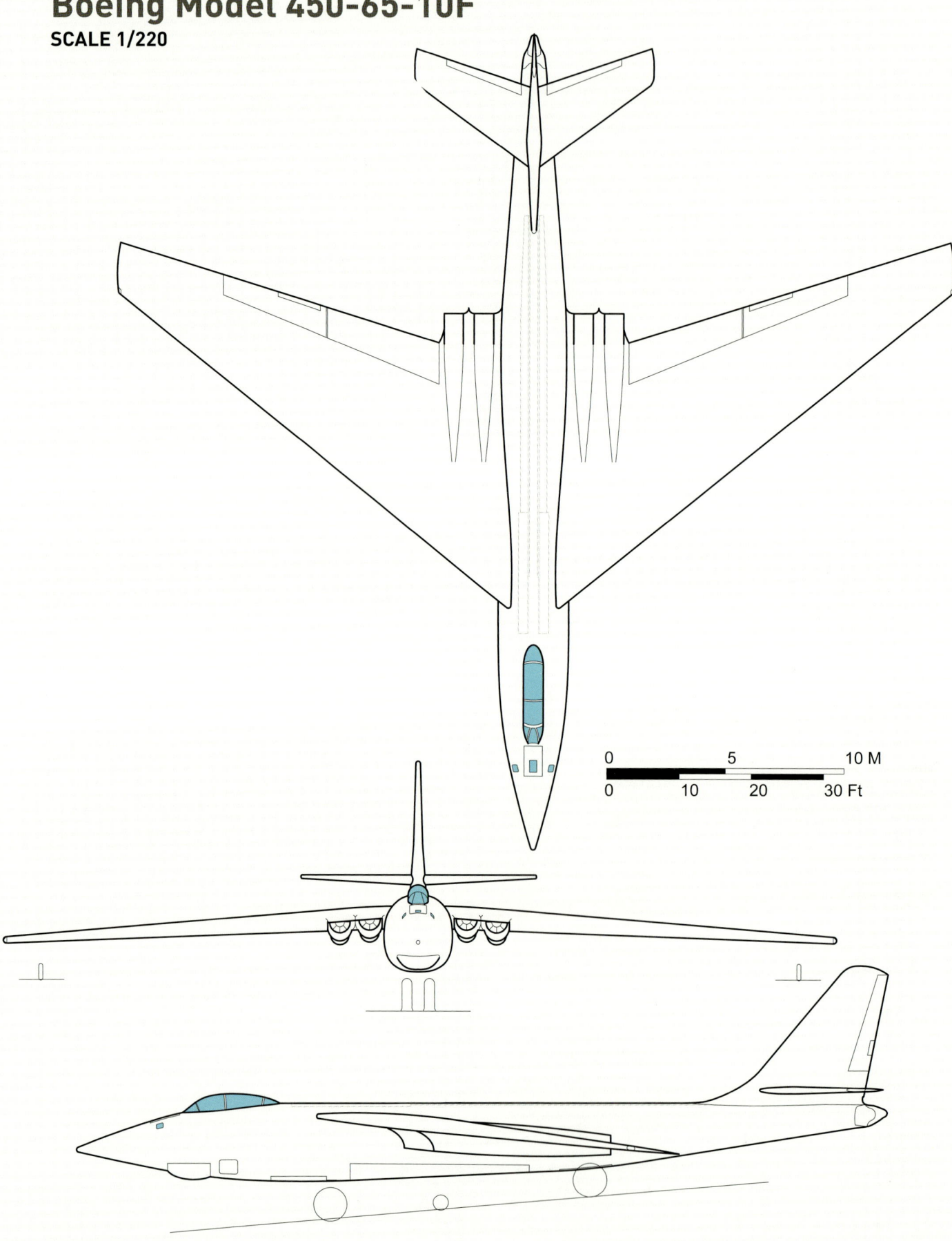

B-47 Development and Projects

four afterburning turbojets located just under the wing roots, with inlets beneath the wings and exhausts just aft of the trailing edges.

A number of variations on the wing and fuselage were studied. Model 450-65-10A had a largely unmodified B-47B fuselage (including the short bomb bay) with a highly swept wing. Model 450-65-10C had the same wing, but added a pointed fairing to the nose; Model 450-65-10F used the same modified fuselage, but added a wing somewhat less swept but with somewhat greater span. Performance data is not available but it seems likely that the -10C version would have had the highest top speed and the -10F the greatest range.

B-58 competitor

A curious episode in the story of the B-47 was its inclusion in the MX-1712 programme, which saw the aircraft proposed as the basis for a Convair B-58 competitor. Running parallel to Convair's B-58 development was Boeing's effort to develop what would eventually be designated the B-59. Boeing began a Phase I contract for project MX-1712 with the USAF in February 1951 to develop a medium-range strategic bomber capable of supersonic dash. Building upon aerodynamic research carried out under earlier studies, the company assigned model number 701 for MX-1712.

Initial requirements for the project included a basic mission radius (with a 200 nautical mile supersonic dash) of 1737 nautical miles and a gross takeoff weight of 200,000lb. Cruise speed was to be 530 knots with a basic mission dash speed of 860 knots and a maximum speed of 1150 knots. The engines were to be Wright J67-W1 turbojets with afterburning and the payload was assumed to be a single 10,000lb nuclear bomb.

Strategic Bombardment General Operating Requirements SAB-51, from December 1951, and Strategic Reconnaissance General Operating Requirements SAR-51 from February 1952 began to lay out the requirements for what would become the XB-59. Boeing programme MX-1965 was an outgrowth of MX-1712 and was seen by the USAF as a conservative backup to the more aggressive Convair MX-1964 (which became the B-58).

The Wright J67s were replaced with General Electric's new J73-X-24A, which was predicted to have improved specific fuel consumption, while the bomb's warhead was reduced in size and combined with a jettisonable fuel tank. This new bomb/tank combination was to weigh 7,700lb, a substantial reduction from the MX-1712 baseline. Other advanced technologies would also be utilized – a high velocity 30mm cannon in the tail for defence, new aluminum and titanium structural alloys, inflight refuelling and new electronics and avionics.

With the great success being demonstrated by Convair, the MX-1965 programme was cancelled in 1953. The ultimate design, Boeing Model 701-299-1, dating from August 1953, would have been the XB-59. But even though MX-1712/1965 was relatively short lived, Boeing drew up a wide variety of designs.

Boeing Model 450-151-31
SCALE 1/250

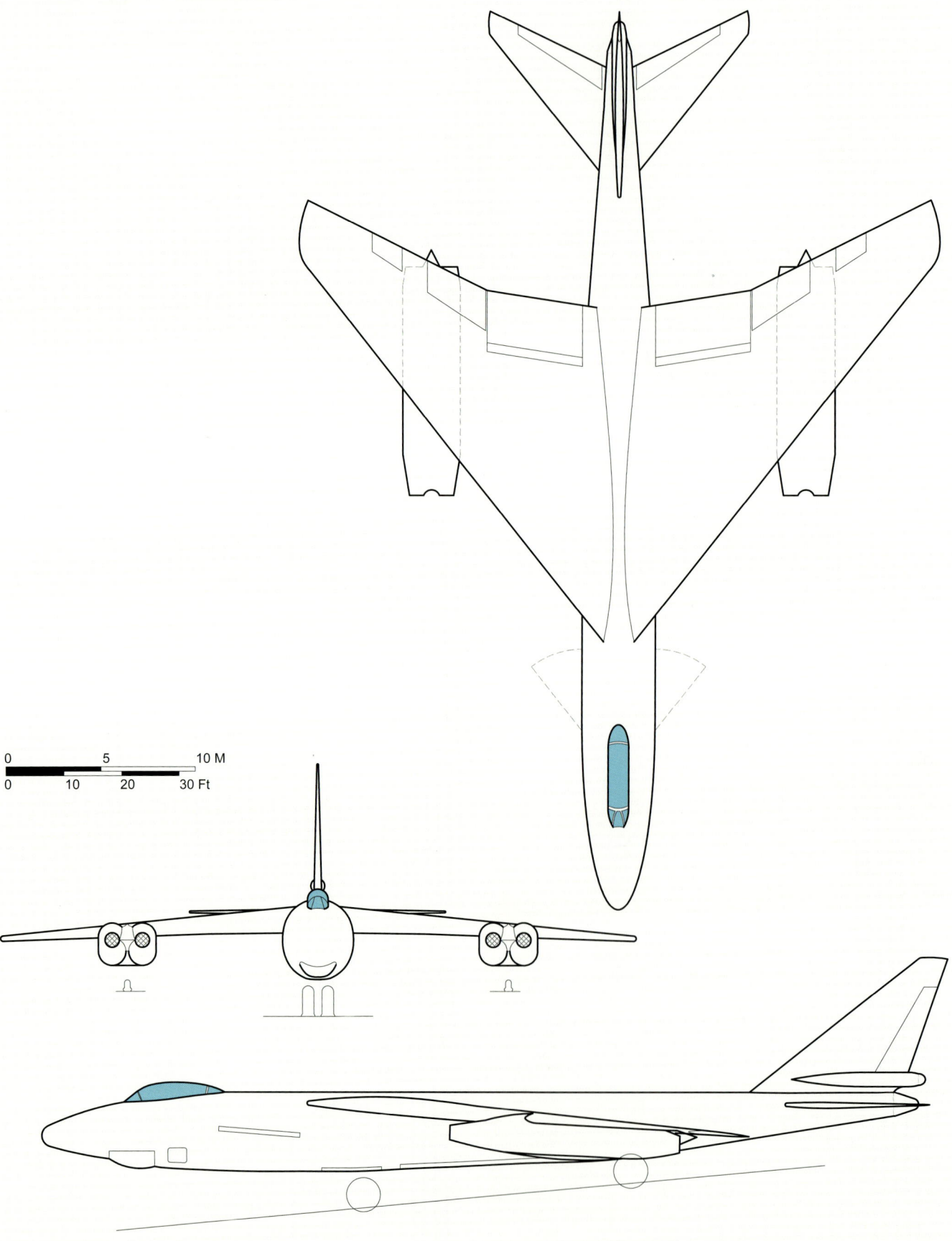

And curiously, while most were all-new, a number of them built in small part or large on the B-47.

Boeing Model 450-151-31

This used a stretched B-47 fuselage married to an all-new shoulder-mounted wing of shorter span and broader chord. Each wing had a single twin-jet nacelle just outboard of mid-span, fixed directly to the wing underside. For added stability at low speed, canards could be deployed from the forward fuselage and stored completely within to reduce drag at high speed. The tail, though similar in appearance to that of the B-47, had surfaces with greater sweepback. While this design had a Model 450 number, clearly associating it with the B-47, it is linked in available documentation with Model 701-238-1, part of the MX-1712 study.

Boeing Model 450-152-27

Model 450-152-27 was, like the -151-31, clearly a modified B-47B. But the fuselage was much closer to the B-47 standard, being about 14ft shorter and much stubbier in appearance. The forward fuselage barely protruded ahead of the wing leading edges and there were no canards, stowable or otherwise. The engine nacelles differed from those of the prior design not only in being held well below the wings via pylons but also in having mirrored vertical ramp inlets rather than simple circular inlets. This design was associated with model number 701-243-1.

Boeing Model 450-154-32

Model 450-154-32, also associated with model number 701-243-1, was a refinement of the previous design. Instead of using the existing B-47 forward fuselage and cockpit canopy – a questionable notion for a supersonic aircraft – an entirely new forward section was designed. This was clearly meant for supersonic speeds, presenting a more sensible pointed nose with sharply angled low windscreens. Data is lacking but it may well be that the -152 was intended as a testbed for the wings and powerplant, a way to get parts of the MX-1712 into the air quickly.

Boeing Model 701-333

The Model 701-333 was another case of a B-47 modified for the MX-1712 programme. Most of the B-47B's structure remained but an almost comically long conical nose was added. The wing and engine arrangement was much more like that of the Convair B-58: a nearly delta wing with four individually-podded engines beneath. With the B-47 fuselage as the basic element, it was a notably larger aircraft than the B-58. The B-59 was intended to have an external weapon, much as the B-58 had, but the B-47 fuselage had an internal bomb bay and virtually no room under the fuselage for an external weapon of any size. Once again, the Model 701-333 was likely intended to be a technology demonstrator rather than an operational aircraft.

Boeing Model 450-139-10 and other miscellaneous designs

In the early days of jet engines, there were really only two ways they worked: the pure turbojet, where all the air went through the combustor to be heated and turned into a stream of high velocity exhaust gas, and the turboprop where that energy was mechanically recovered and transferred to a large diameter propeller, used to claw away at a larger mass flow rate of air. But there was a third method, one that in the decades since has become the standard: the turbofan.

Somewhat similar to the turboprop, some of the power of the core turbojet is used to spin a fan, larger in diameter than the powerplant itself. Turbofans tend to offer a lower top speed but with far lower fuel consumption. In general, the larger the fan, the more fuel efficient the engine. In recent decades turbofans have become so efficient that turboprops have almost disappeared from the market, largely relegated to smaller aircraft or aircraft intended for relatively slow flight.

The Boeing Model 450-139-10 of May 1951 was designed to take advantage of an early turbofan design, the Allison Model 509 'fan jet'. Described in 1950, it was hardly the first turbofan engine designed; the Germans had tested one in 1943. Yet as of 1951, the turbofan was still largely unknown. Boeing designers added the engine to the B-47, resulting in the Model 450-139-10.

The Allison fan jets were noticeably larger in diameter than the turbojets they replaced. Seemingly as a result of this, the two inboard engines were separated into individual nacelles. It may also have been expected that turbofan engines would shed fan blades from time to time and having engines next to each other would be a good way to ensure that a failed engine would damage another. In any event, the otherwise conventional B-47B-like airframe was distinguished by having six separate engines, the inboard four hanging below substantial pylons, the outboardmost engines attached to very short, almost nonexistent pylons. The middle engine pylons were fitted with outrigger landing gear.

Boeing Models 450-148-30 and 450-150-30

Model 450-148-30 and 450-150-30 (also known as Model 704) were slightly different versions of the same quite different B-47. As with the YB-47C, these

Boeing Model 450-152
SCALE 1/250

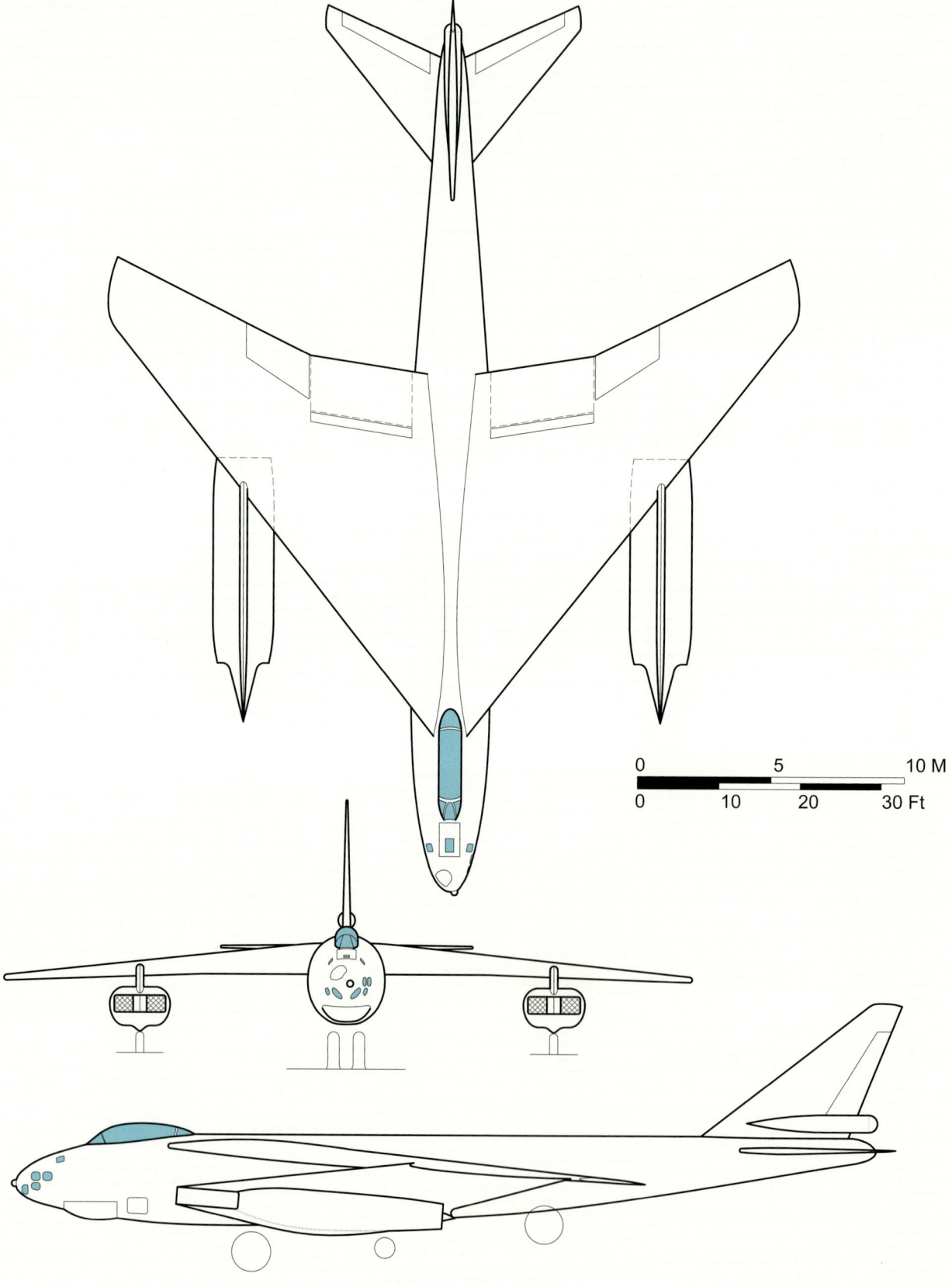

Boeing Model 450-154-32
SCALE 1/220

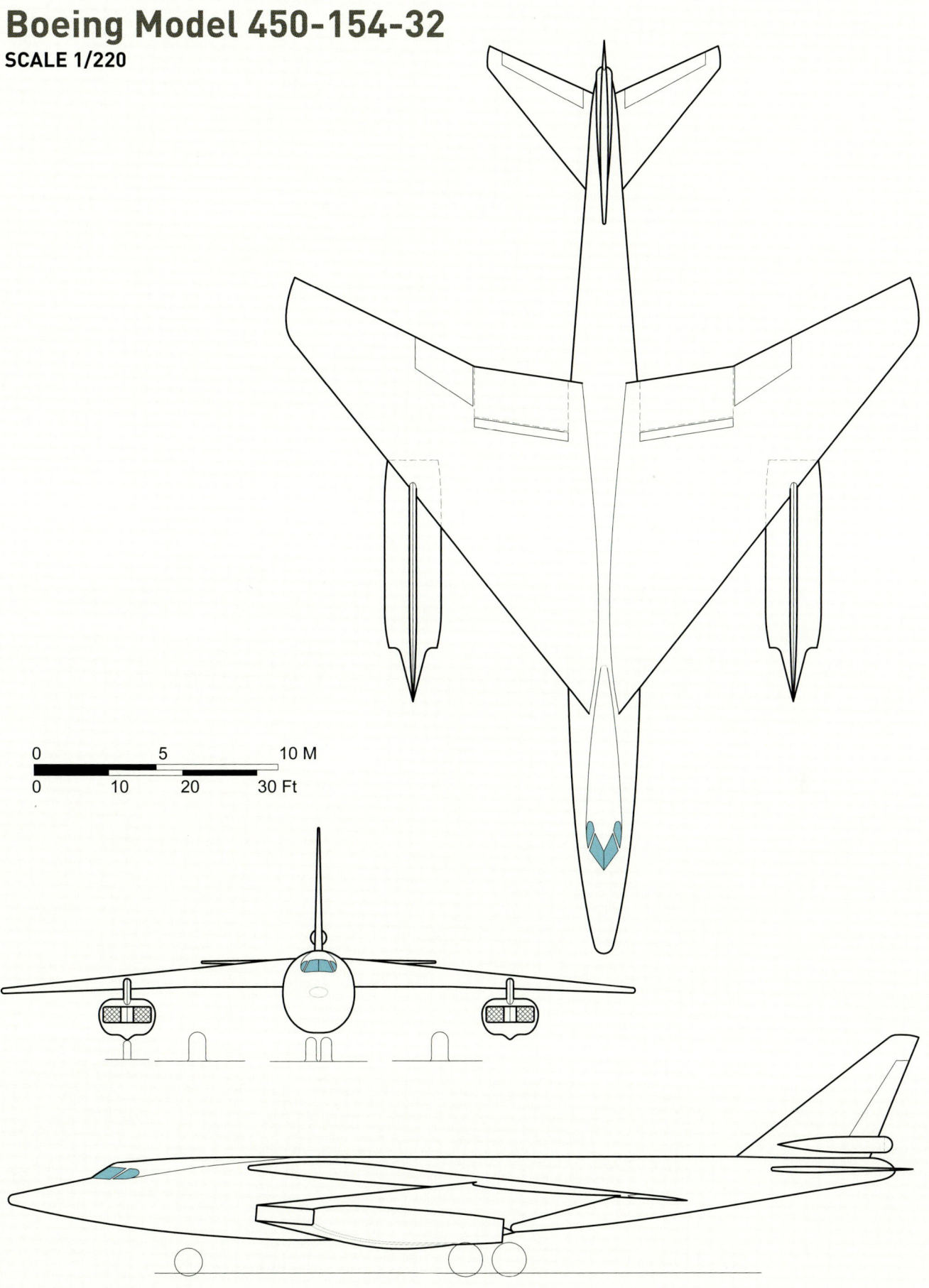

Boeing Model 701-333
SCALE 1/250

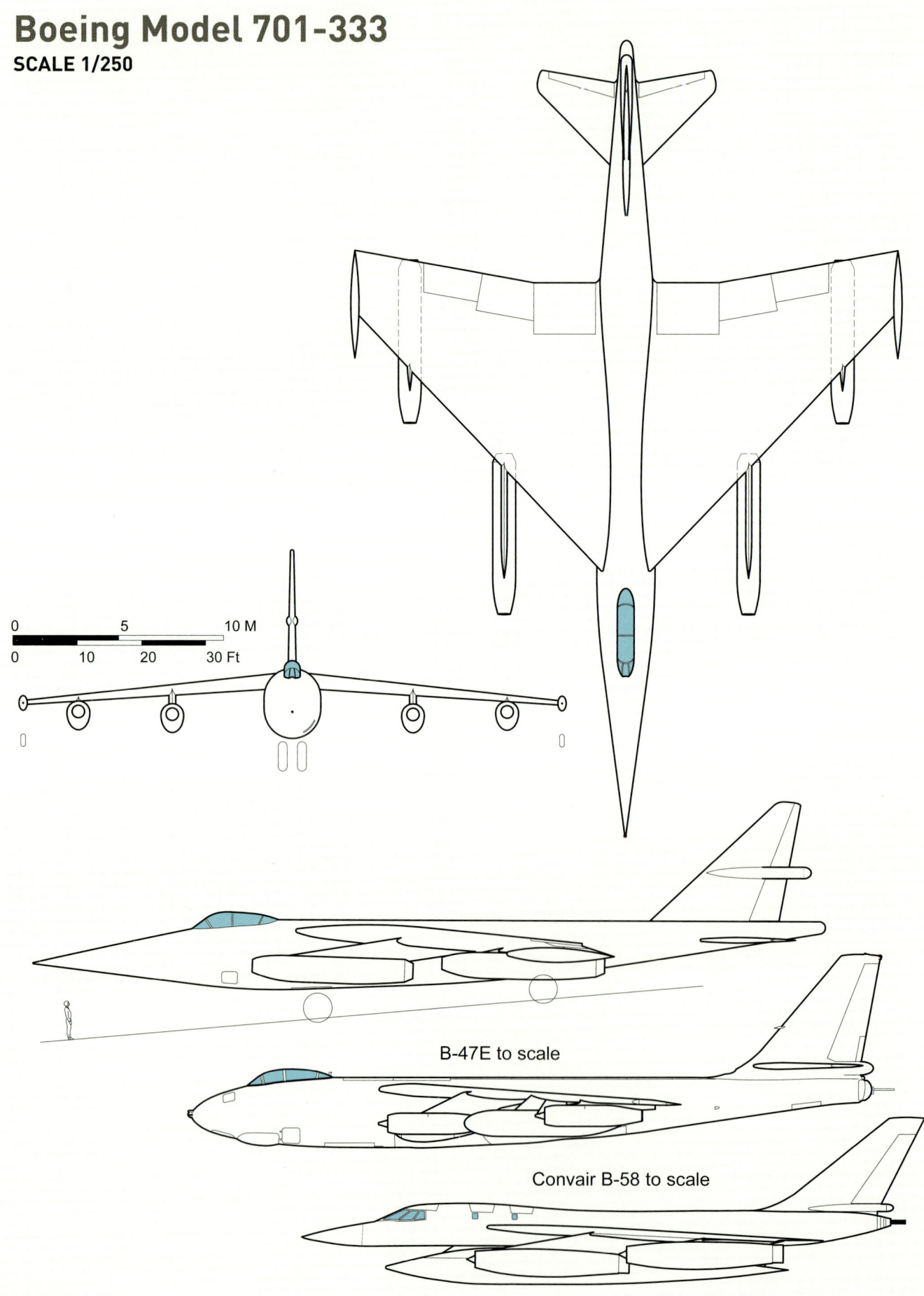

B-47E to scale

Convair B-58 to scale

Boeing Model 450-139-10
SCALE 1/250

Boeing B-47 Stratojet and B-52 Stratofortress: Origins and Evolution

Boeing Model 450-148-30
SCALE 1/220

88

designs had side-by-side seating for the pilot and co-pilot, but in a completely different configuration of cockpit. The change in cockpit was perhaps not the biggest alteration, though… these aircraft were intended for the US Navy. Rather than being a bomber, Model 148/150 was a minelayer. The Class VP minelayer specification was first issued by the Navy in May 1951.

The crew consisted of a pilot, co-pilot, navigator-minelayer, radio-radar ECM operator and a gunner – all in a pressured cabin. The pilot and co-pilot sat side by side; the gunner sat at the rear of the raised, wide cockpit fairing, facing aft and slightly raised. Either the navigator or the ECM operator sat behind the pilots (it's unclear which) and in front of the gunner, while the final crewman sat directly below the gunner. The aircraft featured the long bomb bay originally designed for the B-47 and two radar-guided 20mm cannon in a tail turret, each with 500 rounds.

Model 450-150-30 featured four J57 afterburning turbojets in a configuration similar to the B-47C/B-56. However, the inboard engines were toed inboard by 2.5°. The outboard engines were nearly fixed directly to the underside of the wings, using only a very short pylon; the inboard engines had the longer pylons with offset engines to provide clearance for the outrigger stabilizers.

The Model 450-148-30, however, had a compound propulsion system akin to that of the B-47D, with outboard Wright J67 turbojets and inboard Wright T49 turboprops with 15ft-diameter propellers. The individually podded engines were attached to the wings with very short pylons; the turboprop nacelles were graced with sizable fairings on the underside for the stowage of the outrigger stabilizing gear.

The available diagrams depict neither arrestor hook nor a wing-folding capability, so it seems unlikely that Models 148/150 were intended to operate from aircraft carriers. Both Convair and Martin Aircraft responded to the same specification with jet-powered flying boats; the Martin design winning and being built as the P6M Sea Master. Sadly – for Martin, the Navy and flying boat aficionados – only 16 P6Ms were built. That was the end of the line for American military flying boats. The Boeing design might have proven more successful.

MIT Project LAMP LIGHT

Several important advances in technology, planning and policy have arisen from studies carried out at universities. For instance 'Project Charles', performed at the Massachusetts Institute of Technology's Lincoln Laboratory (a research centre entirely funded by the federal government to study the needs of national defense) in the summer of 1951, set the stage for the air defence of the continental United States through the rest of the 1950s. One important conclusion of the report was the need for a radar system that would see incoming Soviet bombers and even missiles in time to do something about them. This was indeed created: the Distant Early Warning (DEW) line of radar installations in the arctic of Canada and Alaska.

A subsequent study, in 1955, was Project LAMP LIGHT. This expanded upon Project Charles; where the former had proposed a line of early warning radar sites in the far north, LAMP LIGHT suggested similar lines out in the Atlantic and Pacific oceans. Radar systems fitted to offshore oil-rig like platforms, seagoing 'picket ships' and aerial systems were studied as a way to provide warning time for America in the event of an attack from sea using ballistic missiles, bombers and supersonic cruise missiles akin to the Navaho.

But in order for those early warnings to mean anything, the United States would need defensive systems able to take down incoming threats – and the further this could be done away from American shores, the better. The study assumed that aircraft such as the Convair B-58 and the Martin P6M could be modified into high-speed long-range interceptors by sometime in the early 1960s, but an interim solution to the problem was put forward in the form of a highly modified B-47E.

MIT suggested that by 1960 the B-47E could also be turned into a long range interceptor capable of taking the fight up to 900 miles away from United States territory. It would carry eight sizable air-to-air missiles, four in the bomb bay and two under each wing. Each missile had a main stage 172in long, powered by a single liquid propellant rocket engine. Once launched at 30,000ft – at a target expected to be flying at 65,000ft – they would fly at Mach 2 for up to 128 seconds, giving a range of about 54 miles. Guidance of the missile was not firmly determined, but was assumed to likely include a combination of command guidance, infra-red homing and jammer homing.

In order to get the missiles where they needed to be and to find targets for them, the B-47E carrier aircraft would require modifications. The B-47E's standard radar was designed to look downwards for bombing and was of meager assistance in looking upwards and outwards toward incoming Soviet bombers. So the plan was to install a 3ft by 12ft antenna on top of the aircraft in a rotating dish similar to the rotadomes used on AWACS planes. About two tons of electronics would need to be added to the aircraft to support this new system.

The interceptor B-47E would cruise – initially at 23,000ft, and ending at 40,000ft – at 360mph to a

Boeing Model 450-150-30
SCALE 1/220

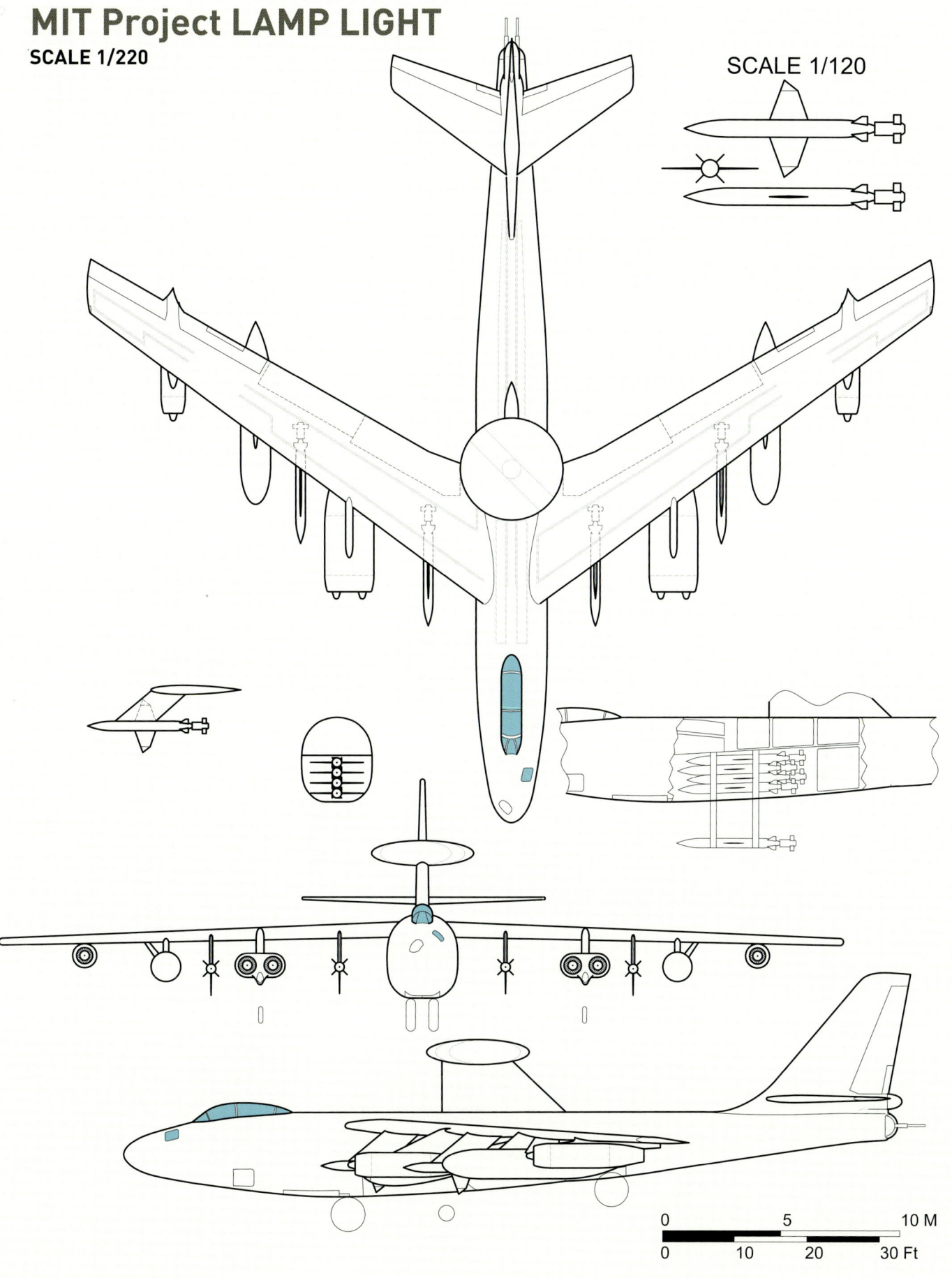

MIT Project LAMP LIGHT
SCALE 1/220

SCALE 1/120

range of 2,700 nautical miles, with inflight refuelling allowing range and duration extensions. The missiles would be stored horizontally in the bomb bay (the outer surface needing some modification so they would fit), but the external missiles would be carried at 90°, with one wing submerged within the pylons. It's unclear why this was done. MIT assumed that these conversions could be carried out by 1960. Three wings of 135 aircraft would be needed and they would work in conjunction with a highly integrated series of radar installations and control computers.

A suggestion was made to use an improved B-47 with larger wing area and J57 engines, though details of this were nonexistent. It sounds, however, like the study participants had had access to information on the Model 450-166-38; the timing certainly seems to support this.

There are reports of Boeing studies on the B-47 as a long-range interceptor, but unfortunately this author has nothing more on them.

The All American 'donkey'

The lack of adequate thrust at takeoff was a constant problem for the B-47; the use of the 'collar' full of 33 JATO units was an expensive and logistically complex solution. So the All American Engineering Company proposed a solution of their own: replace the rockets with a jet-propelled 'pusher' aircraft. All American was a small aeronautical firm that specialized in landing gear, arrestor cable and launch systems, and some unusual projects such as a system enabling aircraft in flight to snatch up packages or people on the ground.

They were not in the aircraft manufacturing business, which might explain the rather bizarre B-47 'pusher' project.

The pusher, apparently nicknamed 'donkey', was little more than two turbojets, a cockpit, a long nose which would mechanically link to the main aircraft and the wings and landing gear needed to recover the vehicle. It would be a lightweight vehicle carrying little fuel and with no need for a pressurized cockpit and anything more than the most basic navigation equipment. The available documentation on this concept, a popular magazine article and a patent, have contradictory depictions of the craft so the diagrams here are somewhat provisional, as is the scale of the craft. No hard data regarding weights, dimensions and performance are currently available.

All American filed the patent in 1956, by which time the B-47 was starting to head off the public stage. It might be thought of as simply an odd little concept from a small and now nearly forgotten company, an idea of little relevance or importance – especially since the B-47 was only crudely sketched in to the patent diagrams. But earlier in 1956 a much more famous aeronautical company, Goodyear Aircraft Corporation, had filed a patent for a somewhat similar concept. The Goodyear concept would have been not only a more substantial aircraft but was also explicitly designed to boost the B-47. The All American 'donkey' could have pushed any number of large jet aircraft as long as they'd had the required modifications, but the Goodyear design was specifically tailored to carry a B-47 on its back.

Not only would the Goodyear design have saved weight for the B-47 by providing takeoff thrust and lift, it would also have allowed the B-47 to save weight by deleting the landing gear. One might be forgiven for wondering how happy the B-47 crews would have been at having to rely on a post-mission rendezvous with a flying undercarriage.

All American Engineering Flying Pusher Catapult
SCALE 1/220

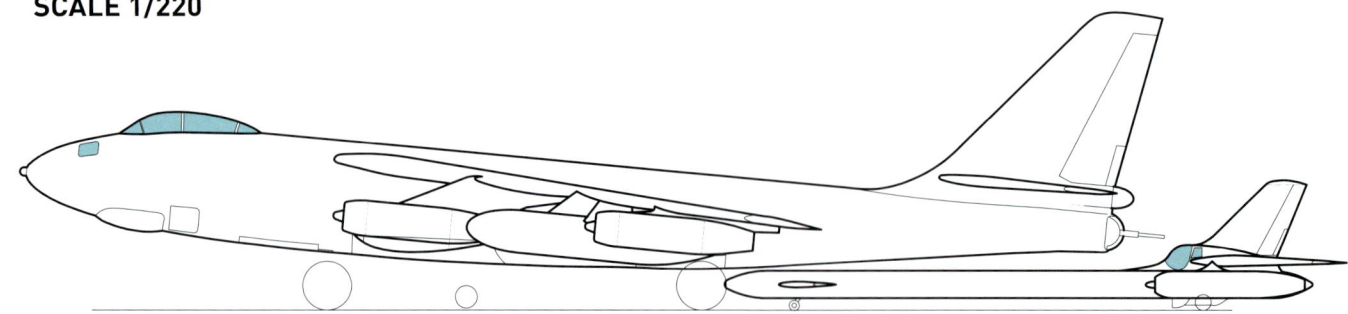

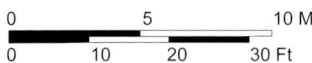

All American Engineering Flying Pusher Catapult
SCALE 1/100

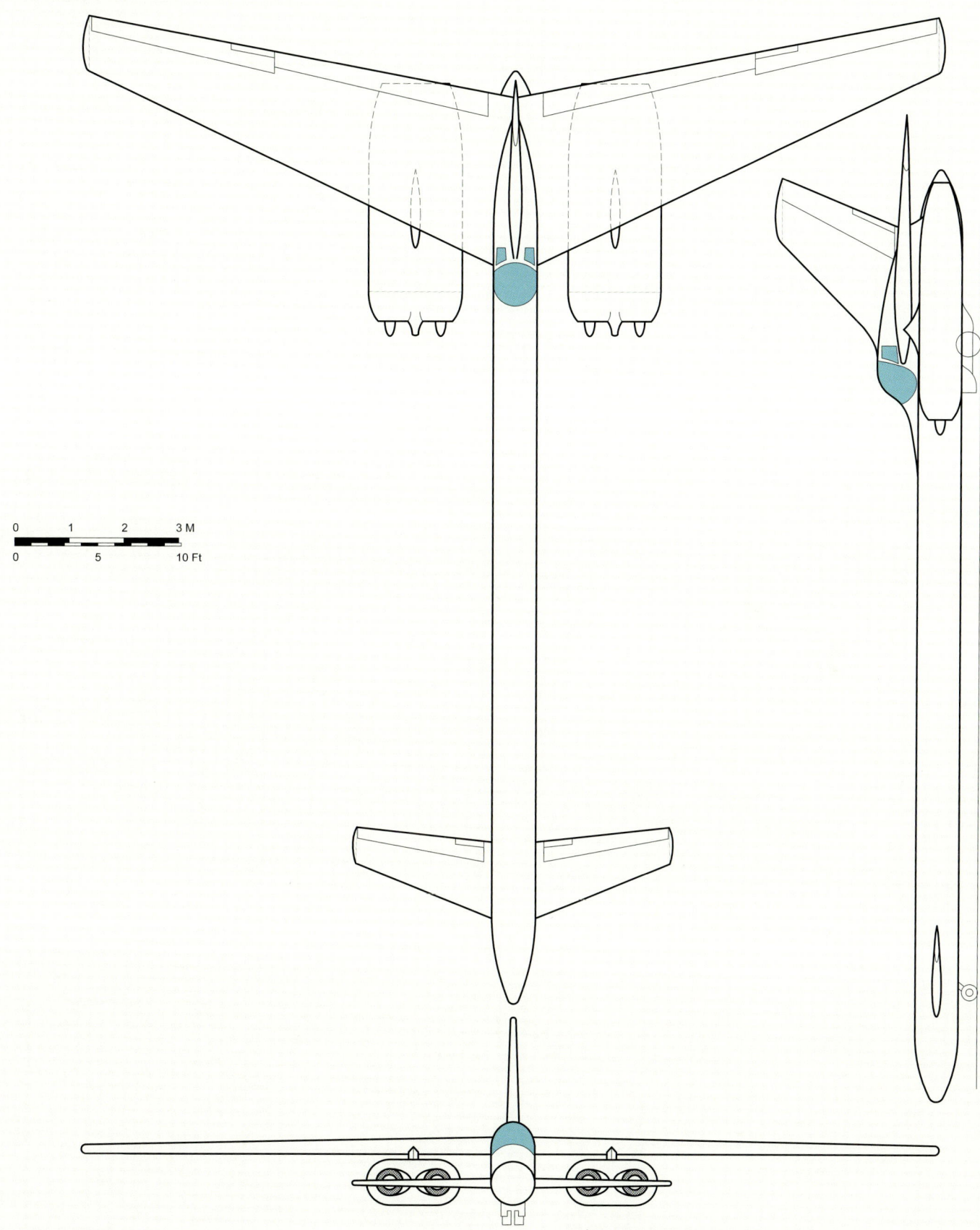

Boeing Model 450-36-10
SCALE 1/250

Extended wings

During the Second World War, the NACA examined the use of fuel tank 'gliders' that would be towed by bombers to extend range, fuel flowing down hoses from the glider to the bomber. There was a meaningful increase in range, but drag and complexity were increased. After the war, aerodynamicist Dr Richard Vogt – a German aircraft designer who had moved to the US under Operation Paperclip – suggested swapping the towed fuel gliders for wingtip-attached gliders, so-called 'floating wingtips'. Instead of droppable fuel tanks being simply attached to the underside of the wing or dragged behind and above as a separate glider, the tank would have its own sizable wing and would attach to the wingtip of the main aircraft.

The attachment point would not be rigid, but allowed to hinge up and down in the roll axis. Little would be needed in the way of active control; under normal flight conditions the 'floating' wingtip tanks would self-adjust. The great advantage here was that tank-wings would add substantially to the main aircraft's wingspan and aspect ratio. This would increase not only wing area, but also the aerodynamic efficiency of the wing.

Boeing Model 450-36-10

Model 450-36-10 from 1950 was essentially Model 450-16-17 with floating wingtips. The engines were changed from Westinghouse J40s to Allison J35s and the wings were modified to accept the wingtip hinges and fuel lines. The wingtips would be able to 'flap' up and down; whether they were meant to be jettisoned when emptied is unclear, but it is probable. The tanks were equipped with dual-wheel landing gear, to be used at least for takeoff. Range was over 1,000 miles greater than for the regular Model 450-16-17. Diagrams depict this aircraft with additional B-47-type drop-tanks under the wings between the inboard and outboard engine nacelles.

Boeing Model 450-57-10

To all appearances, Model 450-57-10 was Model 450-36-10 with the floating wingtips converted into a straight-up extended wing. There was no hinge and no possibility of jettisoning the outer wing panels. The tanks themselves, being bisected by the wings, were also non-jettisonable.

B-36/B-47 Tip-Tow

As bombers grew in range, they quickly outran any practical fighter escort. Until the long-ranged P-51D came along to escort B-17s and B-24s over Europe, many bombing missions ended in catastrophic losses of crews and airframes due to a lack of protection. As bombers grew to truly intercontinental range, fighter escorts would have been simply impossible in the early days of jet technology; the bombers themselves were little more than flying fuel tanks. So the idea of adding 'parasite' fighters to the bombers, that would ride along until needed deep in enemy territory, gained popularity.

Wright Field went forward with practical flight tests in the late 1940s, modifying a C-47 cargo plane wingtip to dock with a single Q-14 Culver Cadet. Unsurprisingly, some difficulties were found in learning just how to go about docking two aircraft

Boeing Model 450-57-10
SCALE 1/250

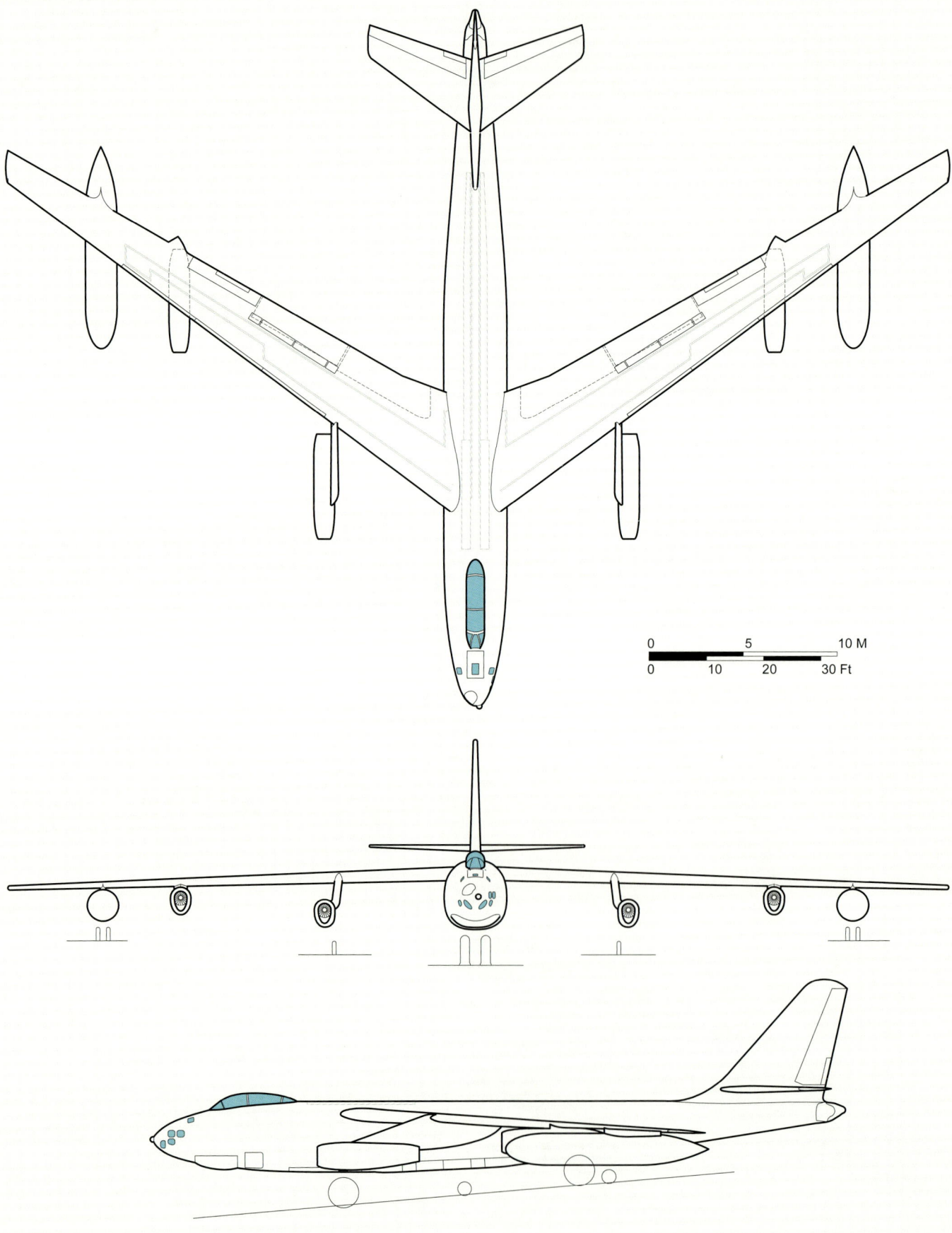

Convair B-36 + Boeing B-47 Wing Tip Tow

SCALE 1/600

Boeing Model 499-1
Long Range High Alt Recon
SCALE 1/275

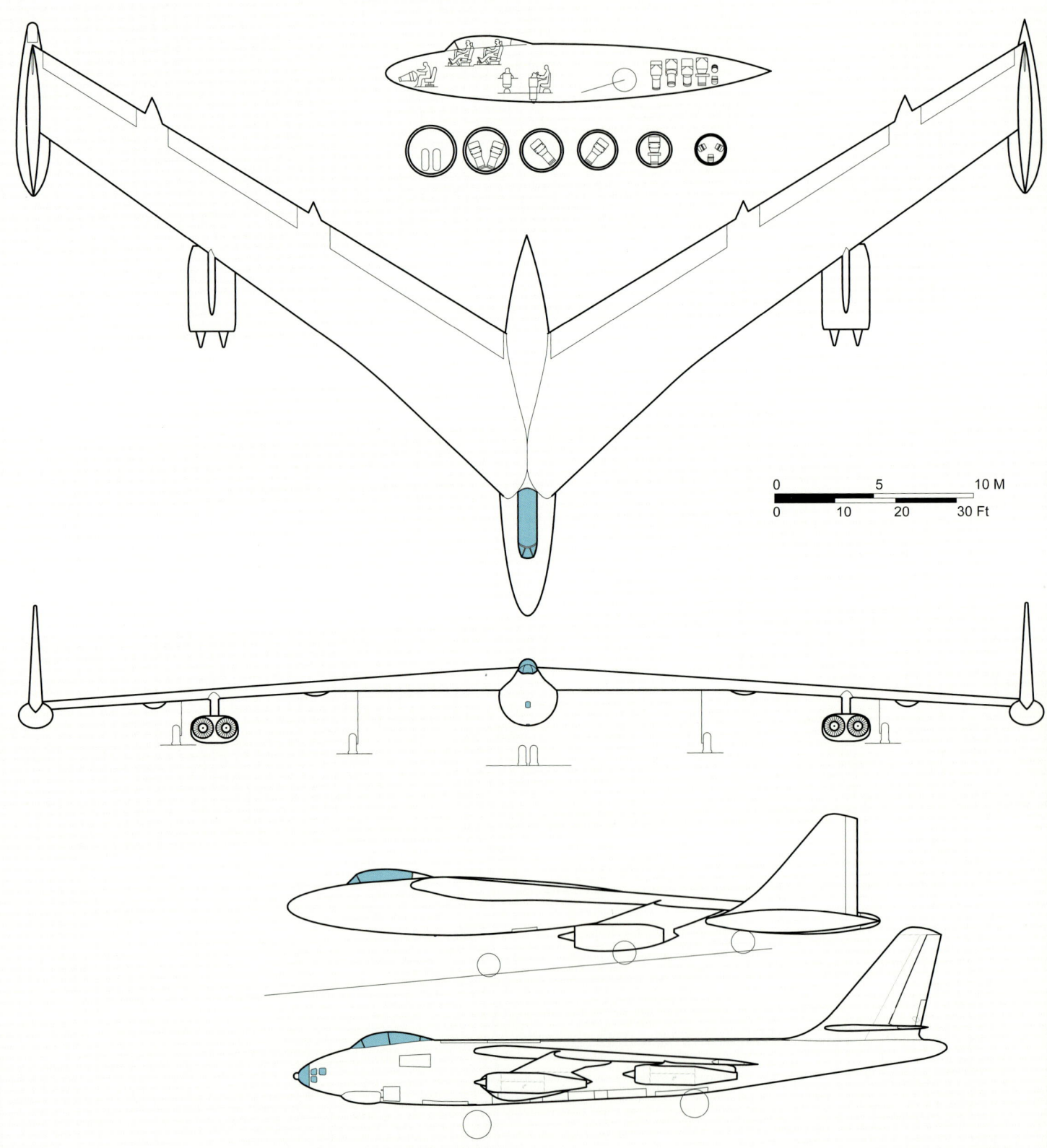

wingtip to wingtip when flight tests began in August 1949, but the wrinkles were ironed out and the two aircraft accumulated more than 28 hours of conjoined flying time during hundreds of flights in late 1950.

The experiments were built upon for project MX-1018, which involved attaching one Republic EF-84D jet fighter to each wingtip of an ETB-29A. Flight tests began in July of 1950 and continued until April 1953 when one of the EF-84Ds, docked to the left wing of the ETB-29A, pitched up and rolled over onto the bomber's wing. The nose of the fighter and the outer wing of the bomber were sheared off and both planes crashed with a loss of all crew, ending the project. However, during the testing programme it was found that the jet fighters could reduce engine power to idle or even shut down entirely and the combination craft would still fly perfectly well. The system was intentionally never flown in rough air though.

Beyond MX-1018 was 'Tom-Tom'. This was similar in concept but with two swept-wing Republic RF-84F fighters docked to the wingtips of a modified Convair RB-36F. This began flight testing in 1955 and continued into 1956, ending when one of the RF-84F's began 'flapping' violently, soon breaking the aircraft coupling mechanisms. In this case all aircraft were able to safely land, but the programme was terminated. Once again, testing showed aerodynamic benefits for the B-36 but challenges for coupling in flight. The hope for 'Tom-Tom' was that it might prove sufficiently useful so as to enter service, providing two fighters to cover the B-36 on its long journey over enemy territory. Exactly how thrilled the fighter pilots would be, stuck in their small cockpits for the better part of two days (the B-36 could stay in the air for around 40 hours), is not a detail that is well documented.

As these programmes were getting under way, the NACA was tasked with studying the concept of a B-36 with a B-47 attached to each wingtip. Whoever thought that attaching two B-47s to a B-36 was a good idea is a detail that seems to have been lost to time... perhaps the notion came from Boeing, perhaps Convair, perhaps the Air Force, perhaps someone at the NACA. In any event, available design information is lean. The lone NACA document presents only the planform of the wings with no additional details. The reconstruction presented here is thus provisional. It is shown with wingtip docking mechanisms taken from the Tom-Tom programme.

For several reasons it is unlikely that this combination aircraft would have lifted off all hooked together. For starters, it can be seen that when level, the B-47s are raised high enough that their landing gear would not have reached the ground when the B-36 was still rolling down the runway. At low enough speeds the B-47s would have tilted far over, perhaps far enough to scrape their outboard wingtips on the ground. And there's the issue of the 462ft wingspan: there are few runways in the world wide enough to handle an aircraft with landing gear that widely spaced.

The probable purpose of this aircraft would be range extension for the B-47s. The B-36 would fly with the attached B-47s idling their engines from the US to near Soviet airspace. The B-47s would then detach and fly to their targets, their tanks full or nearly so, to penetrate deep into enemy territory. The B-36 would presumably then fly home. In this situation the B-36 would presumably swap out its own internal bomb load for more fuel tanks.

The crews of the B-47s would at least be able to get up and move about, unlike the crews of the Tom-Tom F-84s. But even with that fantastic benefit, the idea seems to have existed for only a very brief time.

Boeing Model 499-1

Poorly documented, the Model 499-1 from 1950 was a design for a tailless high-altitude reconnaissance platform. The relationship between the Model 499 and the B-47 is somewhat tenuous, but it clearly shares design heritage with the shape of the forward fuselage, canopy, engine nacelles and wings. Performance data is currently lacking, but it doubtless would have exceeded the B-47's altitude capabilities. It would not have equaled those of the later Lockheed U-2 recon plane, but the payload it carried, four sizable film cameras in the rear fuselage, would have been far greater than the U-2's payload. While not a bomber, it had several crewmen in positions reminiscent of bombardiers; with a crew of five, the workload would have been spread out.

The tailless configuration of the Model 499 was unusual for Boeing, which had not made a great many tailless or flying wing designs.

CHAPTER 3: B-52 Evolution

It is a common misconception that the Atlantic and Pacific oceans provided North America with a natural defence against foreign aggression. Given that the founders of the United States arose from the flames of the French and Indian War of 1754 to 1763, the Continental forces fought British forces on American ground in the Revolutionary War of 1775 to 1783 and again in the War of 1812, it's clear that the United States has never been truly isolated from the effects of distant wars and foreign powers.

Still, up to the dawn of the Second World War many in the US could believe that distance would not only insulate the US from a European war but that distance would be invincible in both directions. Advances in aeronautics had ended that sense of isolation by the early 1940s, but this change in outlook did not come about overnight.

Boeing B-29

Perhaps the first truly modern bomber, certainly the most advanced long-range bomber of the Second World War, was the Boeing B-29. It was all-metal with a pressurized crew compartment and some of the first functional computers, used to help aim the remotely controlled machine gun turrets. And at the peak of its prowess it brought the war to a close by introducing the enemy and their environs to astonishing levels of artificial sunshine. The B-29 was not just another bomber… it ushered in an entirely new age.

That was not exactly the original plan Boeing's designers had in mind for the B-29. Starting out as Model 334A in 1939, then Model 341 in early 1940, it was intended to be an improvement on the state of the art to be sure. Model 341 was refined with improvements in defensive armament and other features to become Model 345 then resubmitted to the Army in May 1940. The Army's 'Superbomber' requirement called for a range of 5,333 miles with a bomb load of 2,000lb, which Model 345 seemed capable of meeting. So in late August 1940, the Army contracted for two prototype XB-29s. Two hundred and fifty B-29s were ordered in May 1941, raised to 500 in January 1942. This was unprecedented as the XB-29s had not yet flown and would not do so until September 1942.

The B-29 was able to meet its requirements by maximizing aerodynamic efficiency. The cockpit, typically raised on pretty much any aircraft up to this point, was fully submerged within the cylindrical fuselage. Defensive gun turrets were unmanned so they could be as flat and aerodynamic as possible, requiring advances in remote control technology. Construction was largely conventional, with a mainly metal structure and fabric-covered control surfaces; the landing gear, unusually for a large aircraft, was a tricycle arrangement.

The story of the B-29 has been often and sufficiently told. But for an aircraft aimed squarely at the heart of Nazi Germany, it is interesting that it was not used in that theatre; only a few were sent to Europe, and then only for propaganda and diversion. Instead, the many B-29s made during the war spent their time turning Japanese targets into rubble and ash. But here they were able to do so only because the United States Navy and Marine Corps had taken islands away from the Japanese, providing runways close enough to the Japanese homeland for the B-29 to carry out its mission.

The B-29 did have the range to reach targets in Europe from bases in North America. What it didn't have was the range to make it back to the United States. The distance from New York City to Berlin is about 3,950 miles; from St John's in Newfoundland, it's about 2,850 miles. Unfortunately the B-29A's basic mission had a radius of 1,843 miles while carrying 10,000lb of bombs. The only way B-29s could strike into Germany if Britain had fallen would be if the US operated from bases in Iceland. And if the Royal Navy was out of commission, US naval support for an Iceland base would have been forced to run an ever-present gauntlet of German U-boats. Additionally, the UK had invaded Iceland in May 1940 to forestall a German invasion; the fall of Britain would have likely ceded Iceland to Germany.

With all this in mind, in April 1941 the United States Army Air Corps issued a request for proposals for a long-range bomber capable to taking the war to Europe from American bases. Requirements were a range of 12,000 miles, a cruise speed of 275mph (with a 450mph top speed), a service ceiling of 45,000ft and a bomb load of 10,000lb. Over a shorter distance, an increased bomb load of 72,000lb was to be carried. Boeing and Consolidated began to work on designs to meet that RFP, but it proved beyond the technology of the day. The RFP was modified to make the aircraft somewhat less demanding in August 1941: range was reduced to 10,000 nautical miles, cruise speed dropped to 240mph and ceiling reduced to 40,000ft.

Boeing Model 360
SCALE 1/300

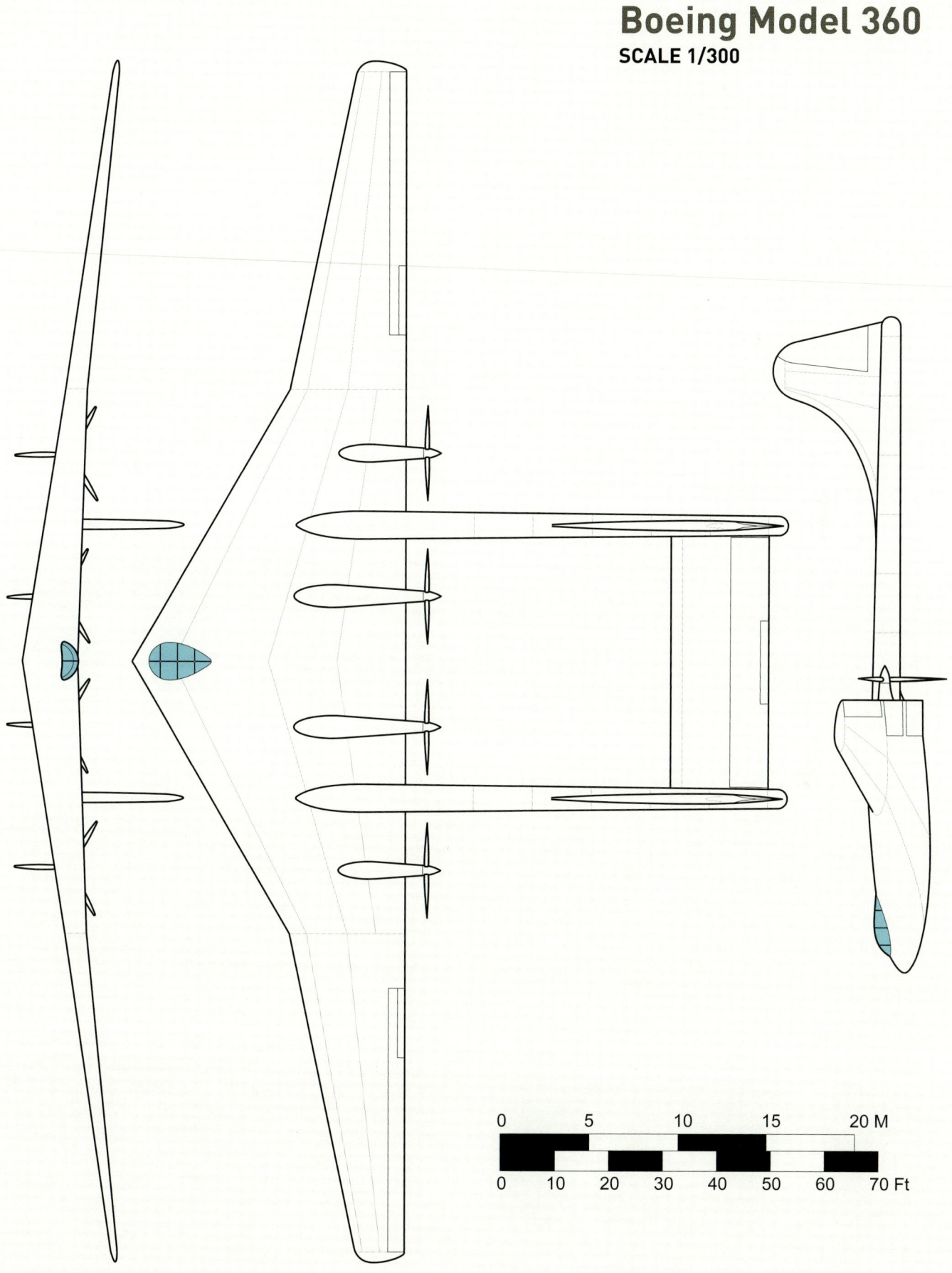

Even with the less exorbitant specification, Boeing struggled to produce practical designs that could meet them. Northrop and Consolidated felt up to the challenge and designed aircraft they felt could meet the requirements, being granted contracts to design and build the B-35 and B-36, respectively. Boeing didn't get a contract but did produce a series of studies – though the available information on these is not overwhelmingly detailed. This indicates that Boeing's work on these designs was not as thorough as the work done by their more successful competitors. Still, these early designs started Boeing down the road to designing true intercontinental strategic bombers.

Boeing Model 360

The Boeing Model 360 was an August 1941 design for a long-range bomber with a maximum payload of 40,000lb. It was an unusual configuration, which at the time would have been called a 'flying wing'. Configured much like the Northrop X-216H 'Flying Wing' from 1929, it featured a long span wing with a thickened centre section and no distinct separate fuselage.

A sizable rectangular horizontal stabilizer and two vertical stabilizers, each similar to that of the B-29, were held well to the rear by two relatively slim booms. Engines buried within the wing centre section drove four three-bladed pusher props. The only crew position depicted is a single sizable cockpit canopy; in all probability though, the airframe would have been festooned with gun positions. Certainly tailguns at the end of each of the booms, and likely turrets both above and below the central wing section. Unfortunately, little else is known beyond some basic dimensions.

Boeing Model 361

Like Model 360, Model 361 is known to this author from a single diagram. It appears to have married the B-29's fuselage and tail to an entirely new wing. The B-29's relatively thin high aspect ratio is gone, replaced with a very thick centre section housing the engines internally to drive tractor propellers via extension shafts. This, presumably, would have allowed in-flight access to the engines for servicing and repair. The wing would also have provided considerable volume for fuel tanks and possible ordnance. The diagram shows the locations of three turrets, two on the upper side of the fuselage and one on the underside, but does not show the turrets themselves. It's not clear if this indicates retractable turrets or if the turrets are simply not depicted. As with Model 360, maximum payload is 40,000lb.

Boeing Model 362

Model 362 from September 1941 mixes the 360 and 361 configurations, with a conventional B-29-like fuselage and tail section with the blended wing/forward fuselage of the 360. The cockpit has a B-29-like greenhouse canopy, though modified somewhat to account for the sizable wing roots at the 'cheeks'. But the biggest change from the prior designs is the actual size of the vehicle… with a span of 300ft, this would have been one of the largest aircraft ever built, with a span exceeded only by the Hughes H-4 'Hercules' and the Stratolaunch 'Roc'. To provide power, eight undefined piston engines were buried within the wings, driving four contra-rotating pusher props.

This resulted in an aircraft that at last met the USAAC's requirements for performance, though at the cost of an immensely large and heavy aircraft. It does not appear that this design was submitted to the USAAC however; it may be that it was simply designed to see just how large an aircraft would have to be to meet the performance requirements based on the aerodynamic, structural and powerplant state of the arts.

Model 362 was to be equipped with a number of turrets. The available diagram shows the locations of five… one on the upper surface of the mid-fuselage, one each atop the wings at about mid-span and one below each wing slightly outboard of the upper turrets. It is possible that there was a lower-fuselage turret or a tail gun, but these are not indicated in the diagram.

The total defensive armament consisted of eight .50 calibre machine guns with 1000 rounds each and six 37mm cannon with 300 rounds each. Offensive payload was baselined at 10,000lb, but exchanging fuel for bombs could give a maximum bomb load of 20 x 4,000lb bombs, 28 x 2,000lb bombs, 76 x 1,000lb bombs, or an impressive 176 x 500lb bombs. The landing gear was of tricycle arrangement, but the main landing gear appears to be very far forward, likely indicating that the central tubular fuselage would not be used to house fuel or bombs, but was merely a structure to hold the tail surfaces well aft.

Boeing Model 365

Boeing's Model 365 from April 1942 was in many ways a scaled-up B-29. The fuselage maximum diameter was increased 2ft from the B-29's 9ft 6in to 11ft 6in. Span and length were increased by more than 50%. And while the configuration was very much like the B-29, it featured some unique alterations. The forward fuselage – otherwise looking very much like the B-29's 'greenhouse' – was fitted with a proboscis-like extension, presumably for the bombardier. This extension looks like a scaled *down* version of the B-29 forward fuselage. The Model 365 had, like the B-29, four piston engine driven propellers, but the four Pratt & Whitney 'X Wasp' engines drove eight-bladed counter-rotating pusher propellers. The engines themselves were still located ahead of the wing leading edge, but the fairings and drive shafts extended well aft over the wing spars.

Boeing Model 361
SCALE 1/300

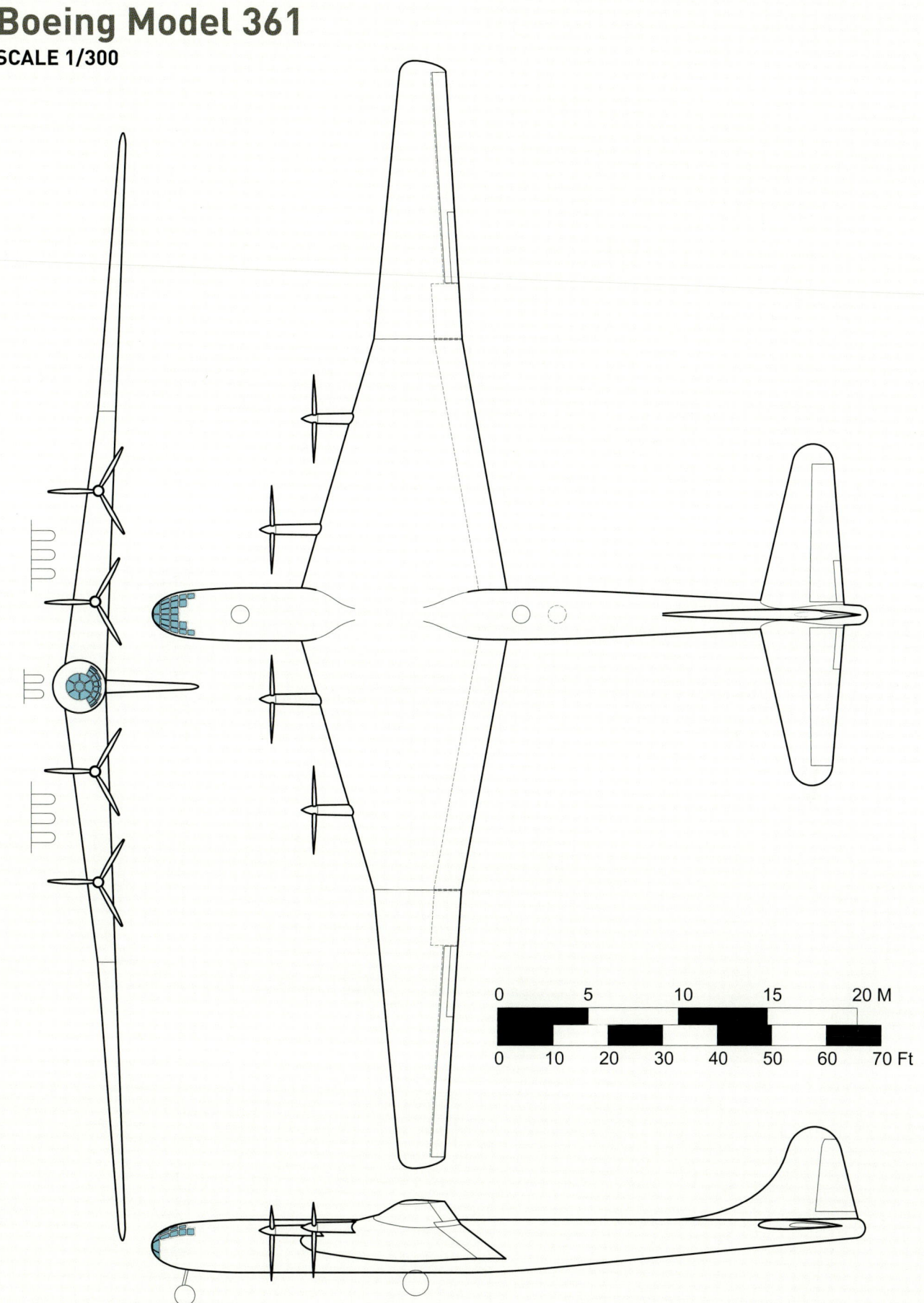

Boeing Model 362
SCALE 1/450

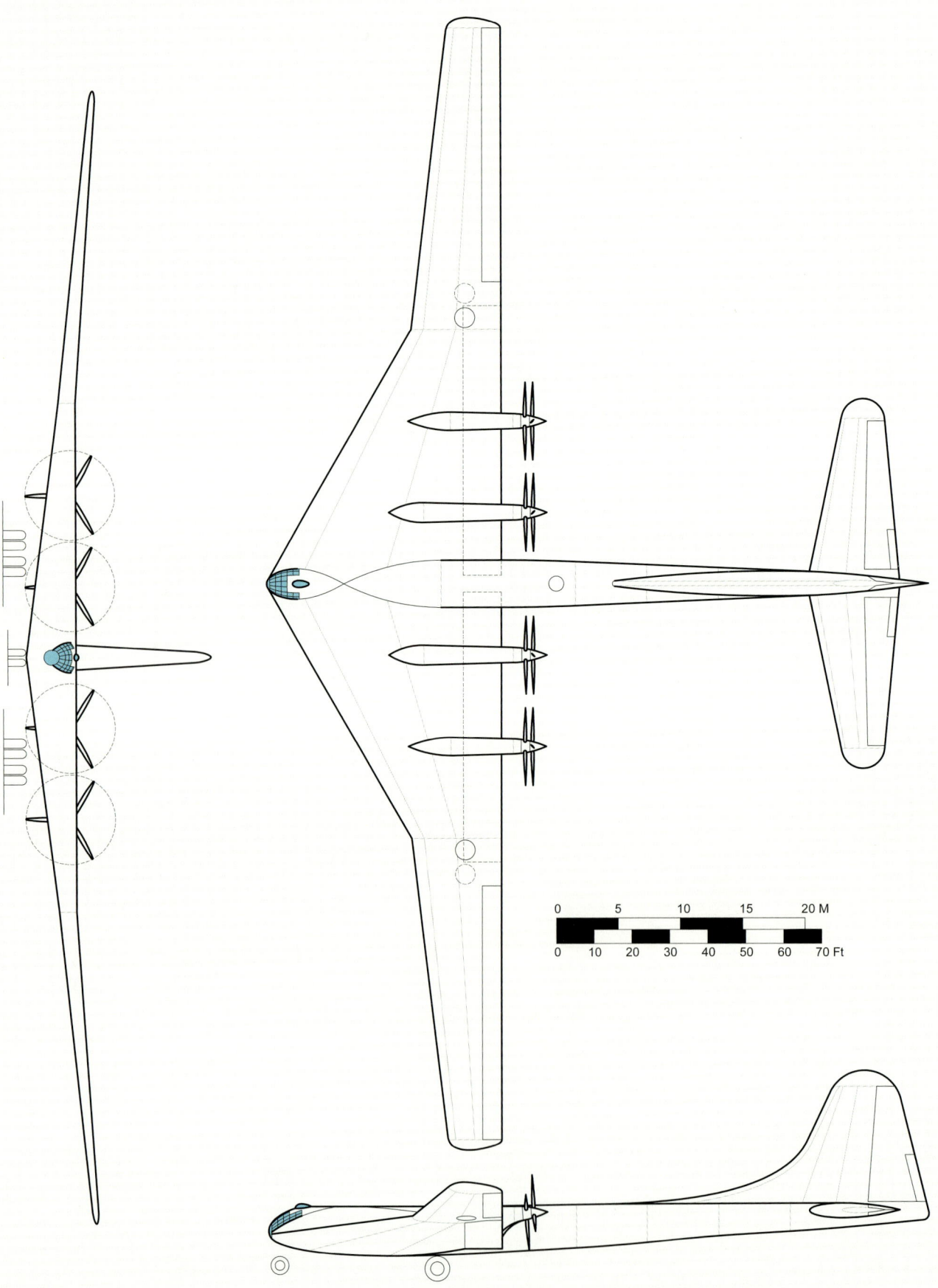

Boeing Model 365
SCALE 1/350

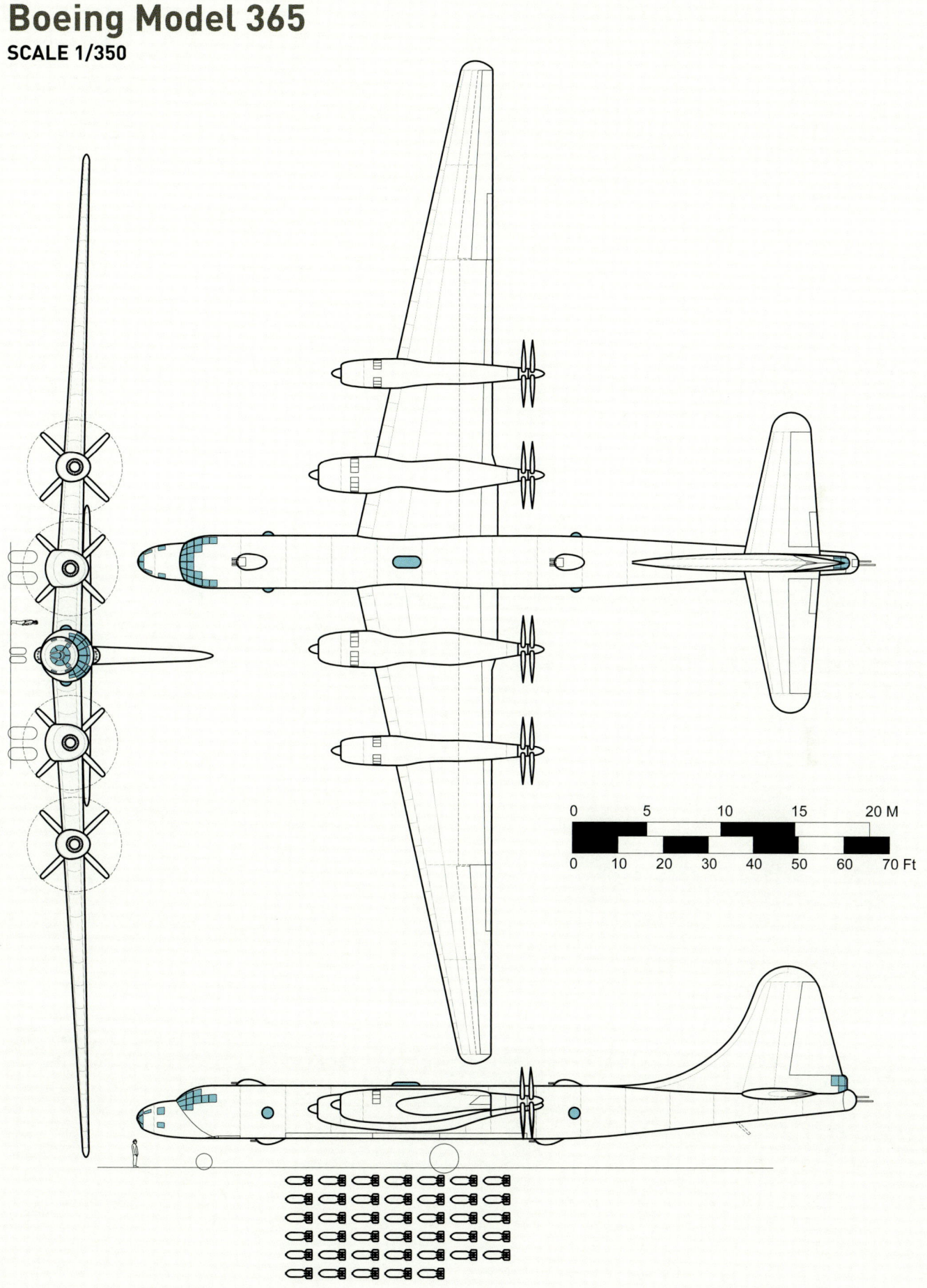

Defensive armament of the Model 365 would have been formidable, with four turrets, two on the upper side of the fuselage, two on the lower, all with four .50 calibre machine guns and a tail turret with two .50 machine guns and an additional two 37mm cannon. Each of the machine guns would have 800 rounds while the cannon would only have 38 rounds each. Maximum payload was to be 32 x 1,600lb bombs, though design payload was 10,000lb. Other bomb loads included two 10,000lb bombs, eight 4,000lb bombs, 24 x 2,000lb bombs, 40 x 1000lb bombs or 84 x 500lb bombs. Range is not given in the available documentation. The design does seem to fall about midway between the B-29 and B-36 in capability.

Boeing Model 384

Model 384 from October 1942 was another 'scaled up' B-29, with an increased length, span and fuselage diameter (to 12ft 3in). The 'greenhouse' canopy was replaced with a more conventional raised cockpit and canopy. The diagram depicts the aircraft with five fuselage turrets (two dorsal turrets, two ventral and one tail), all of which are fundamentally different from the others. The raised cockpit and numerous sizable turrets represents a step backwards in the effort to maximize aerodynamic efficiency.

The aircraft has two separate bomb bays in the fuselage, with a 10ft separation between them. But the biggest difference in configuration between Model 384 and the B-29 is the landing gear: the former is something of a throwback, using taildragger gear rather than nose landing gear. This would lead to the Model 384 sitting with a distinct nose-up stance while on the ground, akin to the British Avro Lancaster.

The nose of the fuselage was a pressurized capsule, with the forward bomb bay directly behind. This would have made installing nose gear difficult. The relatively forward position of the wings meant that the main landing gear would be too close to or even ahead of the centre of gravity of the loaded aircraft. Put together, these issues made a tailwheel a virtual necessity.

Armament was 10 x .50cal machine guns with 800 rounds each and four 20mm cannon with 400 rounds each. It appears that the forward dorsal and ventral turrets each had two .50 machine guns, the aft ventral turret had four .50 machine guns, the aft dorsal turret had two 20mm cannon and the tail had two .50 machine guns plus two 20mm cannon. The offensive load was to be one 10,000lb bomb, two 4,000lb bombs, five 2,000lb bombs, ten 1,000lb bombs or 20 x 500lb bombs. Range was well short of the 10,000 mile requirement.

Boeing Model 385

This August 1942 design had a conventional configuration with six engines. Once again it appears to have taken the B-29 as inspiration, but with everything stretched. The available diagram is fairly rudimentary (and contradictory between top, side and front views), indicating that the design was not well advanced. Payload was nominally 10,000lb, but it could carry a maximum load of 32 x 200lb bombs, 48 x 1000lb bombs or 96 x 500lb bombs. For defence it had 14 x .50cal machine guns with 500 rounds each and a single 20mm cannon with 120 rounds. Information on this design is pretty lean.

Model 384, and possibly Model 385 too, were intended as Boeing's official submissions to the USAAC's RFP. But their performance was well below what was required, as well as being submitted somewhat late. It appears that the Army was underwhelmed with Boeing's tender and no further work was done. However, Boeing was deeply involved in production of the B-17 and development of the B-29; neither the Army nor Boeing were likely too upset that the company would be able to focus on the jobs it already had at hand.

Convair B-36 (Model 35)

Consolidated responded to the April 1941 RFP with their Model 35 design. This was a fairly gigantic aircraft that utilized the best aerodynamics of the day as well as novel metallurgy and design to reduce weight. It featured six of the astonishing Pratt & Whitney R-4360 Wasp radial engines, two tractor and four pusher, with the tractor engines in tandem with two of the pushers. These engines featured a corn cob-like arrangement of four rows of seven cylinders (totalling 28) to generate 3,800hp per engine.

The pusher engines were buried partially within the wings and drove propellers well aft of the wing trailing edges. The wings themselves were aerodynamically nothing spectacular, being fairly straight and certainly having a thick root… but they were vast in expanse, with a span of 232ft 6in.

The B-36 remains the longest-span military aircraft ever put into production, though the Model 35 as originally proposed did differ in other important ways from the B-36 as actually built. The tail surfaces were much more like those of the B-24 than the B-29, with oval vertical stabilizers at the ends of the horizontals. The fuselage was a simple tube, though the nose as originally designed was slimmer than actually built, but broadly conventional. What gave the Model 35 its great range was its sheer scale.

By October 1941, the Army Air Force had decided that the Consolidated design was the best one submitted to the RFP, so the following month a contract was signed calling for the construction of two XB-36s. Model 35 evolved as work progressed – the cockpit section became stubbier; the fuselage became more constant-section cylindrical; the defensive armament arrangement changed time and again.

Boeing Model 384
SCALE 1/300

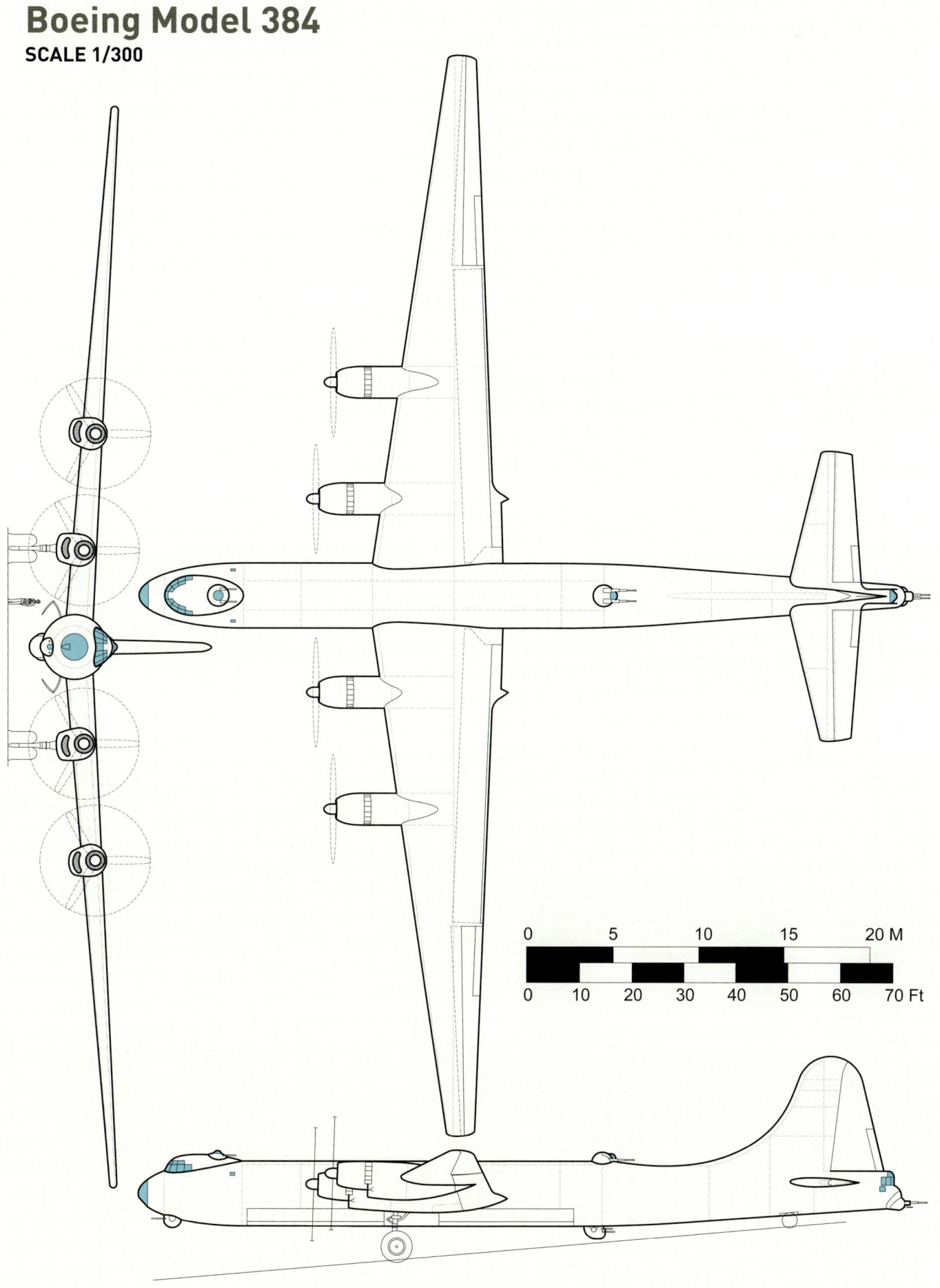

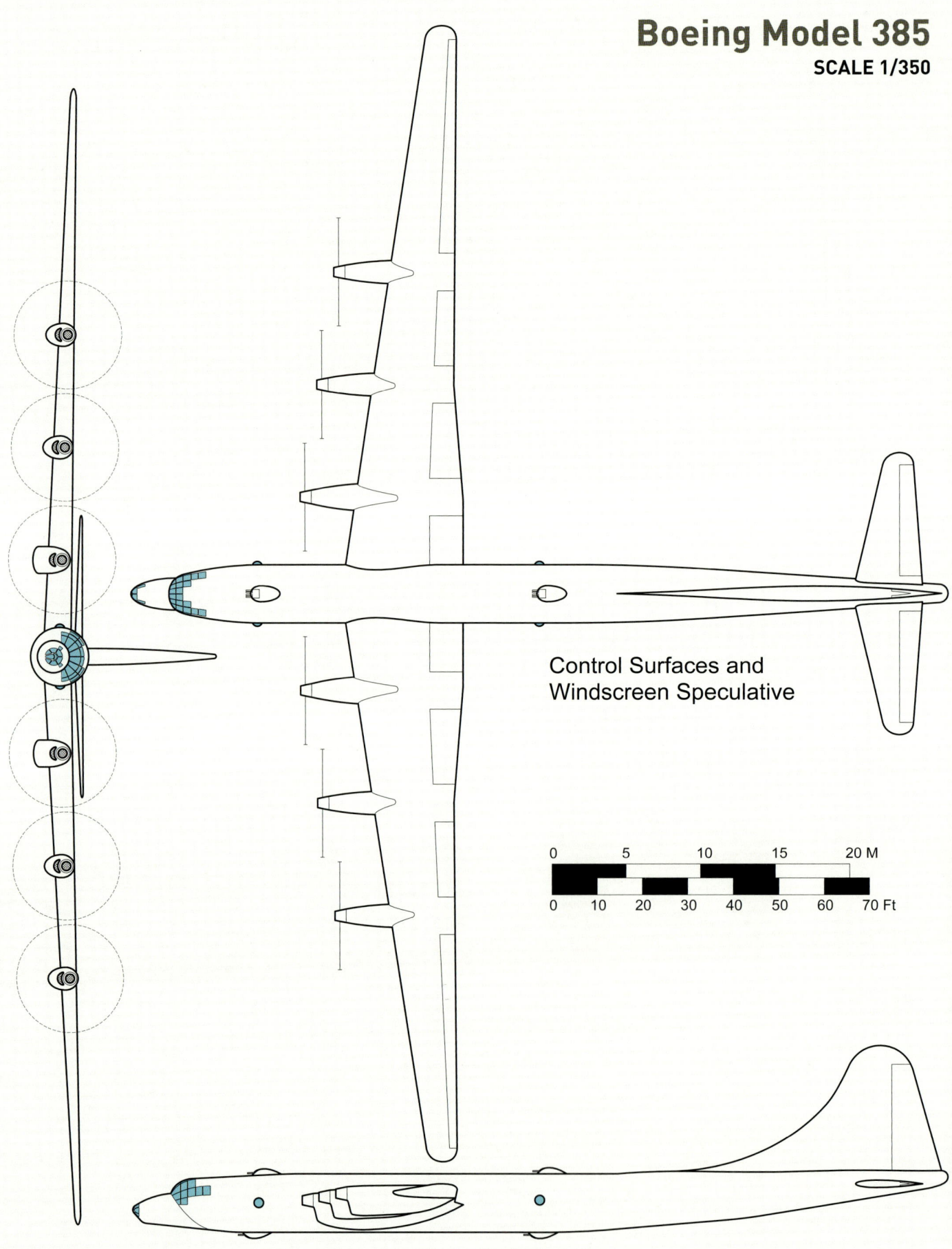

The aircraft eventually changed enough that Consolidated felt that it deserved a new company designation, going from Model 35 to Model 36. Every effort was made to reduce drag; the skin was made as smooth as possible and the six engines become all-pushers. Putting the propellers behind the aircraft ensured that the airflow over the wings would be as clean as possible. The defensive gun turrets, six each with twin 20mm cannon, were stored internally and deployed through doors that slid open, the lowest drag option available. The turrets were controlled remotely using the system pioneered by the B-29.

Progress was slow. In March 1943, well before the first flight of the XB-36, Consolidated merged with Vultee. The new company was often unofficially referred to as 'Convair' until 1954, when it became official. In July 1943, the Army Air Forces ordered a production run of 100 of the B-36 bombers, despite the not-yet-flown XB-36s having run into delays and technical problems. It was now becoming clear that Britain was unlikely to either fall or sign a peace deal with the Nazis – and the war in the Pacific was turning into a horrific slog. The B-29, as well as the B-32 Dominator built as a backup, were also experiencing problems. Troubled as it was, the Army thought that the B-36 might be the only way to strike at the Japanese home islands – hence the production order.

The twin tail became a single vast vertical fin, reducing weight and drag while raising the height of the aircraft. A new directive came in that the bomber needed nose cannon to protect against head-on attack; this necessitated a complete redesign of the forward fuselage, raising the cockpit and giving the aircraft its distinctive and fairly gigantic 'greenhouse' canopy.

The main landing gear was designed to use a single 110in diameter tyre, the largest aircraft tyre in history, and this would fit neatly within the wing. But the single tyre configuration reduced the number of runways the B-36 could land on (without crushing the concrete) to three: Fort Worth, Texas, Eglin Field, Florida and Fairfield-Suisun Army Air Field in California, as well as posing a serious risk to the aircraft if a tyre failed. So the landing gear was redesigned with a four-wheel bogey and more realistic 56in tyres. This arrangement was more practical but also thicker; as a result the wing was redesigned to have fairings both above and below to cover where the wheel projected through the wing outer mold line.

At last the single XB-36 was completed and rolled out, and finally flew on August 8, 1946, a year and two days after the atomic bombing of Hiroshima. The aircraft had missed the war by a large margin and proved to be creakingly slow with a top speed of only 230mph. The XB-36 had the original low-set cockpit, no defensive armament whatsoever and the single large diameter main wheels. Still, the B-36 was all the Army Air Forces, shortly to become the USAF, had as far as intercontinental bombers went. The first B-36A with the final landing gear type, armament positions (though not the actual armament) and raised cockpit first flew on August 28, 1947. The second XB-36 only flew for the first time on December 4, 1947, almost six years after Pearl Harbor. It was built closer to production standard with the redesigned cockpit and full set of cannon, but still had the single main wheel, and was designated YB-36.

The B-36s began entering Strategic Air Command service in June of 1948. Performance was improved compared to the XB-36: after an 8,000ft takeoff run with a 10,000lb bomb load they had a radius of action of 3,880 miles and a maximum speed over the target of 345mph at 31,600ft. This fell well short of the 12,000 mile range they had been meant to deliver, but at least it could almost strike Berlin from New York City – though Berlin was no longer a target by 1948.

The B-36B, first flying on July 8, 1948, improved upon the B-36A in many areas, learning the lessons that the A-model taught. And it did at last come equipped with the defensive armament planned. The B-36B began SAC service in November 1948 and test flights showed that it could carry out a simulated mission with 10,000lb of bombs with a total range of 8,100 miles; truly an intercontinental range. It did so at only 236mph though, a speed fairly described as 'lumbering'. Several proposals for modifying the B-36 with variable discharge turbine or turboprop engines were made but these would have required massive wing modifications and in the end failed to come to fruition. Convair proposed a simpler solution to boost the top speed of the B-36 in October 1948: bolt on some turbojets. The B-36D model was delivered with four General Electric J47 turbojets in twin engine pods – basically identical to those used on the Boeing B-47 except without the provisions for the deployable outrigger landing gear.

It came as no surprise that the additional jet engines substantially improved performance. They reduced the takeoff run by 2,000ft, raised the ceiling to 43,800ft and the top speed at 36,200ft to 406mph. All existing B-36Bs were then converted to B-36D standard.

The B-36D finally gave the United States a truly operational intercontinental strategic bomber. It was able to hit virtually any target on earth, dropping any weapon in the US arsenal. This included not only the T-12 44,000lb bomb but also the 15 megaton Mk 17 thermonuclear bomb. At 41,400lb, it was the heaviest nuclear weapon the US ever produced. It was not a convenient device.

Convair B-36
SCALE 1/400

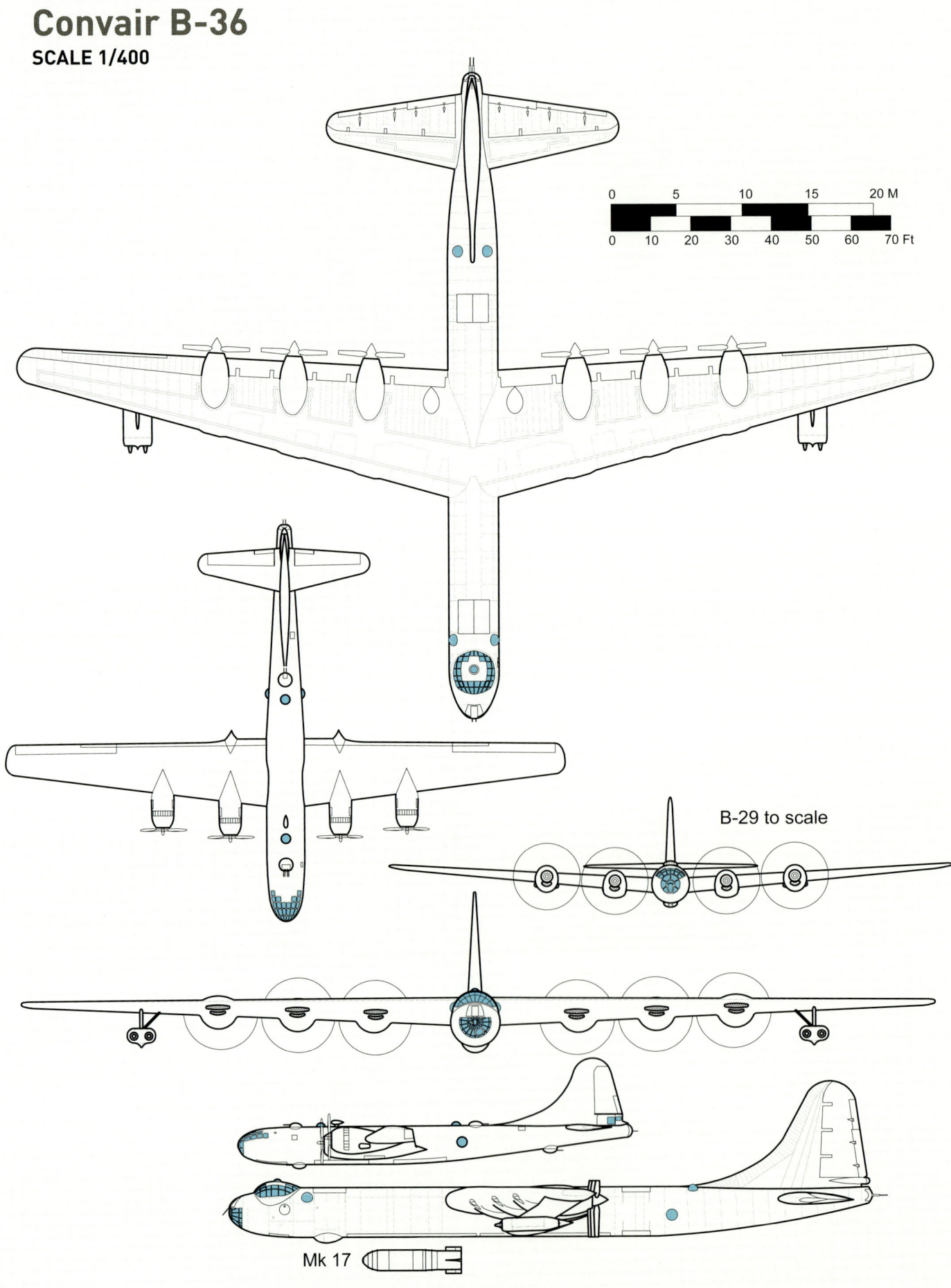

B-29 to scale

Mk 17

Model 444 A
SCALE 1/288

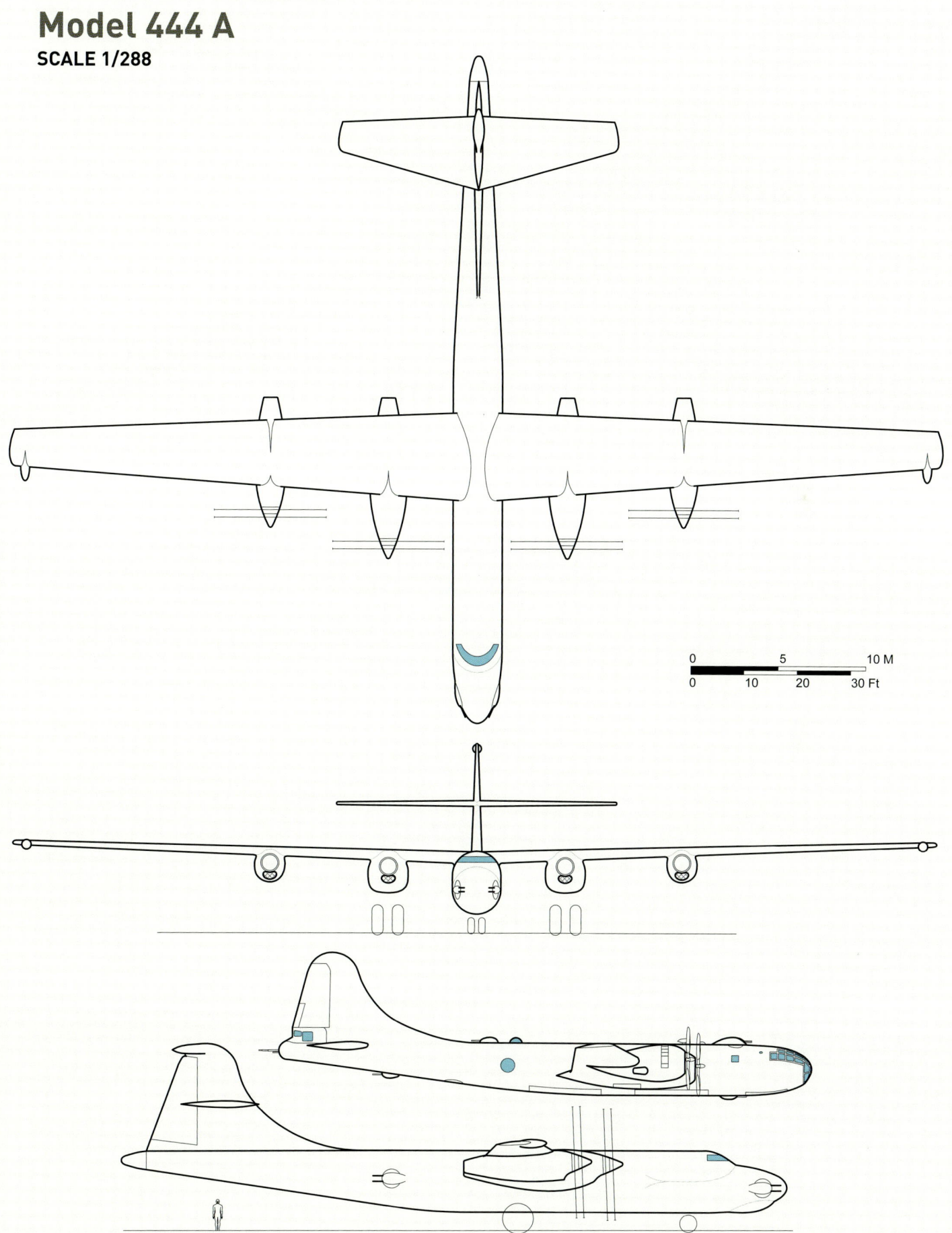

Boeing Model 444A

A May 1945 design, Model 444A was a long-range strategic bomber powered by four large turboprop engines, each with 20ft diameter counter-rotating propellers. Resembling a scaled up B-29, for defence it had four remotely controlled machinegun turrets mounted fore and aft on the sides of the fuselage. Little further data is available.

MCD 392

Aircraft design in the US was not left solely to the aircraft manufacturers during the Second World War. The US Army Air Forces designed aircraft as well – not necessarily to be built, but to illustrate the sort of aircraft it needed or would soon need. One such design was the MCD 392 (Material Command Design 392), produced in 1944 at Wright Field Aircraft Laboratory. This appears to have been more a design study to keep the Wright Field engineers on their toes, rather than an effort to produce a detailed design for production and service.

The MCD 392 was intended to show what kind of aircraft could fulfill a global bombardment role. To that end, it was a gigantic machine, powered by eight Allison V-3420 engines. All these unconventional powerplants were buried in the wing, driving either tractor or pusher propellers, and the thickness of the wing allowed crew to access them for maintenance or repair.

The propellers were contra-rotating, with two engines feeding into a common gearbox to provide power. Each propeller, tractor or pusher, was at the end of a nacelle that extended past the other wing edge. They terminated in ball turrets, each with a quartet of .50cal machine guns. The gunners could access the rest of the aircraft via walkways in the nacelles and wings. Additional ball turrets were located in the nose and tail. This arrangement was expected to provide adequate coverage – though directly above and below, as well as to the side, would have had poor coverage unless those turrets were capable of surprising degrees of traverse and elevation.

Total defensive armament was 16 x .50cal machine guns, each with 1000 rounds, and eight 20mm cannon, each with 300 rounds. The bomb load was only partially contained within the fuselage, the inner portions of the wings also able to hold a substantial payload. Illustrations show a crew of 15.

The MCD 392 seems like an oddity. It was intended to bomb Tokyo, even though that city was already in range of bombers flying from captured Pacific islands by the time it was drawn up. And while its range and payload capacity were vast, its ceiling was shockingly low and its top speed abysmally slow. Its layout, propulsion systems and aerodynamics represented early-war thinking and it had no future in a jet-powered post-war era.

Early Postwar Concepts

When the Second World War ended, the fragile alliance between the Western Allies and Stalin and the Soviet Union quickly fractured. Clearly this was not, as had been hoped, the end of all wars. So the American military and aviation industry kept on designing and innovating, knowing that sooner or later war would come roaring back. And even though the United States had at that moment a massive air force, it was also clear that the vast bulk of it was already well beyond obsolete. Jet engines, turboprops, rockets and ramjets

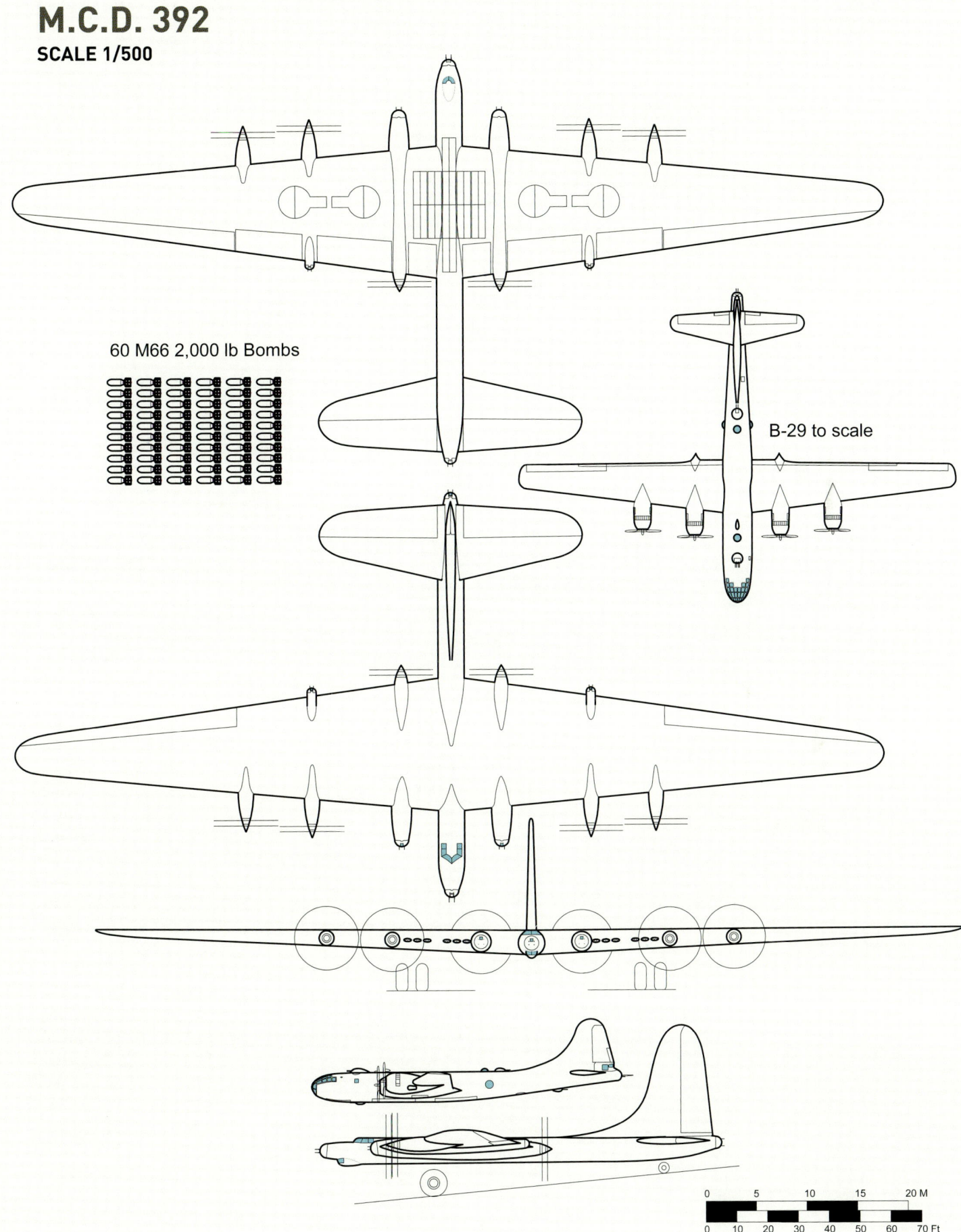

were the way of the future, not piston engines. Every single airplane the US had would need to be replaced… and every American aviation company wanted to be in on that. Designs were produced both to government specification and by company internal desire.

Consolidated Vultee High Performance Patrol Landplane

The US Navy issued a requirement in January 1946 for a heavy land-based patrol bomber and Consolidated Vultee responded two months later with a turboprop-powered design.

The aircraft was intended to fly at high altitude and scan the ocean for enemy vessels using radar in the lower half of the nose. Photographic reconnaissance was a secondary role, along with the launch of 'special weapons' against targets of opportunity. 'Special weapon' generally meant 'atomic bomb' at that time but the use of A-bombs against targets of opportunity – in this case, likely ships or subs – seems vastly expensive overkill. Instead, the term more likely referred to guided munitions such as the VB-6 Felix infrared guided anti-ship bomb or the ASM-N-2 Bat radar guided anti-ship glide bomb.

The inboard profile diagram shows the aircraft armed with a 22,000lb M110/ T-14-type giant bomb; while the smaller 'Tallboy' bombs were used by the British to sink the German battleship Tirpitz, the Tirpitz was essentially a fixed target. Hitting a moving vessel with a 'Grand Slam' would be challenging to say the least… but there are few ships that would not be effectively rattled by the impact of such a device dropped from 40,000ft.

The baseline design was very much a transitional concept between wartime convention and rapidly evolving postwar technology. It had shoulder mounted straight wings, conventional tail unit, propellers and engines on the wings, sharing some traits with the XB-46, then under development. But the engines were what set it apart.

At first glance, the High Performance Patrol Landplane appears to have six tractor engines. And indeed it did have six turboprop engines in separate nacelles, each driving a 14ft 7in diameter four-bladed propeller. But the intermediate and outboard nacelles were also fitted with a separate turbojet engine for takeoff thrust and high-speed performance, bringing the total engine count to ten.

In each nacelle the turboprop engine was mounted aft of the main spars, driving the propeller via an extension shaft; each turbojet was located at the rear of its nacelle, above the turboprop.

At the tail was a remotely controlled turret with four 20mm cannon, each with 400 rounds. The gunner doubled as the radio operator and, as with the rest of the crew, occupied the single pressurized compartment at the front of the fuselage. Additional remotely controlled turrets sat at the wingtips, each with two 20mm cannon with 400 rounds each. It's unclear whether the guns were trained forward or aft, however. Fuel was contained within a series of fuselage cells.

An alternate design was put forward at the same time, deleting the wingtip turrets and going with four more powerful Wright turboprops, each with 18ft diameter contra-rotating propellers. This design was clearly derived from the Heavy Bombardment Airplane of about a year earlier (see below), differing largely in being shorter and only having defensive guns in the tail. It may be that the baseline design was also derived from an earlier Army bomber study; this would explain the Grand Slam.

The High Performance Patrol Landplane was by no means an intercontinental strategic bomber, but it had features that would be useful for such an aircraft. And, importantly, Consolidated Vultee would soon base its B-52 competitor on it.

Turboprop Boeing B-29

A single photograph is known to exist of a large model of a B-29 modified with four turboprop engines, each with contra-rotating pusher propellers. This was likely a wind tunnel model. The date of this design, along with the identity of the engines, is currently unknown, but it's possible that this was an early study for an improved B-29 capable of higher speed.

Boeing Model 461

This 'Patrol Landplane' dated from June 1946 (paradoxically, the Model 461 seems to have been designed after the Model 462). While still very much a Second World War era plane, with straight wings and a stereotypical war-era Boeing vertical tail, the design is starting to trend in the direction of postwar subsonic American bomber design. The cockpit is more modern, the engines are turboprops and the landing gear is located directly under the fuselage, with relatively small stabilizing outriggers further out under the wings. A large radar dome is visible under the nose, for navigation and bomb-aiming.

Weapons load, apart from total weight (39,400lb) is unknown, but by mid-1946 atomic armament can be assumed. At the time, only the 10,300lb Mark III ('Fat Man') bomb would have been available and the Model 461 would have been able to carry three with room to spare. Defensive armament included a tail turret, remote controlled by the crew in the front of the aircraft. Thought was also given to podded turrets on the wingtips, details unavailable. A slightly later artist's impression of a six-engined Model 461 shows nine conventional bombs, probably four thousand pounders.

Consolidated Vultee High Performance Patrol Landplane
SCALE 1/275

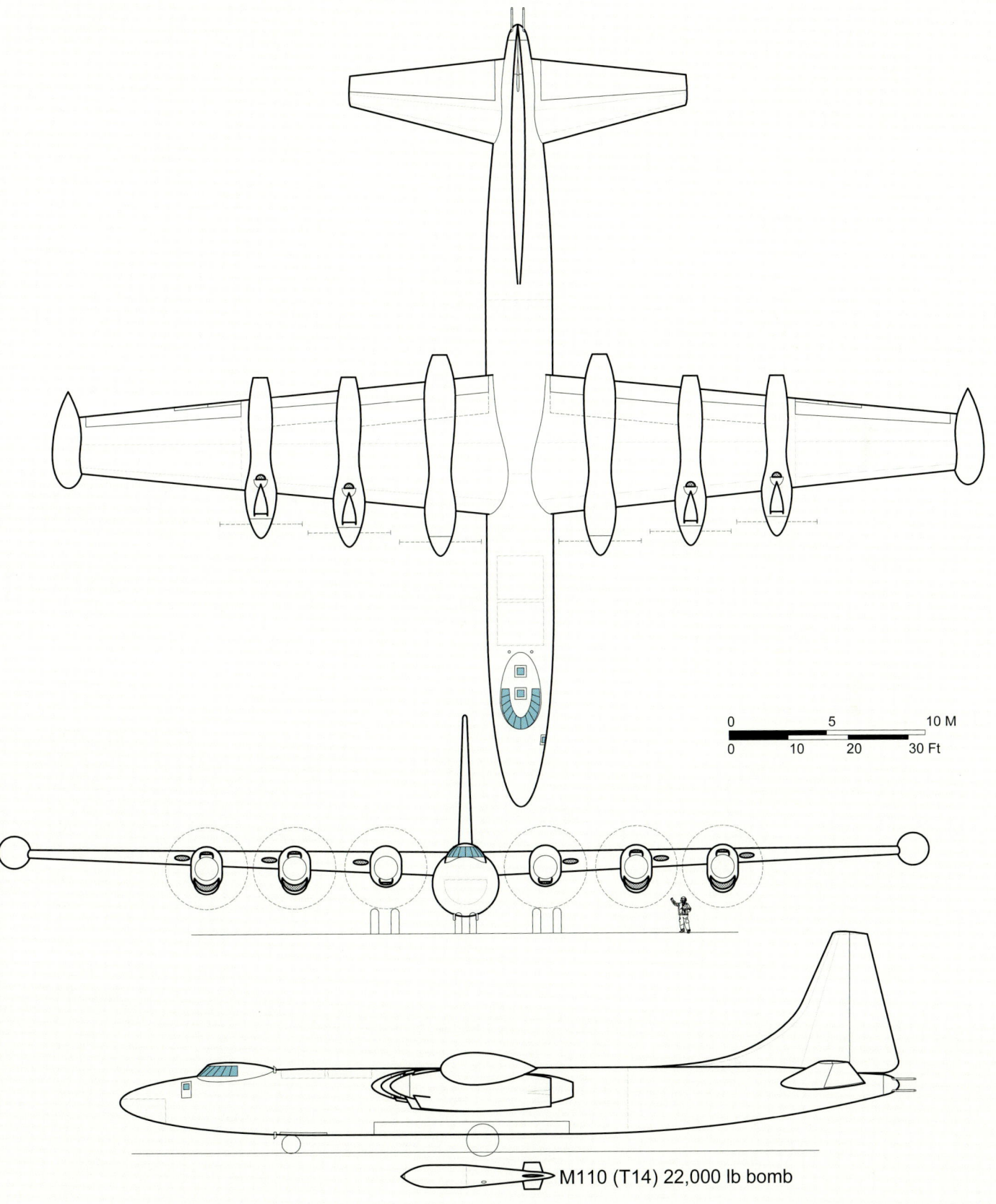

M110 (T14) 22,000 lb bomb

Turboprop B-29
SCALE 1/220

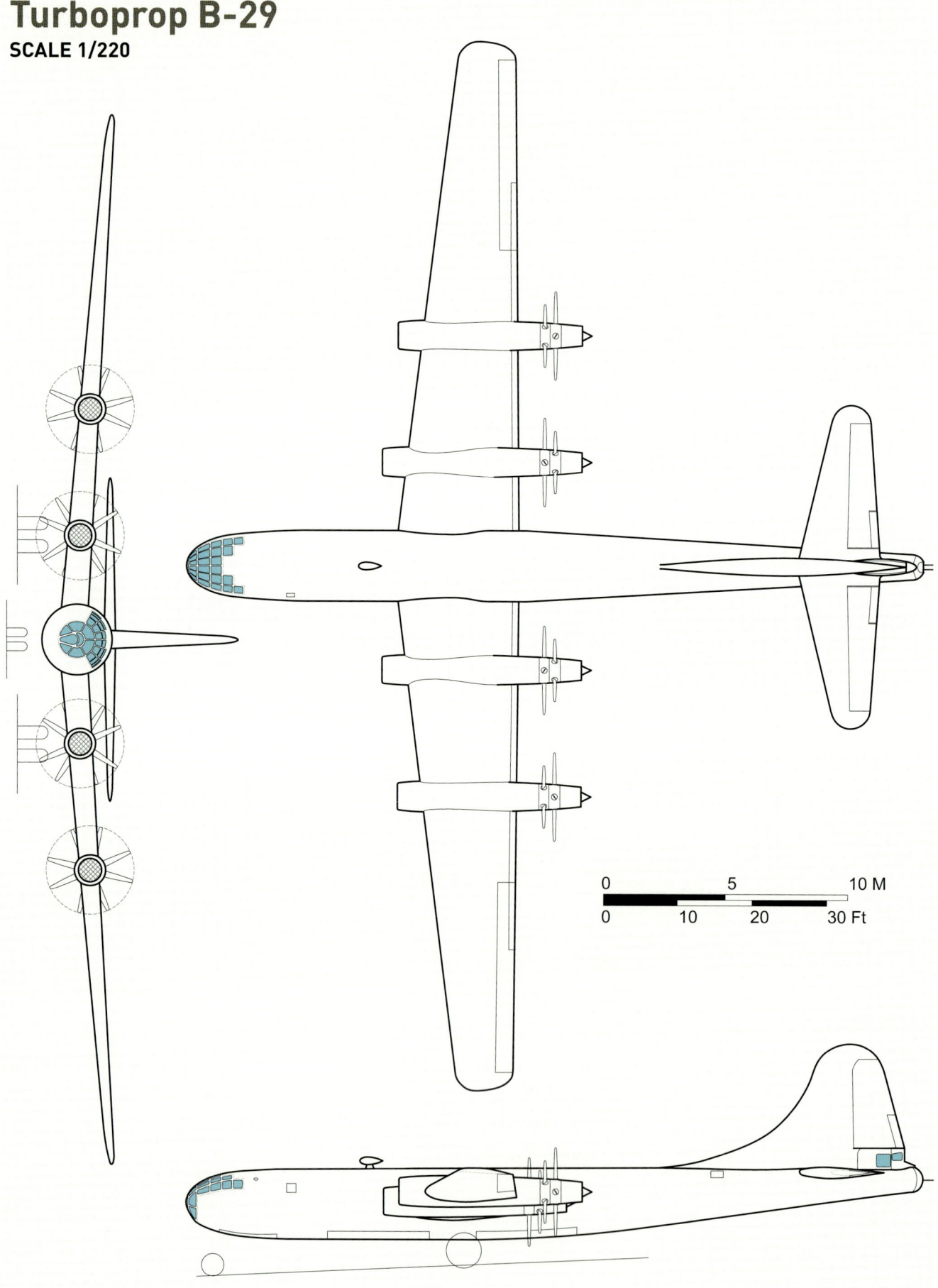

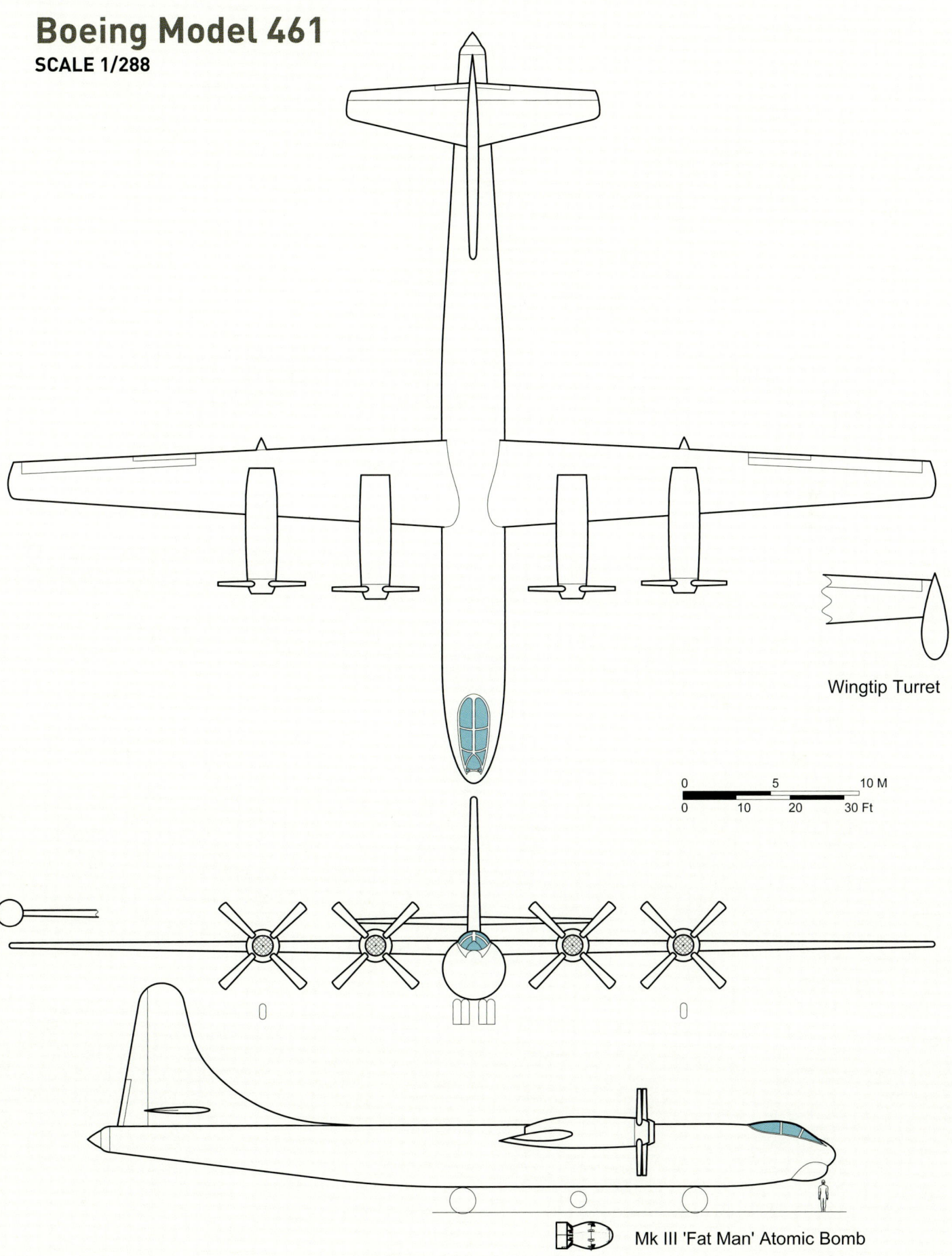

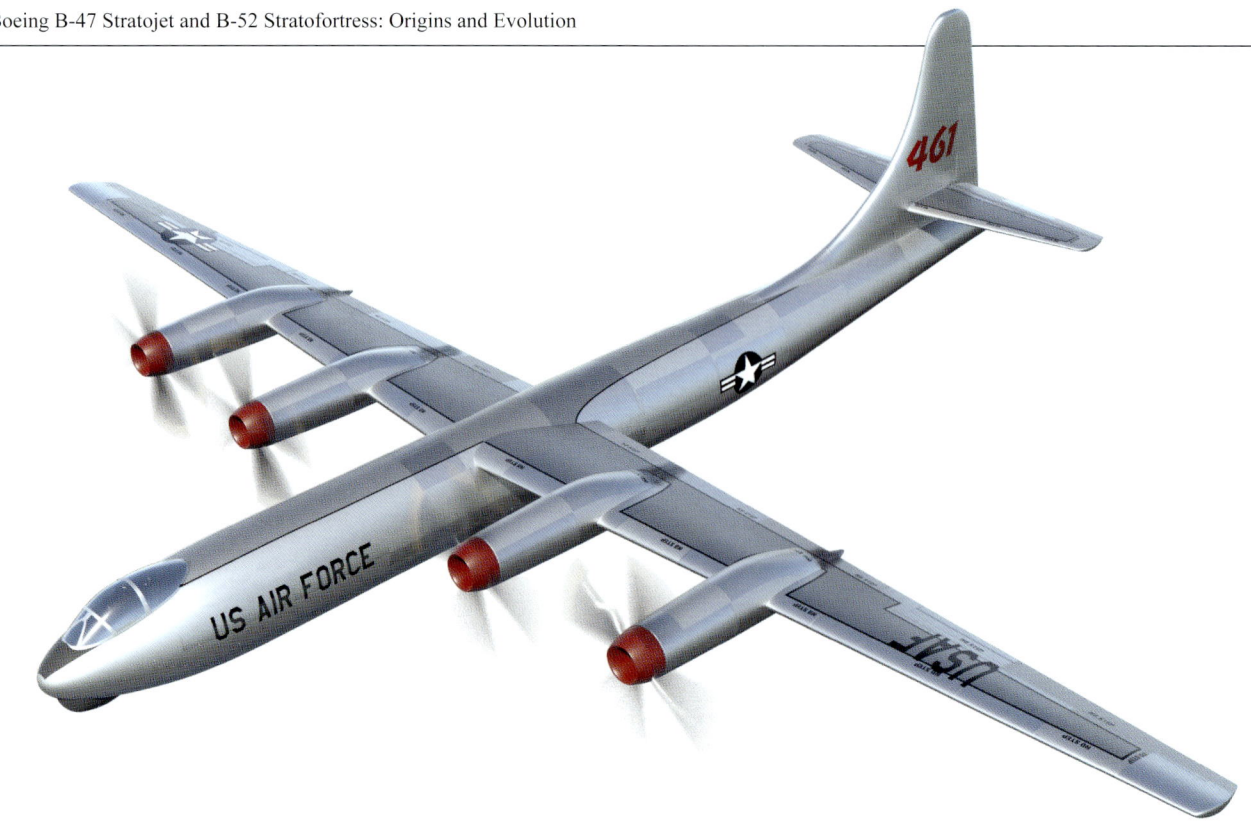

The contest

The Engineering Division at Wright Field recommended in August 1944 that a study for a B-36 replacement be undertaken in 1946. Work on this new heavy long-range bomber would, it was suggested, occur between 1947 and 1949. It would be more advanced than the as-yet unflown B-36, taking advantage of recent developments in turbojet and turboprop engine technologies… the same technologies being put forward for the medium bomber project that would result in the B-47.

The advent of jet interceptors was making the B-36 look vulnerable – and existing jet escort fighters lacked the range to defend it over enemy territory. So a new generation of intercontinental bomber was needed that could fly both high and fast. Prior to April 1945, the Army Air Force published requirements for a new heavy bomber which included top speed of 475mph with half fuel at 35,000ft, 375mph average speed, 40 x 500lb bomb load, 40,000ft ceiling, 15,000ft ceiling on half engines and 5,500 mile range.

This was extremely aggressive for the time. Most aircraft companies declined the challenge, seeing the AAF's requirements as beyond the state of the art. However, one design has been found that was intended to fill that role.

Consolidated Vultee Heavy Bombardment Airplane

Consolidated Vultee designed a turboprop bomber in March 1945 with a view to meeting the AAF's requirements. With the available engines being somewhat disappointing, the design was based on hypothetical engines using data provided by the Wright Aeronautical Corporation. The overall configuration was conventional; it looked like nothing so much as an XB-46 scaled up and given four separate turboprops. As with the XB-46, the wings were narrow chord and unswept; data from German swept wing studies had not yet come in.

The turboprop engines were assumed to be Wright gas turbines with regeneration, producing 5,500hp. Each would drive a 15ft diameter six-bladed counter rotating propeller. The inboard nacelles had a bulbous fairing on the underside into which the dual main wheels would retract, the dual nose wheels retracting beneath the pressurized crew compartment. The forward crew compartment contained the pilot, co-pilot, two radar bombardier navigators, the flight engineer and one radio operator/fire control officer.

Likely the most unconventional design features were the four large diameter remote-controlled turrets on the sides of the fuselage. Their guns, pivoting near the ends of their barrels, were located within the fuselage so that the turrets presented the same low drag no matter where the guns were pointed.

Despite this unusual arrangement, it seems that the four side turrets and the tail stinger would have provided good coverage. All five positions had two .50cal machine guns; the forward turrets had 500 rounds per gun, the aft turrets 800 rounds per gun and the tail had 1,000 rounds per gun. Three fire control operators were housed in a pressurized compartment near the tail, connected to the forward compartment by a long pressurized tube.

There is no evidence that Consolidated Vultee actually submitted the design for consideration – likely being too preliminary to be turned in. It is known from a Consolidated Vultee design summary, rather than a proposal or detailed

Consolidated Vultee Heavy Bombardment
SCALE 1/275

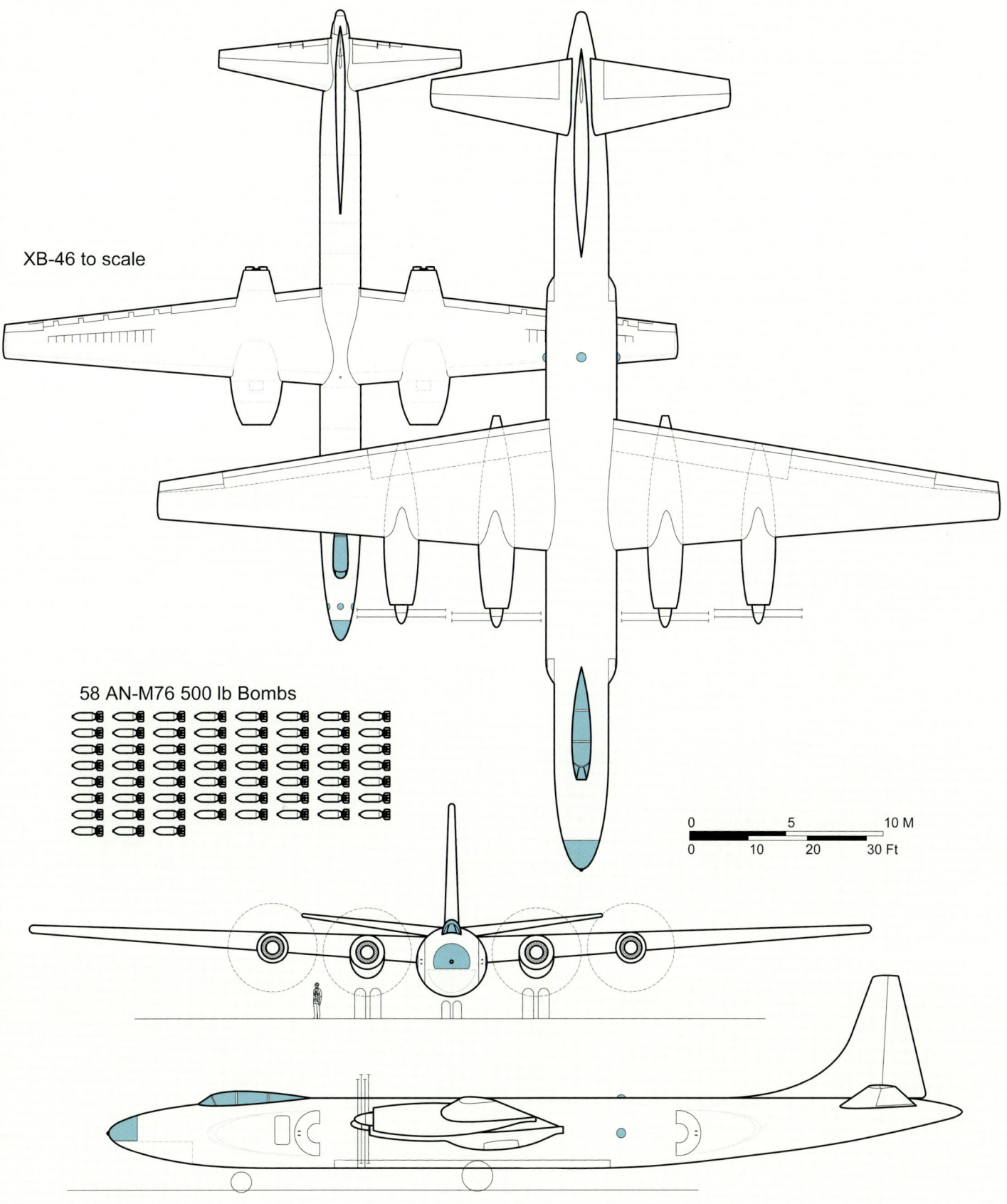

XB-46 to scale

58 AN-M76 500 lb Bombs

engineering schematics. While the performance estimated for the aircraft certainly exceeds the requirements, it seems rather implausible that it would do so in practice. Further examination by CVAC or perhaps AAF staff may well have found it fundamentally flawed.

No other designs are known to have been created specifically to fit the overly aggressive AAF requirements. A rethink therefore produced more modest requirements in January 1946: 450mph top speed at 35,000ft, 300mph average speed, 40,000ft ceiling, 15,000ft ceiling on half engines, 5,000 mile range with 10,000lb of bombs, 7,500ft take-off run to clear 50ft obstacle at gross weight, 4,500ft landing distance over 50ft obstacle at gross weight minus droppable fuel and bombs.

This was still more than could be asked of the turboprops and turbojets of early 1946. It was believed that the development of a turboprop engine able to support such an aircraft would require ten years of development, not the mere handful the RFP called for. Rather astonishingly, the Army wanted an aircraft able to carry – for an undetermined but necessarily shorter range – a single 80,000lb bomb.

The Air Technical Service Command issued another RFP with basically the same requirements the following month... but called only to 'approximate' the previous requirements. This recognition of the need to compromise in order to get a viable aircraft, not necessarily a perfect one, was enough to get airplane manufacturers interested.

Boeing, Consolidated Vultee and Martin submitted their designs in April. Frustratingly, Martin's Model 236 seems to have disappeared from history, with only some scraps of data being available on it. The Consolidated Vultee and Boeing designs are well described, however.

Consolidated Vultee Long Range Heavy Bombardment Airplane

The Consolidated Vultee design was certainly unusual. It was clearly derived from the Heavy Bombardment Airplane of a year earlier and a number of important design features had been carried through – but there was one clear difference. The Long Range Heavy Bombardment Airplane had distinctly forward-swept wings. This was a concept that fascinated Consolidated Vultee at the time; concurrently the company was working on the XA-44 ground-attack aircraft – a rather odd tailless design with forward swept wings and three turbojets. Forward sweep conveys the same high-speed drag reduction as aft-sweep while moving the centre of lift forward. This meant the main spar was further aft – freeing up space in the forward fuselage for equipment, weapons or other features.

Airflow over a swept wing is not purely fore-to-aft, but it is directed spanwise towards the rearmost part of the wing. With a conventional aft-swept wing this means that the airflow heads towards the wingtip, but with a forward swept wing the air heads inboard, towards the wing root. This is a useful feature during landing and other slow speed flight: the forward swept wing will tend to stall first at the wing root rather than the tip as airspeed drops and angle of attack increases. Aft swept wings that stall first near the tips tend to lose aileron effectiveness, since the ailerons are at the wingtips, but forward swept wings retain control.

Unfortunately there is a concern that has always largely negated the value of forward swept wings: as the angle of attack of a wing increases, forces build up on its tips. With a forward swept wing, those forces are trying to bend the wings up and over. In the best case, this will result in the wingtips flexing upwards; in a worse case they take a permanent set, and in the worst case they break. As the tips twist upwards, they generate more lift, which bends them upwards even more. A feedback loop can set in, leading to disaster. In any case, the wingtips twisting up even a little bit can also cause flow instabilities.

Before the Grumman X-29 demonstrated forward swept wings made out of graphite-epoxy composites, wing structures strong enough to withstand the forces trying to warp them tended to be unenviably heavy. A handful of aircraft have entered production with forward swept wings, such as the Hamburger Flugzeugbau HFB 320 Hansa Jet. This 1960s business jet had a forward swept wing for a singular purpose: the spar passed through the rear of the fuselage rather than the front. This allowed for a passenger cabin larger than those contemporary competitors such as the Learjet.

The root stalling advantages of forward swept wings convinced Consolidated Vultee's designers to give their big bomber this unusual feature. The aft location of the spars also permitted the long bomb bay to run through the middle of the fuselage without structural interference. The main landing gear wheels were stored within the fuselage, while the landing gear legs were kept just behind the leading edge of the wing roots.

The wings were bisected by four nacelles. Each of these held a Wright T-35 turboprop up front driving a four bladed propeller. Directly behind each turboprop was a GE TG-180 turbojet; these would be used only intermittently, such as during takeoff and during high-speed dash over the target – where they would add about 115mph to the top speed at 35,000ft. This was considered an interim solution; it was hoped that eventually, when new propulsion systems were ready, the eight engines could be replaced with four.

For defence, the bomber had four of the unconventional flush turrets that Consolidated Vultee had designed into several other bombers, including the Heavy Bombardment Airplane. Each had two 20mm

Consolidated Vultee Long Range Heavy Bombardment
SCALE 1/325

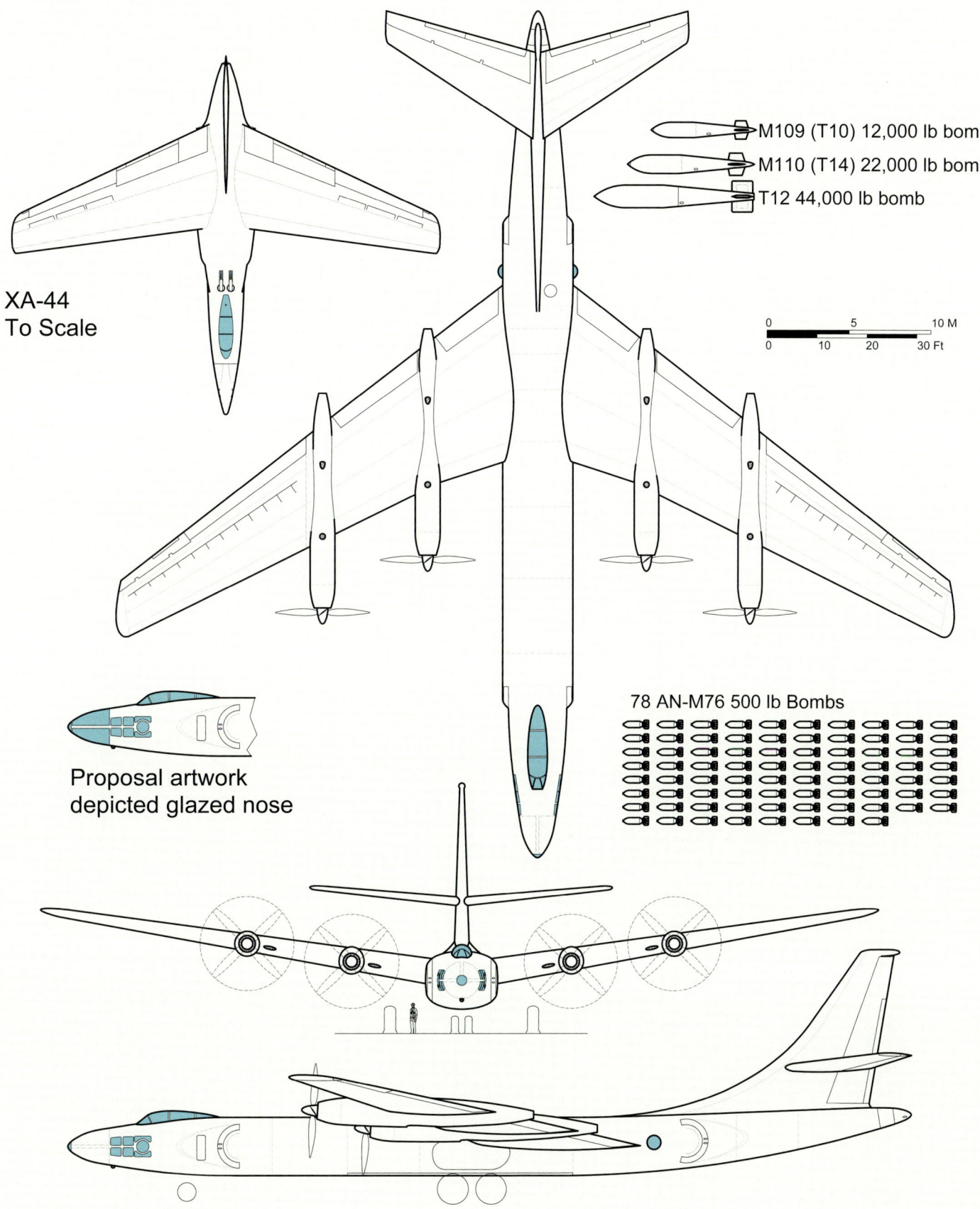

XA-44 To Scale

M109 (T10) 12,000 lb bomb
M110 (T14) 22,000 lb bomb
T12 44,000 lb bomb

78 AN-M76 500 lb Bombs

Proposal artwork depicted glazed nose

Model 462
SCALE 1/375

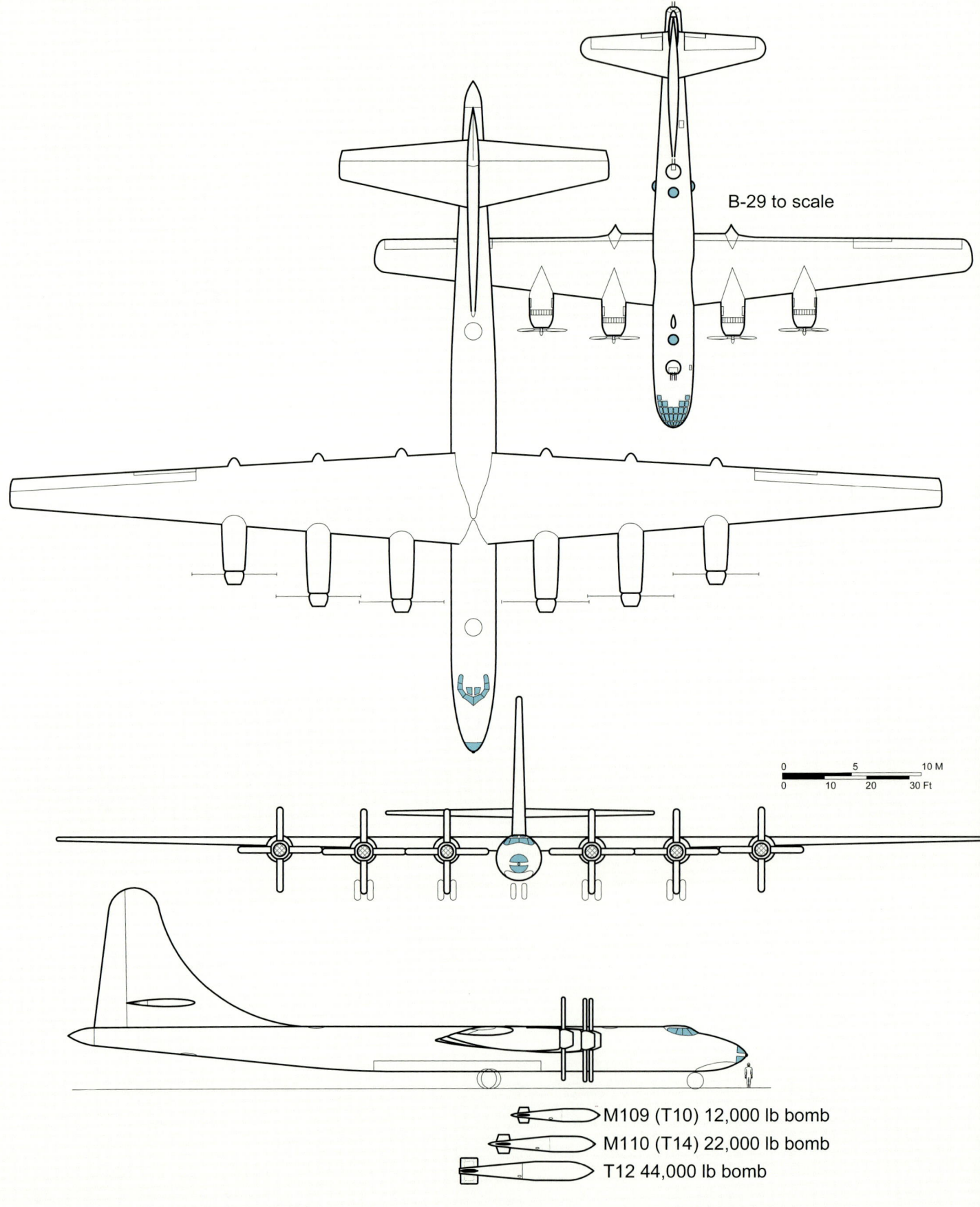

B-29 to scale

M109 (T10) 12,000 lb bomb
M110 (T14) 22,000 lb bomb
T12 44,000 lb bomb

cannon, as did the tail turret. A wide range of offensive payloads were possible, including: one 80,000lb bomb; one 44,000lb bomb; one 22,000lb bomb; two 12,000lb bombs; seven 4,000lb bombs; 18 x 2,000lb bombs; 36 x 1,000lb bombs; 78 x 500lb bombs; and, for the maximum range design condition, 20 x 500lb bombs. Total range at maximum gross weight and with one 80,000lb bomb was just over 4,000 miles, or around 9,000 miles with a 10,000lb payload.

The crew were split into two locations. The pilot, co-pilot, flight engineer, radar bombardier-navigator and optical bombardier-navigator sat in a pressurized compartment at the front of the aircraft, while three gunners sat side-by-side in a pressurized compartment near the tail. The crew could pass fore and aft through a pressurized tunnel, much as they would be able to do in the B-36.

The Long Range Heavy Bombardment Airplane did not pass Army Air Forces muster. Doubtless the dubious nature of forward swept wings had more than a little to do with it, along with the fact that Convair was at that moment already deeply involved with development of the B-36.

Martin Model 236

Martin's entry into the competition seems to have vanished utterly from the records. The only known description of it is a summary of data from the three competitors; it had a gross weight of 275,000lb, intermediate between the Consolidated Vultee and Boeing designs, with a span of 195ft and a wing area of 3,930sq ft. The configuration, number and type of engines are unknown.

Boeing Model 462

Boeing's April 1946 response to the long range heavy bomber RFP was Model 462, a large aircraft of conventional layout with six turboprops hanging under the shoulder-mounted straight wing. While it bore some faint resemblance in parts to the B-29, unlike several earlier aircraft it was not obviously a direct descendant of that aircraft. The cockpit was entirely different, with a slightly raised canopy for a pilot and co-pilot sitting side by side and a relatively long, sloping forward fuselage leading to a Plexiglas nose cap for the bombardier.

The aircraft had six Wright T-35 turboprop engines, each generating around 5,500hp while turning 20ft diameter four-bladed propellers. The forward sections of the engine nacelles were all identical, from propellers to engines to fairings, connections and mountings. This would simplify and speed maintenance; a single assembly could be bolted on to any of the six positions. The inboard and intermediate nacelles containing dual-wheel main landing gear bogies. A retractable nose wheel completed the landing gear.

Four turrets with two 20mm cannon each were provided for defence, submerged within the fuselage until needed. Two were directly above/below each other just aft of the pressurized crew compartment, one was atop the rear fuselage just ahead of the vertical stabilizer's leading edge extension, and the last was below the rear fuselage closer

to the tail. A fifth defensive position, a tail stinger, had four 20mm cannon. Two gunners sat to the rear of the crew compartment, equipped with a total of four periscopes... two above the fuselage, two below, offset to right and left.

Possible bomb loads included a single 44,000lb bomb; two 22,000lb bombs; three 12,000lb bombs; nine 4,000lb bombs; 24 x 2,000lb bombs; 50 x 1,000lb bombs or 84 x 500lb bombs. A 10,000lb 'special bomb' (atomic and possibly speculative on the part of Boeing engineers, though it's likely that they received guidance about the size and weight of atomic bombs of probable interest to the Army Air Forces, such as the Mark III 'Fat Man') of 61in diameter and 150in length was also considered.

The mighty 44,000lb T-12 'Cloudmaker' bomb, an enlarged version of the 22,000lb 'Grand Slam' developed by the British during the war, would have largely filled the bomb bay. As with the Grand Slam, the Cloudmaker was designed to penetrate deep underground before detonation, using a thick steel case to penetrate dirt and concrete before detonation. The B-36 bomber was the only aircraft that could actually carry the T-12; none were used in action as the bomb was not available until 1948. But in 1946, the ability to carry such a massive bomb indicated not only that specific capability, but the capability to haul other as-yet undeveloped massive weapons... such as the new atomic bombs that were being designed and developed just as fast as the early American atomic industry could manage.

General Laurence C Craigie, chief of the Engineering Division at Wright Field, recommended on May 23, 1946, that the Army Air Forces accept the Model 462 for a Phase I development contract. The following month, the Air Materiel Command at Wright Field informed Boeing that they had won the competition and on June 15 the new aircraft was officially designated XB-52. A letter contract for Project MX-839 (the XB-52) was approved on June 28, awarding Boeing the Phase I effort to create detailed designs, wind tunnel models and tests and a full-scale mockup.

What should have been a time of exultation for those working on the XB-52 was promptly rained on. The aircraft design was massive and failed to meet the range requirement. This opened it up to attack by those in the Army Air Forces who didn't care for it, attacks that would go on for several more years and would threaten the whole project with cancellation.

Boeing Model 462-5

Model 462 was a very large aircraft, with a span only 9ft shorter than that of the B-36, and a length only a foot shorter. Empty weight was lower than that of the B-36, but gross weight was substantially greater.

While most performance requirements were met, armament weight and turboprop inefficiencies meant that the operating radius fell short at only 3,570 statute miles (3,100 nautical miles). Some thought was given to increasing gross takeoff weight to 480,000lb by adding droppable external fuel tanks; this would have required rocket assist for takeoff.

The Model 462 design continued to evolve through multiple iterations; Model 462-5 from September 1946 moved the turboprops from below the wing to above. Importantly, the main landing gear was moved from the engine nacelles to the fuselage and smaller outrigger stabilizer landing gear were added – forming the arrangement that would eventually be used on the B-52. Putting the landing gear in the fuselage meant widening it and saying goodbye to a purely circular cross-section. Unconventionally, 'blisters' that form-fitted the outboard wheels were added, since even with a wider fuselage the wheels just didn't fit. Despite all the tinkering, though, Model 462 was simply not going to meet range requirements.

By October, complaints about Model 462's performance shortfall had prompted Boeing to greatly modify the original concept, resulting in Model 464. This was generally similar to the Model 462 but smaller, lighter and with only four engines. A number of early 464 variants were studied and within a few months Boeing had one that fit the bill.

Boeing Model 464-16 and 464-17

Boeing unveiled Models 464-16 and 464-17 in December 1946. Boeing's redesign efforts had been aided by new data from Wright Aeronautical on an improved T-35-5 turboprop. With greater power (8,900hp) and better specific fuel consumption, Model 464 was able to fill the mission requirements with only four engines. Range was at last where it was supposed to be.

Model 464 had a conventional configuration, with long straight shoulder-mounted wings and a tail much like that of Model 462. The turboprops were mounted to the leading edges (with the bulk of the nacelles below the wings), each powering 22ft-diameter counter-rotating propellers.

Models 464-16 and 464-17 were externally largely identical but their interiors were quite different. The -16 was specifically designed as an atomic bomber, with a 15ft long bomb bay just large enough to carry a single 'special' weapon. Model 462's retractable turrets were completely deleted, the vacated space now filled with fuel cells, bringing total capacity to 36,500 gallons and giving a range more than adequate to meet requirements. In effect, it was a flying fuel tank wrapped around a single atom bomb, dependent wholly on speed, altitude and luck to avoid enemy interception. Model 464-17 kept the turrets and a long bomb bay, making it a general purpose bomber capable of carrying a Cloudmaker or a factory's worth of smaller bombs. Fuel capacity was 20,200 gallons, substantially less than that of 464-16; the fuselage

Boeing Model 462-5
SCALE 1/350

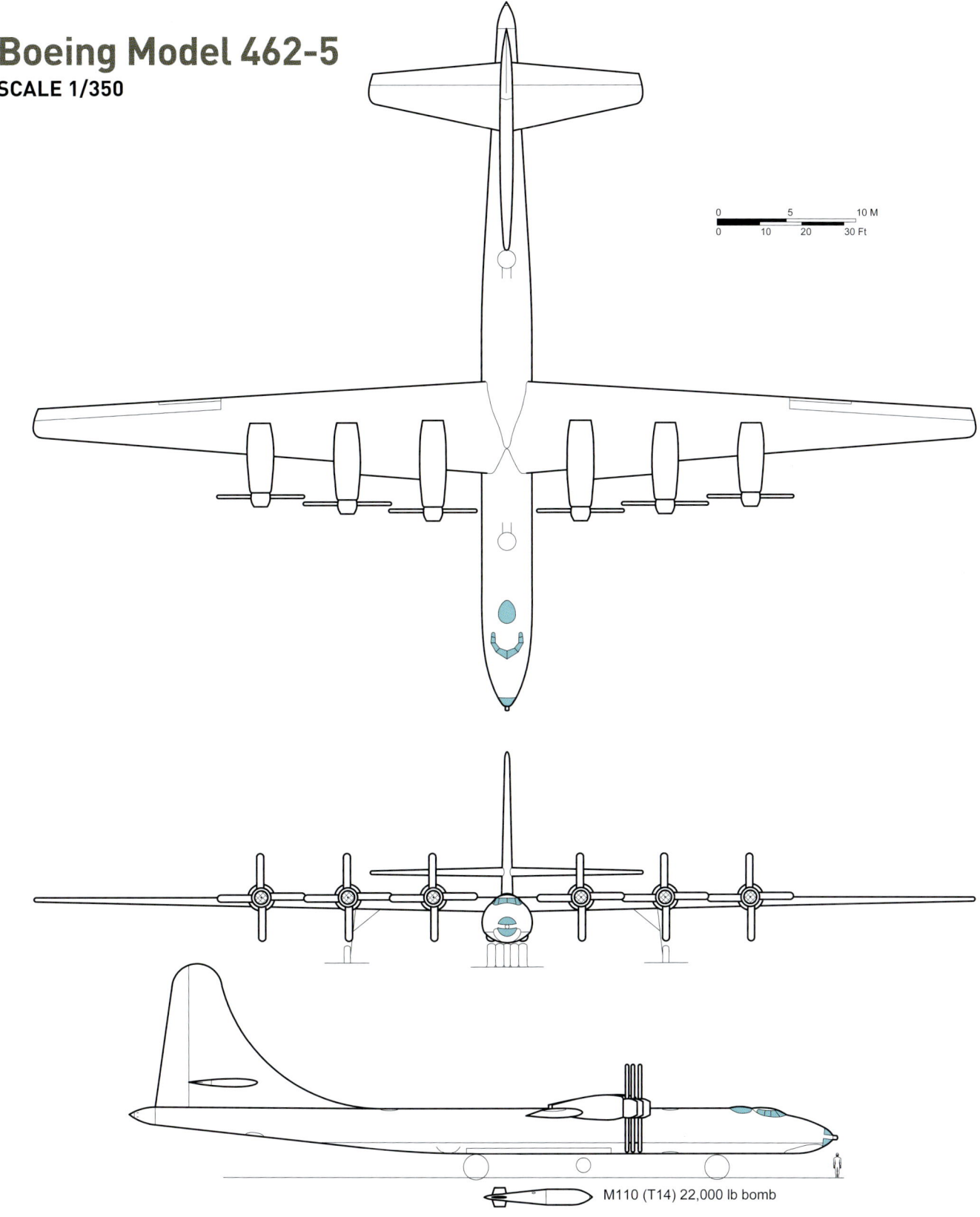

M110 (T14) 22,000 lb bomb

was slightly larger in diameter than the 464-16's, 130in vs 126in. The Army Air Forces preferred Model 464-17 to Model 464-16 due to its greater flexibility.

At 400,000lb takeoff weight, Model 464-17 could carry 10,000lb of bombs on a mission radius of 3,800 nautical miles. This was still short of the hoped-for 4,350 nautical mile radius. However, adding drop tanks and increasing takeoff weight to 480,000lb gave a 5,000 nautical mile radius. Model 464-17 could carry 90,000lb of bombs with a mission radius of 1,950 miles. Maximum speed at 35,000ft would be 382 knots. The Air Force requested in April 1947 that Boeing examine the XB-52 as a GAM-63 RASCAL missile carrier. Boeing found that, with some modifications

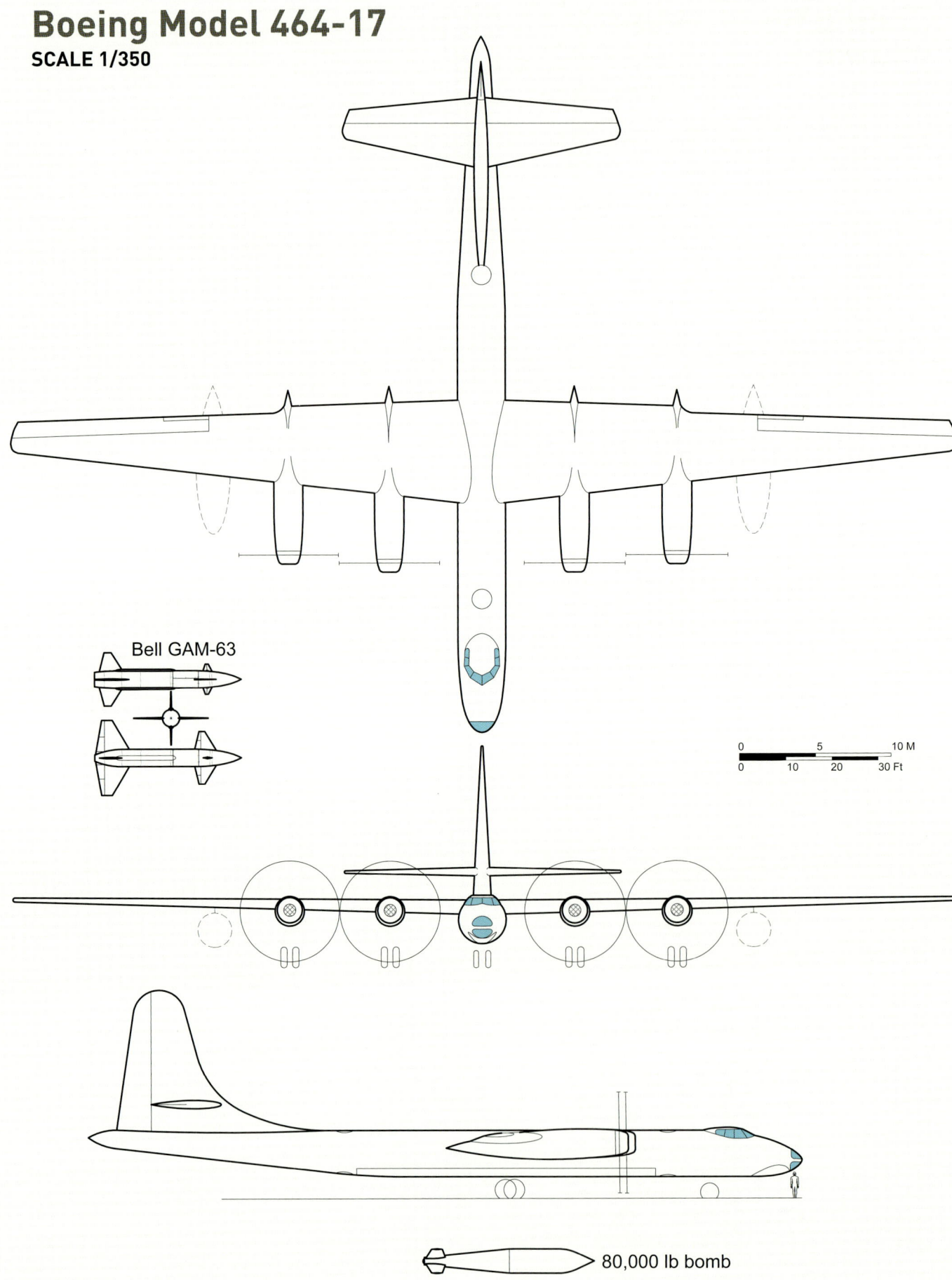

Boeing Model 464-18
SCALE 1/288

Control Surfaces Undefined

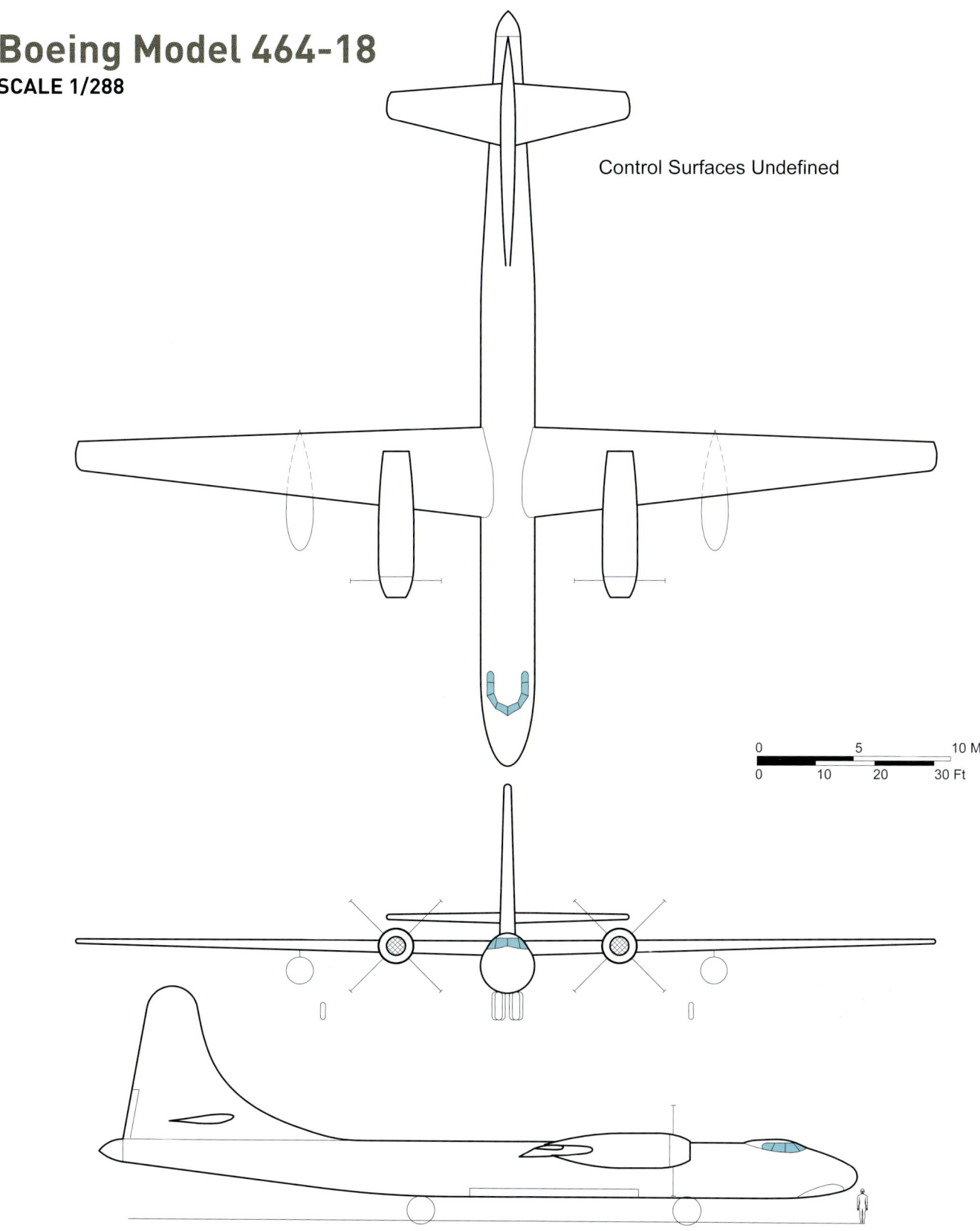

to the RASCAL, one could just barely be carried internally, though it would likely be easier to carry two externally – most likely on pylons either between the inboard engines nacelles and the fuselage or between the inboard and intermediate nacelles. No diagrams of these layouts have been found by this author, however.

Defensive armament consisted of four body-mounted retractable turrets and one tail turret, with a total of 12 x 20mm cannon, much as with the Model 462. Studies also looked at the use of manned wingtip gun turrets. A Phase II proposal was issued for Model 464-17 in January 1947.

Boeing Model 464-18

A smaller version of 464-17 with just two turboprops was drawn up in mid-late 1946 as Model 464-18. Bomb load was reduced from a maximum of 90,000lb to just 10,000lb. Range, thanks to the addition of large underwing drop-tanks, was only slightly decreased compared to the -17. Armament was four 20mm cannon. Distribution of the guns is unclear, but it is probable that two of them were in a tail turret. Little data is available on this design and since it went the wrong way with regards to range, it was probably a rather academic exercise and went nowhere.

Boeing Model 464-25

Offering slight improvement on 464-17 was Model 464-25 of July 1947. The wings got a leading edge sweepback of about 14°, the props got different spinners, the nose was slightly recontoured with a flush-topped cockpit and defensive armament was cut to a tail turret plus two retractable turrets (one above the forward fuselage, one below), all three armed with 0.50cal machine guns. Perhaps the biggest step towards the B-52 from the -17 to the -25 was the change in landing gear: -25 had four main wheels within the fuselage, with small outriggers located in the outboard engine nacelles.

Fuselage fuel load was reduced slightly (7,700 gallons from 9,460 gallons), but wing fuel tankage was greatly increased (19,200 gallons from 10,740 gallons). Additionally, external fuel capacity was increased from 19,200lb to 68,700lb. With a decrease in bomb load from 90,000lb to 12,000lb, suiting a single Mark III 'Fat Man' atomic bomb (drawn as such in the original Boeing diagram), this brought gross weight to 400,000lb.

Boeing Model 464-27

This short-lived continuation of the B-52 design evolution, designed on October 20, 1947, followed on from Model 464-25. It featured wings swept back slightly more, a refined fuselage and a return to the raised cockpit canopy. The tail surfaces were substantially changed too but the basic data for the Model 464-27 was not much different from that of the 464-25. Model 464-29 was also produced at this time and differed from 464-27 mainly in being 17in longer and in having a vertical stabilizer trailing edge that was straight up and down rather than slightly aft-swept.

Dark clouds formed on the horizon in November 1947. The baseline Model 464-17, at nearly 500,000lb gross takeoff weight, was seen as too slow and heavy

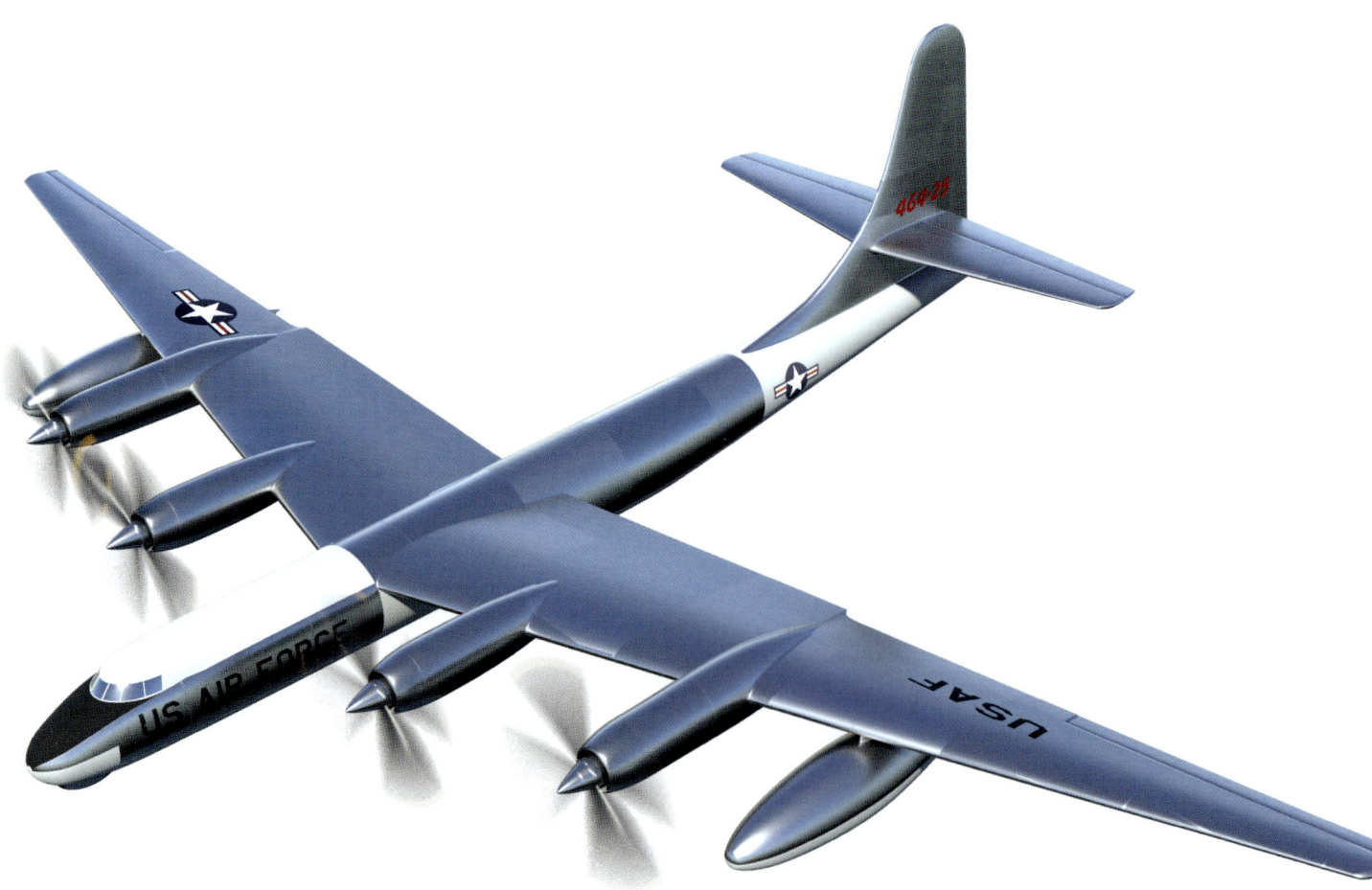

Boeing Model 464-25
SCALE 1/350

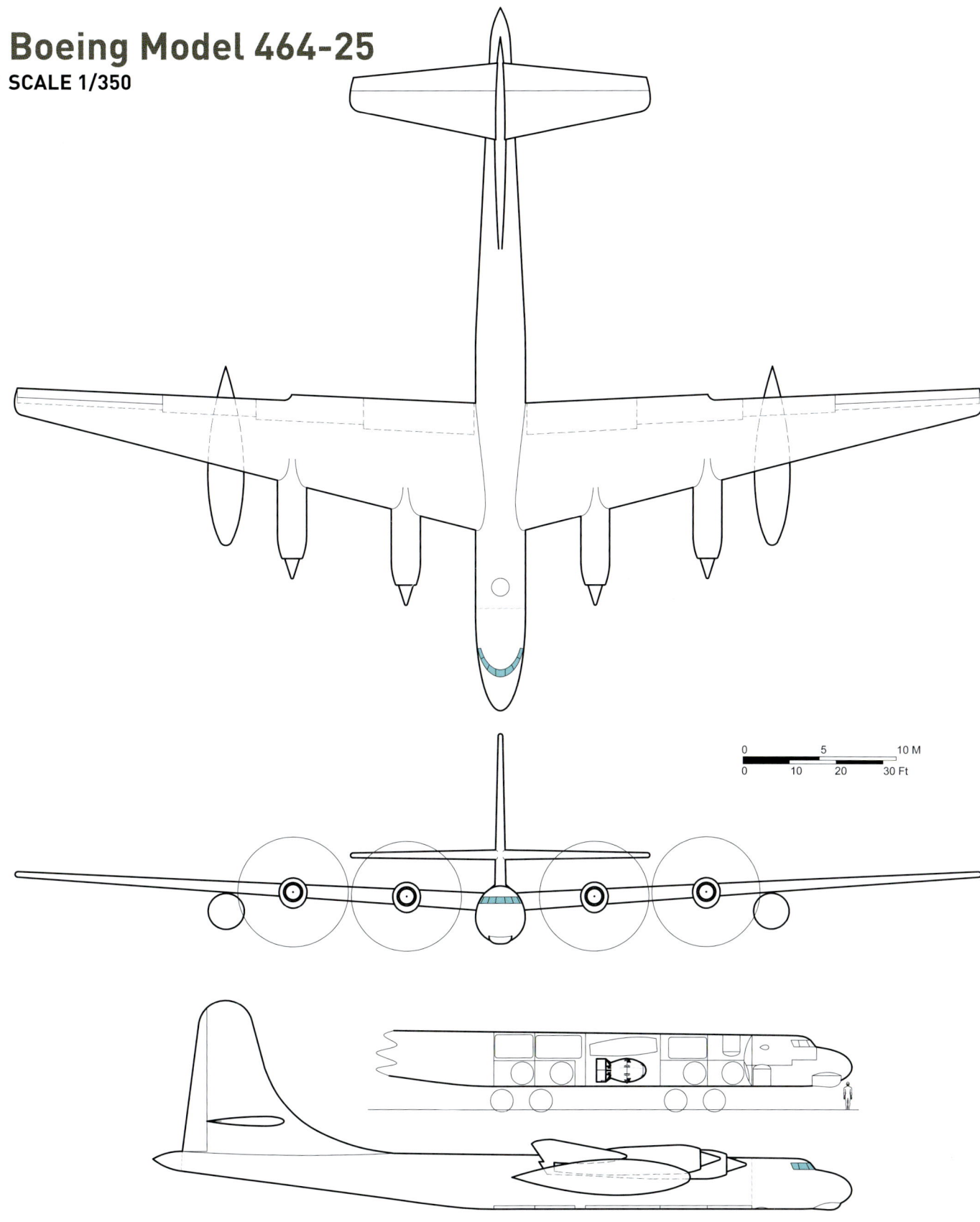

by many in the Air Force. It was also too expensive to procure in useful numbers. All staff present at a conference in the office of the Director of Research and Development in Washington, D.C. agreed that the 464-17 should be cancelled.

Turbojet development had reached a point where some believed the mission required a pure jet-bomber, not one powered by turboprops. But should Boeing be ordered to redesign the XB-52 or should the whole contract be scrapped and a new competition started?

In mid-December, the Director of Chief of Staff/Materiel ordered the Air Materiel Command to cancel the contract and share the following characteristics among the American aviation industry for re-bid: 500mph+ top speed at tactical altitude, 500mph average speed, 35,000ft tactical operating altitude, 8,000 statute mile range at design weight conditions, 9,000ft takeoff distance over 50ft obstacle with RATO, 9,000ft landing distance over 50ft obstacle without droppable fuel and bombs, and a 10,000lb average bomb load.

At the same time, in flight refuelling was becoming a reality. Postwar experiments using a trailing hose snagged by the plane attempting to refuel had been successful, though the process was clunky and fuel flow rate was poor. Boeing, however, was developing the 'flying boom' system, which put most of the work into the hands of an operator in the tanker aircraft; tests with modified B-29 aircraft had been successfully carried out in September 1948, and promised a practical means of greatly extending the range and duration of the XB-52.

Boeing and its allies in the Air Force defended their efforts all through December 1948 while others argued to cancel the contract. But finally, at the end of December the Air Force was authorized by the Air Materiel Command to keep going with the Boeing programme. The atom bomb itself was also in flux as the Mark III design was due for replacement by the Mark IV with improved aerodynamics; the 'Fat Man' configuration had proven to be erratic, missing the target by more than 2,000ft during the Able shot of the Crossroads operation at Bikini Atoll on July 1, 1946. The Mark IV configuration was due in early 1948.

Boeing Model 464-33-0

Designed in January 1948, Model 464-33-0 was a refinement of the earlier Model 464-27. Some changes were subtle – the wings had slight anhedral rather than dihedral, the wing sweep was increased, and the tail surfaces were distinctly aft-swept and more modern in appearance. However, while the designs looked similar, the -33 was a substantially smaller vehicle than the -27 and had a much lower gross weight. This was possible largely thanks to the incorporation of inflight refuelling into the design. Fuel could be loaded through a flying boom at a rate of 400 gallons per minute. Wingspan dropped from 205ft to 185ft – the span the eventual B-52 would have. Gross weight dropped from more than 400,000lb to a much more manageable 260,000lb.

As with the -27, the undercarriage was 'bicycle' main gear and smaller stabilizing outriggers deploying from the outboard turboprop nacelles.

Boeing Model 464-34-3

A minor modification of the 464-33, this design primarily involved a change in cockpit layout, though it also had a substantial increase in gross takeoff weight. The wide, dome-like cockpit of the earlier vehicle – somewhat like that of the B-36 – was replaced with a long slim tandem cockpit of the kind used on the Boeing B-47. In some ways, the Model 464-34 was the first of the design series that began to look like the actual XB-52.

Boeing Model 464-35

Model 464-35 of January 1948 was a major design iteration and was, for a time, looking like it would either be the XB-52 as built or the last design before cancellation. Just after it had survived a cancellation attempt by some in the Air Force, Boeing's project faced a new threat from Secretary of the Air Force Stuart Symington, who wanted to cancel it in favour of Northrop's YB-49 Flying Wing. This is somewhat ironic, given Symington's stormy relationship with Northrop. As it turned out, the Flying Wing had serious stability issues and ended up crashing during a test flight in June 1948. Once that happened the Flying Wing programme, already troubled, was doomed; it never again threatened the B-52.

Wind tunnel testing had produced an important change in wing geometry. The wing had a root thickness of 16%, but it tapered down to 10% at the tip. The thicker wing root did not appreciably affect drag characteristics but did improve overall structural efficiency. The thin, high-speed wings designed for the early Model 464 and B-47 aircraft had to be heavily reinforced internally since they were so thin; this made them heavy and precluded the use of wing tanks. Thick wing roots were more spacious and could be used for extra fuel storage.

Model 464-35 had a tail turret for defence, but sources vary on whether it would be crewed or remote-controlled. The diagram included here depicts what seems to be the main line of thinking, that the tail gun would be controlled remotely. While the payload was relatively low at 15,000lb, it retained a long bomb bay capable of carrying a physically bulky payload. By this point, though, the idea of dropping giant conventional bombs had become almost passe; atomic weapons were what a long-range bomber would carry. Still, the larger bay gave options.

The Strategic Air Command weighed in on the B-52's cockpit layout during February 1948, based in large part on a series of mockups built for the Model 464-17. A tandem arrangement was preferred; a cockpit like that of the B-47 made ejection seats easier to accommodate and produced lower drag than a side by side cockpit. Also in February, the Air Force

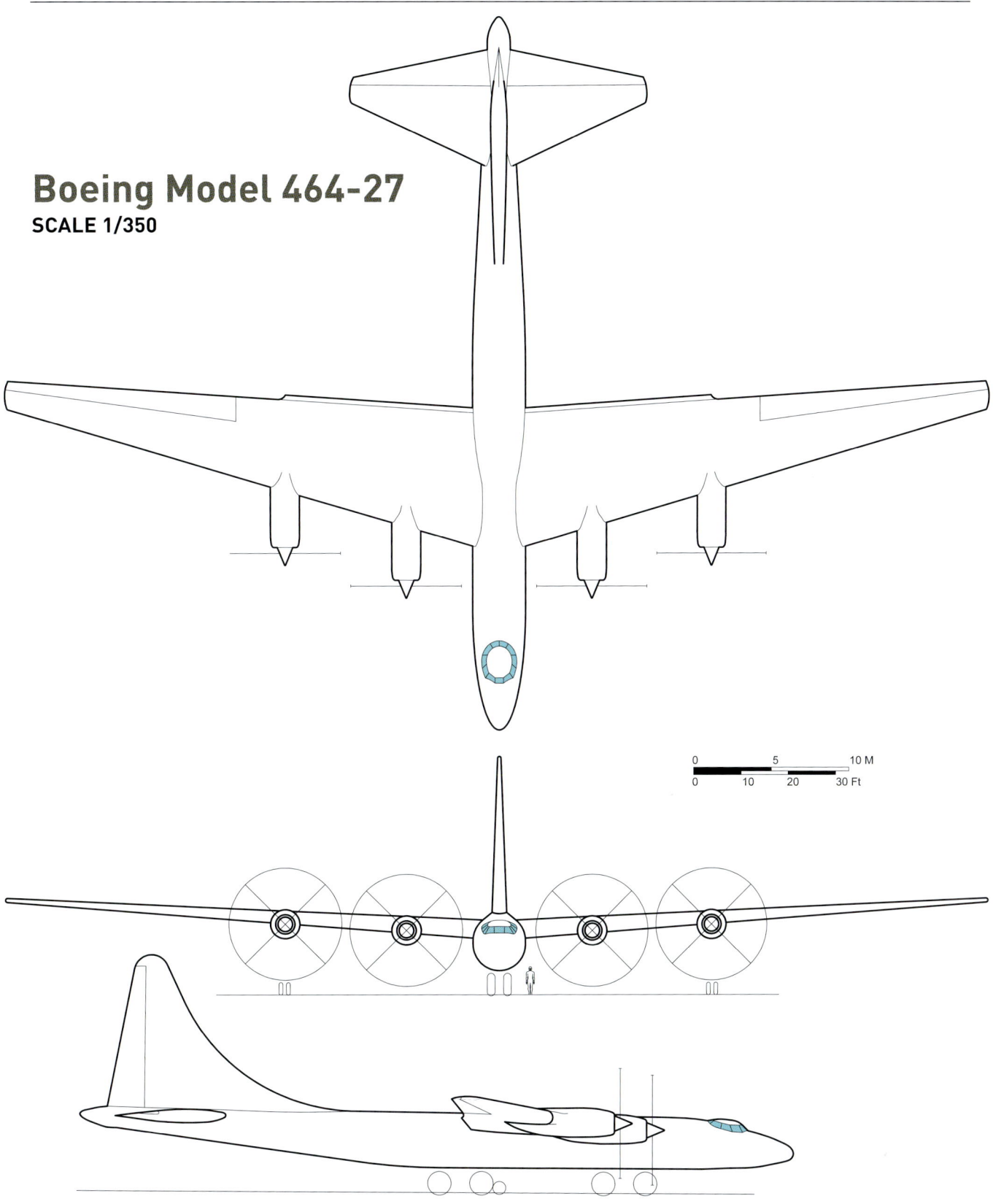

Boeing Model 464-27
SCALE 1/350

definitively decided against reopening the competition; at last Boeing could rest easy in the knowledge that its long-range heavy bomber would not be taken away. The contract between Boeing and the Air Force was amended in March to make Model 464-35 the baseline.

Discussions between engine and propeller manufacturers in May 1948 indicated that it could take as long as four years to perfect the turboprop propulsion system required for Model 464-35. Additionally, the Air Force personnel involved evidently just didn't like propellers, preferring jets instead. Given that the late 1940s was entering firmly into the Jet Age, with jet fighters well along the path of replacing prop fighters, and the B-47 jet bomber in flight testing, this

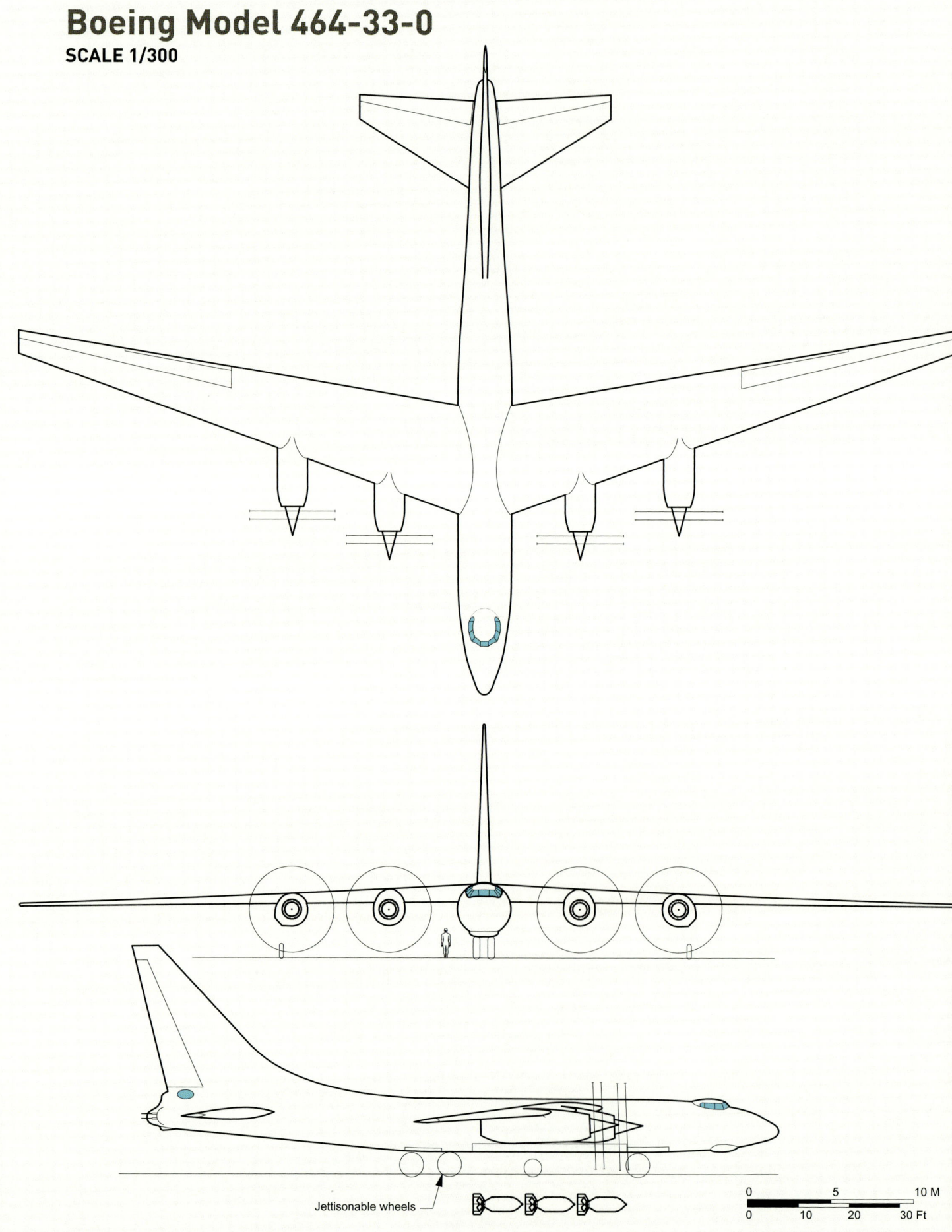

Boeing Model 464-33-0
SCALE 1/300

Jettisonable wheels

B-52 Evolution

Boeing Model 464-34-3
SCALE 1/300

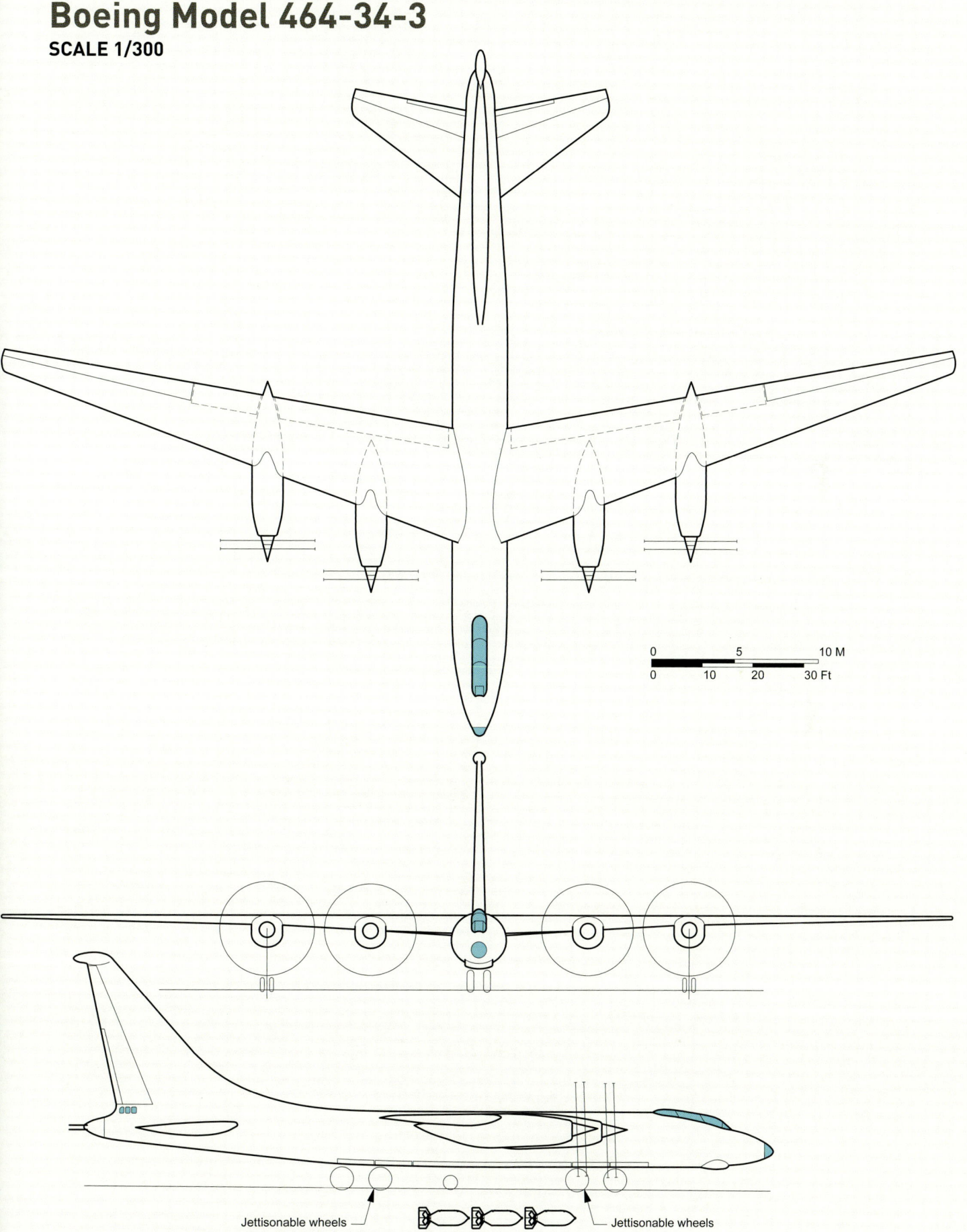

Jettisonable wheels   Jettisonable wheels

133

preference was becoming less a matter of taste and more a matter of simple practicality. Boeing certainly took note of this, and while retaining the baseline 464-35 for the time being, produced another design.

Boeing Model 464-40

The Model 464-40 preliminary design, drafted in April 1948, was submitted by Boeing in July 1948, following a USAF request the previous month. Its airframe and equipment were essentially those of Model 464-35 but with eight Westinghouse XJ40-WE-6 turbojets. The engines were in pairs, with nacelles similar to those of the B-47 and with B-47-like pylons holding them below and ahead of the wing leading edges.

This at last cemented the basic configuration of what would become the B-52. The design was not carried out with much detail and was not considered entirely ready for further development, but it was an important step in determining the usefulness of jets for strategic bomber missions. The aircraft was substantially faster than the turboprop version but had notably shorter range. Still, it was considered adequate, especially in light of mid-air refuelling. Studies indicated that if the aircraft had multiple external fuel tanks spaced correctly under the wings, it could be refuelled in flight to a weight of 450,000lb without greatly increasing the empty weight. This would give the Model 464-40 a range of around 9,000 nautical miles.

The design was seemingly intended to be built not as a production aircraft but as a modified version of the Model 464-35 XB-52; an XB-52 turboprop airframe would be either rebuilt with jet engines or taken from the assembly line and completed with jets. The available diagram of Model 464-40 is subtly different from the 464-35 in a number of areas, however. As interesting as the Model 464-40 was, though, 464-35 continued to remain the baseline.

Boeing Model 464-41

Designed in April 1948, 464-41 returned to the all-turboprop configuration from the all-jet 464-40. Little information is available it, but even though it didn't have the final B-52's engine arrangements it did have a fuselage much more like that of the final XB-52. Rather than being circular in cross-section it was slab-sided, with rounded corners. The wing had a lower sweep than that of the B-52, but the cockpit canopy was refined from prior designs and somewhat resembled the canopy that would be used on the first two prototype B-52s. The tail extended quite a bit beyond the stabilizers and was fitted with ventral and dorsal sighting bubbles for the gunner.

Boeing Models 464-46 and 464-47

A brief and minor pair of designs from sometime around September 1948, Models 464-46 and 464-47 were similar to the earlier Model 464-40 but with six rather than eight turbojets and with the 464-41

Boeing Model 464-35
SCALE 1/300

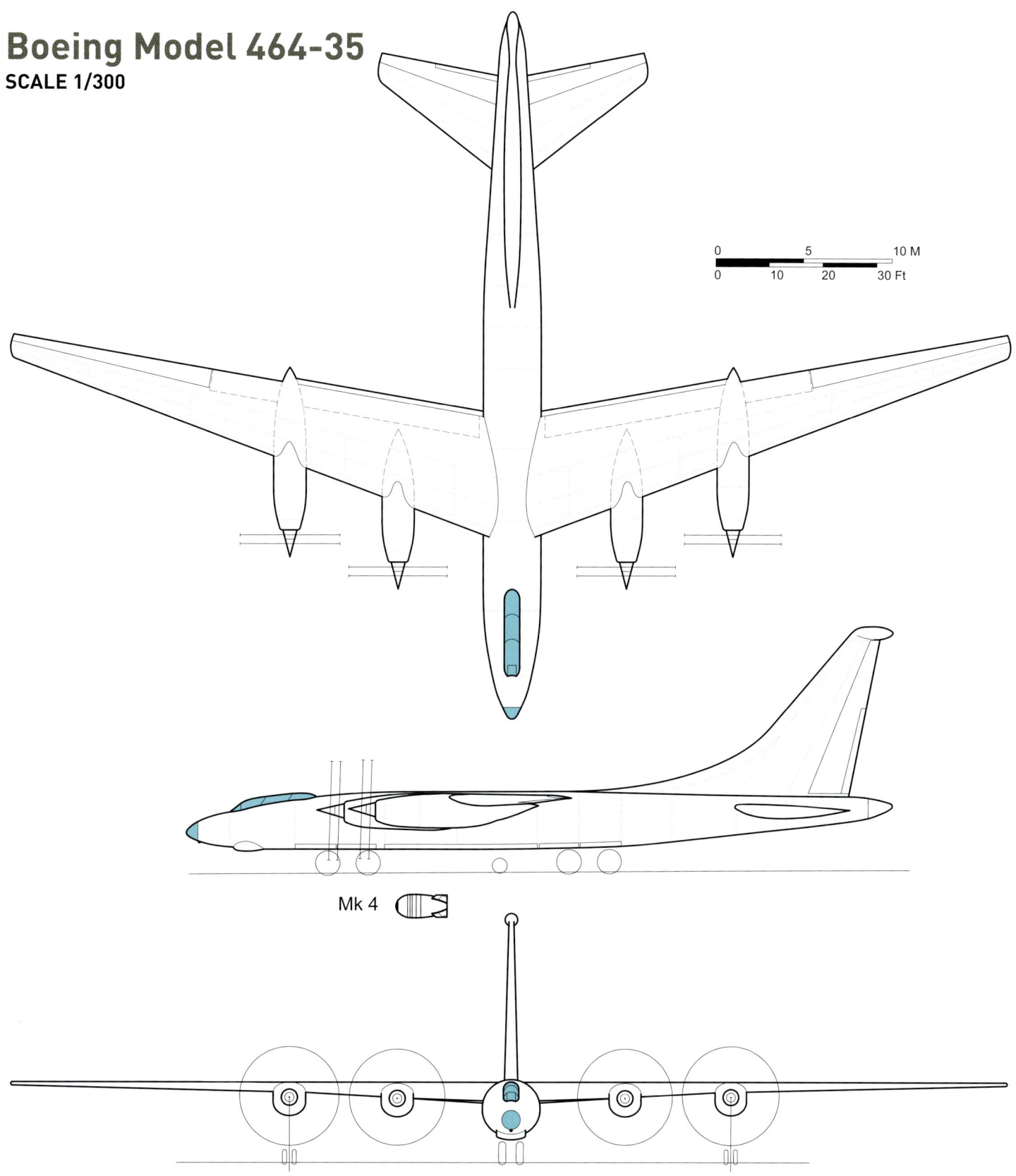

fuselage and stabilizers. While 464-46 and 464-47 had the same number of engines as the B-47 and put them into two pods of two jets and two pods of one like the B-47, the 464-46 and 464-47 swapped the inboard and outboard engine pods. The inboard pod had one engine while the outboard pod had the pair. Why it was done this way is unknown, though likely for structural reasons. Model 464-46 had six Westinghouse XJ40-10 engines while 464-47 had six Pratt & Whitney XJ-57 engines… the type that would go on to power the B-52. Performance data is unavailable for either design.

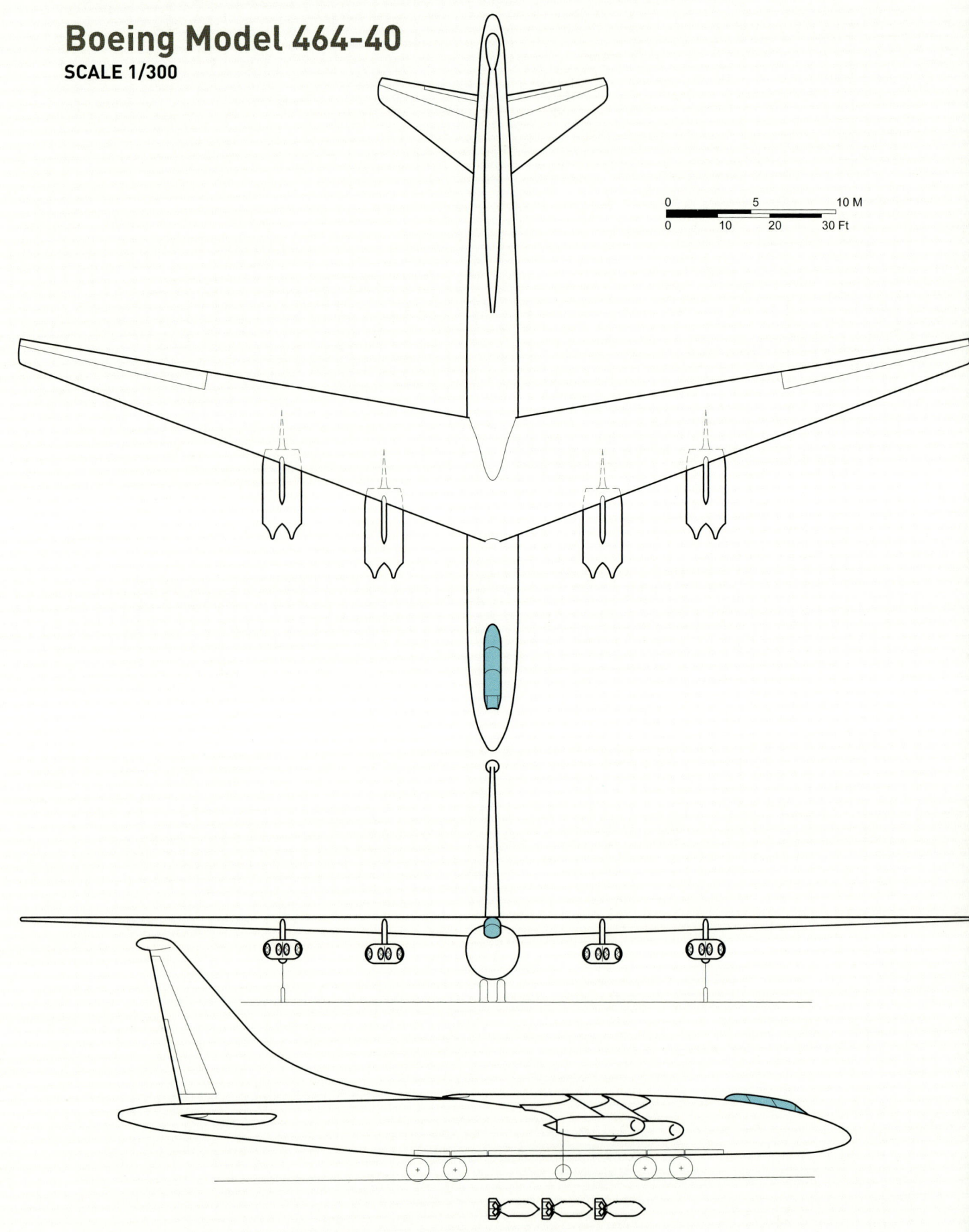

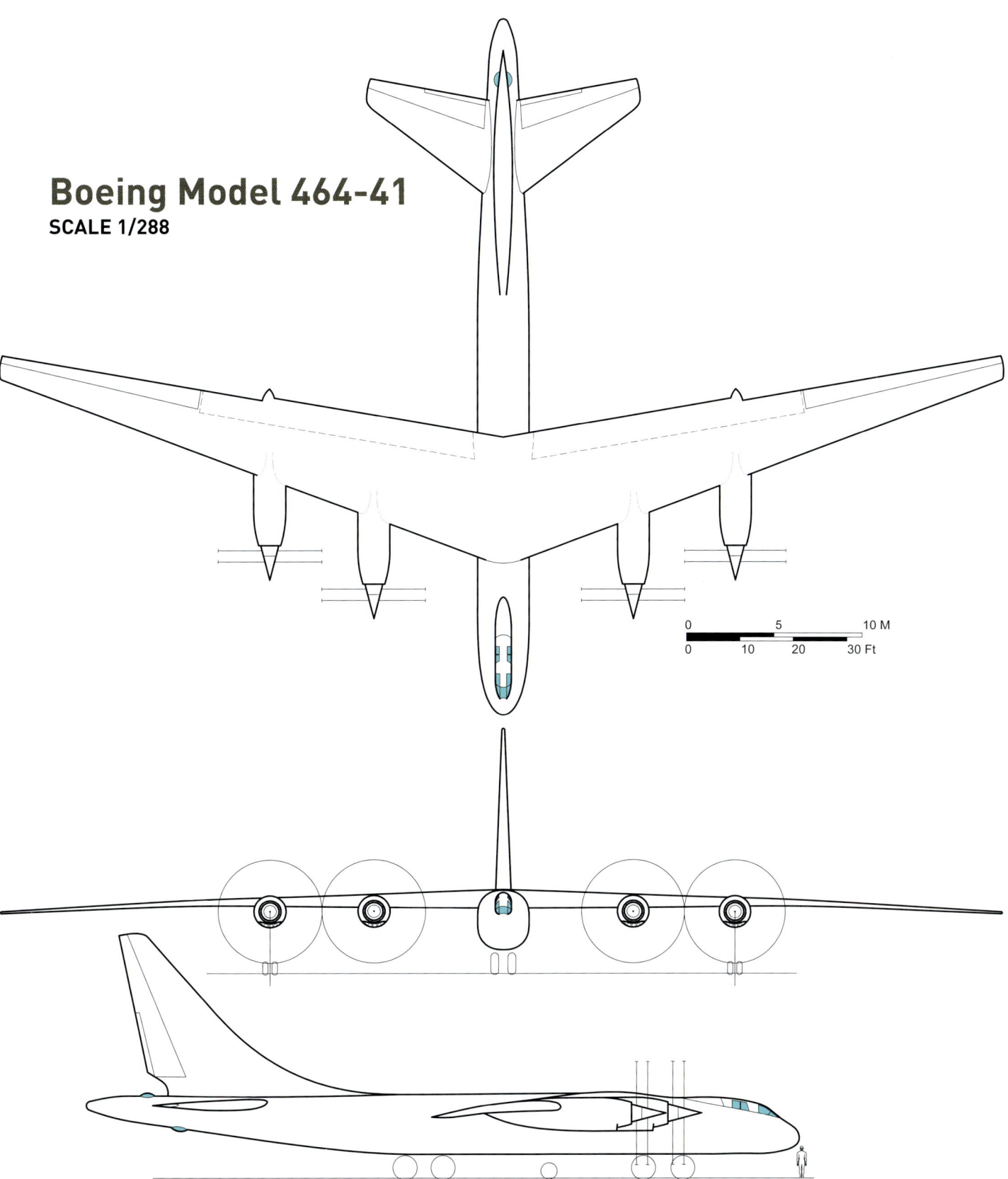

Boeing Model 464-41
SCALE 1/288

Boeing 'Hotel' design

One configuration iteration has gained a measure of fame as the 'hotel' design. A Boeing team arrived at Wright Field on October 21, 1948, to discuss the B-52 (specifically Model 464-35) with Air Force officials. At this time, the turboprop Model 464-35 design was still the baseline. During the ensuing meeting, the Boeing team were asked to redesign the B-52, taking the latest advances in aerodynamics and propulsion into account. Turboprops weren't working out and flight testing of the B-47 had shown that wing sweep could provide greater aerodynamic benefits than expected.

Boeing Model 464-46, 47
SCALE 1/300

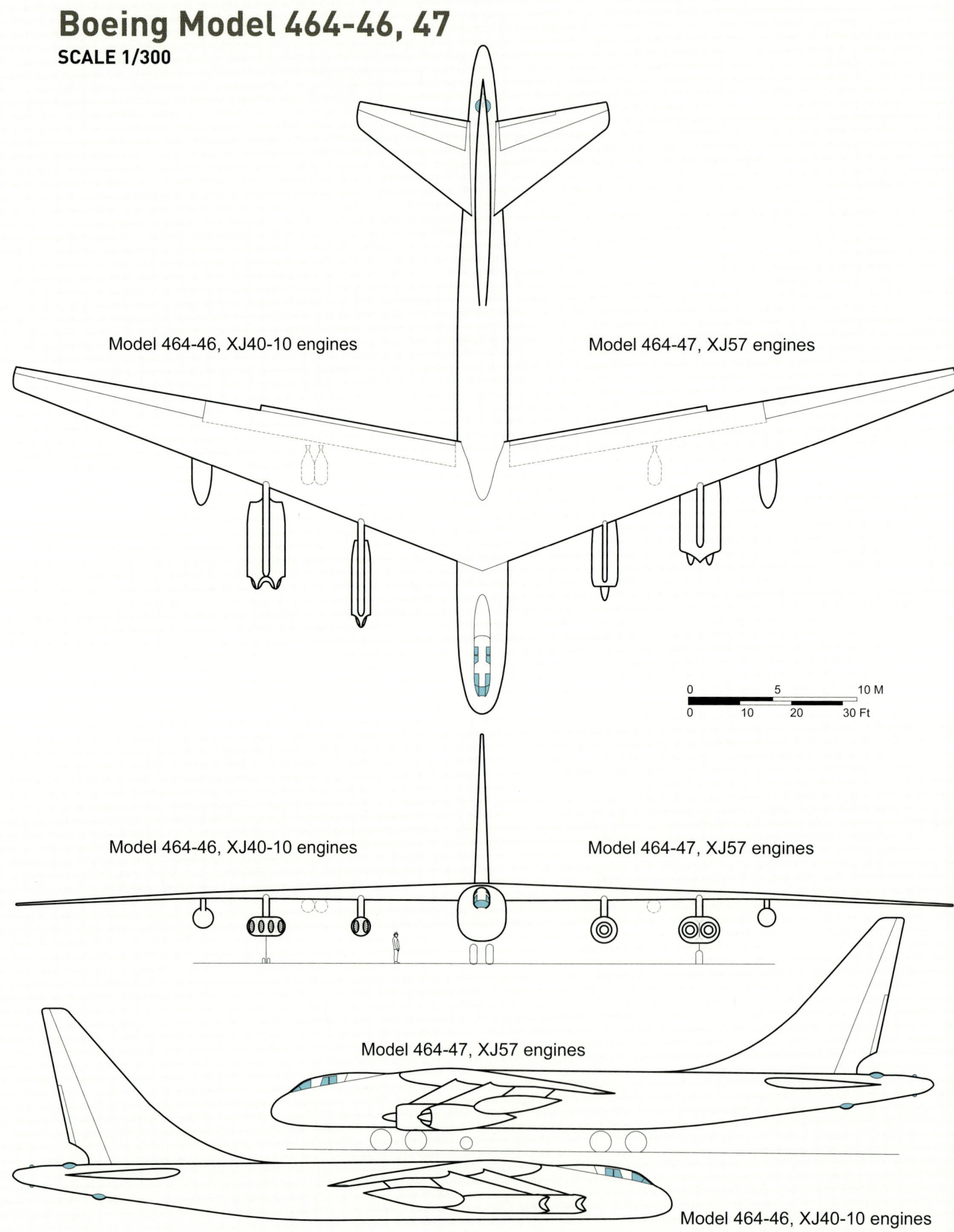

As the story (simplistically) goes, the Boeing team retired to their motel rooms and over a weekend produced a new, more advanced design. The story usually has it that these Boeing officials whipped up a whole new airplane over a busy weekend, producing a 33-page proposal with a general arrangement, inboard profile and carved balsa wood display model basically from thin air and determination. But there had already been numerous designs produced since 464–35, so the hotel-bound Boeing designers had plenty to work from.

The resulting 'Hotel' configuration (also referred to as the D10,000 design due to the report number) was similar to the previous major design, 464-47, but with many refinements. It had much the same configuration but with increased wing sweepback; as a result of considerable wind tunnel research already performed by Boeing, the 20° 52' leading edge sweepback of the 464-47 was further swept to 35° (at the quarter chord). This was the same angle of sweepback that the B-47 had, providing a direct source for comparison. Wingspan was the same at 185ft, but the wing area was substantially raised by increasing the tip chord. The wings were also given substantial anhedral.

The six engines of the 464-47 were replaced with eight in an arrangement similar to that of 464-40, but the new design used Pratt & Whitney XJ-57 turbojets. It was much like the B-52 as actually built, but understandably rather crude given the rapidity of its off-site genesis. The only major difference in shape between this layout and the actual XB-52 was that the fuselage forward of the wing was about 14ft shorter.

In fact, the canopy of the tandem cockpit extended further aft than the leading edge of the wing root. The tail surfaces were much closer to those of the B-52 than to those of Model 464-47. The new configuration was presented on October 25, to considerable approval.

Boeing Model 464-49

The penultimate major design in the B-52's development was Model 464-49, a refinement of the 'Hotel' design. The configuration was quite close, but with the details worked out. Visually there were two clearly changes: the vertical stabilizer finally took on its ultimate form. The D10,000 design had featured a vertical stabilizer leading edge that gradually curved into the fuselage since before Model 464. With Model 464-49, though, the leading edge was straight from top to bottom. Secondly, the inboard nacelles were moved markedly outboard from their 'Hotel' design position. The outboard nacelles, however, basically stayed put.

The 464-49 used four pairs of main wheels for the landing gear, with small outrigger stabilizers dropping from the outboard engine nacelles. An important difference from the B-52 was that the second and third mainwheel pairs would be droppable after takeoff in order to reduce weight. The five crew were pilot, co-pilot/engineer, navigator, bombardier and tail gunner. The plane's sole defensive armament was two .50cal machine guns in the tail. Given the high altitude and speed expected of the craft, a tail-on attack was considered the only feasible way for enemy fighters to take it on. Model 464-49 was also capable of inflight refuelling.

Boeing D10,000
B-52 Proposal
SCALE 1/300

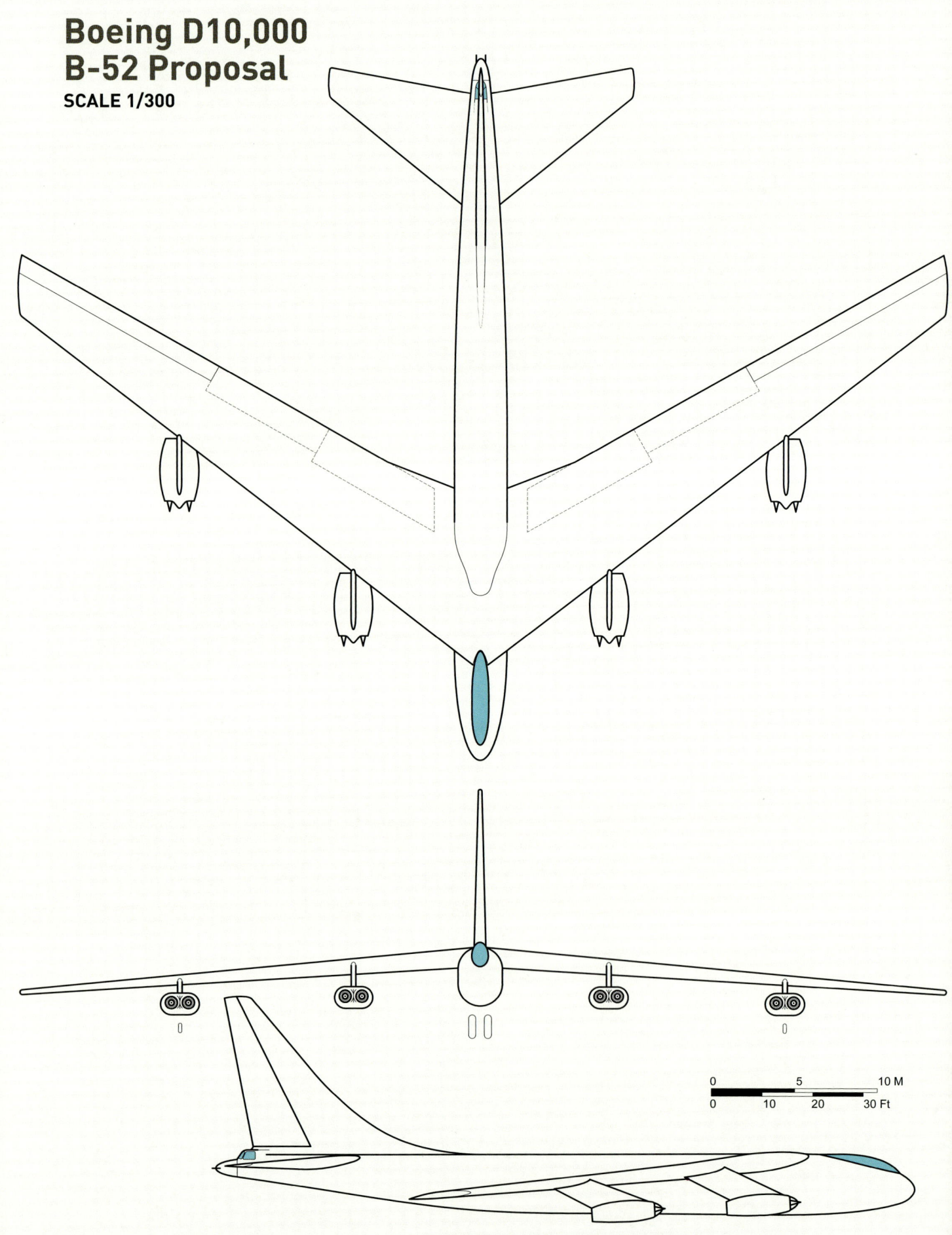

Boeing Model 464-49
SCALE 1/300

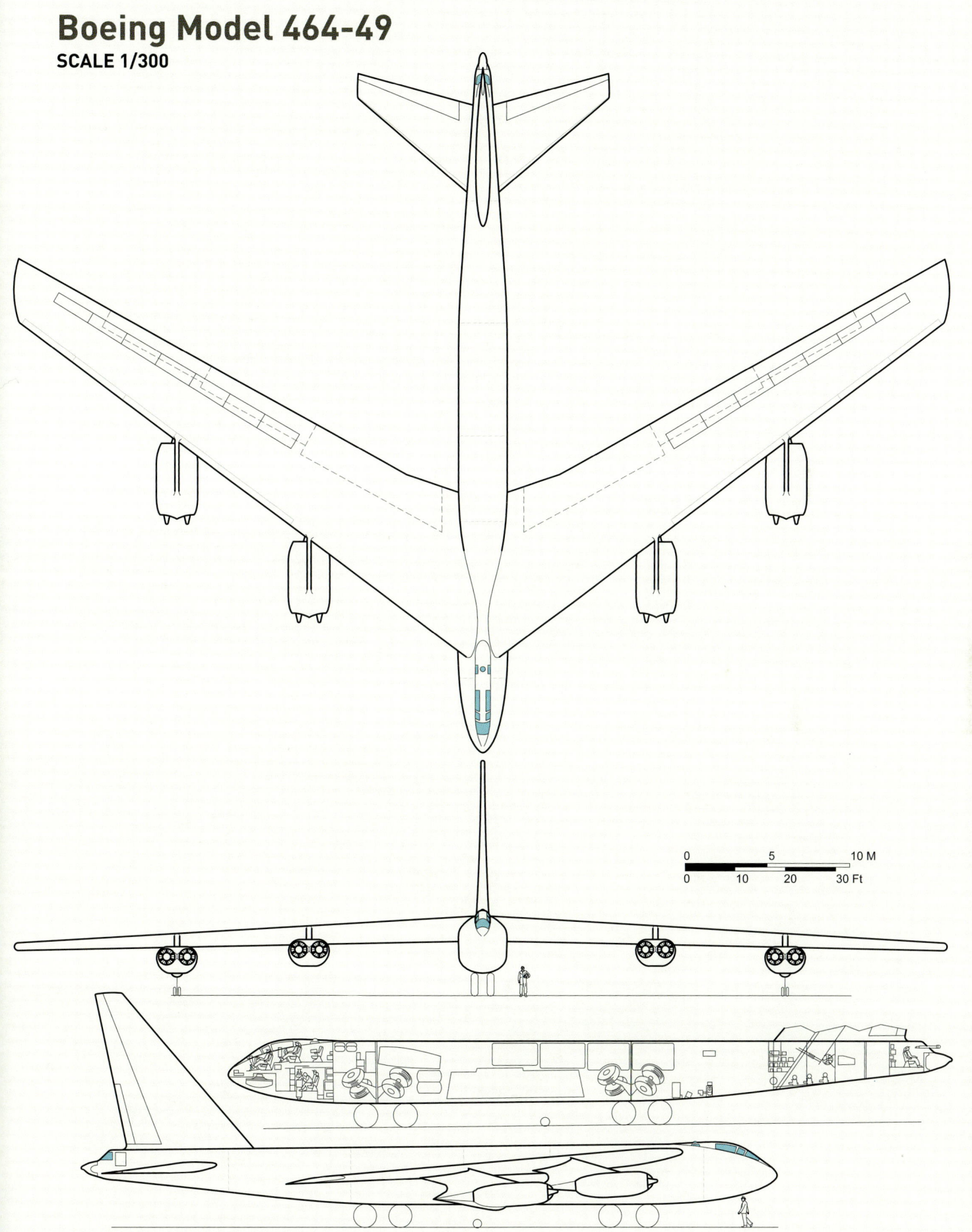

464-49 was sufficiently impressive that in January 1949, the USAF cleared Boeing to proceed with the all-jet design over the turboprop 464-35. At the same time, Pratt & Whitney received additional USAF funds to continue work on the XJ57 engine and carry it through to component testing.

The XB-52 actually built was Model 464-67 but though there is a substantial gap in the designation list between 464-49 and 464-67 (Model 464-54 was the Phase II version of 464-49), the differences between the designs appear to have been minor. A mockup review of Model 464-52, then known as the XB-52, was held at Boeing's Seattle facility in April 1949. The B-52 design not only evolved, it competed. Several other companies put forward designs to fulfill the long range strategic bombardment role.

The Boeing B-55 diversion

Curiously, one of the companies that worked the hardest to replace the Boeing B-47, and to produce a bomber that would compete against the Boeing B-52, was Boeing itself.

Before the B-47 had even flown, the Bombardment Branch at Wright Field was looking towards a future where the B-47 would be replaced. Given the rapid pace of advancement in both jet propulsion and aerodynamics, it was a reasonable assumption that aircraft then on the drawing board would be obsolete in a handful of years. And the fleet of Boeing B-29s and B-50s then in service was already looking elderly. So in October 1947, the Air Force held a competition to design a replacement for the B-29, B-50 and B-47; requirements included a 2,000 mile radius, a 10,000lb bomb load and a gross weight under 200,000lb. Boeing won the competition with Model 474, officially dubbed XB-55, and received the contract to build it in July 1948.

Boeing Model 474

Model 474 was, more or less, an enlarged B-47. Along with the increase in scale, the turbojets were replaced with four Allison T-40-A2 turboprops. The wings had less sweep at 20°, which makes sense given that the turboprops made the aircraft slower than the B-47. Range, however, was notably greater than that projected for the B-47.

Defensive armament was a little unusual. Along with the expected stinger turret in the tail, there was another blended into the leading edge of the vertical stabilizer root, and a third under the chin. All seemed to be equipped with four 20mm cannon, certainly a potent punch for the time. The tail gun was manned by a gunner who sat at the base of the trailing edge of the vertical stabilizer, looking out through bubble windows on either side; the gunner for the rear fuselage turret had a larger single bubble canopy at the base of the leading edge of the vertical stabilizer. The forward gun was operated by the radio operator or the navigator (it's unclear). Total crew complement was ten.

The T-40-A2 presented some difficulties however. The six-bladed contra-rotating props would need to split the engine's power through a gearbox, and there was some concern that the sheer power of the engine would tear the drive shaft apart. Allison estimated at

Boeing Model 474
XB-55
SCALE 1/250

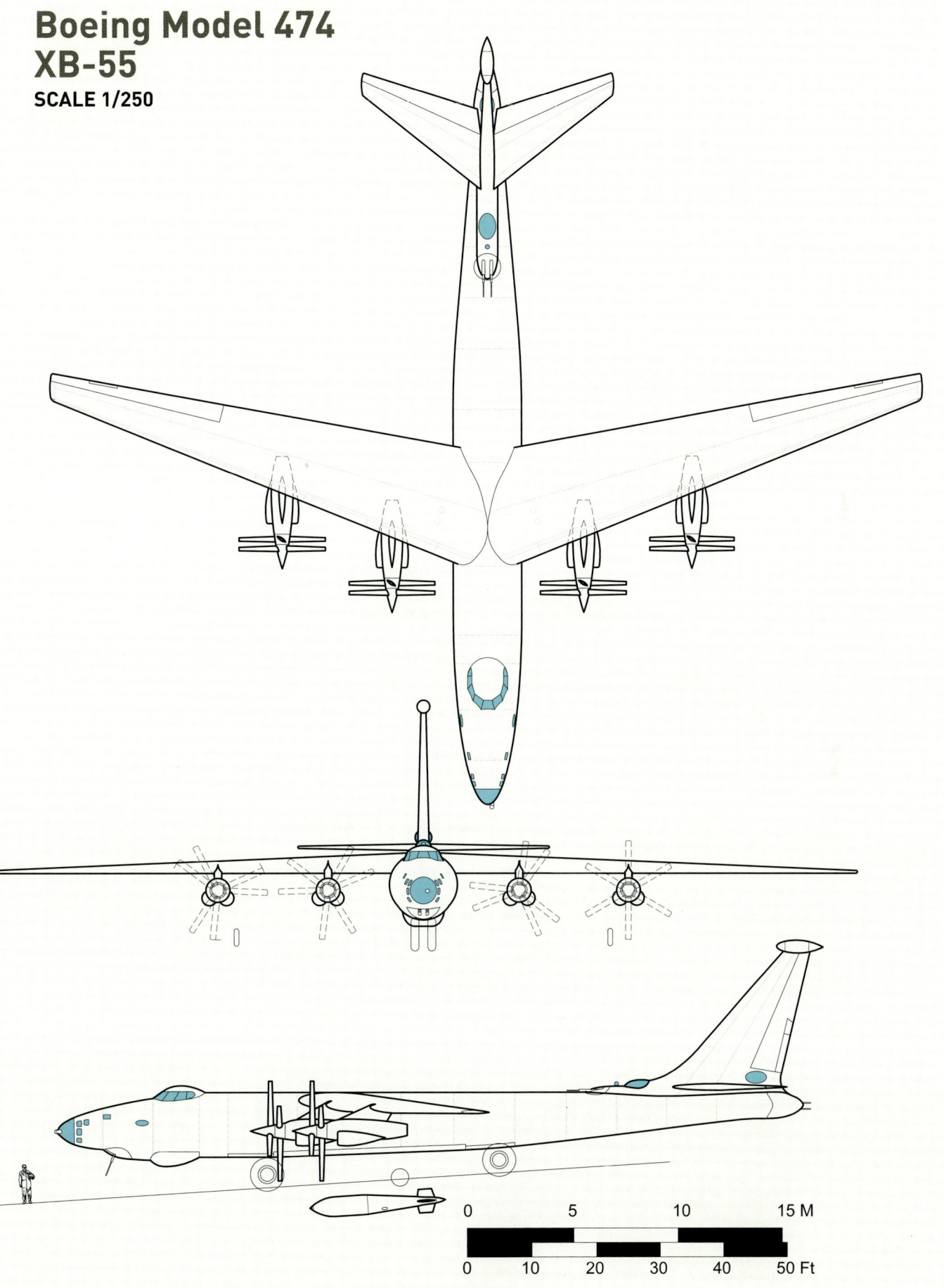

143

least three years to get the whole propulsion package running properly. This was a problem.

Boeing Model 479

In January 1948, shortly after Boeing won the competition, but well before being granted the contract to build the XB-55, it produced Model 479. This was the logical next step in the development of the XB-55 and resulted from an Air Force request to examine a turbojet version, bypassing any turboprop-related issues.

It was similar to Model 474, with an engine arrangement just like that of the B-47: two engines in a nacelle below a pylon under the wing, with a single engine attached to the underside of the wing further out. This was the third turboprop-powered Boeing bomber that had made the switch to turbojets while still on the drawing board.

Wingspan was increased by 15ft and area by 1,000sq ft. Defensive armament was reduced to just the tail turret; Model 479 would depend on speed for survival. The nose was shortened substantially. The wing featured different thickness ratios, being thinner at the wingtip, the advantages of which are described in the section on the 464-35. The 'second generation swept wing' as Boeing called it was first introduced here in the Model 479 XB-55. The XB-52 design team would incorporate the feature in Model 464-35 and take full advantage of it in Model 464-49.

The US Air Force cancelled XB-55 development in January 1949. This was partly due to lack of funds in 1949 and partly because LeMay wanted Boeing to concentrate its efforts on the B-52.

Boeing Model 483-11

The XB-55 project had a number of offshoots, unusual designs linked to it in ways that remain somewhat ill-defined. The Model 483 series is included as part of the B-55 design family; whether these were designed that replaced Model 479, or were simply studied alongside it is not currently clear. Only a few examples of Model 483 are known today, and then only via fragments.

Model 483-11 had a layout somewhat similar to the B-47 or the Model 479, but equipped with only four jet engines. Wing sweep was substantially greater than that of Model 479 or even the B-47 at 41.5°. This done to help improve its top speed. While performance data for 483-11 is not available, some is available for Model 483-36. As Models 483-11 and 483-34 are known from layout drawings, have very similar configurations and are almost exactly the same size (the two differ only in the location of the engine nacelles; the -11 has them spaced 50ft apart; the -34, 72ft 4in), it's reasonable to assume that Model 483-36 was also quite similar.

This aircraft was intended to reach Mach 0.95. An aircraft capable of that would probably be able to go supersonic in a dive and stay there after pulling out into level flight. Whether staying supersonic would break things on the aircraft, such as skin panels, compressor blade or wing spars, is another question.

Boeing Model 483 Delta

A very different configuration also lays some claim to the XB-55 mantle... a tailless configuration with near-delta wings. This design is known from a single simple diagram, devoid of any real data except a scale reference. It had four engines, just like the other known Model 483 configurations, contained in two twin-engine nacelles, each pylon suspended under the wings. No other data is known.

Boeing Model 484-102

The Model 484 designation was something of a catch-all for jet bomber designs. Most Model 484 designs ended up as part of the development of the supersonic MX-1712/MX-1965/XB-59 project, but some were subsonic and related to the extended XB-55 project. A few of these have survived as at least basic three-view diagrams.

Model 484-106 dates from approximately 1948 and was a geometrically straightforward delta wing design. Known solely from a single small set of diagrams, it was fitted with a single large delta vertical fin for stability. Four turbojet engines were buried within the wing, two on either side of the central bomb bay. Each engine had its own long inlet from the leading edge and exhaust duct to the trailing edge.

Unconventionally, the design used a four-poster landing gear arrangement. The main gear had a pair of wheels that would extend only a short distance, while the forward gear had much longer legs. This resulted in the bomber having a very 'nose up' stance on the ground. A fairing ran along the underside, starting with the ground-mapping radar and stretching to the tail. The crew of four were located in the forward fuselage under a long, narrow canopy somewhat like that of the B-47. No defensive armament is indicated in the available diagram.

With the exception of the large dorsal fin, Model 484-106 was a virtually timeless design in appearance. Boeing and Rockwell would produce very similar concepts 30 years later for 'stealth' bombers. The cockpit was less obvious and the fin was gone, but otherwise the basic configuration would hold up.

Model 484-102 was quite similar, but had a single straight trailing edge, forming a triangular planform. The vertical stabilizer also had an unswept trailing edge. Model 484-102 had something of a tail-dragger tricycle undercarriage arrangement, the tail gear being a pair of wheels that extended only a very short distance.

Boeing Model 479
XB-55
SCALE 1/275

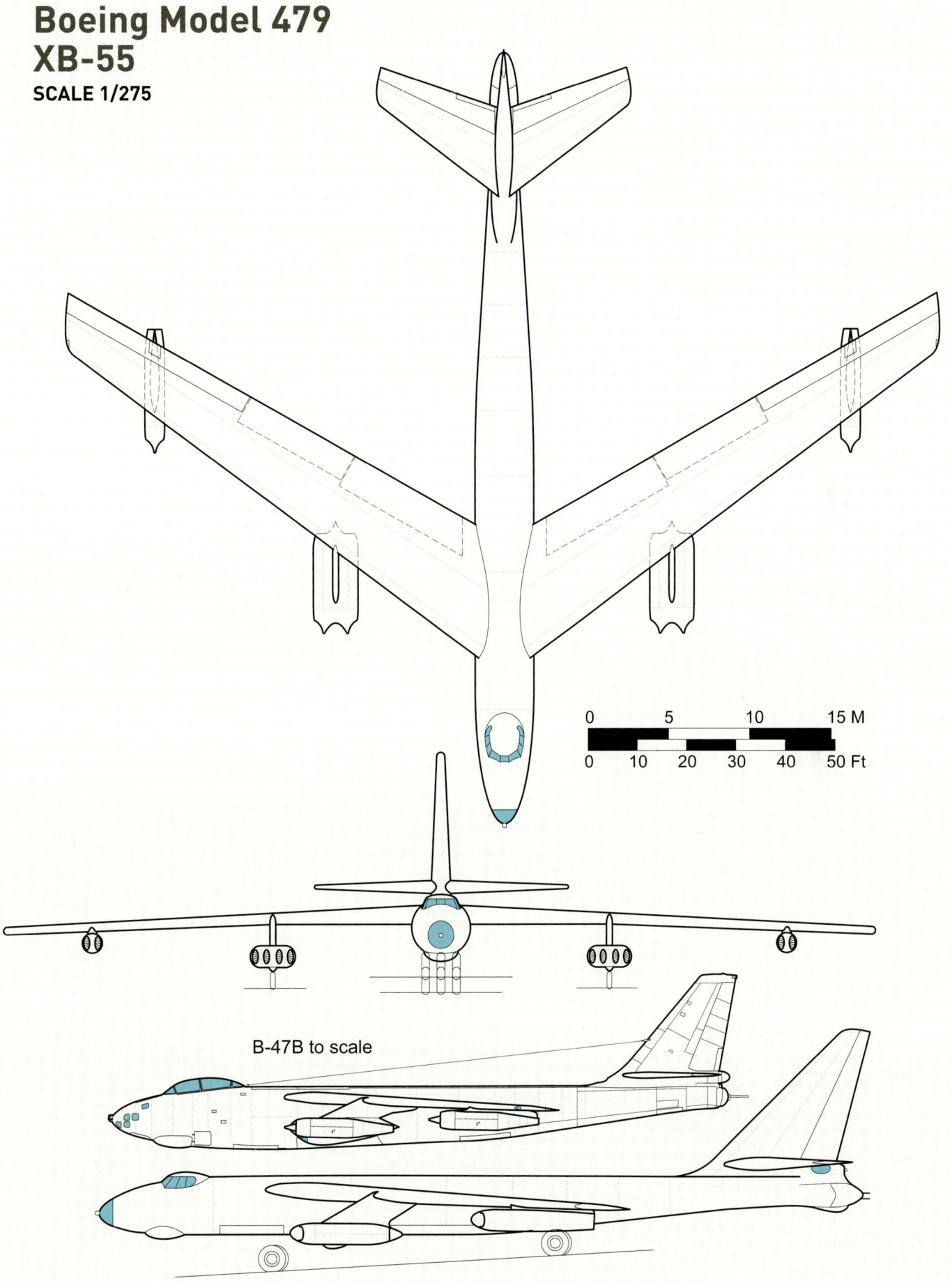

B-47B to scale

Boeing Model 474 + 479
XB-55
SCALE 1/200

Unknown bomb type
M109
0
5
10
15 M
0
10
20
30
40
50 Ft

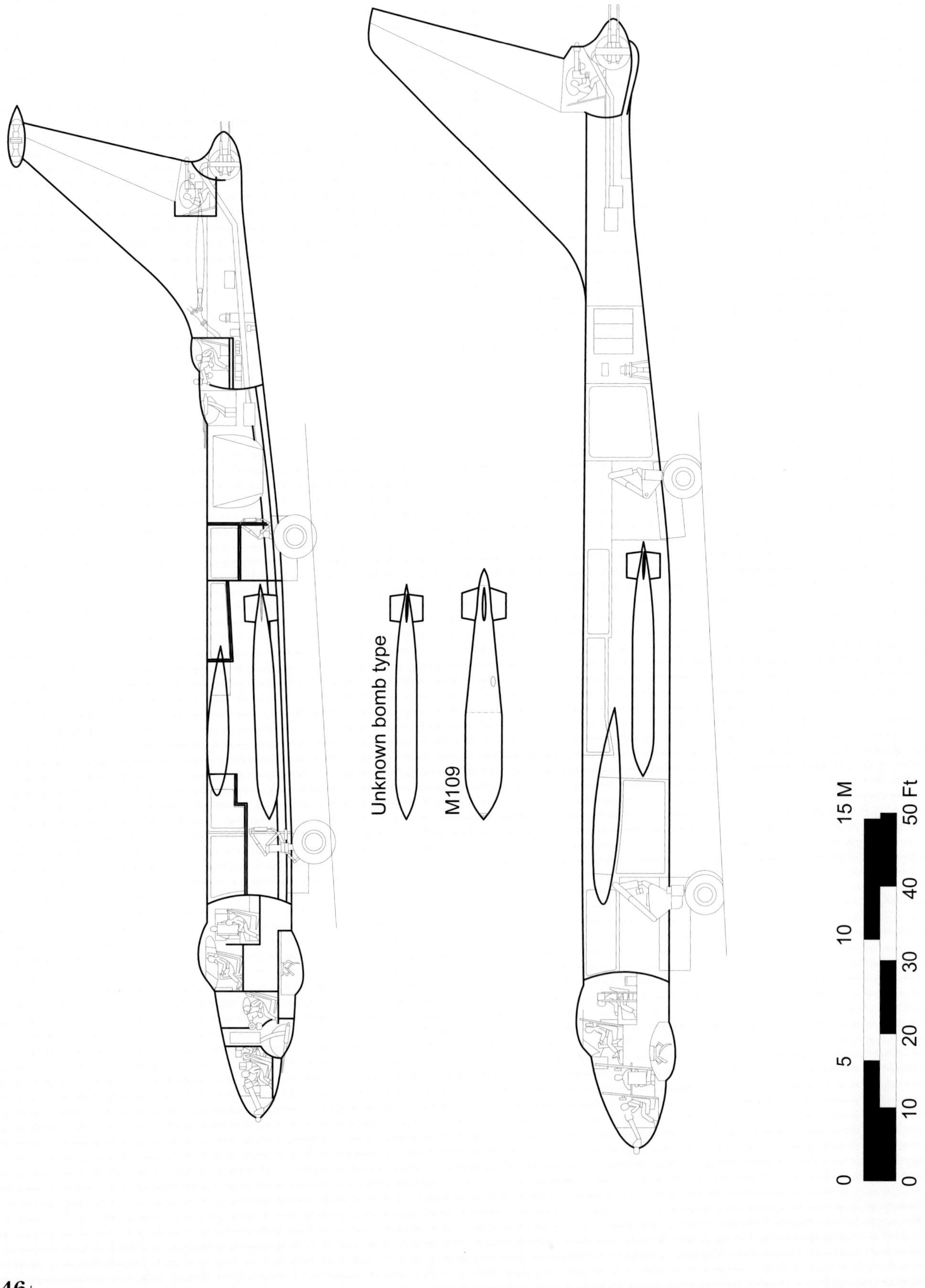

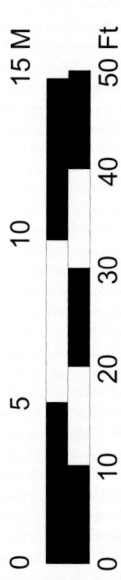

Boeing Model 483-11
XB-55
SCALE 1/250

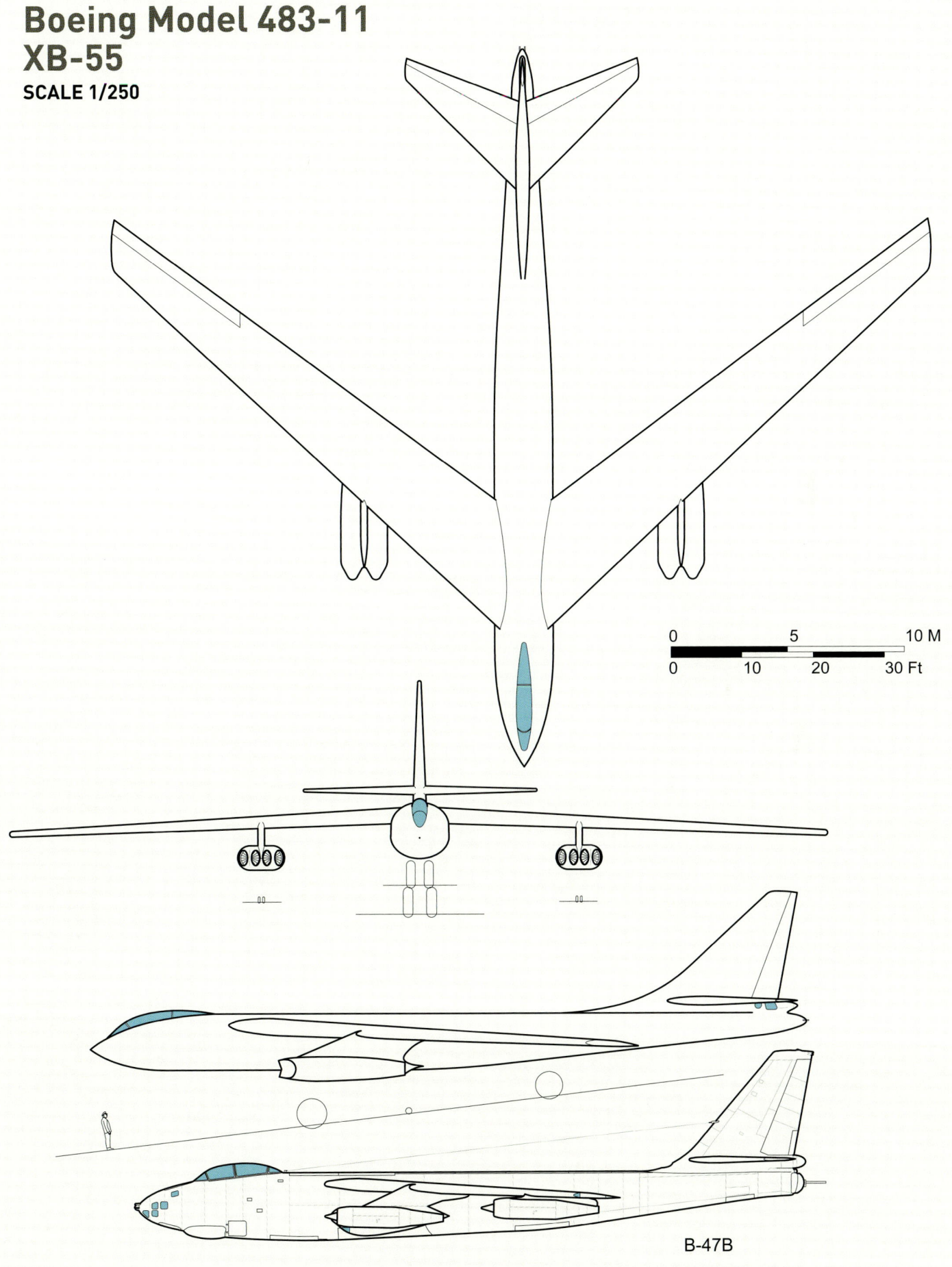

B-47B

Boeing Model 483 Delta
XB-55
SCALE 1/200

A-12 Avenger II to scale

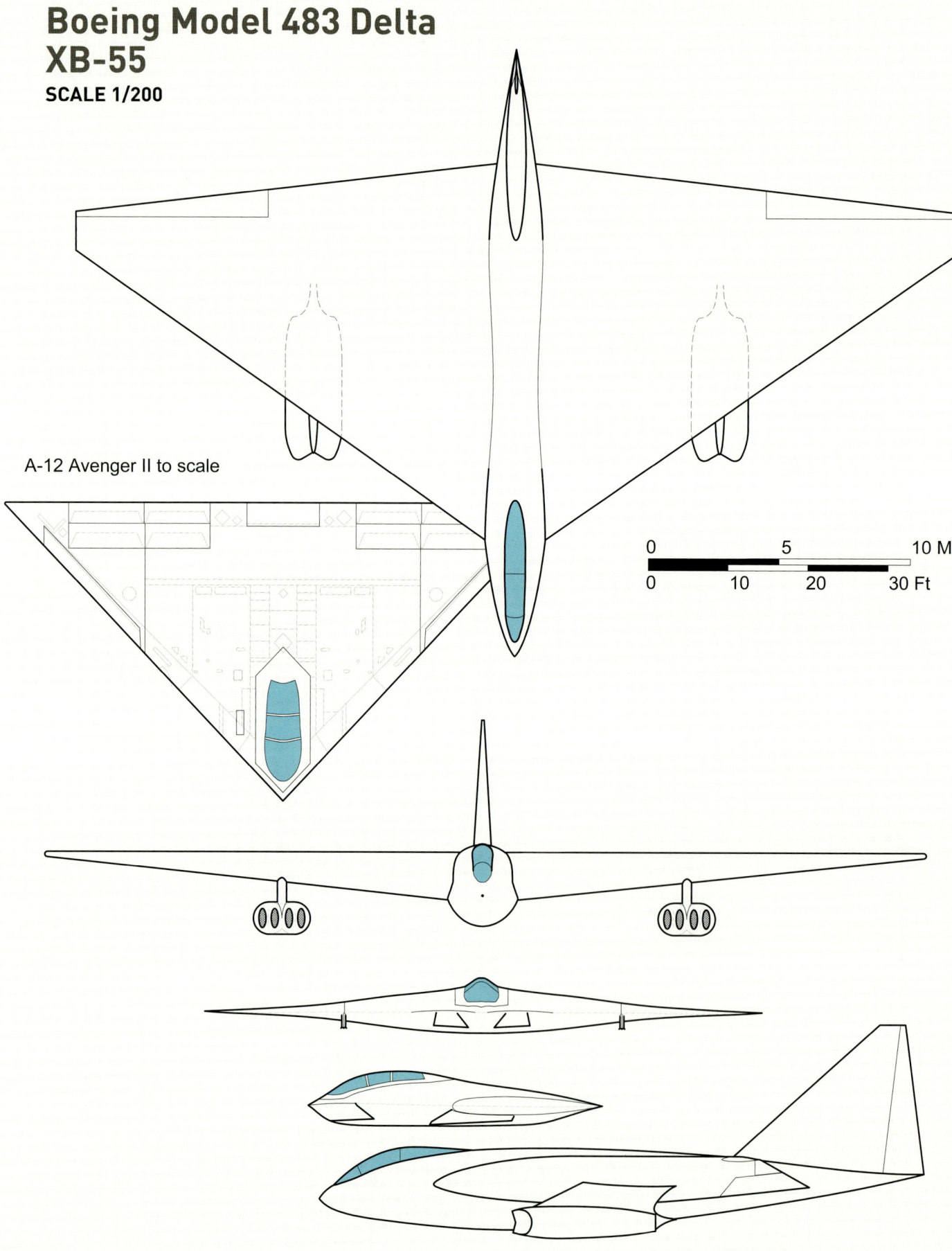

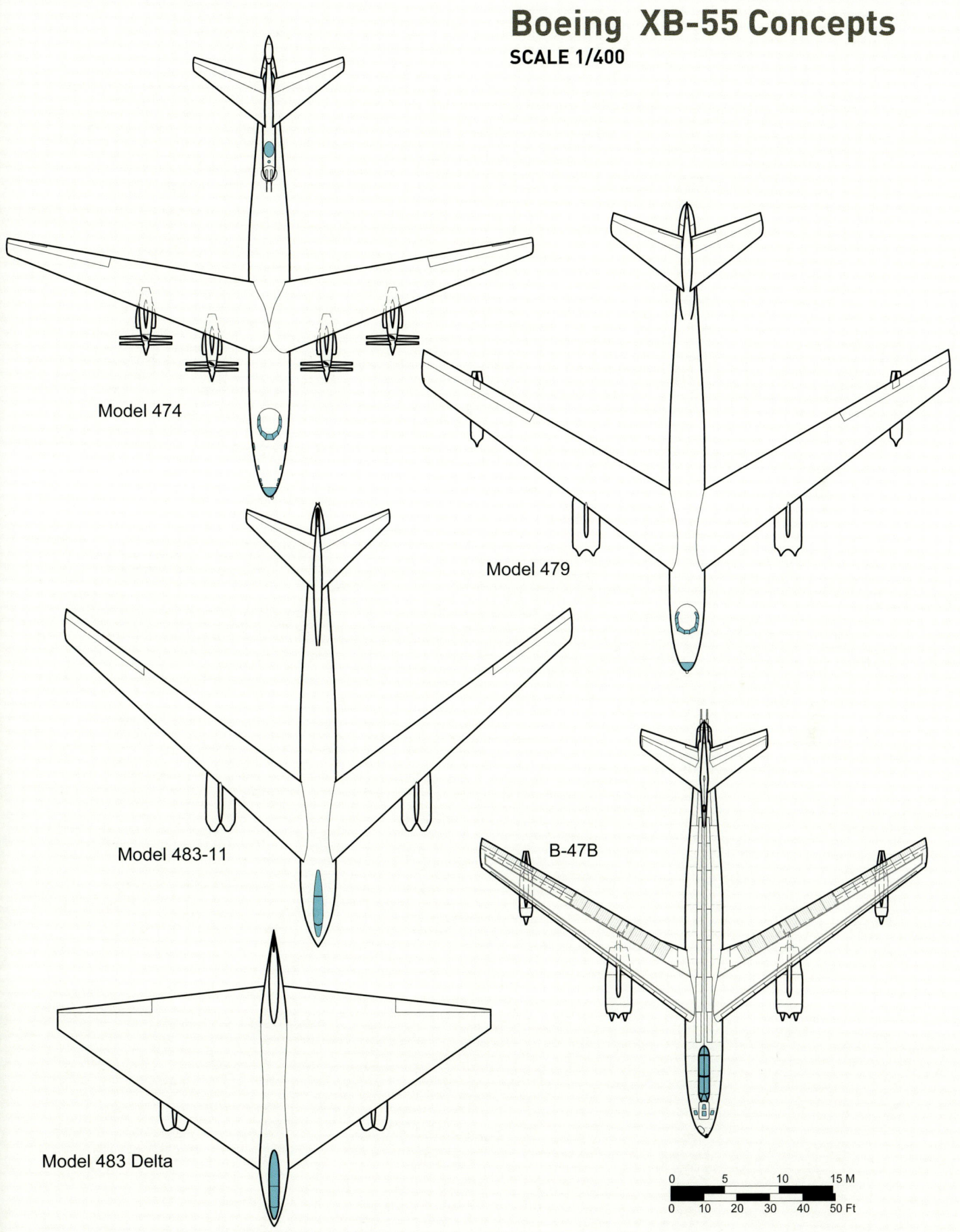

Boeing Model 484-230

Another early Model 484 design, 484-230 is known from a single diagram dated March 1, 1949. It was also a flying wing, but larger and more elegant than the seemingly simplistic Model 484-106. Looking not unlike a wartime German project, Model 484-230 was a swept wing with two widely-spaced vertical stabilizers and six turbojets in two banks of three on either side of the centrebody. Where the wing of the Model 484-106 had a constant thickness ratio, the wing of -230 was markedly thicker in the centre section.

The diagram seem to indicate a crew of four – two pilots, a navigator/bombardier and a gunner in a tail position much like the tailgunners position on the early B-52s. Unlike Model 484-106, 484-230 utilized a standard tricycle landing gear arrangement. There was a modest fuselage and a single central bomb bay. Small swept vertical stabilizers were fitted above the outer wing panels.

Model 484-301 was similar in configuration, but was somewhat larger and used an extended forward fuselage. The inlets were built into the leading edge of the wing rather than underneath.

Later competitors

While no competition was held for a design that would stop B-52 development and production, other manufacturers nevertheless made occasional proposals for bombers of a similar role. Some were vague and simplistic; others were the products of massive effort.

Douglas Model 1064

One of the prizes brought back from Germany after the war was data on delta wings generated by Messerschmitt Me 163 designer Alexander Lippisch. As with swept wings, delta wings showed great promise for high-speed flight, and American aeronautical companies set out to exploit this new potential. Douglas designers were no exception and they sketched out the Model 1064 series in 1947.

A number of different designs were produced, all based on the theme of a turbojet-powered delta-winged strategic bomber. The designs varied somewhat, but were all of a sort: a simple, fairly featureless delta wing housing the engines and with a central fin. The inlets of some were beneath the wing, some above, some in the leading edge. Detailed vehicle and programme data is not currently available.

A number of the designs were geometrically very similar to the Douglas D-571, the basis for the F4D Skyray fighter. The D-571-1 as designed for the US Navy in June 1947 was a geometrically much simpler configuration than the Skyray as built and looked very much like a wartime German design. It seems Douglas took some of the experience it gained on the Navy's dime designing the D-571 and tried to turn it into something it could sell to the Air Force. But as all that is known to exist of the Model 1064 series is a few scattered diagrams; it seems likely that the concept was not finalized, nor seriously proposed to the Air Force.

The design reproduced here is presumed to be the final Model 1064, a close replica of the D-571 configuration but scaled up. Had work on the design continued, it could perhaps have played a role decades later when stealth designing was more in vogue.

Fairchild M-121

One of the more unconventional designs of the post-war era, Fairchild's M-121 was a concept for a ten-engined strategic bomber that would spend a good fraction of its operational career as a biplane. The M-121, which existed in some form as early as February 1948, was intended for intercontinental strategic bombing like the B-52 but took a markedly different design approach.

It had a roughly conical fuselage, modestly swept main wings at the rear of the fuselage, canards up front and, most unconventionally, a vast drop tank in the form of a second wing mounted above the fuselage. This would provide additional lift while taking off, a process further assisted by sitting the aircraft on a jet-powered rail car during the takeoff run. This would allow the undercarriage to be sized only for landing, greatly reducing weight. The wingtank would be jettisoned after the fuel was consumed, reducing weight and drag, with the remaining wing's area being sufficient for cruise and landing.

The engine arrangement was equally as unconventional. The ten jet engines were clustered in a circle at the rear of the fuselage, surrounding a cylindrical bomb bay. Bombs, like those of the North American A-5 Vigilante, were ejected straight to the rear. This method of bomb ejection had been studied by Cornell university in 1948 and shown to be practical; the Cornell reports included some basic configurations of bomber aircraft that could utilize rearward ejection, including a few tailless configurations that, if stared at a bit cross-eyed, could be seen to have a familial relationship to the Fairchild M-121.

Advantages of the system included zero time required to open and close bomb bay doors, since there weren't any, and the lack of sudden aerodynamic changes as the bomb bay opened up to the airstream. Disadvantages included a sudden shift of the aircraft's centre of gravity to the rear as the atom bomb slid aft; this would be the moment when the canards were put to the test. And if something went wrong and the bomb got stuck prior to jettison, the aircraft could be in serious trouble.

Boeing Model 484-106
SCALE 1/175

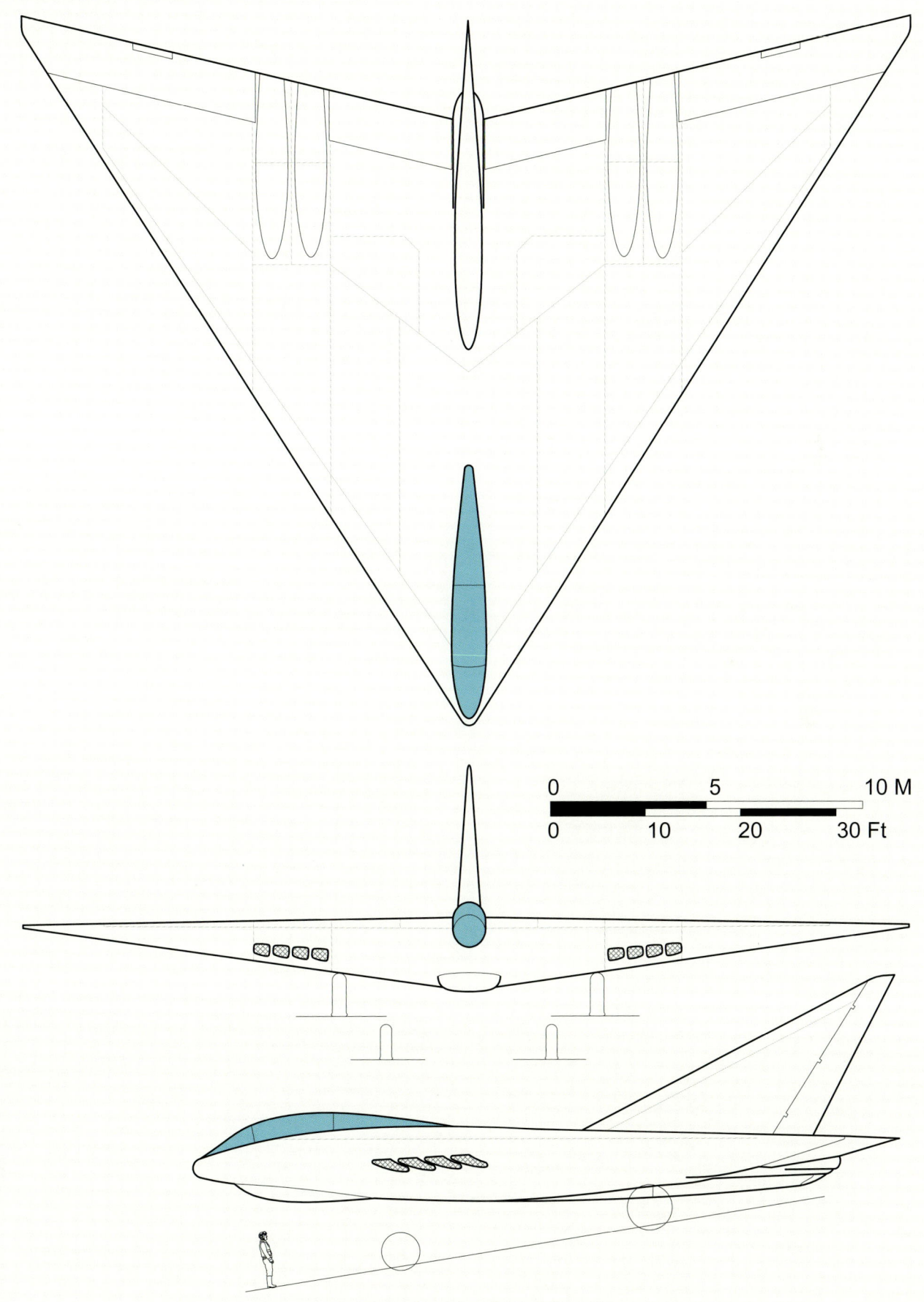

Boeing Model 484-230
SCALE 1/200

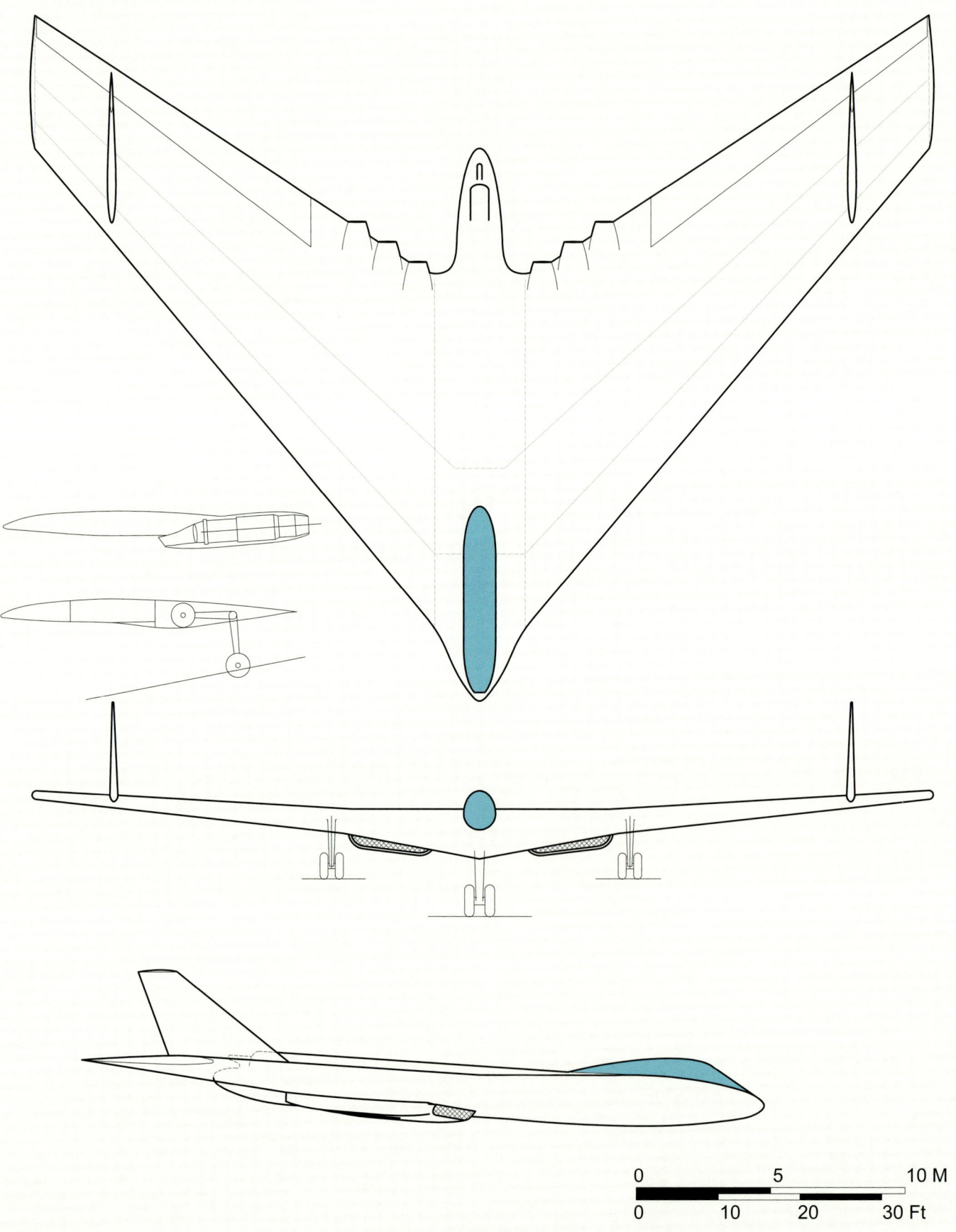

Douglas Model 1064
SCALE 1/200

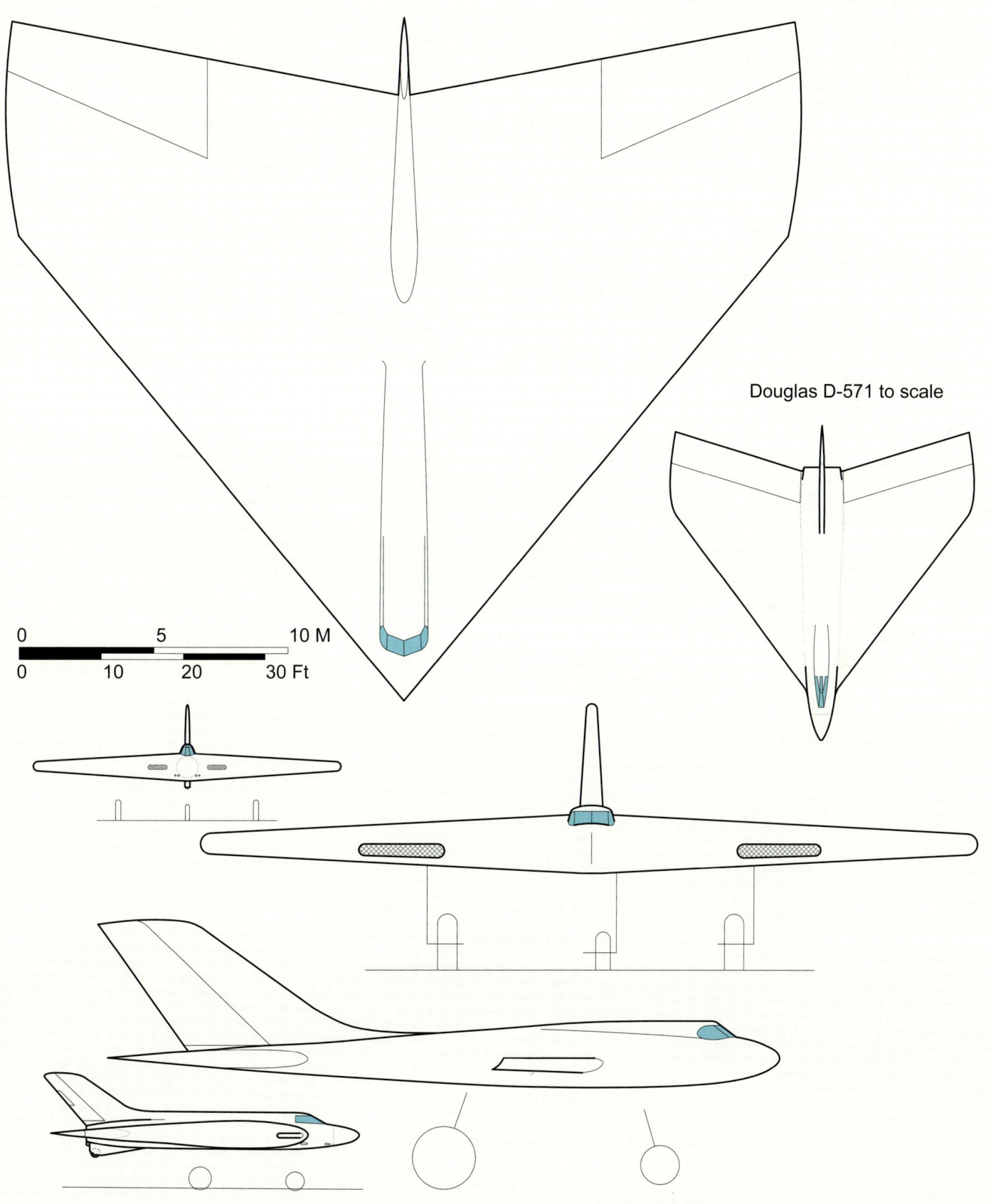

Douglas D-571 to scale

The aerodynamics of the M-121 were obviously very different from those of the B-52. The fuselage had a relatively large cross sectional area, but this also encompassed the frontal area of the turbojets. The blunt tail would normally be an aerodynamic nightmare, creating vast amounts of base drag, but the engine arrangement served to create something of a plug nozzle or a 'fluid faring' according to Fairchild. This has been proven an effective way to reduce base drag on blunt-tailed vehicles such as proposed single stage to orbit rocket boosters; the physics would work the same here. However, Fairchild also suggested that the M-121 could achieve great range by shutting a few engines down during flight and running the rest at high power, where the greatest fuel efficiency could be had. This would have played havoc with the base aerodynamics though.

The cockpit, covered in a many-paned greenhouse canopy, occupied the extreme forward portion of the nose with downward visibility hampered somewhat by its positioning in relation to the canards. Much was made of the 'spacious interior' available to the crew of four: pilot, co-pilot/radio operator, navigator/bombardier, engineer/crew chief.

The pilot sat in a centrally positioned seat quite near the tip of the nose; from here the canards were behind him and his field of view was good. In an emergency, the forward portion of the fuselage would become an escape module. The canards would be jettisoned, stabilizing fins would deploy and then the cockpit would separate from the aircraft via a 'breaker cylinder'. Parachutes would deploy to slow its descent then the crew would bail out and use their own individual parachutes. The system, in retrospect, seems a bit clunky compared to just punching out with ejection seats.

Fairchild proposed a variety of missions for the M-121. Since the forward fuselage was designed to break away, it made sense that different cockpit modules could be attached for different roles. The basic M-121 long-range bomber had no defensive armament, so one possibility was an escort fighter version. A raised cockpit would replace the flush-canopy of the original and the crew complement was reduced from five to four. The pilot sat back away from the nose, which became a turret with four M-24 20mm cannon.

The terrain-mapping radar would be eliminated and replaced with 24 missiles, either 5in high velocity aircraft rockets (unguided volley-fire missiles) or Falcon air-to-air missiles. For added speed in combat, a four-cylinder 16,000lb thrust liquid rocket system would be added to the rear of the now-useless bomb bay. This aircraft could accompany the bomber version wherever it might go.

A similar cockpit module would replace the armament with cameras, creating a long-range reconnaissance platform. A trainer version would be largely indistinguishable from the bomber and by modifying the aircraft structure appropriately it could be converted into a tanker with a flying boom, just the thing for extending the range of M-121 bombers to 9,750 nautical miles from their bases.

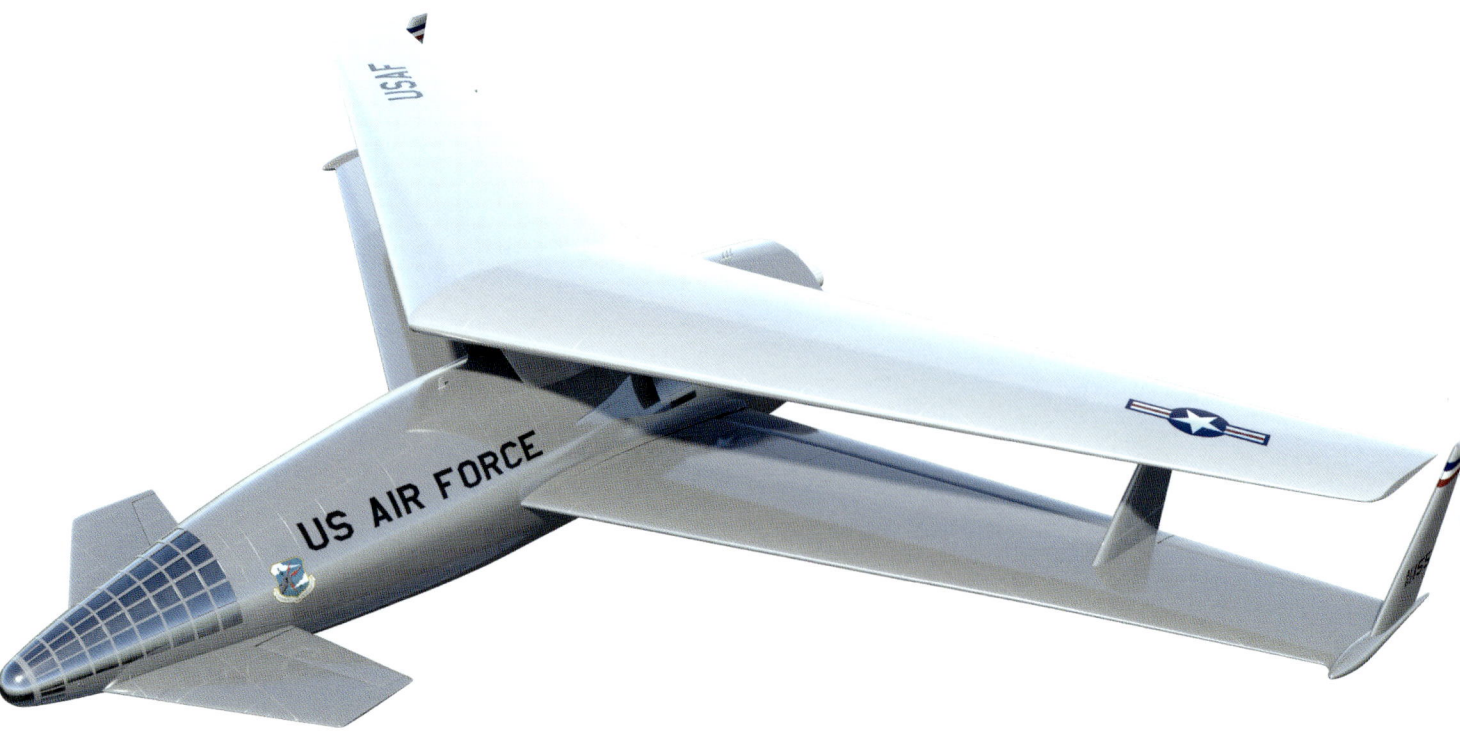

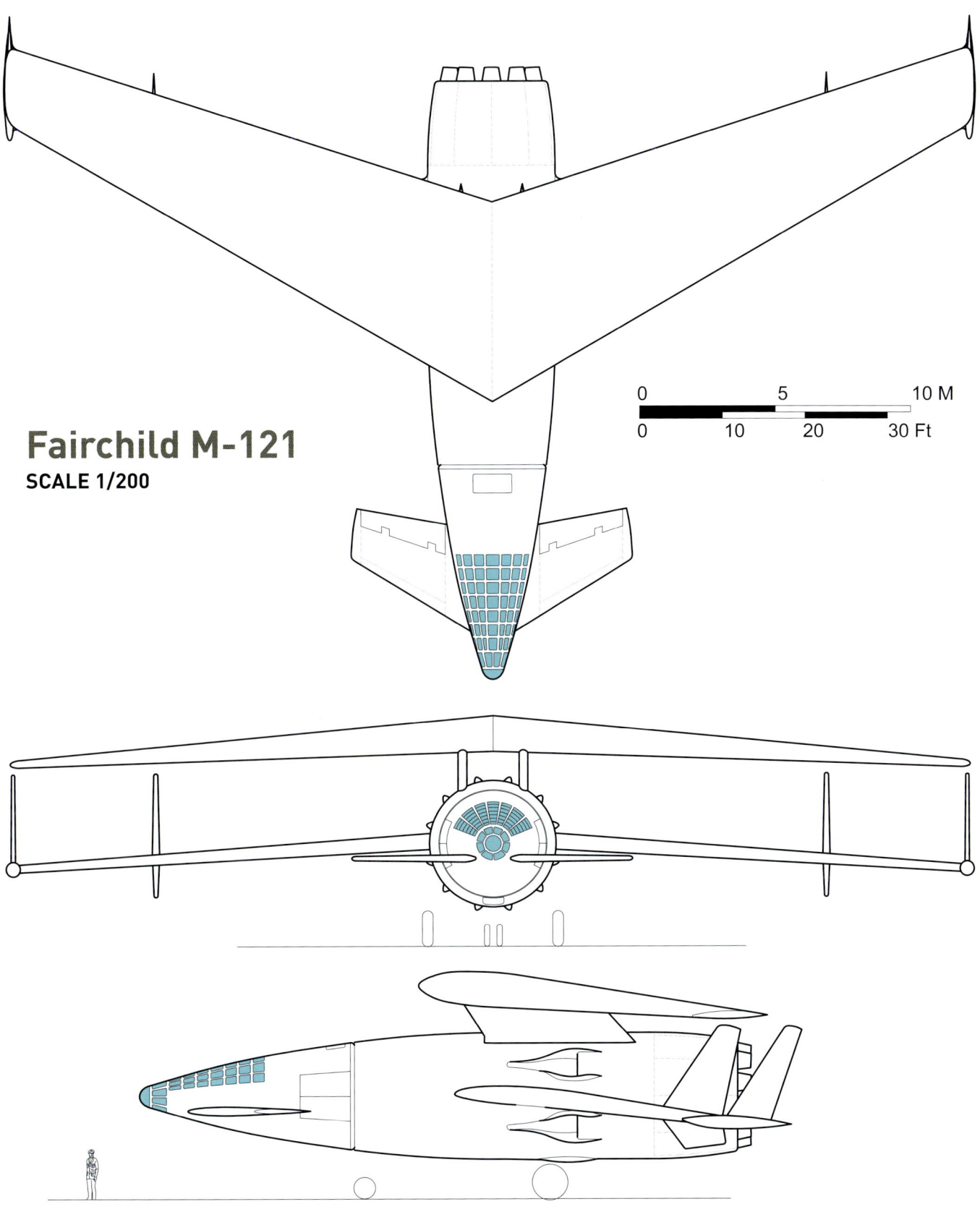

Fairchild M-121
SCALE 1/200

A deficiency of a design like this is that it is essentially locked into the engines it started off with; new engines could be swapped in, but they would have to fit within the confines of the original units. Larger diameter or longer engines would require massive modifications to the airframe.

The M-121 had been presented to the Bombardment Branch at Wright Field in September 1948, where it was received positively. It was seen as innovative and potentially useful, but it was understood that cancelling the Boeing effort and proceeding with Fairchild's design would set the adoption of a new

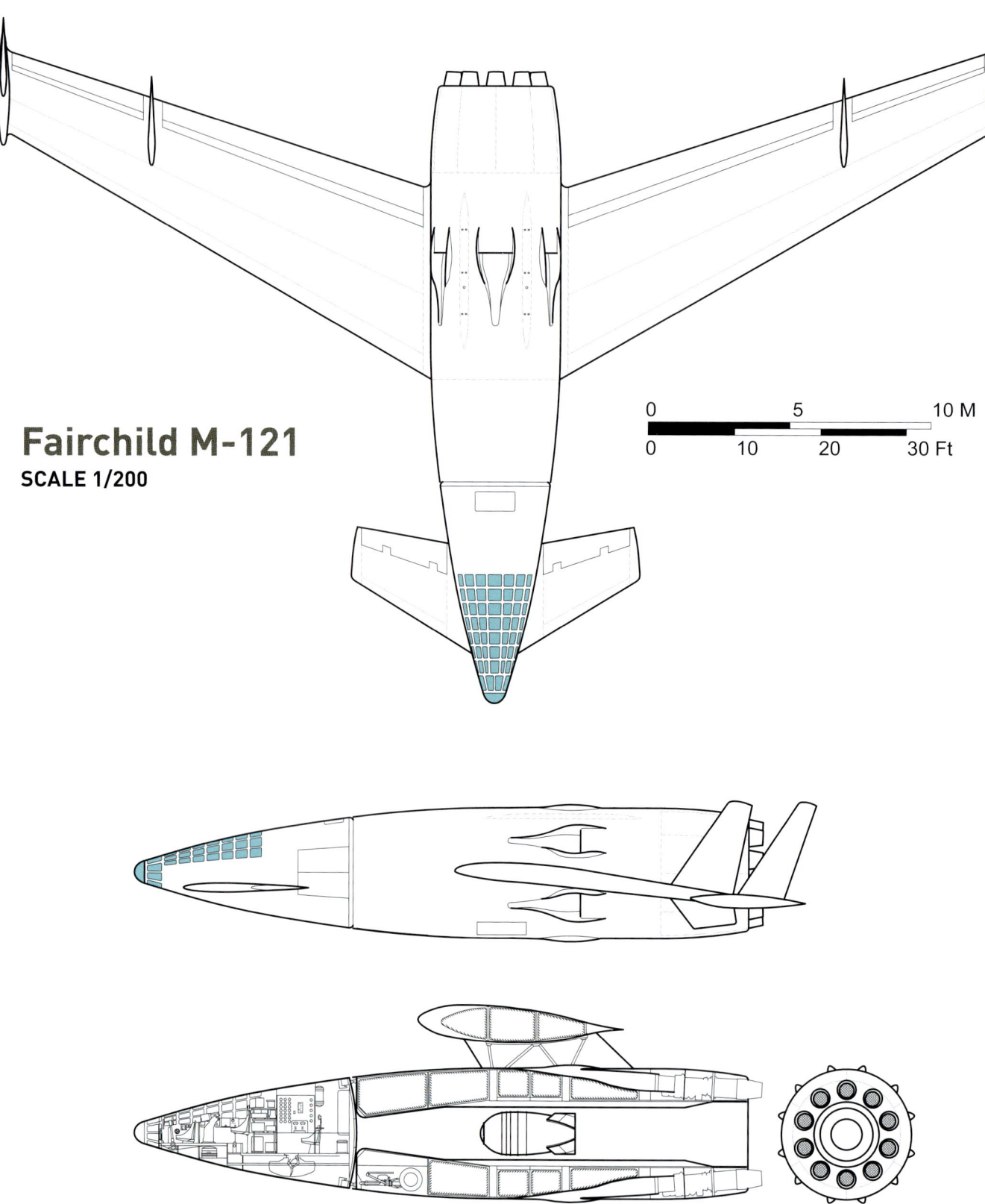

Fairchild M-121
SCALE 1/200

bomber back by at least three and as many as six years. It was also recognized that the M-121's performance was based on a lot of assumptions, not all of them necessarily wholly grounded.

The aircraft would be limited in growth potential and would require a new fuel tank wing on every flight, as well as the special jet-powered launching rail system. The Bombardment Branch concluded that if such a radical departure in aircraft design was to be adopted, the practical thing to do would be to open the competition again and see what new ideas other companies might have. The idea was floated to keep working on the Boeing XB-52, but open to industry a new competition specifically for radical new designs.

But no money for such a competition was procured and the M-121 faded away.

Douglas Model 1211-J

The one design that came closest to actually competing with Boeing for the B-52's role was the Douglas Model 1211-J. In many ways an American version of the Soviet 'Bear' bomber, the 1211-J was a large conventional bomber with swept wings and turboprop engines.

Douglas began examining advanced long-range jet bombers at least as far back as April 1947 with the 'Very Long Range Bomber' concepts. These early designs were generally fairly conventional in layout with long, thin, straight wings and two jet engines in the wing roots. Other designs such as Model 1112 from May 1947 used advanced radial engines to drive pusher props, like overgrown versions of the XB-42 'Mixmaster'. By 1948 Douglas had produced Model 1155, a four-jet, straight winged strategic bomber that looked much more like a jetliner than a bomber… in no small part because it utilized wing and fuselage components from the Douglas DC-6. None of these seem to have been entered into official USAF competitions; if they were, nothing came of them.

By August 1949, Douglas had embarked on Model 1211. This was not some minor design study; instead, Douglas spent a considerable effort through early 1951 designing at least 40 versions. It was a turboprop-powered long-range strategic bomber in the mold of Boeing's Model 464-35 and was clearly meant to replace the Boeing effort in Air Force plans.

The first Model 1211 design was another straight-winged vehicle, though a somewhat spindly one. But the wing, spanning some 160ft, almost immediately gained an appreciable aft sweep, along with increased span. The apparent primary design, Model 1211-J from January 1950, has certainly received the most notice over the years, having even been discussed and illustrated in *Aviation Week's* January 29, 1951, issue. It is unusual in the extreme for secret projects to get press; in all likelihood information about Model 1211-J was leaked to the press in order to raise awareness and interest in the design in the hope that this would nudge the Air Force into throwing open the B-52 competition again. Whether this might have been done by Douglas staff looking for a new contract, or Air Force staff looking to dump the Boeing design, is not known.

Model 1211-J was one of a great many variations on a theme. The 1211 line included smaller planes, larger planes, turboprops, turbojets, wings with less sweep and greater. Model 1211-J was the one that seems to have been presented to the Air Force for consideration. Its fuselage was narrow in diameter and extremely long, while the wings had an almost sailplane-like aspect ratio of 12. Four unspecified turboprop engines were spaced along the 227ft, 6in wingspan, the nacelles submerged within the wing.

It appears that turbojet engines were included at the rear of the nacelles, something that had appeared in some of the earlier long-range bomber studies, raising the total number of engines to eight. According to contemporary Douglas art and display models, the exhausts of the turbojets were capped with conical fairings; this would indicate that they weren't to be used for takeoff or cruise, only for high-speed dash over the target.

The aircraft had not only a bicycle landing gear arrangement but also main landing gear within the inboard engine nacelles. This was a necessity given the 322,000lb gross takeoff weight. The size of the aircraft, its aerodynamic quality and the sheer mass of fuel it could carry gave it substantial performance, with a range of 12,000 miles and a ceiling above 55,000ft. This at least exceeded the Boeing design's performance of the time… but by this point Boeing had moved on to pure turbojet power and was able to fly faster than the Douglas design. And speed was life, both in the battlefield and in the halls of government power.

The plane would have range for global bombing missions but was substantially slower than the B-52; consequently, missions would take much longer. A crew of nine therefore included a second shift to take over at some point in the mission. It does not appear that the crew were given ejection seats.

Douglas also suggested Model 1211-J as a carrier for a wide range of missiles, recon aircraft and defensive fighters. Two X-3-derived recon planes could be carried, one under each wing between the inboard and outboard engines; alternatively, two F4D-1 Skyray fighters (or even two XF-92s) could be carried in the same positions to provide cover for the bomber while over enemy territory. All of these parasite aircraft could be recovered in flight. The wings of the Model 1211-J were too thin to permit the pilots of the parasite craft to leave their planes in flight, so they would need to remain in their cramped cockpit for the entire duration of the mission.

Another option was two early-configuration Northrop XSSM-A-3 Snark missiles carried in the same positions, giving the bomber a substantial standoff capability. Model 1211-K appears to have been a largely identical craft, but with a single North American Aviation Model 704 XSSM-A-2. This was an early design for what would become the Navaho supersonic cruise missile. Due to space limitations, the missile could not conceivably be carried underneath, so it was positioned above the fuselage on a rack that would raise and tilt it for launch. An alternative to the parasite aircraft or standoff Snark missiles were gigantic pods. Filled with fuel, they could greatly

Douglas Model 1211-J
SCALE 1/400

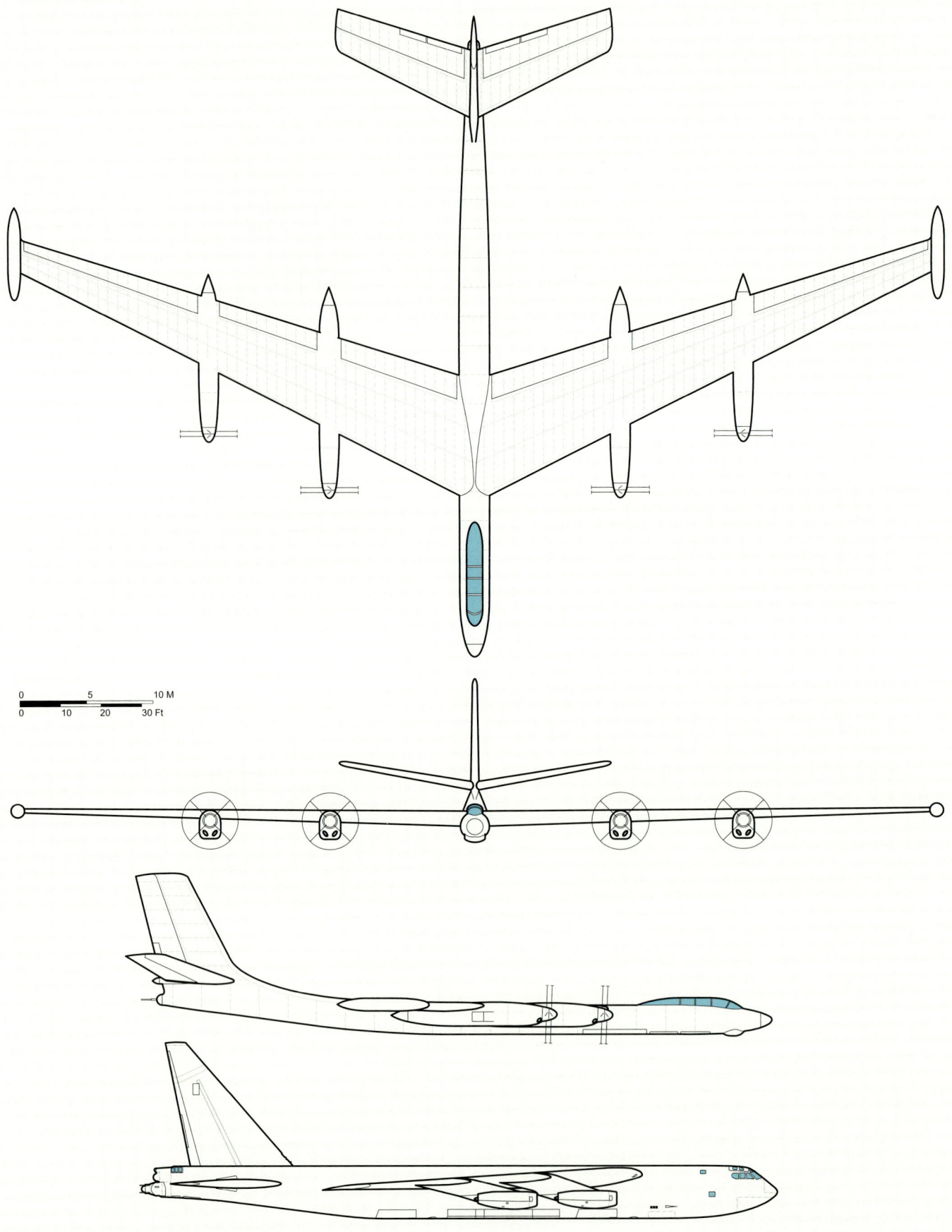

increase the range of the aircraft; or they could be filled with unspecified equipment to aid in specialized missions – probably reconnaissance and/or electronic warfare. The pods would also carry the personnel need to operate the equipment.

Despite Douglas' best efforts, the Model 1211 series faded away around early 1951.

RAND Corp G-2

The RAND Corporation seems to have had it in for the Boeing B-52 almost from the beginning, with recommendations for cancelling the programme dating back to April 1947 (when it was still Project RAND). The RAND think-tank was specifically created to review Air Force programmes and make recommendations regarding research and development priorities, so reviewing the B-52 programme was certainly within its purview. This included not simply critiquing Boeing's evolving designs but also producing preliminary concepts of its own. As early as late 1947, RAND was tinkering with flying wing concepts for long-range strategic bombing.

Data on the RAND designs is so far a bit lean, but one of them has come to light. The 'Model G-2' flying wing dates from June 1949 at the latest. Based in large part on Northrop studies, the G-2 could come in either turbojet (TG 180 or TG 190) or turboprop (Northrop-Hendy T-37) versions. The wing design was simple and straightforward; well-swept and mated to a minimal fuselage, with four engines embedded within. The turboprop version drove 20ft 5in diameter eight-bladed counter-rotating pusher props.

The fuselage would have defensive gun turrets in the nose and tail, with a raised cockpit just behind the nose turret. It had a single bomb bay within the central fuselage; unlike the Northrop designs, it did not have separate bomb bays strung along the wing (this is likely due to the fact that by 1949 a long-range bomber would be expected to carry a single nuclear weapon, not a mass of smaller high explosive or incendiary bombs as was the case for Northrop's Second World War era designs). It had a tricycle landing gear and inflight refuelling capability. The turboprop variant seems to have received the greatest interest; the fact that it was expected to be able to reach a target 4,500 miles from its base while the turbojet version could only go 2,500 miles to drop its bomb doubtless had much to do with that.

As RAND was a think tank and not an aircraft manufacturer, the G-2 bomber never left the drawing board.

Convair B-60

Of all the aircraft that tried to dethrone the B-52, only one came anywhere close… and even then not very close. In late 1950, successful development of the Boeing B-52 was still far from a certainty. On the other hand, Convair's B-36 was then well along in production. The company had proposed to modify the B-36 with turboprops to boost top speed, though by keeping the standard straight B-36 wing the maximum speed achievable would remain far below that of the B-52.

Yet Convair had another, somewhat more radical, idea: keep the wings, but sweep them back and hang turbojets under them. This concept, dubbed the B-36G, was proposed, unsolicited, to the Air Force in August 1950. It wouldn't be as fast as the B-52 but it would be far cheaper… and far more likely to fly. Very little about the Convair proposal was radical. As a backup in case the B-52 failed, The Air Force approved of the idea as a backup in case the B-52 failed and authorized Convair to convert two B-36Fs (taken unfinished from the active assembly line) to B-36G configuration in March 1951. Shortly afterwards, the aircraft were officially redesignated YB-60 due to the massive changes in configuration.

The wings were chopped off and re-attached with a sweep of 35° and given new wedge-shaped wing roots, new squared-off wingtips and new leading and trailing edges. The piston engines were removed and replaced with four jet engine pods, each with two Pratt & Whitney J57-P-3 turbojets; these were the same nacelles and pylons that Boeing was using on the B-52. This had precedent, as the nacelles used on the B-36D and subsequent models were taken from the Boeing B-47. The same basic B-36 landing gear was used and consequently still required aerodynamic protrusions above and below the wings.

The change in weight and balance compared to the B-36 meant that the YB-60s did not rest as securely on their landing gear however. So a fourth set of gear was added in the form of a deployable tail wheel. While taxiing, this would convert the aircraft landing gear into a four-point system. The tail wheel was steerable to aid in ground manoeuvrability and would remain deployed while the aircraft began its acceleration down the runway, retracting prior to aircraft rotation.

The fuselage was stretched and a new pointier and more aerodynamic nose was added; the tail surfaces were replaced with modern swept surfaces. It was intended that the nose turret and the upper rear turrets would be removed; the tail turret, the upper forward turrets and the lower rear turrets would have been retained in an operational B-60. While less well armed than the B-36, the B-60 would certainly be far more heavily defended than the B-52 with its single tail turret and four .50ca machine guns.

One of the two YB-60s was completed and did successfully fly, but its B-36 heritage hampered it greatly. Unlike the elegantly thin wings of the

RAND Corp G-2
SCALE 1/300

Northrop B-49
SCALE 1/300

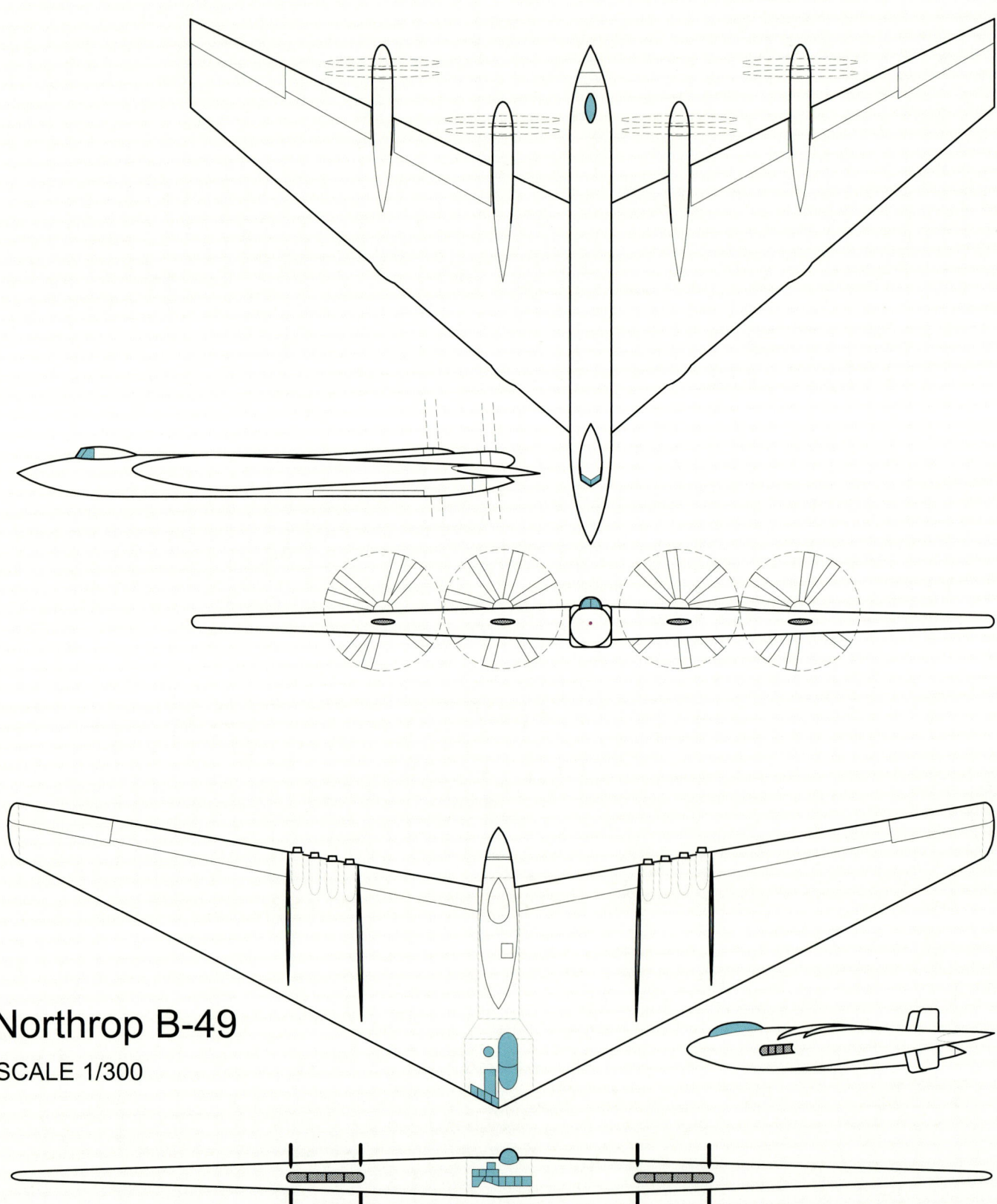

Convair B-60
SCALE 1/350

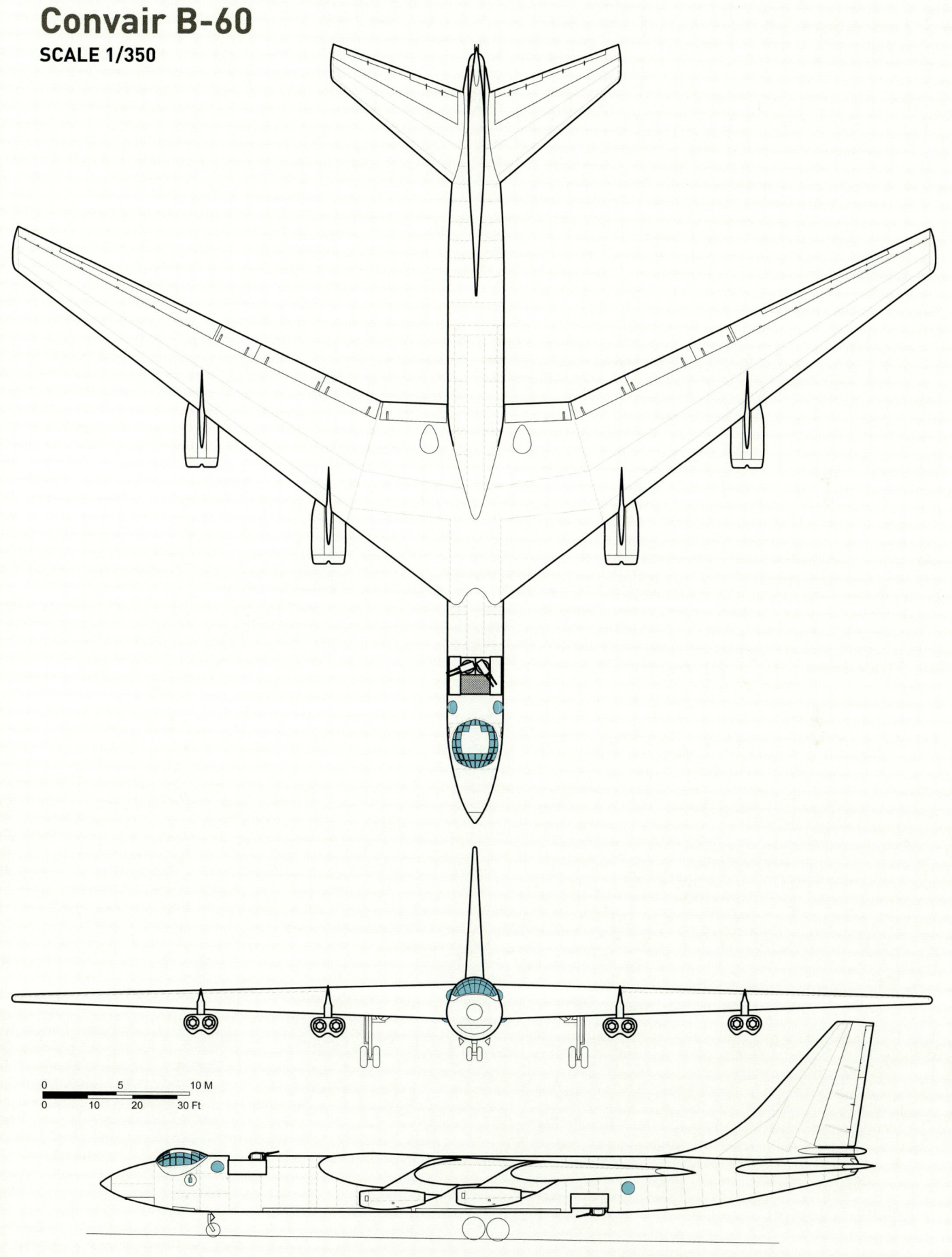

Convair B-60
SCALE 1/350

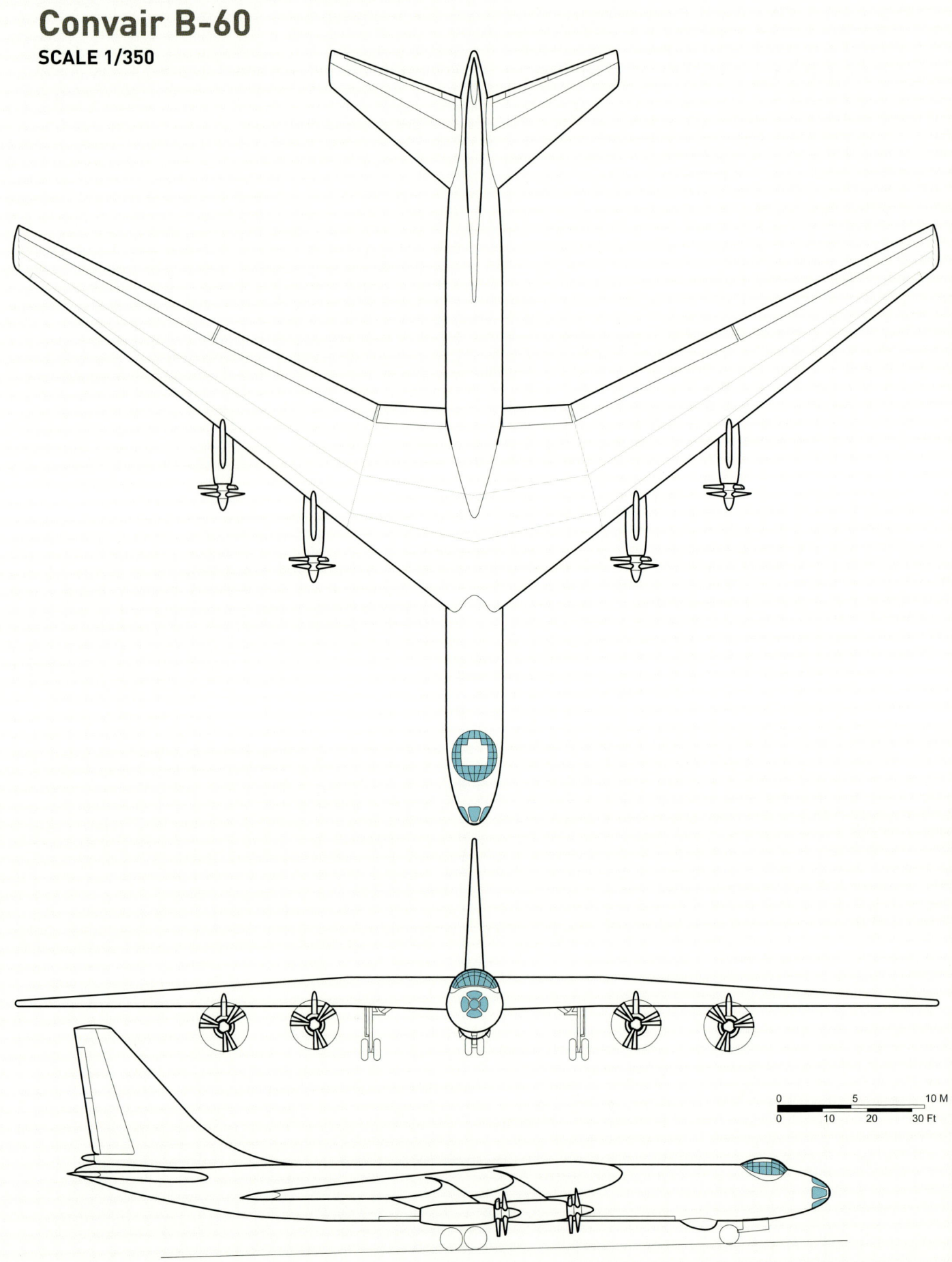

B-52, the B-60s wings were still those of the B-36; designed for slower speeds and less flexibility, they were far thicker and much draggier. Overall the B-60 was loaded with aerodynamic issues and wound up being about 100mph slower than the B-52.

The cost advantages that the B-60 offered, due to parts commonality with the B-36, did not offset the performance difference. The B-52 and B-60 both flew for the first time in April 1952, the YB-52 flying only three days before the YB-60; this demonstrated if nothing else converting the existing B-36 into a very different aircraft was a much faster process than developing an entirely new one. Flight testing, though, showed great differences between the two aircraft and while they were not officially competing, it was quickly clear that the B-60 was not a winning design.

The single completed YB-60 flew less than two dozen times, racking up a litany of failures and malfunctions, some which could be blamed on the B-60 design or construction, others leaving the B-60 blameless. Its flight record was unspectacular and its demonstrated top speed depressingly slow – not materially faster than the maximum speed of the B-36.

It had trouble with rudder flutter, control forces were too high, aileron effectiveness was marginal and it had a tendency to Dutch roll (a yaw-roll coupling mode). These were fixable but troubling. But the poor top speed was its most unfortunate feature. In addition to thick draggy wings and a fat fuselage, the B-60 also had the raised greenhouse canopy of the B-36, while the B-52 had a lower drag cockpit that was flush with the top of the fuselage. Photos of the YB-60 in flight show that its thick, sturdy wings did not flex like those of the B-52.

After a mere 66 hours of flight testing, the YB-60 was grounded for good and the whole project was cancelled on January 20, 1953. In July of that year, the aircraft were stripped for useful parts and equipment then promptly hacked to bits for the scrap yard.

The YB-60 as built was not necessarily the B-60 that would have been put into production. The B-60 would most likely have had four T35 turboprops, the same as had been planned for earlier iterations of the B-52. The fact that the B-60 was slower than the B-52 would have been an acceptable tradeoff for the improved fuel and range performance; and the unfortunate aerodynamics of the B-60 would have had somewhat less impact had it been intended to have a lower tops speed.

Each turboprop would have had 11ft diameter contra-rotating props with tip that reached transonic speeds. This would have provided good high-speed performance, at the expense of brain-melting acoustics as discussed in the section on the B-47D. An operational B-60 would also have had in flight refuelling capability… but it was not to be.

Boeing Model 481-100

Among the Boeing designs that offered some competition to the company's own B-52 was Model 481-100. At the end of January 1948, Secretary Stuart Symington ordered that work on the Boeing XB-52 (at that time the Model 464-17) be stopped because the flying wing concept was showing great promise. It was thought that conventionally configured designs ought to be abandoned in favour of this new layout. Boeing of course disagreed and argued – successfully in the end – that the XB-52 was going to provide the capabilities the Air Force wanted, and would do so years before a new flying wing type of aircraft could be developed.

But if the XB-52 was to be cancelled and replaced with a flying wing, Boeing wanted to be the designer and builder of that flying wing. So on February 2, 1948, a design was produced at the Seattle division for a flying wing that would meet the latest B-52 performance requirements. The Model 481-100 was not a pure flying wing though; it had a substantial central fuselage as well as large wingtip vertical stabilizers. The fuselage was a practical necessity: the nearly-pure flying wings of Northrop had been designed prior to the atom bomb and could not internally carry the large atomic weaponry that the B-52 would be required to.

The Model 481-100 had a fuselage with a substantial bomb bay. The nose projected well ahead of the leading edge of the wing and bore a resemblance to the forward fuselage of the B-47. At the rear of the fuselage was a defensive gun turret with a separate pressurized cabin for the gunner. Four Westinghouse XJ-40-WE6 turbojets were semi-submerged beneath the inboard section of each wing, inlets being a long, low common scoop.

The Model 481-100 was a fairly large aircraft with 4,500sq ft of wing area, a 185ft span and a gross takeoff weight of 320,000lb. Performance data is not available, but apparently the range and speed capability of this design were not so spectacular that Boeing was ready to abandon the Model 464 and go with this new design.

The size and shape are quite close to those of the Northrop B-49; it may be that this was Boeing's attempt to redesign the competing vehicle. The central fuselage would certainly have aided the aircraft in carrying the larger atomic bombs, while the underslung inlets reduced the length of the inlet ducts and thus decreased weight and internal drag. Model 481 also had the variable thickness ratio that Boeing introduced during the XB-55's development which helped to make the B-52 wing so successful, while the B-49 had a constant thickness ratio.

Boeing Model 481-100
SCALE 1/300

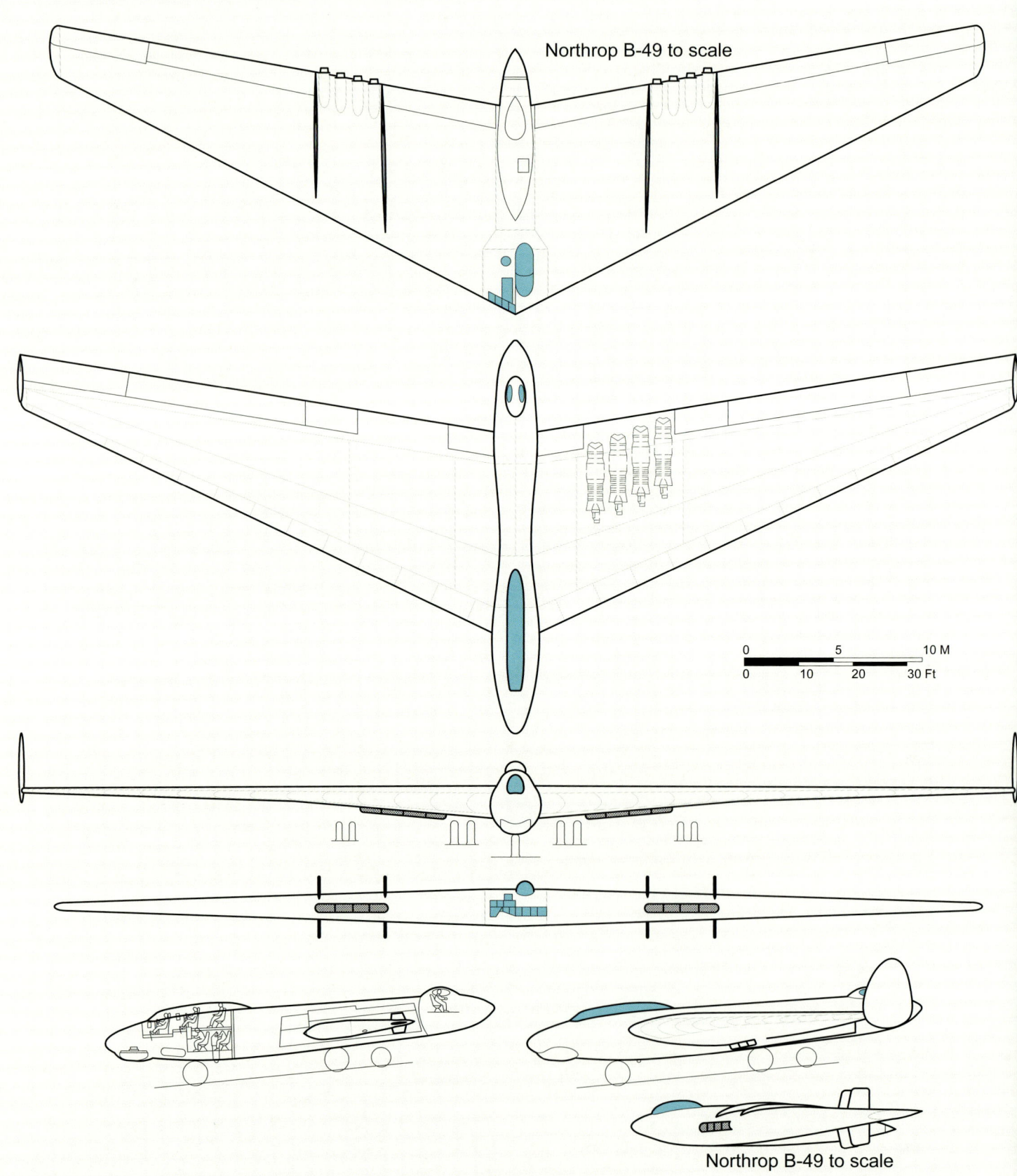

CHAPTER 4: B-52 Development

Boeing XB-52 (Model 464-67)

The winning design for the XB-52, Model 464-49, transitioned to Model 464-67. While largely the same, there were some notable differences, most obviously the extension of the forward fuselage. Where 464-49 had the rear of the cockpit canopy behind the leading edge of the wing roots, 464-67 put the cockpit well ahead of the wing. The relatively vast expanse of spoilers on the wings were scaled down and the engine nacelles were reshaped. With those changes and an Air Force 'letter of intent' for B-52 tooling in March 1951, Boeing was ready to begin constructing two Model 464-67s.

These prototype B-52s were given the designations XB-52 and YB-52… X for 'experimental' and Y being the designation for 'prototype'. Typically an 'experimental' aircraft is built before a 'prototype', but in this case while the XB-52 (serial number 49-230) rolled out on November 29, 1951, and the YB-52 (serial number 49-231) followed on March 15, 1952, the YB-52 flew first on April 15, 1952. This was due to the XB-52 suffering damage during pneumatic system pressurization testing which required extensive repairs.

The XB-52 followed the prototype into the air on October 2, 1952. The first flight of the YB-52 lasted two hours and was powered by prototype YJ57-P-3 engines. Despite the difference in designations, the XB-52 and the YB-52 were essentially identical.

The prototype B-52s were largely similar to the production aircraft in appearance. An immediately distinguishing feature of both aircraft, though, was the cockpit. A tandem fighter-style canopy somewhat similar to that used on the B-47 was employed; it was low-drag and gave the pilot excellent visibility.

The prototypes pioneered the landing gear layout that the rest of the B-52 fleet would employ. Somewhat similar at first glance to the bicycle arrangement used by the B-47, the gear used by the B-52 was quite different. Four separate dual-wheel bogies were stored within the B-52 fuselage, but instead of deploying straight down they deployed out to the sides, twisting around so that the bogies stored fore-and-aft ended up side-by-side. This gave the B-52 not a bicycle arrangement, but a quadricycle. The B-52 would comfortably sit level on its main landing gear and not tip to one side or the other. It still employed smaller outrigger gear near the wingtips, but this was to keep the wingtips from striking the ground during heavily laden takeoffs or bumpy landings.

Additionally, the forward bogies could rotate up to 20° side to side, allowing the B-52 to do something unique: land while 'crabbing' into the wind, the fuselage of the aircraft pointed well off the axis of the groundpath of the flight. This would permit safe landings in high winds.

The prototypes had flapperons, ailerons and spoilers on the main wings. The ailerons were

Boeing XB-52
SCALE 1/300

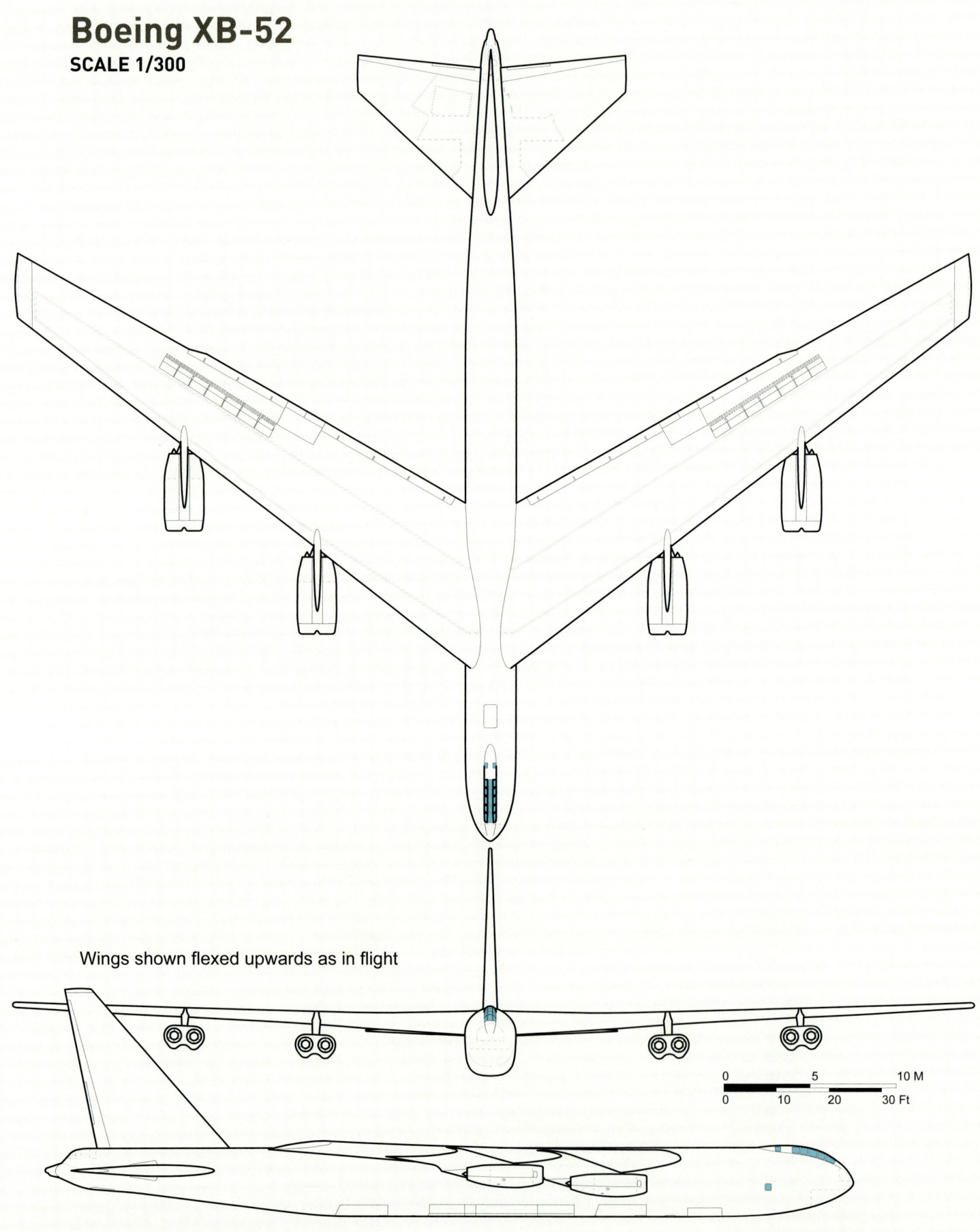

Wings shown flexed upwards as in flight

relatively small and located far from the wingtip; in fact, just outboard of the inboard engine pylon. A wingtip location for the ailerons would have given them more authority, but that would have put them in a much thinner section of the wing, a section much given to flexing. The inboard location was sufficient for the manoeuvring that the bomber was expected to perform.

In any event, the spoilers were to take care of the bulk of the control needs of the aircraft, and the ailerons would eventually find themselves redundant. Unlike the production aircraft that followed, the prototypes did not have the capability for inflight refuelling. Neither did they, initially, have the external fuel tanks that generally graced the outer wings of production model B-52s, but such tanks were eventually added later in the testing phase.

The horizontal stabilizers were all-moving, but this was meant for trim stabilization. Actual control was via slim elevators along the trailing edge. The elevators had, through the B-52F, trim tabs. An important but rarely noted feature not only of the prototype B-52s but of all B-52s that followed was the folding vertical fin. The fin was, at least until the G-model, a vast structure; too tall by far to allow the B-52 to fit within standard hangars. So it could fold over 90-degrees, greatly reducing the effective height of the aircraft. Unlike naval aircraft with wings that fold to fit in the limited space on board aircraft carriers, the folding fin is not a self-contained system – an external crane is needed to lay it over and raise it back up again.

The prototypes were essentially hand-made at the Boeing Seattle factory. Production methods were not used as the jigs were not finalized; the equipment and instruments employed were also often not what would become standard. Neither prototype was fitted with defensive weapons; the tail turrets were represented by static fairings, with the frames of the proposed gunners' canopies represented with painted-on lines.

The YB-52 was donated to the US Air Force Museum on January 27, 1958, having flown for 783 hours. It was on display for a time but due to a 'beautification' scheme orchestrated by First Lady Lady Bird Johnson, both the XB-52 and YB-52 were scrapped sometime in the 1960s. Exactly how the official museum of the United States Air Force was 'beautified' by converting one of the most beautiful aircraft ever built into razor blades and soda cans is not adequately explained in the available literature.

Boeing B-52A (Model 464-201-0)

More than a year before the YB-52 flew, on February 14, 1951, the USAF signed a letter contract with Boeing for 13 B-52A aircraft. These would also be built at the Boeing Seattle facility, in a much more production-like process. As described previously, General LeMay disliked the tandem B-47-style of cockpit as it made communication between pilot and co-pilot somewhat more difficult than when they were seated side-by-side. At a B-52 mockup meeting held in Seattle on March 1 and 2, 1951, LeMay expressed his preference for the side-by-side cockpit arrangement.

The unbuilt YB-47C was projected to have an all-new cockpit featuring side-by-side seating and cockpit design was incorporated directly into the B-52A. This change, coupled with a slight extension to the forward fuselage, was the most obvious change from the prototypes to the A-models. A few weeks later, on March 21, the contract was revised: ten of the 13 B-52As would instead be built as B-52Bs, leaving only three B-52A aircraft. A definitive contract in December 1952 called for three B-52As and 17 RB-52B aircraft.

Boeing B-52A
SCALE 1/300

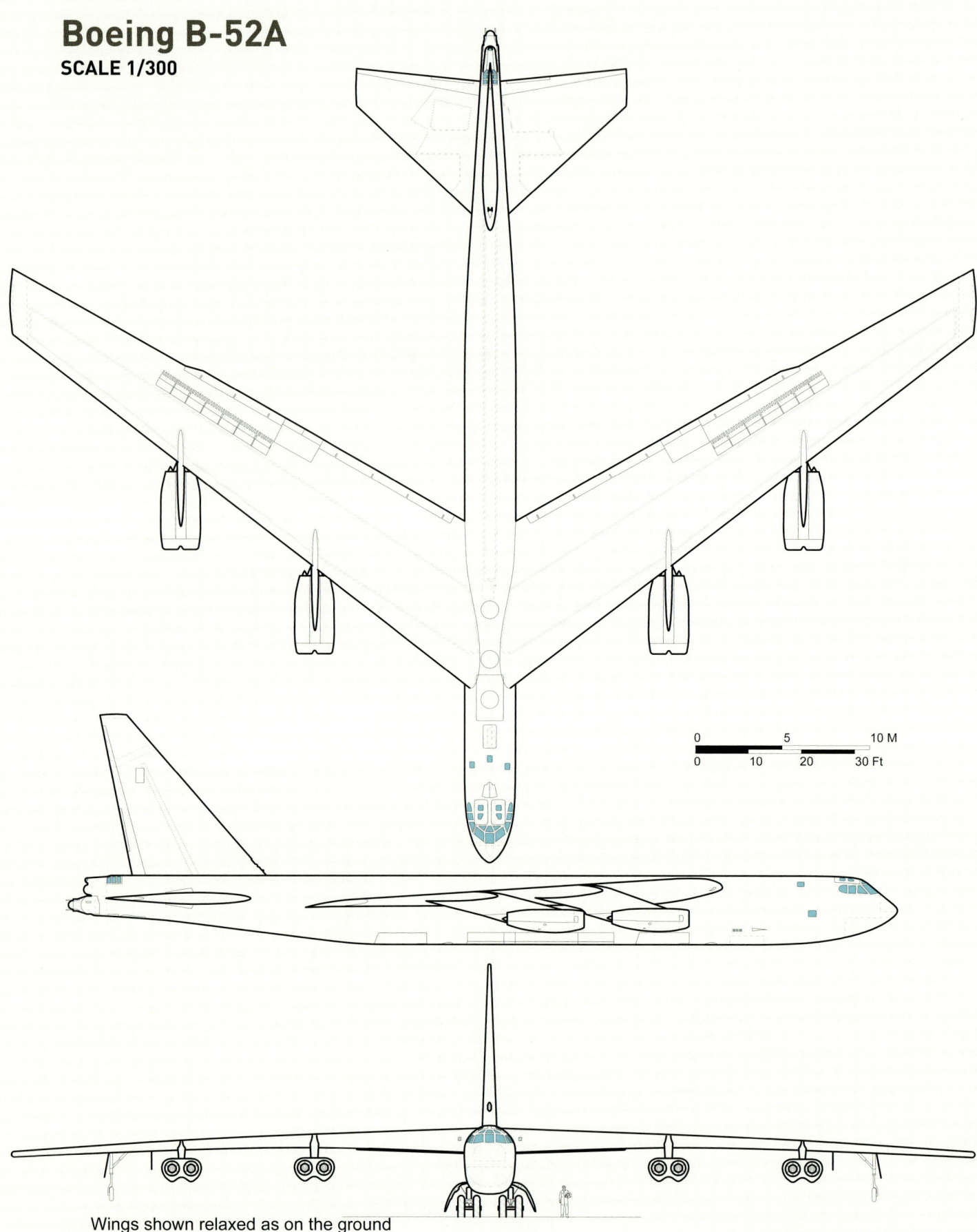

Wings shown relaxed as on the ground

Boeing B-52A
SCALE 1/288

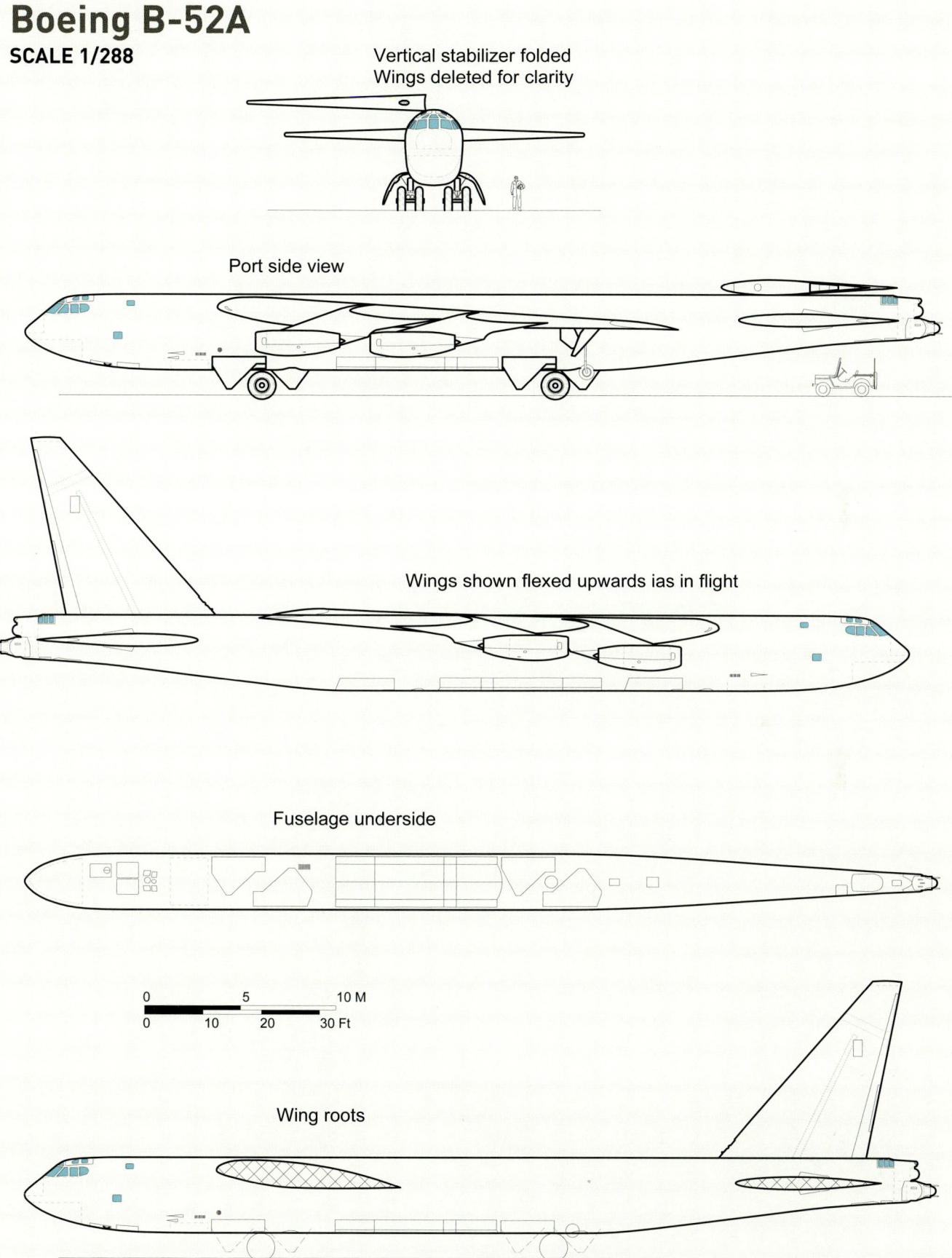

Vertical stabilizer folded
Wings deleted for clarity

Port side view

Wings shown flexed upwards ias in flight

Fuselage underside

Wing roots

The YJ57-P-3 engines used on the prototypes had proven to be temperamental and were replaced on the B-52A with uprated J57-P-1W engines. These included water injection, used during takeoff to briefly increase thrust (and negate the need for the rocket assistance that the B-47 had required). While the water injection did both cool the engine and increase thrust, it also led to the distinctive smokey exhaust trails that B-52s left on takeoff.

The first B-52A was rolled out in Seattle on March 18, 1954, accepted by the USAF in June 1954 and first flown on August 5, 1954. The B-52As were built to production standards on production jigs but would never see operational service. Instead they would serve as test aircraft, with the B-52 airframe being put through its performance paces. The B-52A could be fitted with 1,000-gallon external fuel tanks, but often flew without them. The type was also equipped with the inflight refuelling receptacles behind the cockpit which would become a standard feature for all B-52 models, allowing the aircraft to roam the entire globe. The B-52As were not equipped with bomb aiming systems.

Each had six crew (one more than the prototypes): two pilots, two bombardier-navigators, one ECM operator and one tail gunner. All three B-52As were fitted with a functional tail turret complete with four .50cal machine guns, each with 600 rounds. The turret had an A-3A fire control system that could automatically aim and fire the guns using a built-in radar unit. The tail gunner was in a separate pressurized compartment at the extreme rear of the fuselage with no access to the forward compartment – there being no pressurized tunnel as there was on the B-36. In the event of a need to bail out, the entire rear fuselage aft of the gunner's station would be jettisoned; the gunner could then undo his straps and simply fall away, no ejection seat needed.

Of the three B-52As, one (52-0002) was scrapped in 1961, another (52-0001) was grounded and turned into a teaching aid and another (52-0003) was converted into the NB-52A carrier aircraft, described later.

Boeing B-52B (Model 464-201-01, -03, -04)

The first operational B-52s were the B-52Bs (Model 464-201-03) and the RB-52Bs (Model 464-201-01 and –04). Largely identical to the B-52A, the B type was at last fitted with the full suite of equipment that would allow it to aim and drop bombs.

Initiated in April 1952, the first B-52B rolled out in July 1954. The first flight occurred on January 25, 1955, followed by acceptance of the first B-52B in June 1955. The B-52B was initially equipped with the K-3A bombing/navigation system from the B-36, though there was a problem: above around 35,000ft, in particular at 45,000ft where the B-52 was expected to operate, the system provided very poor visibility of the ground.

Since the bombardier could not see the targets clearly enough to identify them, the K-3A turned out to be virtually useless and was replaced with the MA-6A system. This allowed the B-52B to carry and deliver nuclear weapons such as the 8,500lb Mark 6 (a fission bomb with a yield of up to 160 kilotons) and the 17,600lb Mark 21 hydrogen bomb (with a theoretical but never tested yield of 18 megatons). It could also carry 27 x 1,000lb conventional bombs, or one 50,000lb bomb. A B-52B earned the distinction of being the first aircraft to drop a hydrogen bomb when, on May 21, 1956, a Mark 15 was dropped in the Redwing Cherokee test over Bikini Atoll. Detonating at an altitude of 4,330ft, this bomb generated a yield of about 3.8 megatons. Unfortunately, a navigation error

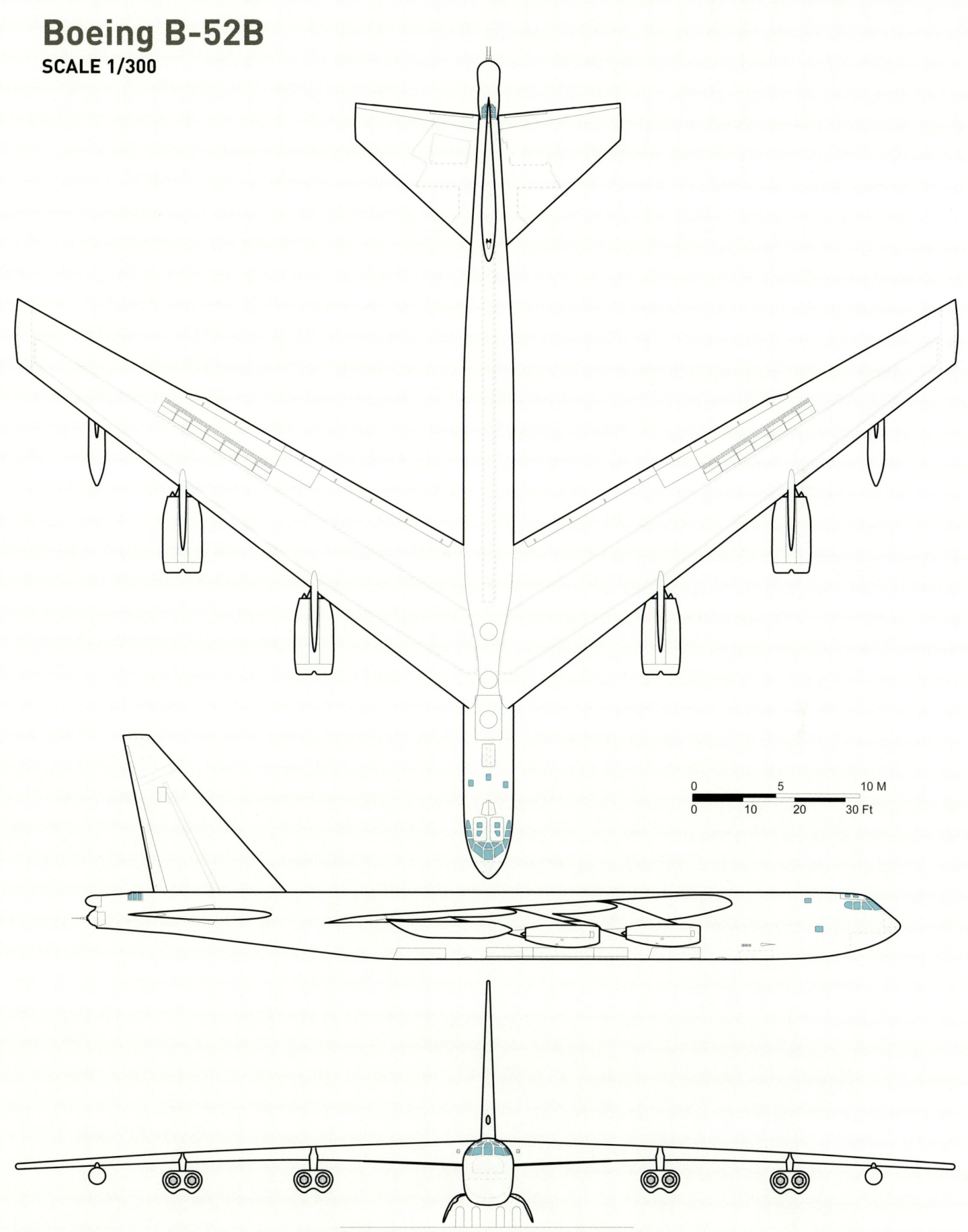

led to the bomb being dropped about four miles off target, hitting the wrong island and ruining much of the attempt to obtain data.

A better result was obtained when, in January 1957, five B-52Bs set off from Castle Air Force Base on a round-the-world nonstop flight. Forty-five hours and 19 minutes later three of the aircraft had covered 24,235 miles (one spare had failed to refuel in flight due to a frozen-over receptacle; another had made a scheduled diversion to and stop in England). Mission duration was within two minutes of prediction. This demonstrated that the B-52 could reach anywhere on earth in a reasonable time.

For defence, the B-52B series was initially was equipped with the same four-gun turret fitted to the B-52A. But another tail stinger was developed with two M24A1 20mm cannon with 400 rounds reach. These would hit far harder than the .50cal machine guns, and theoretically at greater range. To aim them, the MD-5 Fire Control System featured a larger radar in a distinctive and bulbous housing above the guns. Unfortunately this system proved a disappointment and the B-52B series ended its production run fitted with the four machine gun tail.

The RB-52B variant had special provision made in the bomb bay to accept a large cylindrical pressurized reconnaissance capsule. Several variants of this were available, all including seats for two reconnaissance operators. The pod had its own fixed fairing on the underside that replaced the bomb bay doors. In the event of an emergency, the operators were equipped with ejection seats; but given their position these necessarily fired downwards – making ejection at low altitude a risky proposition. The general purpose pod was used for photo reconnaissance, coming equipped with a bank of large (up to 36in) cameras and 24 M-120 photo-flash bombs.

An electronic intelligence capsule with antennae for receiving and instruments for recording enemy radio and radar transmissions was produced, as well as a weather reconnaissance version. At this stage in the development of the B-52 there was debate about whether the aircraft should be primarily a bomber or a recon platform; it was a debate that would rage for several years.

All B-52Bs were built in Seattle. There were 23 B-52Bs and 27 RB-52Bs. Every B-52Bs had been retired by 1966 except for one: the NB-52B.

Boeing B-52C (Model 464-201-6)

The next B-52 variant, the C-model, was initiated in August 1953. In many respect the B-52C was little changed from the B-52B, but there were some differences of note. The B-52C was built in fewer numbers than any other operational variant, only 35 coming off the production line in Seattle. First flight of the model was on March 9, 1956.

The first of the external changes was the replacement of the 1,000-gallon external tanks with much larger 3,000-gallon tanks, bringing the total fuel capacity of the aircraft to 41,700 gallons. The other visual change was in the paint: the B-52C was the first model of the B-52 to be painted with 'thermonuclear white' or 'anti-flash white' paint on the underside of the fuselage, wings, engines and tail. The upper surfaces remained largely bare metal.

The purpose of this was not camouflage; at high altitude, black or dark blue would have made much more sense. White, in contrast, would have clearly stood out. But a high altitude thermonuclear bomber would likely still be well within sight of the nuclear detonation when the time came, and tests with high-altitude aircraft showed that bare-metal skins absorbed a distressing amount of the visible

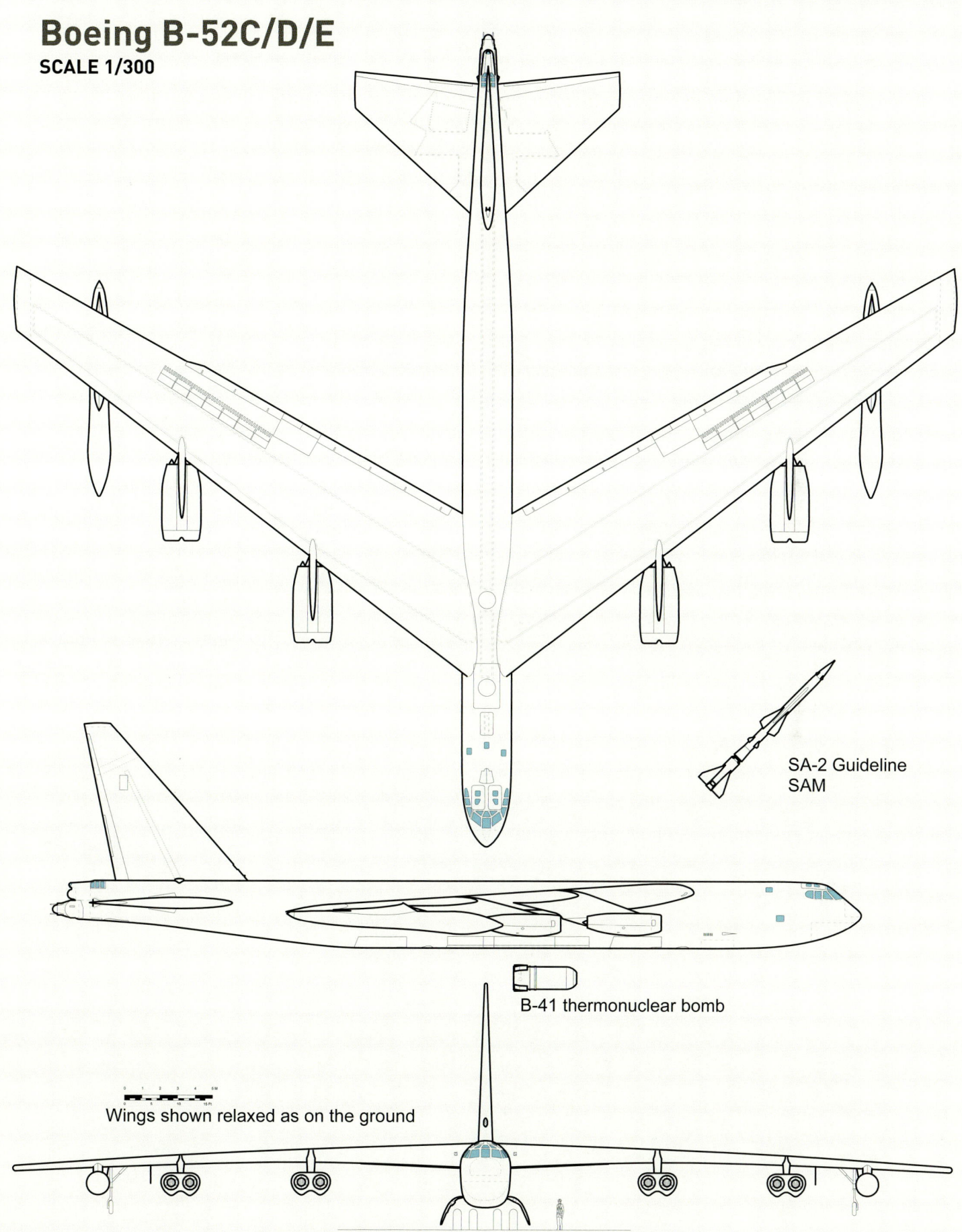

and infra-red radiation from the flash. Enough energy was absorbed that the thin aluminum skins heated up substantially almost instantly, causing warping and sometimes tearing as the skins tried to expand against an underlying unyielding structure. The white paint, using a classified composition, reflected much more of the flash and kept the structure cool… or at least cool enough to not take thermal damage. The B-47 would also find itself with a similar paint job. In contrast, while the British strategic 'V-bombers' – the Avro Vulcan, Handley-Page Victor and Vickers Valiant – also used anti-flash white paint, they tended to cover the entire aircraft, not just the underside. The North American Aviation XB-70 was similarly covered in overall anti-flash white.

The B-52Cs were mostly provided with the same four-machine-gun turret and A-3A fire control system that had been given to the B-52A and the first and last of the B-52Bs. However, the very last B-52C was fitted with an MD-9 fire control system, which proved to be much more reliable. The bombing/navigation system was upgraded to the AN/ASQ-48 package. With the larger fuel tanks, the gross weight was increased but the engines remained the same. The B-52C was designed to accommodate the same reconnaissance capsules that characterized the RB-52B, but the 'R' was never added to the B-52C designation. The final operational B-52C was retired to the boneyard at Davis-Monthan in September 1971, while one lone B-52C used as a flying testbed was retired in July 1975.

Boeing B-52D (Model 464-201-7)

The design of the B-52D was initiated at the same time as the B-52C, in August 1953. With the advent of the B-52D model, Seattle began to share manufacturing responsibility with another facility, Boeing's Wichita, Kansas factory. The first flight of a Wichita B-52D was on May 14, 1956, while the first Seattle B-52D flew on September 28, 1956. Wichita delivered 69 B-52Ds while Seattle delivered 101.

The B-52D was visually little different from the B-52C, sharing the same tail guns and large external fuel tanks. However, reconnaissance capsules were not included in planning for this variant, so the bomb bays did not have the features needed to accommodate them. The B-52D was to be a dedicated bomber. For this role it was capable of carrying the latest nuclear weapons such as the Mk 28, Mk 48, Mk 51 and Mk 57. It was also modified to be able to carry AGM-28 'Hound Dog' cruise missiles. The tail guns were the same four machine guns as had been featured previously, but the B-52D used the MD-9 fire control system incorporated into the last B-52C. This proved to be a much more reliable and capable system.

The B-52 had begun life as a high-altitude bomber, its survival based on flying higher than enemy interceptors could reach. That seeming invulnerability was fading fast by the late 1950s, and the B-52D was the first model to incorporate low-level bombing. This would put the bomber down in the 'clutter' of terrain, difficult for air defence radar systems to spot until the aircraft was right on top of them. But low-altitude high-speed flying presents difficulties, including being a very rough ride. While in later decades advanced technologies would be first tested and then incorporated that would reduce the turbulent ride at low altitude, in the late 1950s the new flight mode meant that the B-52D structure needed strengthening.

The 'Hi-Stress' programme, begun in the very late 1950s, took aircraft as they approached 2,000 hours and again at 2,500 hours and strengthened areas such as the aileron bay, fuselage bulkheads and numerous wing structures. Begun in 1972 and completed in 1977 was the 'Pacer Plank' programme which replaced many of the skin panels on the B-52D, not only getting rid of cracked or potentially cracked skins, but making the aircraft slightly more aerodynamic due to smother skin.

Starting in December 1965, the entire operational B-52D fleet went in for 'Big Belly' modifications to make them more capacious bomb trucks. Main bomb bay capacity was increased to either 84 x 500lb bombs (Mk 82) or 42 x 750lb bombs (M117), with a further 24 of either bomb being carried on external racks attached to pylons under the inner wing. These modifications did not change the B-52D's ability to deliver nuclear weapons, but did raise the total payload capability to around 60,000lb. The bombs were loaded into the bomb bay in pre-loaded 'clips', greatly speeding the rearming process.

This capability was turned on North Vietnam (and targets in South Vietnam) starting in 1965, with Big Belly modified B-52D and B-52F bombers laying into the jungle. Carpet bombing raids cleared large sections of forest and collapsed underground bunkers and tunnels, and filled the North Vietnamese Army and the Viet Cong with dread, as the bombs came raining down largely without warning. A hundred raids were performed in 1965… a number raised to 5,000 in 1966, 9,700 in 1967, and 20,560 in 1968. This expansion in missions was made possible due to the original base of operations in far-off Guam being replaced with a base in Thailand, merely an hour from the target.

These missions continued into 1968, expanding to bomb North Vietnamese military targets in Laos and Cambodia. The Johnson administration cancelled the bombing of targets in North Vietnam just prior to the elections of 1968, but bombed the hell out of everyone else nearby; a policy that Nixon deescalated but did not end. There were about 19,500 sorties in 1969, 15,100 in 1970 and 12,550 in 1971. By this point, everyone was largely sick of the war and just wanted to be done with it. The Nixon administration wanted to pull US forces out of frontline fighting and turn the war over to the South Vietnamese; the North Vietnamese, utterly unsurprisingly, took advantage of that and launched fresh offensives.

So the Nixon administration ended the Johnson administration's bar on bombing North Vietnam directly, and began a series of bombing raids that hearkened back to the bomber raids of the Second World War. Ports, rail yards and industrial facilities of all kinds in North Vietnam came under heavy bombardment. An unfortunate first occurred for the B-52D on November 22, 1972, when a North Vietnamese surface-to-air missile struck and damaged one, forcing it down. Fortunately it was able to reach friendly territory and the crew escaped.

When the North Vietnamese delegation walked out of the Paris peace talks in mid-December 1972, Nixon ordered a massive concentrated strategic bombing campaign against North Vietnamese infrastructure. Over an 11-day period some 741 sorties destroyed much of North Vietnam's power generation and distribution infrastructure, a good fraction of its petroleum reserves and its desire to stay away from the peace talks. Soon after, they were back at the table in Paris and quickly reached an agreement to end hostilities.

Unfortunately, surface-to-air missiles brought down 15 of the bombers involved – the only B-52s lost to enemy action. The Soviet Union had provided a great many SA-2 'Guideline' surface-to-air missiles and to defend against this threat the B-52Ds were equipped with an extensive suite of electronic countermeasures during the 'Rivet Rambler' programme of 1967 to 1969. This included the addition of the AN/ALR-18 automated receiving set; the AN/ALR-20 panoramic receiving set; the AN/APR-25 radar homing and warning set; four of the AN/ALT-6B or AN/ALT-22 continuous wave jammers; two AN/ALT-16 barrage jammers; two AN/ALT-32H high- and one AN/ALT-32L low-band jammers; six AN/ALE-20 flare dispensers with 96 total flares; and eight AN/ALE-24 chaff dispensers with 1,125 bundles of radar disrupting chaff.

On the other hand, the December 1972 raids did also result in the only times when B-52s brought down enemy aircraft: B-52 tailgunners managing to shoot down two MiG-21 jet fighters. The last B-52 raids of the war occurred on August 15, 1973; by the end, B-52D, F and G bombers had dropped some 2.63 million tons of ordnance on South East Asia.

Incredibly, the total destructive power contained within those tens of thousands of sorties, conducted over eight years, would have been equalled or surpassed by a single bomb of the type that the B-52 was actually *meant* to carry.

Perhaps ironically, the B-52Ds with the Big Belly modifications used to bomb targets in Vietnam dispensed with the anti-flash white paintjob in favor of a gloss black underside. Here the risk of damage to the underside of the aircraft from a thermonuclear flash was minimal, but the risk of taking a missile or cannon fire from a North Vietnamese MiG was very real. The use of carpet bombing raids in Vietnam has been argued about for decades, both in terms of the moral justification for it and the actual tactical value it provided; but eyewitness testimony – including from this author's own father – can attest to the fact that a number of B-52s, each laying down dozens of high explosive bombs on a stretch of jungle, resulted in landscapes that were hard to ignore or forget.

The last B-52D was retired in October, 1983.

Boeing B-52E (Model 464-259)

Design work on the B-52E was initiated in May 1953. Kansas began to take the manufacturing lead, as 58 were manufactured in Wichita and 42 in Seattle. The first flight of a Seattle-made B-52E was on October 3, 1957, while the first Wichita B-52E flew on October 17.

Where the B-52D had had to be modified for the low level bombing role, the B-52E was designed and built for that from the outset. It remained a purely nuclear strike vehicle throughout its career, never being altered for carrying conventional bombs, although it certainly could be loaded with a combination of high explosive, cluster and incendiary bombs as well as mines. The nuclear payload included two Mk 21s and two Mk 15s.

The primary differences between the B-52E and the B-52D were invisible from the outside, relating to the navigation equipment installed. In order to carry out the new low-level nuclear bombardment mission, the AN/ASQ-38 navigation system was added, along with new instruments and systems for bomb navigation and terrain avoidance.

Most B-52Es were retired by March 1970, although several went on to have interesting careers as technology and propulsion testbeds.

Boeing B-52F (Model 464-260)

The last B-52s built in Seattle were F-models, with 44 made in Washington state and 45 in Wichita. They were largely identical to the B-52Es that came before except for one noticeable difference. As well as having uprated J57-P-43W, 43WA or -43WB engines, they also had alternators added to the port sides of each engine nacelle. These resulted in a conspicuous bulge. An additional scoop inlet was introduced under the main turbojet inlet, providing cooling air for the alternators and engine oil. These new 'hard-drive' alternators replaced pneumatically-driven alternators that had been installed in the fuselage next to the wing roots.

The design of the B-52F began in May 1954, with the Seattle plant manufacturing 44 and Wichita 45. The first Seattle-made B-52F flew on May 6, 1958, the first Wichita one on May 14. While the B-52F was not included in the 'Big Belly' modification programme, it was used to drop conventional ordnance on the North Vietnamese. When the modified B-52Ds began showing up with their greater ordnance loads, the B-52Fs were withdrawn and returned to the nuclear strike mission. For this role they could carry the Mk 28, Mk 41, Mk 53 and Mk 57 bombs as well as four ADM-20 Quail decoys and two AGM-28B Hound Dog cruise missiles. The final B-52Fs were withdrawn from service in 1978.

Boeing B-52G (Model 464-253)

The B-52G was the first model since the B-52A to feature a major change in appearance, as well as being the first to be built entirely at the Wichita facility. The design was initiated in June 1956 with first flight on August 31, 1958. A total of 193 were built.

The G-variant was intended to be a major advance, with a large range of changes to airframe and avionics. As the Model 464-253/B-52G was being designed, the future of strategic bombing looked far faster than the B-52: the Convair B-58 'Hustler' was in advanced development and the WS-110A programme (which would lead to the North American XB-70) had just been announced. The skies would be filled with bombers reaching Mach 2 or Mach 3, powered by engines burning boron-based 'zip' fuels or even having nuclear reactors. The B-52 would look archaic fast if it was not improved, and the Model 464-253 was intended as an important step in that direction.

The obvious design change for the B-52G was the vertical stabilizer: some 8ft was cut off the tip, reducing height, area and, importantly, weight and drag. The loss of stabilizing area was made up for with the addition of a new hydraulically powered stability augmentation system. The shortened fin had been previously tested on the very first B-52A and resulted in impressive weight savings.

The vertical stabilizer change was the most obvious aspect of an overall programme to reduce structural weight. The ailerons, which had never been particularly useful, were deleted entirely, losing the weight of the hinges, actuators and control systems. Lighter materials were used throughout the aircraft structure, the flaps got lighter drive motors and the original pressurized tail gunner compartment was removed. The gunner remained but was now on an ejection seat next to the ECM operator in the forward compartment, using radar for targeting rather than visual sighting through windows.

Within the wings, the rubber bladder fuel tanks were removed and the fuel poured directly in, creating 'wet wings'. This both saved dry weight and increased fuel capacity. As a result of the greatly increased wing fuel capacity, the 3,000 gallon tanks used on the preceding models was replaced with a much smaller 700 gallon tank. The new wings were lighter than those of the previous models but while this was great for performance, they were structurally less resilient to the effects of low-level flying. Fatigue-related issues plagued the wings of the B-52G and subsequent H-model, requiring occasional repairs and modifications.

Another change to the wings was a built-in ability to carry the AGM-28 'Hound Dog' cruise missile under pylons attached to the inboard wings. While earlier models of the B-52 would be retrofitted with these pylons, the B-52G was designed and built with the requisite structural modifications, hardpoint power, plumbing and control systems from the outset.

The nose of the B-52G was changed several times. The initial version was noticeably different from the nose that had graced the B-52A through B-52F… instead of two separate radomes, one smaller one above the nose and a separate larger one below, a single-piece fibreglass fairing covered the whole nose. This new nose was longer and blunter, changing the profile of the forward fuselage. Where before the lines of the windows carried further down the nose, with the B-52G there was a distinct step where the new nose met the bottoms of the windows. There was also a slight discontinuity in the lines where the new nose was attached at the rear of the old lower radome.

Early thinking had involved the use of afterburning J75 engines on the H-model. Details on this are sparse, but it was doubtless intended to result in a single engine on each pylon (see the later description of the XB-52 with two J75 engines). While this would have decreased the total number of engines and increased thrust, it was not incorporated; the B-52G used the same engines and nacelles as the B-52F, including the distinctive alternator fairings. However, the water injection system used for increased thrust at takeoff was improved and the capacity increased to 1,500 gallons.

The B-52G was a nuclear bomber…a role that it held for many years, but not without incident. On January 17, 1966, A B-52G collided with its KC-135 tanker; both aircraft broke apart in the air and crashed near Palomares, Spain. The B-52G had been

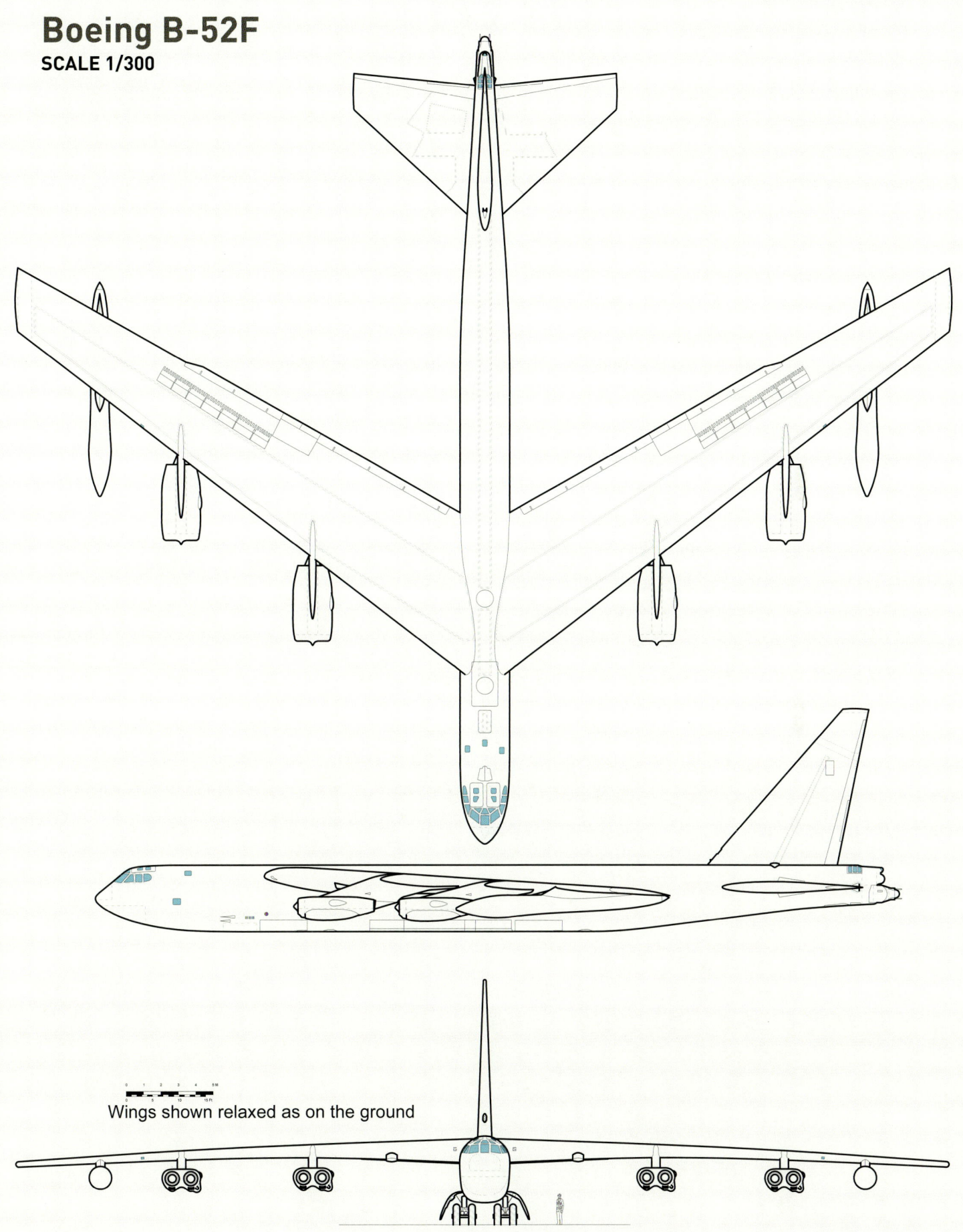

Boeing B-52F
SCALE 1/300

Wings shown relaxed as on the ground

carrying four B28FI nuclear weapons; three impacted on land, one in several thousand feet of water. The pair that hit the ground had their chemical explosives detonate, spreading plutonium over a wide area and necessitating a major cleanup operation. And then on January 21, 1968, a B-52G again carrying four B28FI bombs caught fire in flight, leading the crew to bail out. The aircraft crashed onto the ice in North Star Bay, Greenland, and the resulting fire caused the high explosive components of all four bombs to explode, once again scattering plutonium hither and yon, requiring another major cleanup.

Beginning in October 1971, the B-52G and B-52H fleet underwent modification to carry a new weapon: the AGM-69A Short Range Attack Missile (SRAM). This was a relatively small solid rocket-propelled weapon that could be carried almost like a conventional bomb. Smaller and lighter by far than the Hound Dog cruise missile, the range of the SRAM was much shorter – 110 to 785 miles. Once modified, a B-52G could carry 20 of them… eight in a rotary launcher in the bomb bay and six on a pylon under each wing. Armed with a 200 kiloton thermonuclear warhead, the SRAM would have been used for not just attacking primary targets but also for SAM suppression. Its Mach 3 maximum speed meant it would reach out far enough ahead of the bomber to destroy SAM sites before they could fire. The SRAM entered service in 1972 and was retired in 1990.

The B-52G entered the war in South East Asia in 1972 and seven were lost; six to enemy action and one to a crash in the ocean shortly after takeoff from Guam.

Between 1971 and 1976, the B-52G and B-52H fleet underwent another round of modifications. This time the fibreglass nose cones were replaced with slightly longer, slimmer noses. More importantly, the aircraft were also fitted with electro-optical sensors and improved electronic warfare capabilities. These manifested in the form of variously sized nose blisters.

Two large side by side blisters on the underside each contained a turret with a large camera. The starboard turret contained a Hughes AN/AAQ-6 forward-looking infrared (FLIR) camera; the rectangular window was made of germanium and was opaque to optical light. The port turret contained a Westinghouse AN/AVQ-22 'starlight' low light level TV camera behind an oval window. When not in use the turrets rotated 180° 'backwards' to protect them from dust, bugs and rain; also, while stowed, the windows could be cleaned with sprays of 500psi hot water. When in use, they could rotate 20° to either side and provide views of the terrain forward and below for piloting and bombing. These were useful not only for low-level night flying, but would have been handy had actual nuclear war broken out. The pilots would be able to close off their windows with opaque flash curtains to keep from being blinded by nearby nuclear detonations, while continuing to fly using the electro-optical system to see.

At roughly the same time, the 'Rivet Ace' programme added a series of electronic countermeasures blisters onto the B-52G and H. Along each side of the nose was a small fairing for the AN/ALQ-117 active countermeasure system; another AN/ALQ-117 was later installed just above the tail turret, replacing a TV camera (the gunner was then reliant upon radar for targeting).

Along the top of the nose, just in front of the canopy, was the fairing for the AN/ALQ-122 (multiple false target generator), the AN/ALR-46 (digital radar warning receiver) and the AN/ALR-20A (countermeasures receiving set, with further sets located in the port wingtip and next to the tail

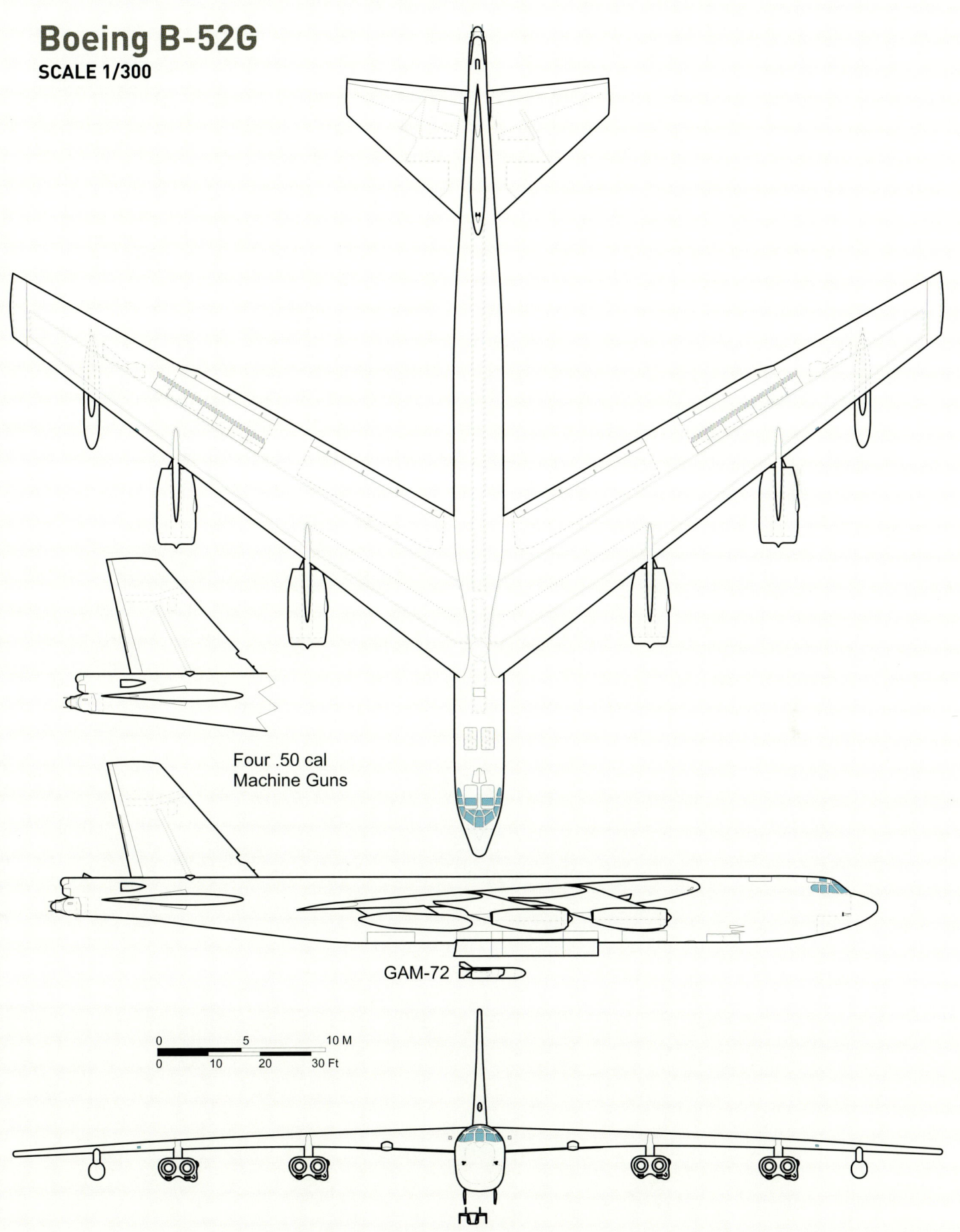

Boeing B-52G
SCALE 1/300

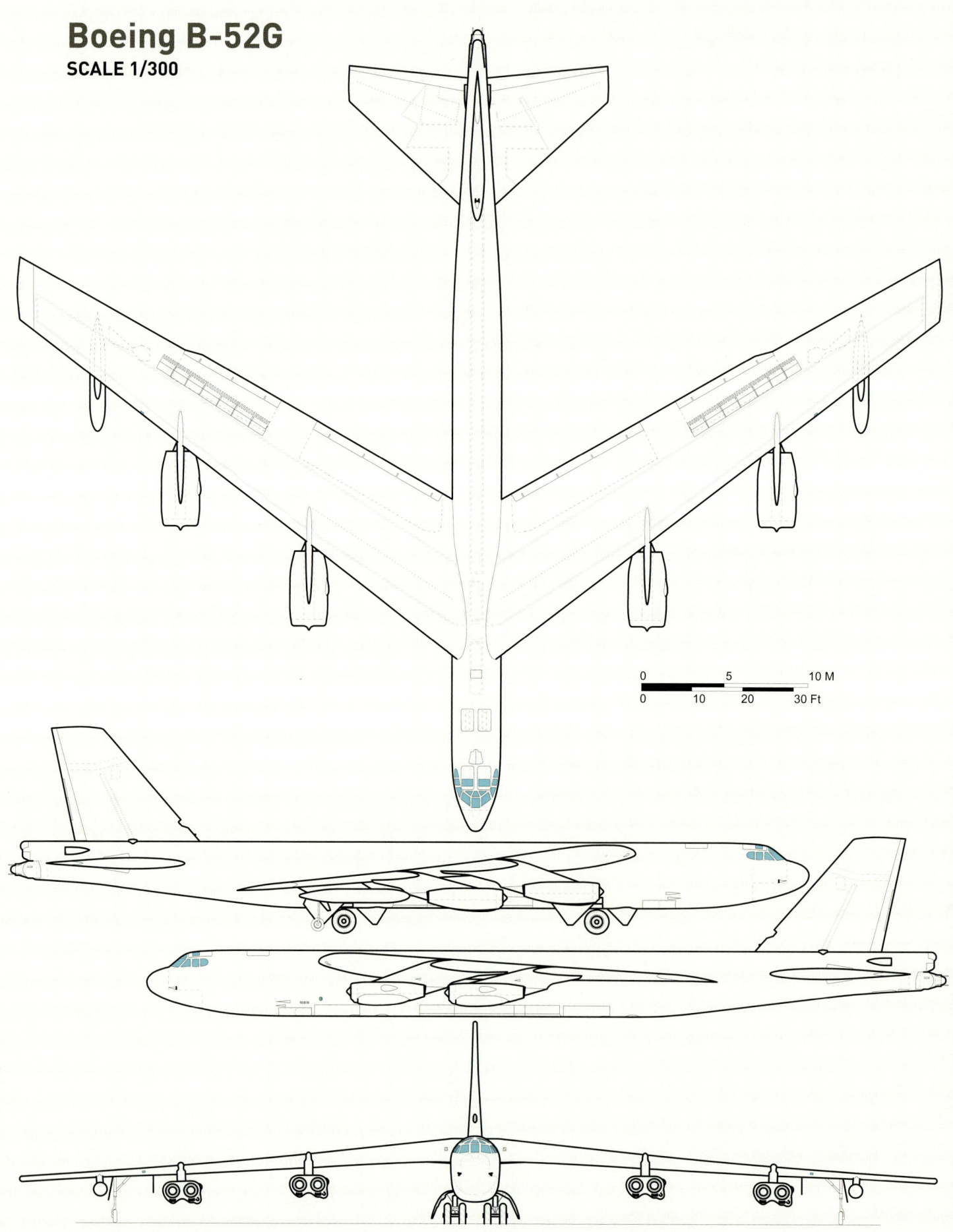

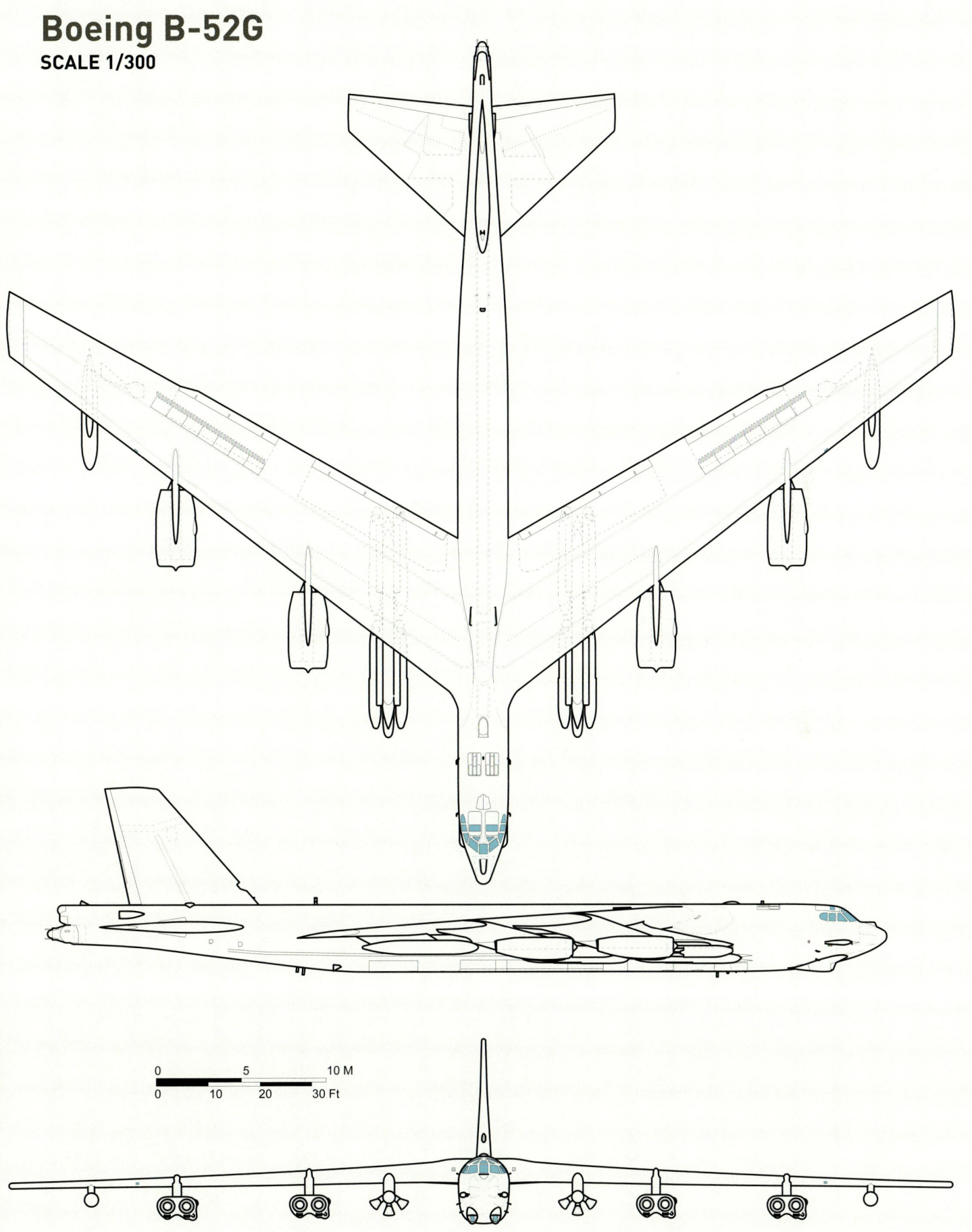

Boeing B-52G
SCALE 1/300

turret). Further modifications and additions appeared elsewhere on the G and H models: the AN/ALQ-153 rear-looking pulse Doppler radar system (used to distinguish missiles and aircraft pursuing the B-52) was added approximately halfway up the vertical stabilizer, and four AN/ALE-24 chaff dispensers were added to the underside of each wing in the position formerly occupied by the ailerons actuators. Rivet Ace also included a 40in extension to the rear fuselage, lengthening the tail in order to install further equipment.

B-52G retirement began in 1989. Sent to Davis-Monthan, 'retired' aircraft were cut or chopped into large sections that could be easily identified by Soviet reconnaissance satellites. However, the following year surviving B-52Gs began to see action again. When Iraq invaded Kuwait, kicking off the Persian Gulf War, only the B-52G and B-52H remained in service. And while the Gulf War was seen at the time as nearly a video game, given how high-tech the American weaponry used was – stealth fighters, cruise missiles guided by GPS, bombs flying through windows, daily briefings filled with the latest exciting footage taken with night-vision scopes from great distance etc. – the B-52 was ready.

Now in their thirties, older than many of the crew who flew in them, the B-52s had been updated with the latest tech and were supported by the latest systems. When the United States kicked off Operation Desert Storm in mid-January 1991, F-117 stealth strike aircraft took out much of the Iraqis' air defence capability, and seven B-52Gs flew all the way from Barksdale Air force Base in Louisiana in order to launch 35 cruise missiles at Iraqi military communications sites and power systems. It was a testament to the effectiveness of the US Air Forces in taking out air defence systems that none of the B-52s in that raid, nor in any of the raids that followed, were shot down. Several did take damage from enemy action however, and one crashed in the Indian ocean after a raid due to electrical system failures.

Following the initial raid, B-52s were used to drop large numbers of conventional unguided bombs onto Iraqi troop and weapons concentrations during the Gulf War. In January and February 1991, some 1,645 strike missions were flown, doing great damage not only to Iraqi military infrastructure but also to the psyche of many Iraqi service personnel who witnessed and survived the B-52G raids.

The Gulf War was a great success for the B-52G, but just a few months later came perhaps one of the worst events in the history of the aircraft – in December 1991, the Soviet Union dissolved. While this was unquestionably a good thing for the world as a whole, for the B-52 it was the loss of a major reason for its existence. Unsurprisingly, retirement of the B-52G continued and was complete by 1995.

Boeing B-52H (Model 464-260)

The B-52H was the final version to see production and is the only version still in service. The design was initiated in January 1959, and as with the B-52G, the type was manufactured entirely in Wichita, Kansas. A total of 102 were built with the first example flying on March 1, 1961. The final B-52H was delivered to Minot Air Force base on October 26, 1962. As such, at the time of writing the youngest B-52 in service was approaching 60 years old… likely three times as old as some of its crew.

The most obvious change from the B-52G was the use, for the first time in the B-52's history, of entirely new engines. Gone were the reliable but inefficient J57 turbojets. In their place were Pratt & Whitney TF33-P-3 turbofans. These engines, the military version of the JT3D engines used on such early jetliners as the 707 and the DC-8, were derived from the J57 but with the first three stages of the compressor replaced by a two-stage fan with a 1.4:1 bypass ratio. This made the engine larger in overall diameter and the resulting nacelle draggier… but thrust was increased and fuel consumption decreased. Water injection was no longer needed to boost thrust at takeoff, saving not just weight but also operational headaches. The improved engines greatly increased the unrefuelled range of the aircraft.

Another visually apparent change from the B-52G was the removal of the four .50cal machine guns in the tail. In their place was a single 20mm M61 'Vulcan' Gatling gun. Capable of a respectable 6,000 rounds per minute, this weapon – eventually the standard used on nearly all USAF and US Navy fighters – had far greater hitting power than the earlier machine guns, as well as greater range. As with the B-52G, the gunner was located in the forward crew compartment. The cannon was equipped with 1,242 rounds – enough for perhaps 12 and a half seconds of firing (short bursts recommended).

The gunner used the AN/ASG-21 fire control system, reliant solely upon radar to aim. But a decision was made in 1991 to remove the tail gun entirely from the B-52H. The days when the threat to a bomber was a fighter wandering up behind it, guns blazing, were long over; realistic threats came in the form of surface-to-air missiles and air-to-air missiles, neither of which the tail gun could really counter. So the guns and the gunners were removed, the hole in the tail covered with a flat plate.

The B-52H was intended from the outset to be equipped with GAM-87 Skybolt air launched ballistic missiles. When these were cancelled, the AGM-28

Boeing B-52H
SCALE 1/300

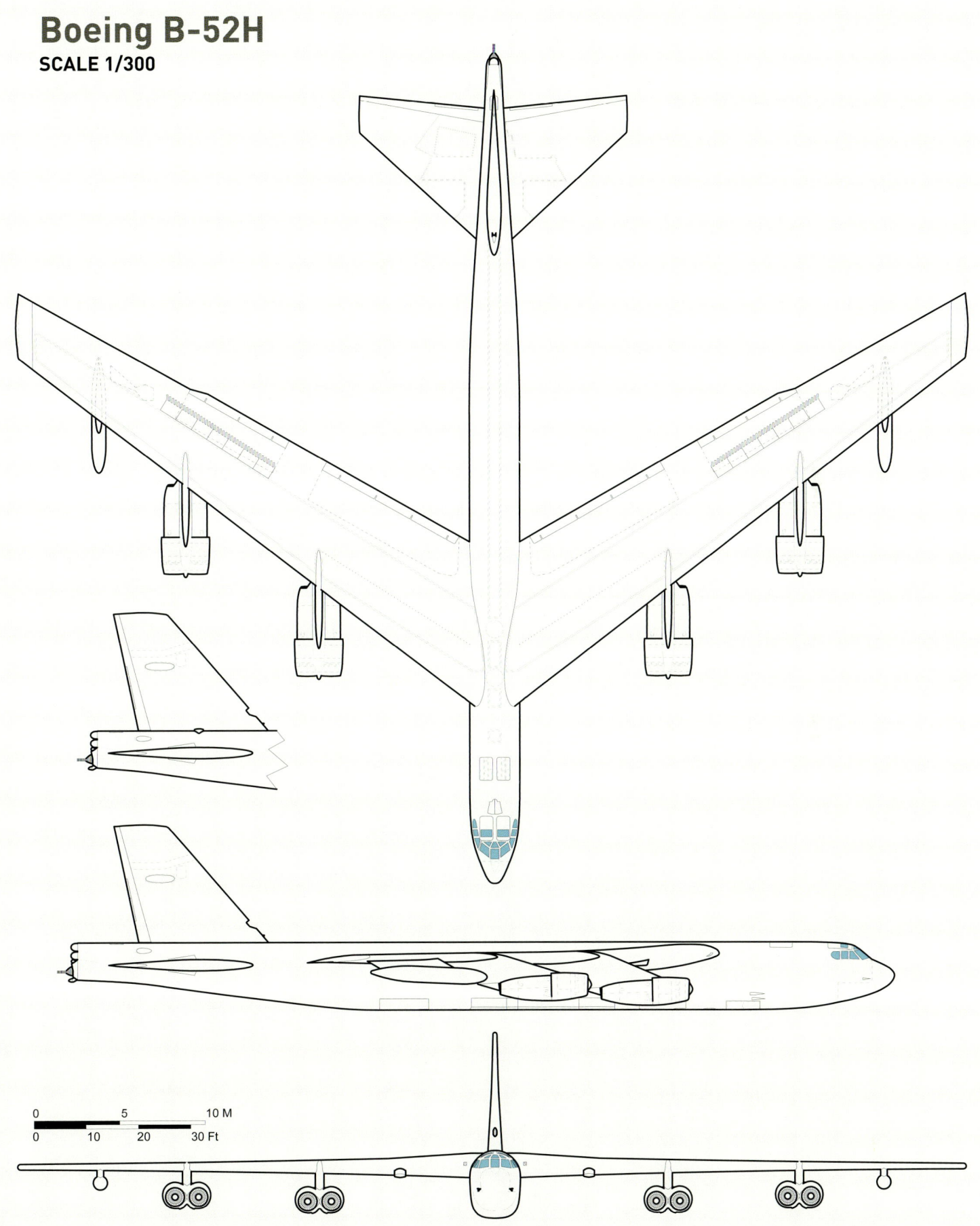

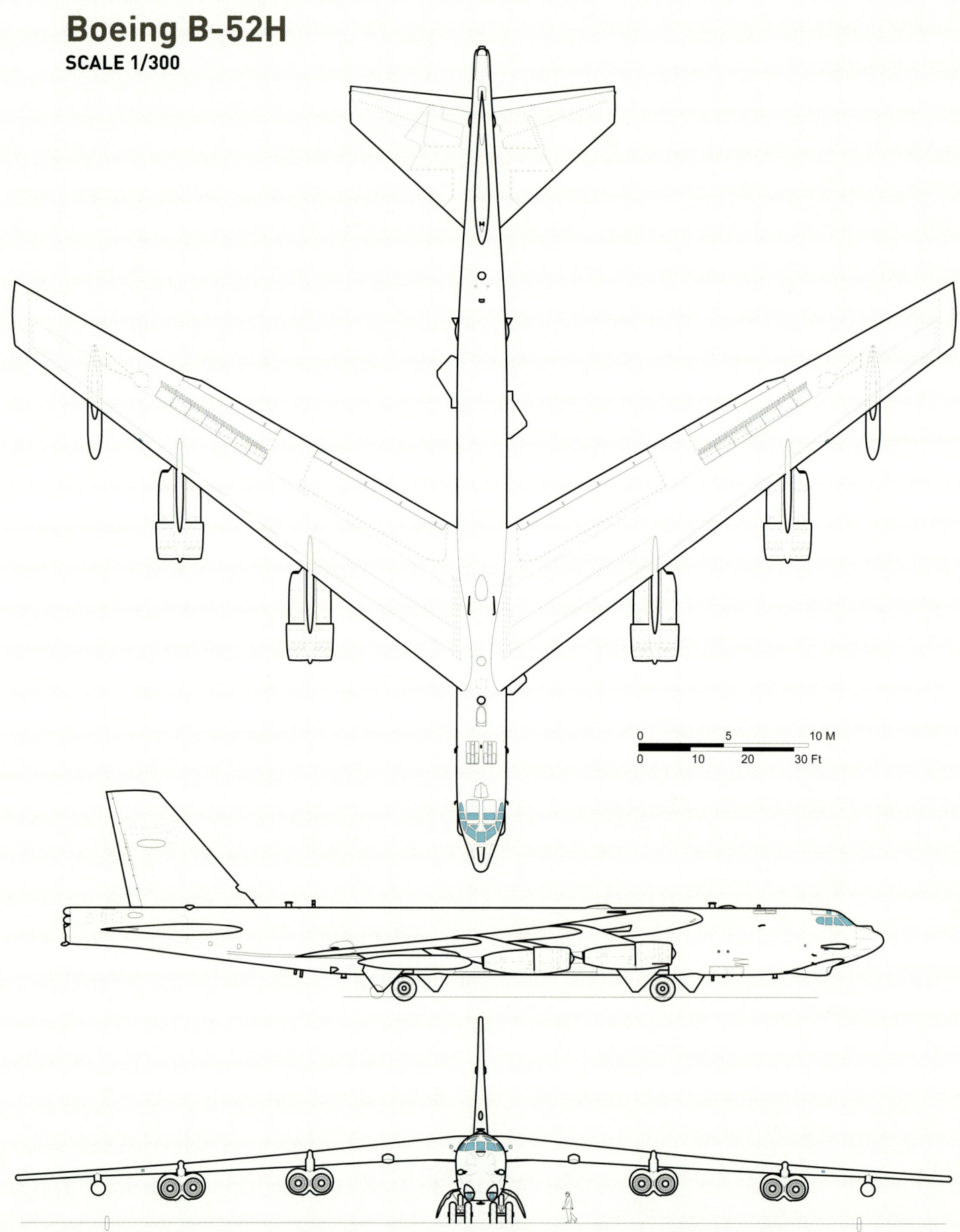

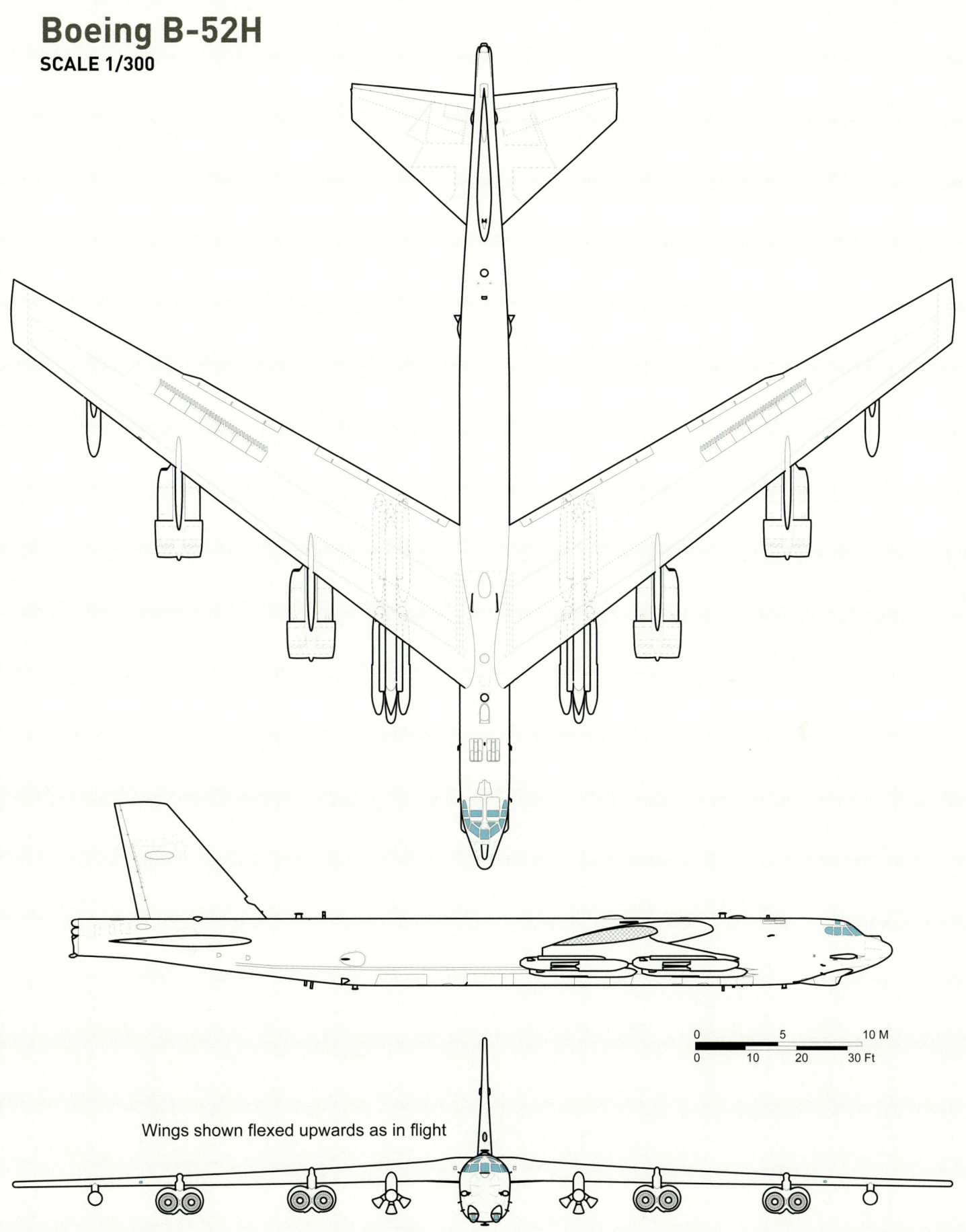

Boeing B-52H

SCALE 1/300

0
5
0
10
20
30 Ft
AGM-86B
B-61
B-83
GBU-57 MOP

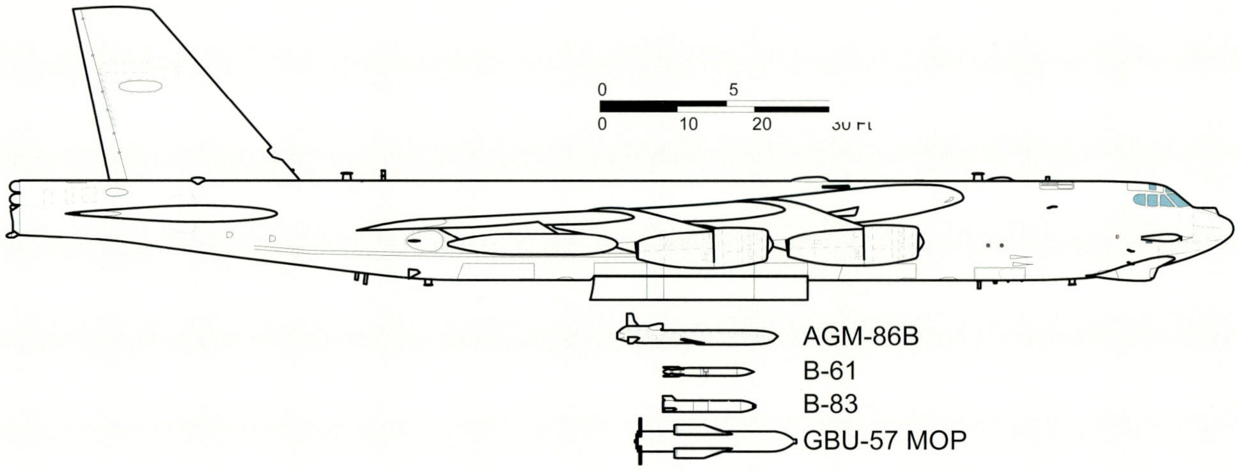

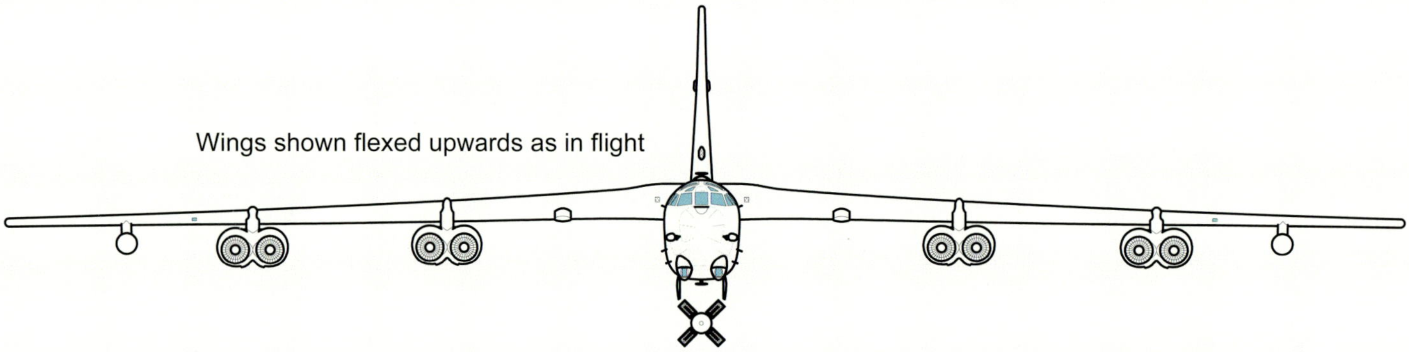

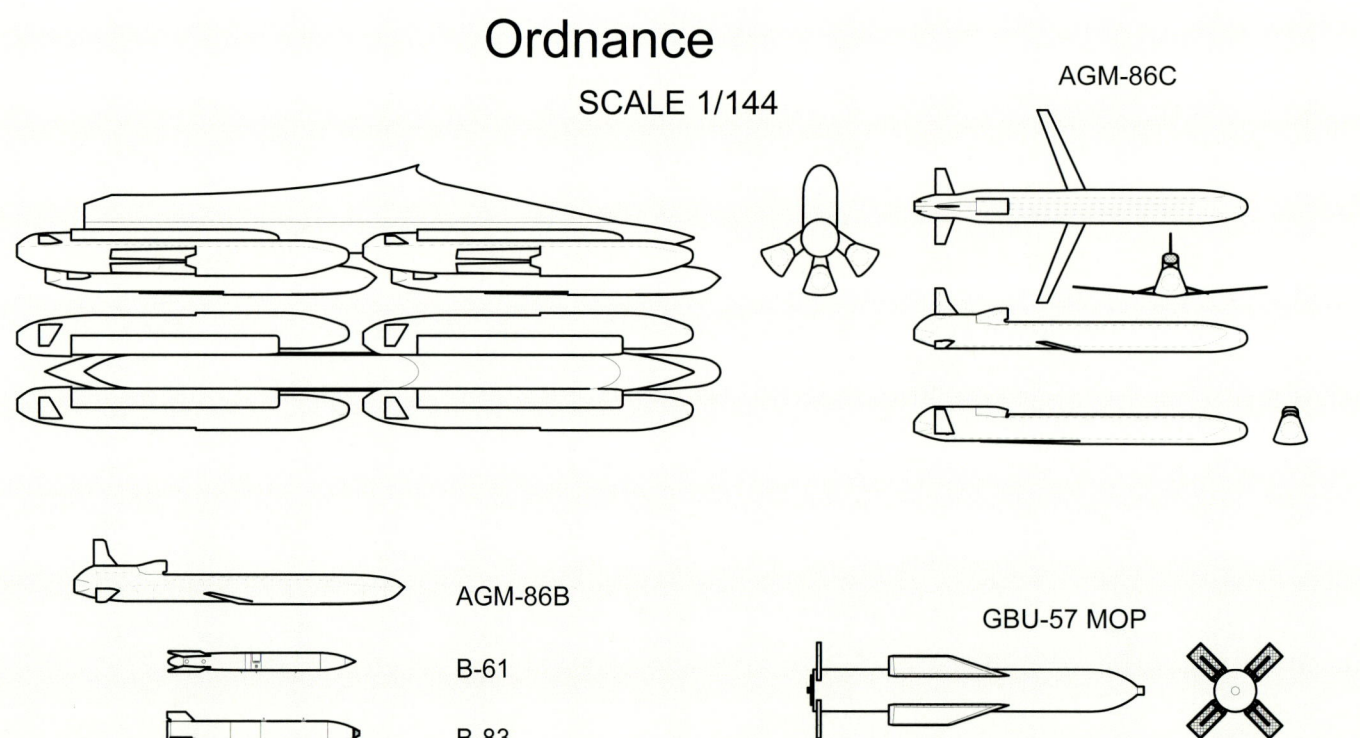

Wings shown flexed upwards as in flight

Ordnance

SCALE 1/144

AGM-86C

AGM-86B
B-61
B-83

GBU-57 MOP

Hound Dog was the fallback until the AGM-69A SRAM became available, as well as nuclear bombs in the bomb bay. Over time a large range of other weapons were developed for use on the B-52H – including the AGM-86 Air Launched Cruise Missile, the AGM-129 Advanced Cruise Missile, the GBU-57 Massive Ordnance Penetrator, a wide array of conventional unguided and laser guided bombs (guided by a laser beam projected from the AN/AAQ-28(V)1 'Litening II' or AN/AAQ-33 'Sniper' pod suspected beneath the starboard wing), glide bombs such as JDAMs, JSOWs, and SDBs, the AGM84 Harpoon anti-shipping missile, the AGM-158A Joint Air-to-Surface Standoff Missile (a turbojet powered cruise missile akin to the ALCM, but smaller) and the ADM-120 Miniature Air Launched Decoy (a modernized 'Quail', in essence).

The B-52H was originally delivered with the same single-piece, somewhat blunt nose that graced the B-52G. But it was replaced with the same pointier nose and electro-optical vision system that the B-52G received, along with the electronic warfare systems and blisters of the 'Rivet Ace' update. However, the B-52H was granted even more bumps and protrusions. Atop the fuselage between the wings (about two-thirds of the way back along the chord) and along each side of the aft fuselage are blisters for the AN/ARR-85 Miniature Receive Terminal, a communications system designed to survive in a nuclear war environment.

Atop the fuselage between the wings (just aft of the leading edges) is a circular GPS receiver; ahead of that is a raised fairing containing the AN/ASC-19 AFSATCOM (satellite communications system). Along the underside are a number of blade antennae for electronic countermeasures.

The total population of operational B-52s by the mid-1990s had been reduced to 85 B-52Hs and Strategic Air Command itself was done away with on May 31, 1992. The B-52 now shared nuclear strike duties with the Rockwell B-1 and the new Northrop B-2 – but there seemed to be nobody around to drop nuclear weapons on. After a brief lull in the wake of the Soviet Union's dissolution, a genocidally bloody civil war erupted in the former Yugoslavia.

Beginning in 1992, Serbia, Croatia and Bosnia decided that the time for fellowship was over and into 1995 they fought the first real war Europe had seen since the Second World War. The hatred didn't end with the Dayton Accords either; soon enough Serbia was fighting to retain the city-state of Kosovo. NATO decided that enough was enough and sent in armed aircraft and ground forces to put a stop to it. On March 24, 1999, NATO began air operations against Serbian forces. This involved the B-52H, the B-1 and the B-2. The B-52s launched cruise missiles, rained down Mk 82 unguided bombs and even dropped leaflets warning of further B-52 raids. As in Iraq, troop concentrations and columns of armour proved vulnerable targets for B-52s dropping simple gravity bombs. Serbia finally threw in the towel on June 10, 1999, bringing the air campaign to an end.

Once again, the end seemed near for the B-52; nobody left to nuke, and few enough targets to carpet bomb.

And then came September 11, 2001. The attacks on the Pentagon and the World Trade Center led directly to one war and indirectly to another. Operation Enduring Freedom began on October 7, 2001, with B-52s operating from the British island of Diego Garcia in the Indian Ocean bombing al-Qaeda and Taliban sites throughout Afghanistan. A new capability was on display: the turbofan engines of the B-52H – at this point around 40 years old – allowed it to loiter, heavily laden, for extended periods at high altitude. Using co-ordinates provided by forward air controllers or spotters on the ground, the B-52 could pick out individual targets and put GPS guided bombs right onto them… and it could keep that up for hours, one bomb at a time out of a clear blue – or black – sky.

And then came Operation Iraqi Freedom. Saddam Hussein had worn out his welcome; while he was not involved in the September 11 attacks, nobody wanted to give him the opportunity to be an aggressor again. With contradictory and confused intelligence on whether or not he was working on weapons of mass destruction again, the controversial decision was made to go ahead and make sure that he wasn't by removing him from power. So operations began on March 19, 2003, using a much smaller force of aircraft than had been deployed more than a decade earlier. Nevertheless, the combination of B-52s, B-1s and B-2s, together with strike fighters of various breeds, gradually erased the Iraqi military. The war was officially over by mid-April; Saddam was captured in December. While counter-insurgency operations continued for years, the bomber's role was over relatively quickly in Iraq.

Not so in Afghanistan. There were few massed troop concentrations, no columns of armour, no major industrial infrastructure to turn into splinters and gravel. But there were targets aplenty in a war that dragged on far too many years (culminating with Afghanistan falling to the Taliban again, negating two decades of effort, expense and death). The B-52 was used in fits and starts throughout the nearly 20-year conflict, with sorties right up to the end.

CHAPTER 5: B-52 Projects

Boeing Models 464-72 and 464-74
The B-52 began as a turboprop-powered bomber but evolved into a turbojet-propelled design before the configuration was finalized. Turbojet engines had overtaken turboprops for well and good, it had seemed. But even after the Model 464-67 had settled the design, there were some in the Air Force as late as April 1950 who felt that turboprops were still worth considering for the future. B-52s, it was felt, could be retrofitted with turboprops; they would be slower, but range would be increased.

Either as a response to Air Force interest or from sheer cussedness, Boeing designers created Models 464-72 and 464-74 in February 1950, using the airframe of the Model 464-67/XB-52 married to turboprop engines. Both configurations had four turboprop engines derived from the Pratt & Whitney JT 3A turbojets, but utilized them differently.

The configuration of 464-72 was very similar to that of the B-47 derived Model 450-30-10 (designed at the same time; it's probable that the concepts were produced side by side) with engines mounted ahead of the wings in long, slim nacelles that terminated in pusher propellers. The engines were connected by a drive shaft that ran aft to 15ft diameter four-bladed supersonic propellers. These would have had all the problems described for the XB-47D and related designs using supersonic propellers… damage to nearby aircraft structures as well as converting ground crews brains to Silly Putty.

Model 464-74 used the same engines and basically the same propellers, but used them as tractors rather than pushers. Again, the engines were mounted ahead of the wings, now connected almost directly to the propellers. The nacelles were mounted to the wing leading edge and the prop spinners doubled as the engine air inlets. Both Models 464-72 and 464-74 had exhaust nozzles on the upper surface of the nacelles. Each also had four 'slipper' tanks projecting forward from the outer wings.

No performance data is available for either design. It may be assumed that maximum speed was less than that of the jet-equipped B-52, while range ought to have been greater. But since inflight refuelling quickly became routine for the B-52, the added range provided by turboprops would seem to have been a little superfluous while the reduction in top speed could have made the aircraft much more vulnerable.

Extended wings
Just as Boeing studied extended wings and floating wingtips for the B-47, these were also studied for the B-52. Several of the resulting designs are unsurprisingly very similar to their B-47 counterparts. In November 1949 the Bombardment Branch of the Engineering Division at Wright Field requested that Boeing study floating wingtip tanks as a means of extending the range of the B-52, then falling somewhat short. Boeing did so, and seems to have thrown the B-47 into the mix.

Boeing Model 464-79-0
A design known from a single drawing and some data, Boeing Model 464-79-0 was a modification of Model 464-67 (complete with the originally intended tandem cockpit of the XB-52) with floating wingtips. The wing was impressive, with a span 60ft greater than that of the standard B-52. There was a hinge roughly 64.5ft out from the centreline on either wing (about 5ft was chopped off the original wingtips) and 35ft wingtip extensions were added. This outer wing panel was free to 'float' on the hinge, flapping up or down 35°. This would have granted Model 464-79 the same aerodynamic benefits described for Model 450-36-10.

While the B-52 had a single external fuel tank on each wing outboard of the furthest engine nacelles, Model 464-79-0 had two such tanks, one on either side of the hinge line. The tanks furthest out were substantially larger than the inboard tanks and were fitted with horizontal stabilizers at their tails, with an area of 73.5sq ft each. Model 464-79-0 had the B-52 fuselage landing gear as well as similar 'outrigger' stabilizing gear just inboar d of the smaller inboard external fuel tank. But it also had dual-wheel landing gear integrated into the outboard fuel tanks to keep them off the ground.

Boeing Model 464-149
Model 464-149 was seemingly Model 464-79 but with a permanently fixed outer wing panel. The wing was simply extended to a span of 230ft. The new outer wing was not attached at the same line as the hinge on the 464-79, parallel to the direction of motion; instead

Boeing Model 464-72
SCALE 1/350

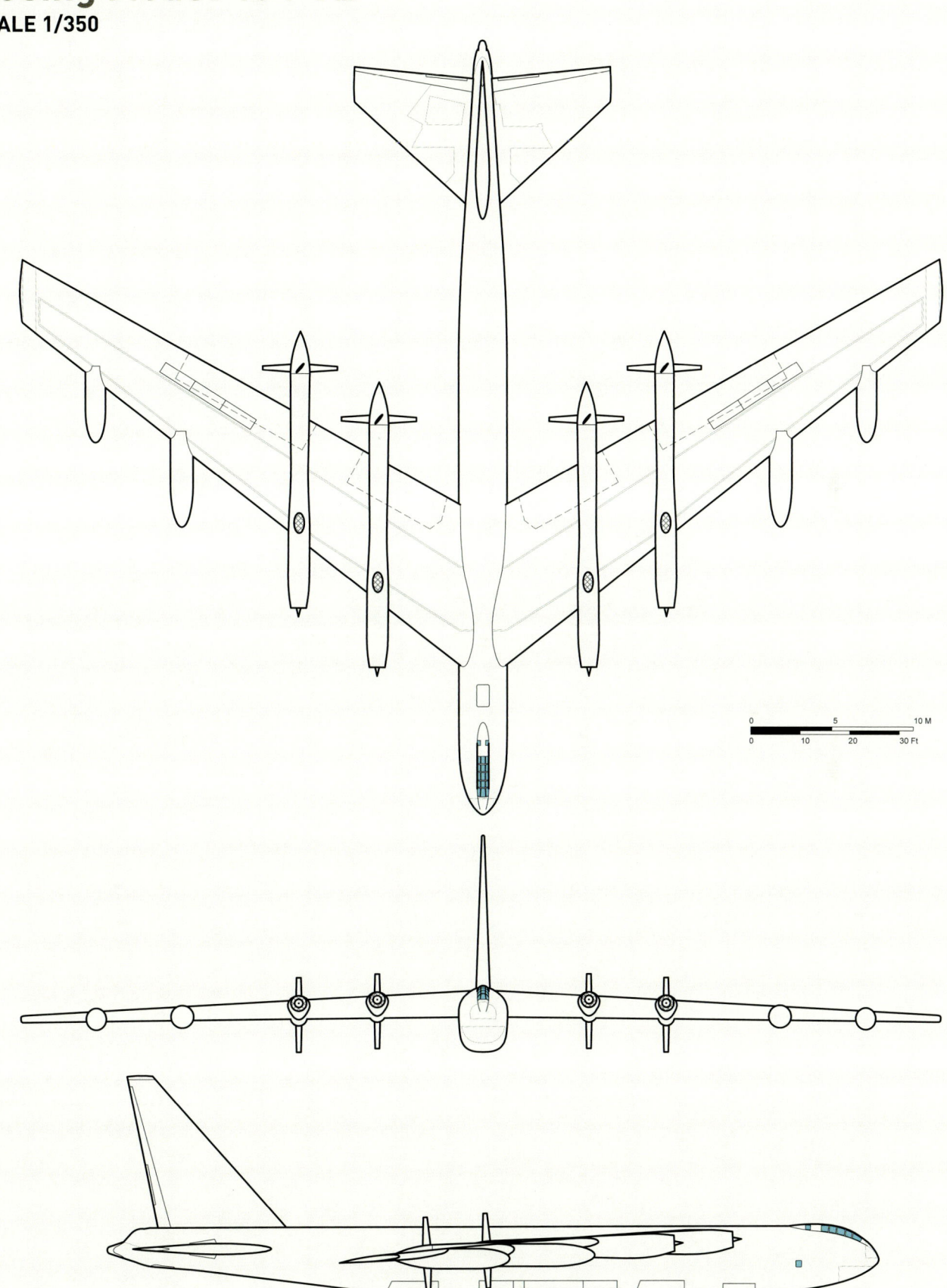

Boeing Model 464-74
SCALE 1/350

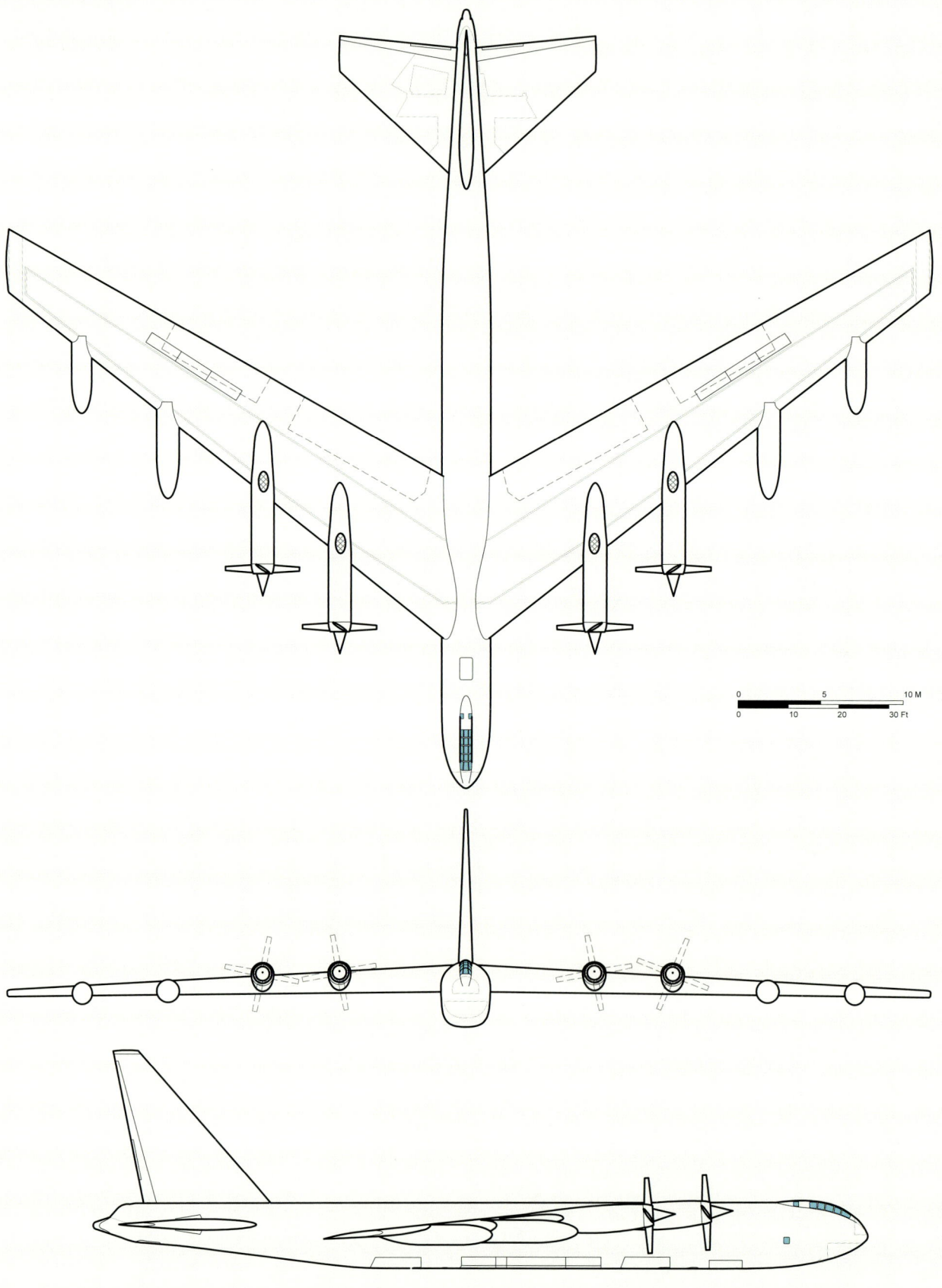

Boeing Model 464-79-0
SCALE 1/400

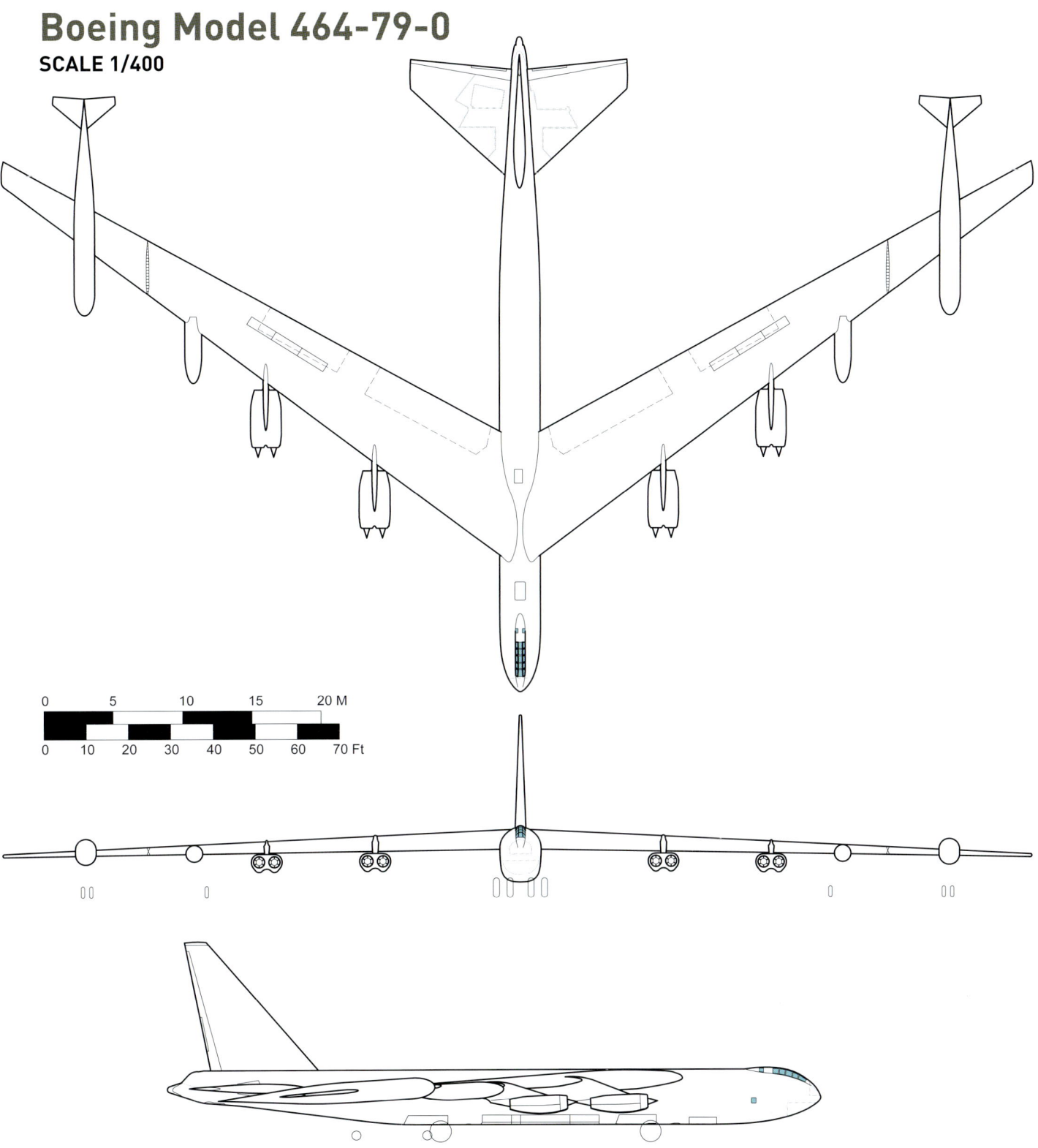

the wing was cut at an angle and the new extension spliced on. Obviously, the intention almost certainly was that the wing would be built that way from the ground up rather than a shorter wing being surgically altered; a modified wing would always be heavier and more structurally suspect than one built correctly the first time.

Gone were the two differently sized external fuel tanks; now there was only a single mid-sized tank permanently affixed to the underside of the outer wing. The normal B-52 outrigger gear was used for ground stability and to keep the nearby external fuel tanks from contacting the runway. No performance data is available.

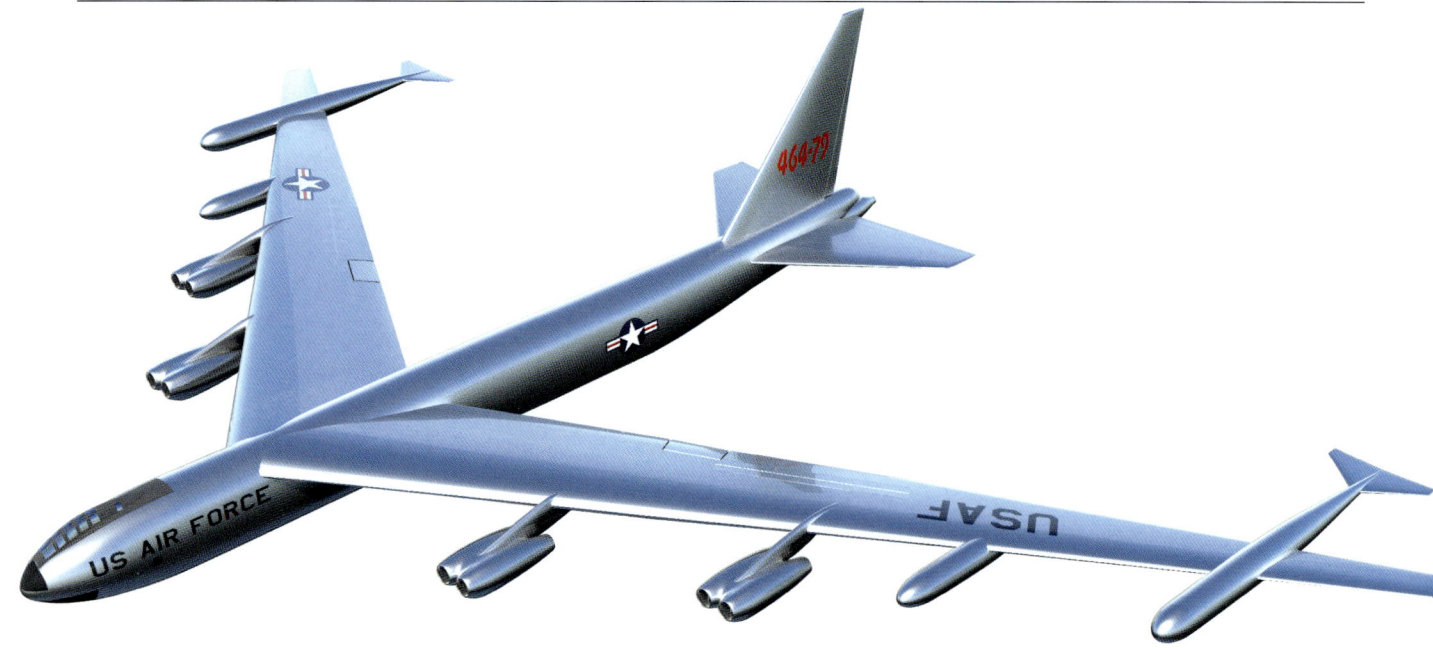

Boeing Model 464-EXT 1 and 464-EXT 2

Two designs known only from general arrangement diagrams are Model 464-EXT 1 and Model 464-EXT 2 – slight modifications on the theme established by Model 464-149. 464-EXT 1 was very similar to 464-149, but with a somewhat larger external tank under the outer wings, held by a short pylon, rather than integrated into the wing lower surface. The tank came complete with its own two-wheel landing gear to assure a safe ride on the runway. The outrigger gear was deleted from the wing itself.

464-EXT 2 had two somewhat smaller external tanks attached by pylons below the outer wings. These also had their own landing gear. Once again there is no performance data for these designs.

Boeing Model 724-15

Discussion of floating wingtips for the B-52 brings up another Boeing design which used that concept. While not derived in any meaningful way from the B-52, Model 724-15 was a design meant to supplant the B-52; it was Boeing's initial entry into the competition that led to the B-70 Valkyrie Mach 3 bomber.

The USAF had begun a programme to develop a new chemically powered (specifically non-nuclear powered) strategic bomber in 1954. The Weapons System 110A concept was meant to eventually replace the B-52 with something far harder for the Soviets to kill. The WS-110A needed to be able to operate from the continental US and strike targets within the Soviet Union. Performance requirements included not only the range to strike far distant targets, but also a minimum cruising speed of Mach 0.9 and a Mach 3 dash for at least 1,000 nautical miles – preferably 2,000 miles.

In July 1955, the USAF held a WS-110A bidders conference (also incorporating the WS-125A nuclear powered bomber studies) at Wright Field; only North American Aviation and Boeing decided to submit WS-110A proposals. Both companies presented their designs to date in mid-1956, and both featured the same gimmick: short-span supersonic bombers, optimized for Mach 3 flight, were married to long, high aspect ratio jettisonable wing panels equipped with large aerodynamic fuel tanks.

Boeing's Model 724-15 of April 1956 was of remarkably similar in layout to its North American competitor. A large supersonic bomber with four engines and highly swept rather short-span wings was equipped with two long-span wing extensions, each with its own fuel tank. The wings themselves were also full of fuel, giving the bomber an extra 22,400 gallons of fuel per wingtip for takeoff and long-range subsonic cruise as well as triple its own basic wing area.

When the time came for high speed dash, the wing panels would be simply jettisoned, leaving a clean high-speed configuration capable of triple-sonic flight. The jettisonable wingtips – similar in concept to those used on the Boeing Model 464-79-0 – had their own model number, 724-1003, and were large structures in their own right, about two-thirds the size of a B-47.

Model 724-15 had a cockpit canopy somewhat similar to that of the final B-70 in that a fairing would tilt up or down to present either a smooth low-drag surface or improved forward vision for takeoff and landing. The nose as a whole was fixed, unlike the drooping noses of most supersonic transports.

As with the North American design, 724-15 was a clumsy solution to the problem. It required runways that were not only long but also incredibly wide; the span from centreline to centreline across the fuel pods was 172ft. Runways would have needed to be around 200ft wide. Additionally, in order for aircraft to simply

Boeing Model 464-149
SCALE 1/400

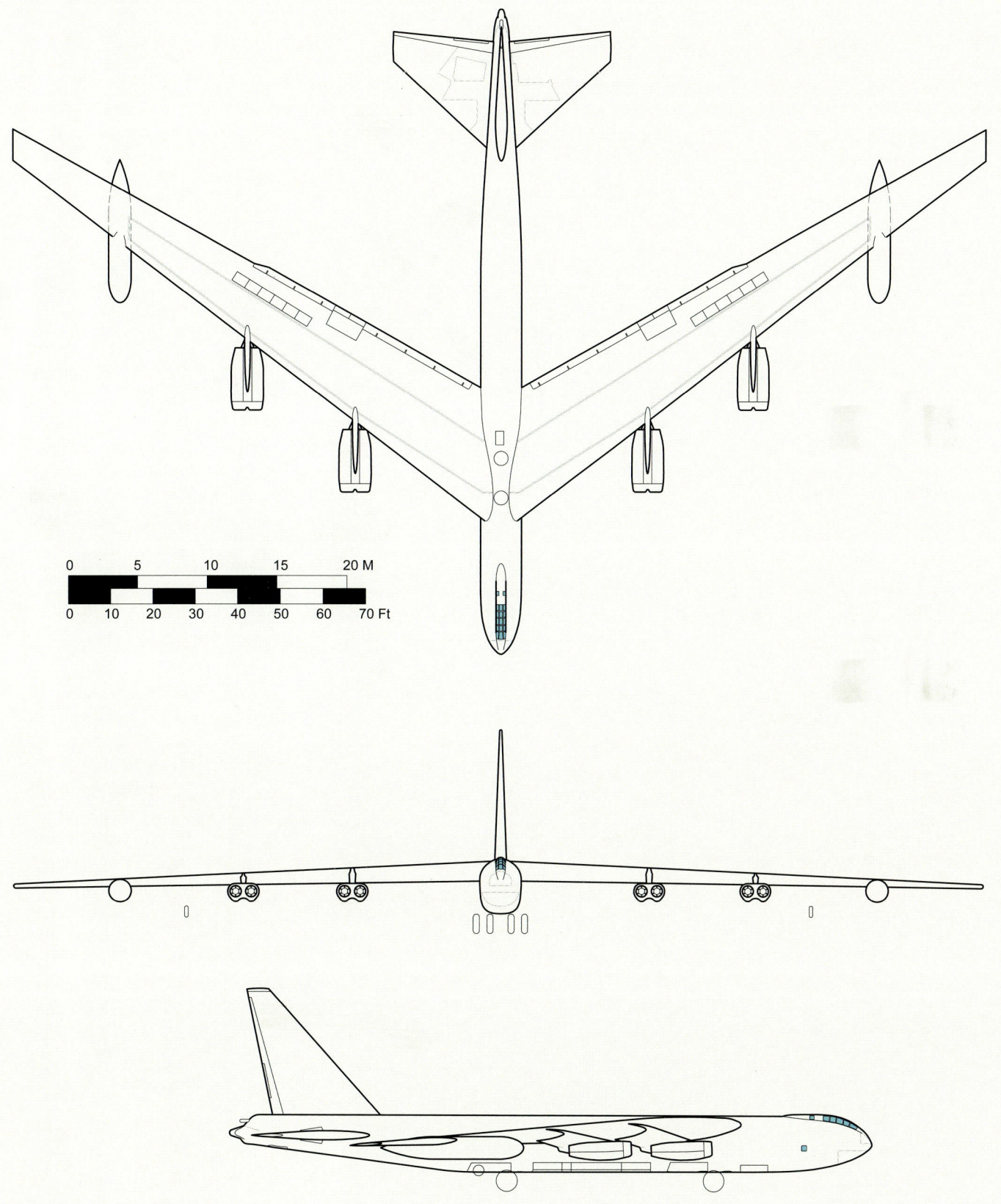

Boeing Model 464-EXT #1
SCALE 1/400

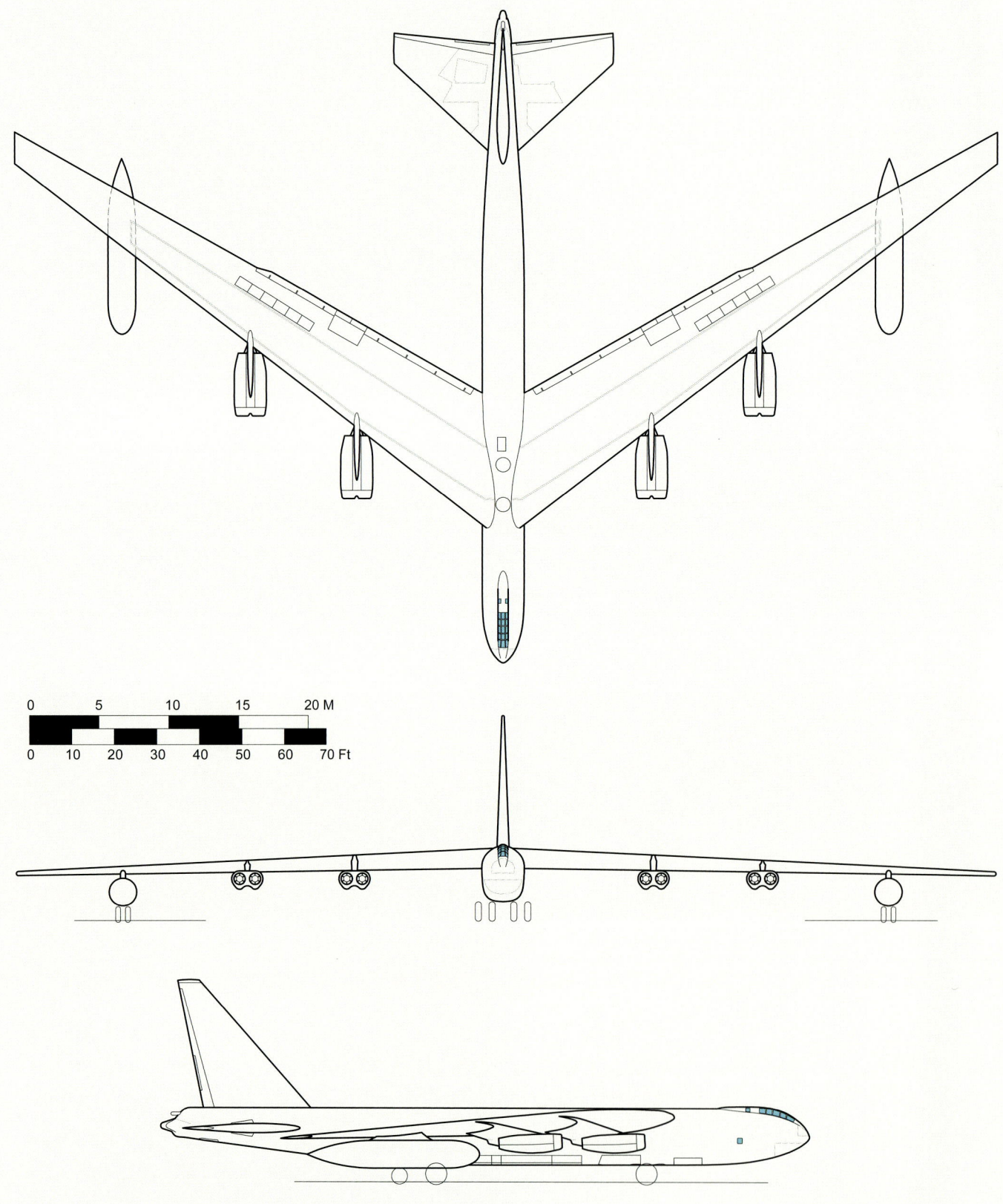

Boeing Model 464-EXT #2
SCALE 1/400

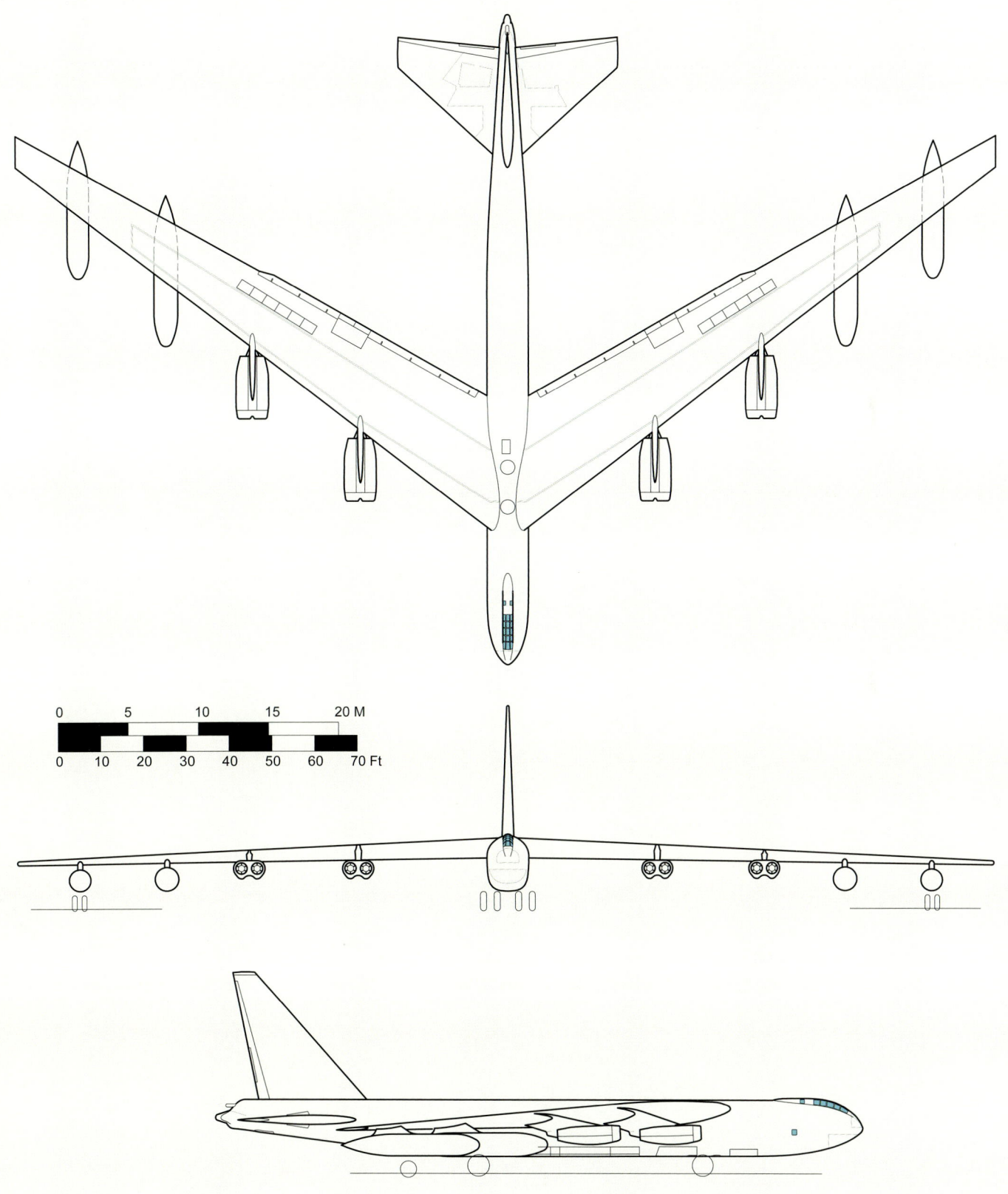

Boeing Model 724-15
SCALE 1/450

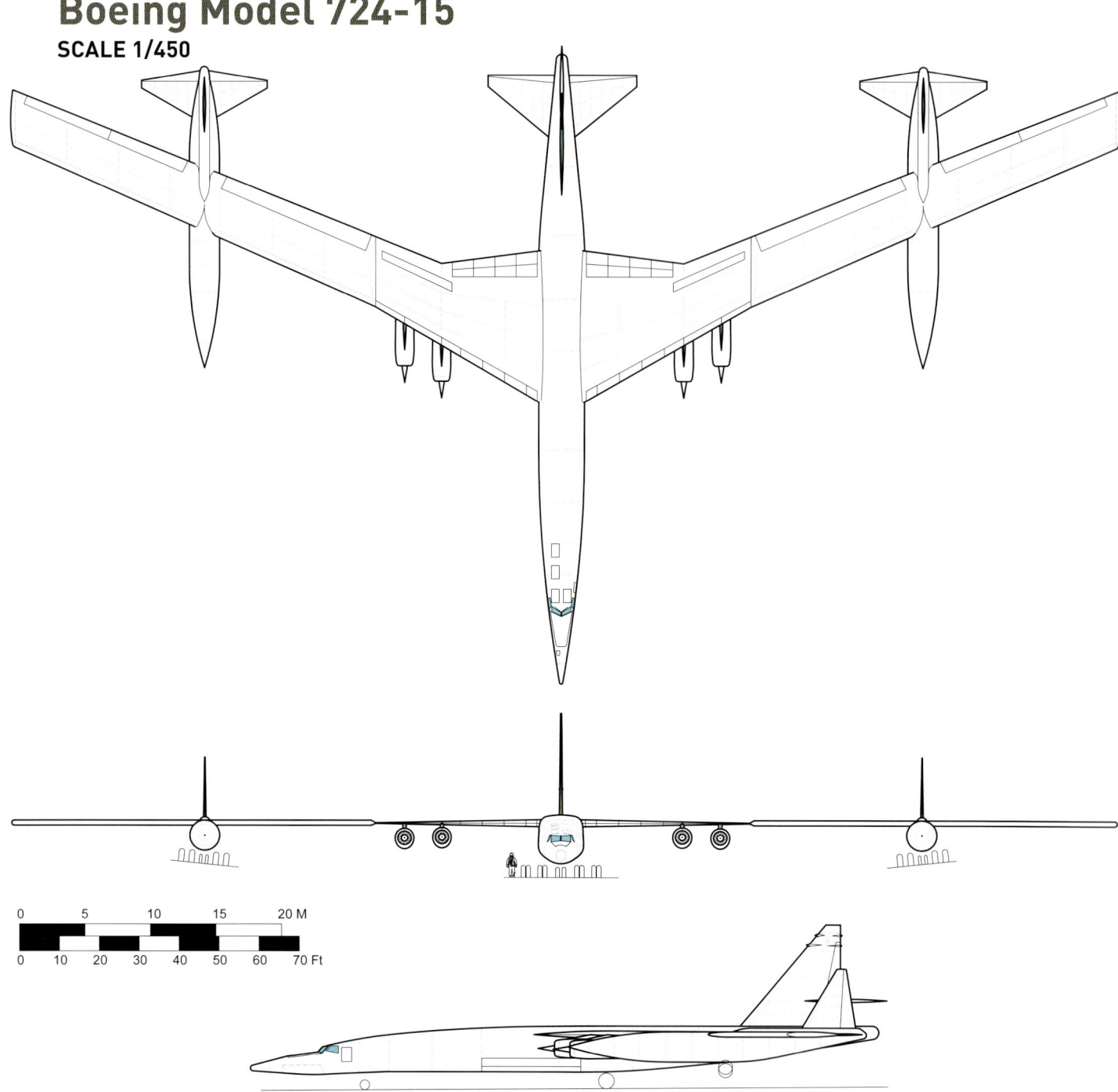

practice at supersonic speeds, the large and expensive jettisonable wingtips would have had to be expended every time.

The baseline payload was a single undefined 10,000lb 'special' weapon – a large hydrogen bomb. Other weapons loads were considered, including a single 25,000lb hydrogen bomb. During a standard mission, Model 724-15 would cruise at 460 knots for 1960 nautical miles at an altitude of 32,100-36,100ft. It would then jettison the wingtips and begins its dash, cruising at 1725 knots while ascending to 63,100ft. After dropping its weapon it would turn around and fly another 1,070 miles at 1,725 knots while climbing to 67,600ft. Then it would descend to 37,800ft and slow to 518 knots for the subsonic cruise home.

As with the North American submission, the Boeing design was rejected outright. General Curtis LeMay considered the designs to be more like three-ship formations than proper aircraft. As a result, both North American and Boeing went back to the drawing board.

Boeing Models 464-238 and 464-239

Another way to improve the B-52's range performance was to simply scale up the wings. This

Boeing Model 724-15
Model 464-79-0
Model 450-36-10

SCALE 1/450

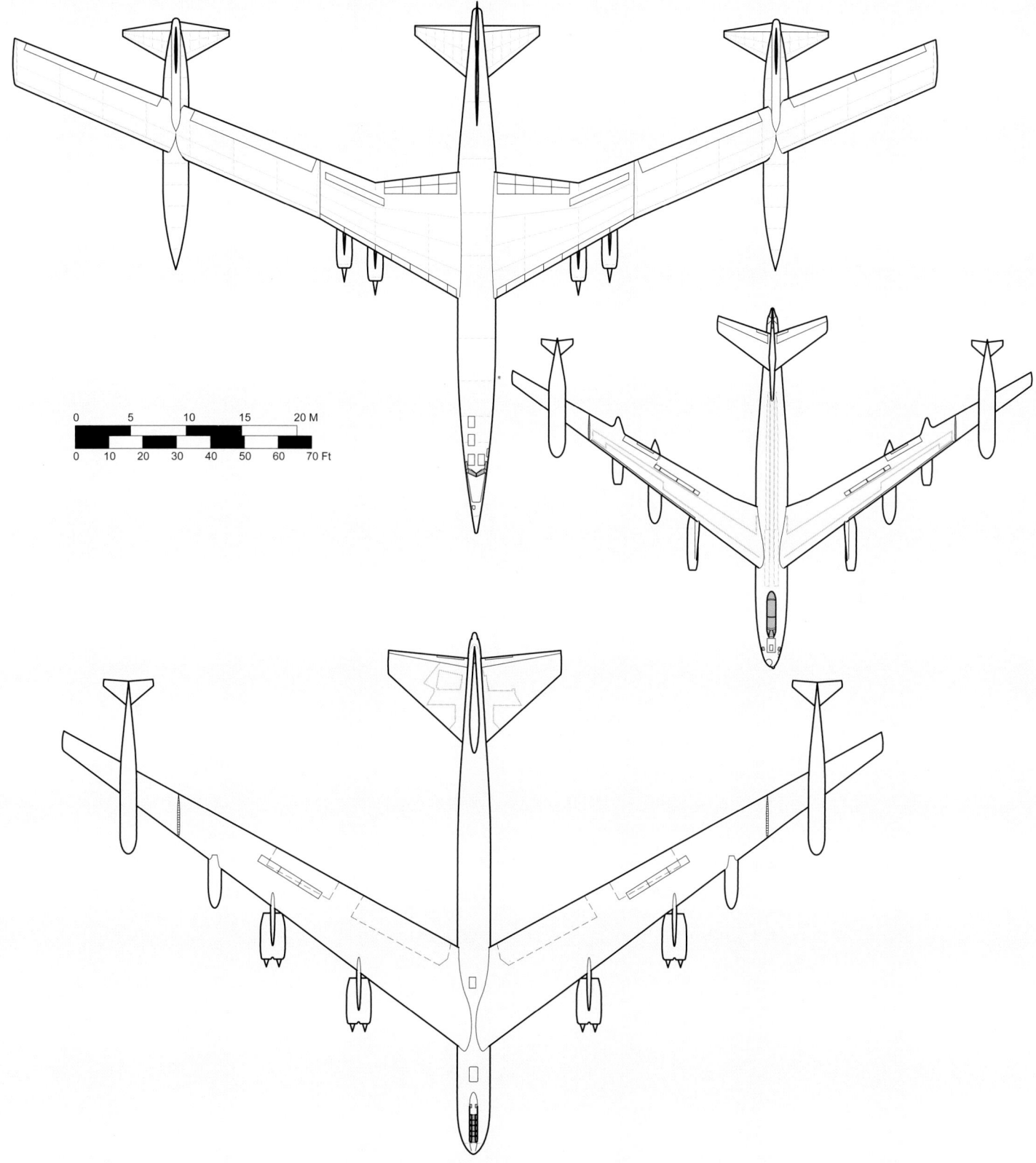

Boeing Model 464-238
SCALE 1/350

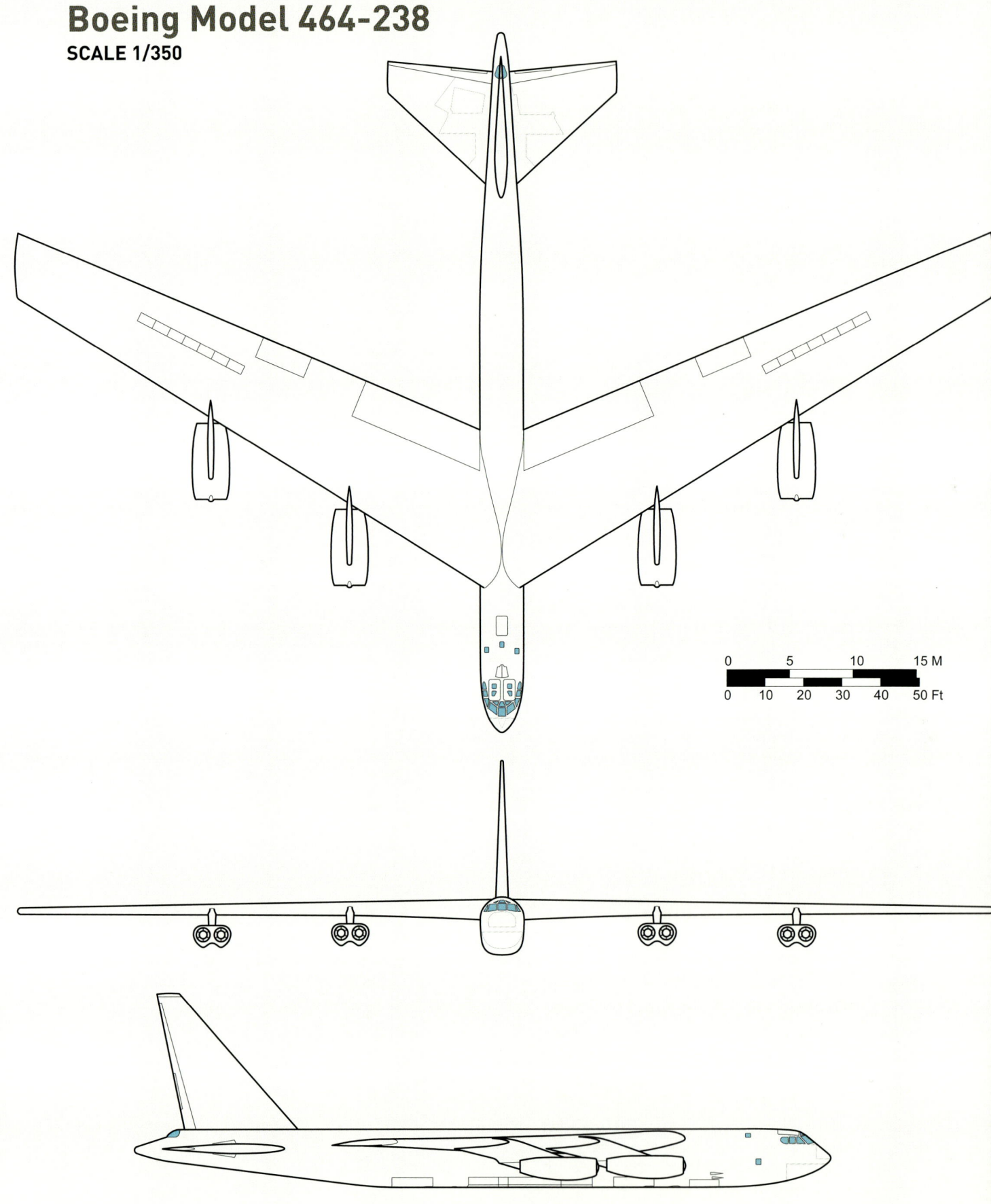

was the aim of Models 464-238 and 464-239, known to this author solely through two top-view diagrams dated April 1954. The 464-238 had a wing scaled up to a span of 218ft and a wing area of 5,600sq ft; otherwise the fuselage design seems to have been largely unchanged from the B-52A. The engines were changed to Wright J67s.

Model 464-239 had a wing of the usual 185ft span, but the chord was widened to increase wing area. The standard B-52 had a theoretical chord of 30ft 11in at the centreline; the Model 464-238 was 36ft 6.7in and the Model 464-239 was 37ft 1.2in. No further data is available. Given the high dash-number of these two aircraft, it is clear that Boeing's designers were kept busy dreaming up modifications and alternate versions of the B-52, though most have faded into the mists.

Boeing Model 713

Boeing used the Model 713 designation as a catch-all for advanced bomber concepts. Dozens of designs were included, beginning in June of 1954; many were supersonic, many were nuclear powered, and more than a few were both. Those designs fall outside the scope of this work, but several were clearly inspired by the B-52, if not actually directly derived from it. The configurations of the Model 713-1-133 and 713-1-138 series were much like the B-52… only more so. In some ways, much more so.

Model 713-1-133 came in three variants, A, B and C, all dating from mid-December 1954. They, and the apparently largely identical Model 713-1-138 from February 1955, shared a common wing, fuselage and tail, all reminiscent of the B-52. The cockpit contours and fuselage cross section were very much like those of the B-52. Where 713-1-133 differed from the B-52 was in scale. Its span and length were more than a quarter again as long as the B-52; the wing area was more than doubled. While the nominal bomb load remained the standard 10,000lb, gross weight was far beyond that of the B-52.

The differences between the Model 712-1-133 variations were in the engines. The -133A had 12 or 14 Pratt & Whitney J75 engines (the data says 14; the diagram shows 12), while the -133B had 14 General Electric X-84 engines. The X-84 was an unbuilt turbofan engine design of variable size and performance; various sources list thrust levels of 20,000 to more than 24,000lb of thrust. The 133C was similar to the -138 in that they each had 16 X-84 engines. This would have produced something along the lines of 320,000 to 384,000lb of thrust, compared to the 74,000lb that the early B-52s could produce. This would be a remarkably high power output for a subsonic bomber, producing a thrust to weight ratio of more than 1:2 at gross takeoff weight… and more than 1:1 at empty weight. The -133 series and the -138 had the same weights, but the -138 had an extra 12ft of wingspan and 1,000sq ft of wing area, bring it to a total of 10,000sq ft (929sq m) – vast by any reckoning.

No performance data is available. It can be assumed that range, and probably ceiling, would have been impressive. As nuclear power was common in the other designs in the Model 713 series, it's safe to assume that the 713-1-133 and -138 were meant to show competitive performance from a purely chemical system.

Boeing Model 464-197

Possibly the fastest B-52 derivative concept ever designed, Boeing Model 464-197 is known from a single diagram and some basic data. It was produced for the MX-1965 program which led to the B-59, the same programme that led to the Model 450-65-10 series, Model 450-151-31, Model 450-154-32 and Model 701-333 described in the B-47 chapter. But beyond that, little is known about the 464-197 other than it was a B-52 heavily modified into a supersonic-capable configuration.

The 464-197's fuselage and tail were much like those of a standard B-52B, with the exception of the nose. The relatively blunt nose of the B-52 was replaced with a pointed nosecone and the windscreen was reshaped for lower drag at high speed. The biggest difference was that the long, narrow wings of the B-52 were replaced with entirely new shorter-span, longer-chord wings of about 45° sweep (40° at quarter chord), a thickness ratio of only 4% and substantially greater wing area. The vertical tail was reduced in size somewhat. It retained the main quadricycle landing gear of the standard B-52, with outrigger stabilizing gear deployed from the outboard nacelles. The aircraft still had eight engines in four underslung pods, but the pods were entirely different in shape and size from those of the standard B-52.

While the standard B-52 pods had quite simple inlets, those of the 464-197 had vertical ramps. The engines inside were Wright J67s, which was to be a license-built version of the Rolls-Royce Olympus engines that powered the British Vulcan bomber. But while the Vulcan had four of these afterburning engines, Model 464-197 was to have eight, enough to power the aircraft through the sound barrier. The nacelles were substantially longer than standard B-52 nacelles in order to provide space for the afterburners. The performance potential of the design is unfortunately not available.

It's not clear if Model 464-197 was meant to be an operational bomber for series production or possibly a test aircraft for some of the MX-1965 design concepts. However, it was to have a 10,000lb bomb as standard, with the ability to carry a 50,000lb bomb (the data seems to indicate that this would be a single

Boeing Model 464-239
SCALE 1/350

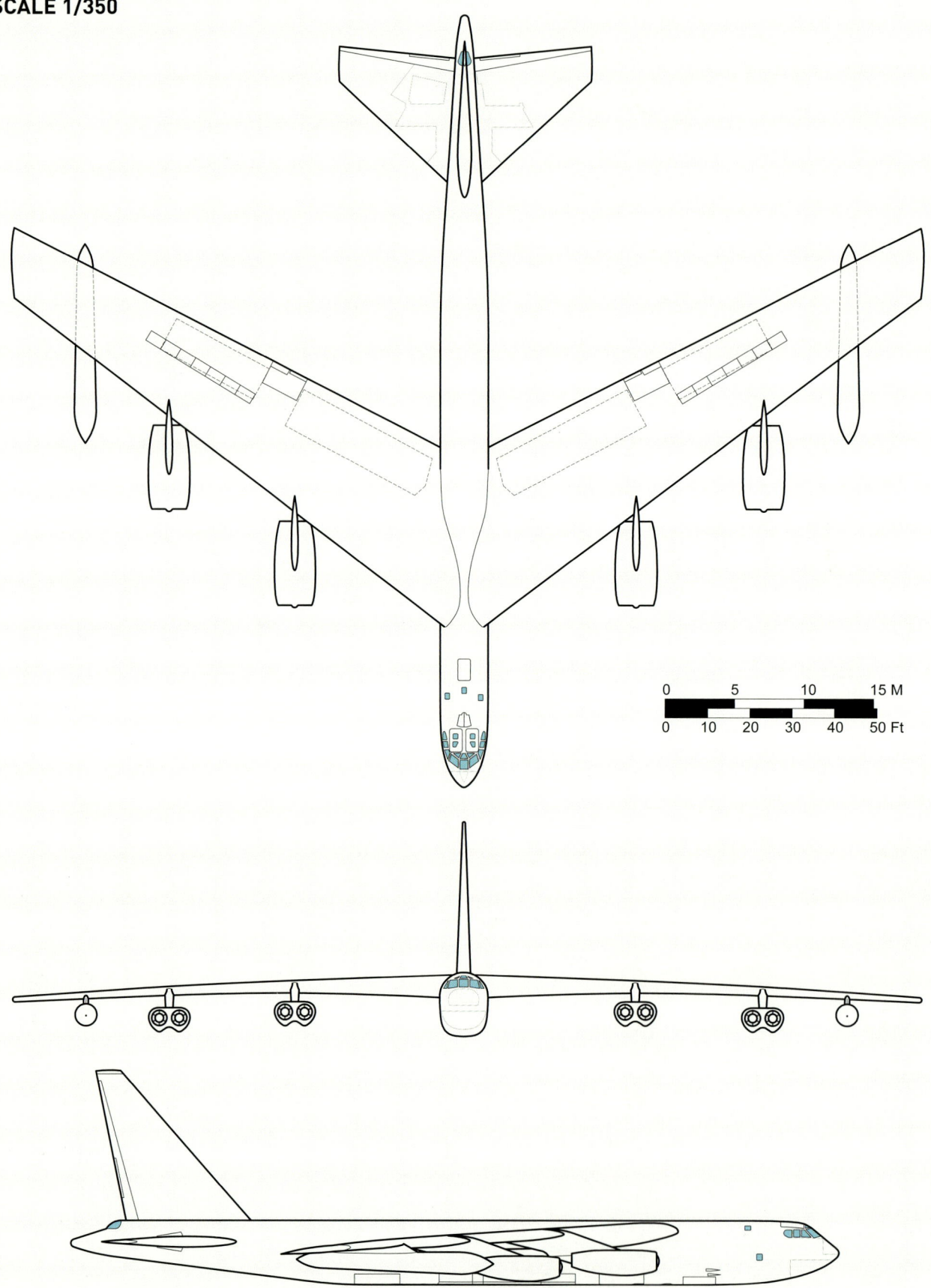

Boeing Model 713-1-133
SCALE 1/400

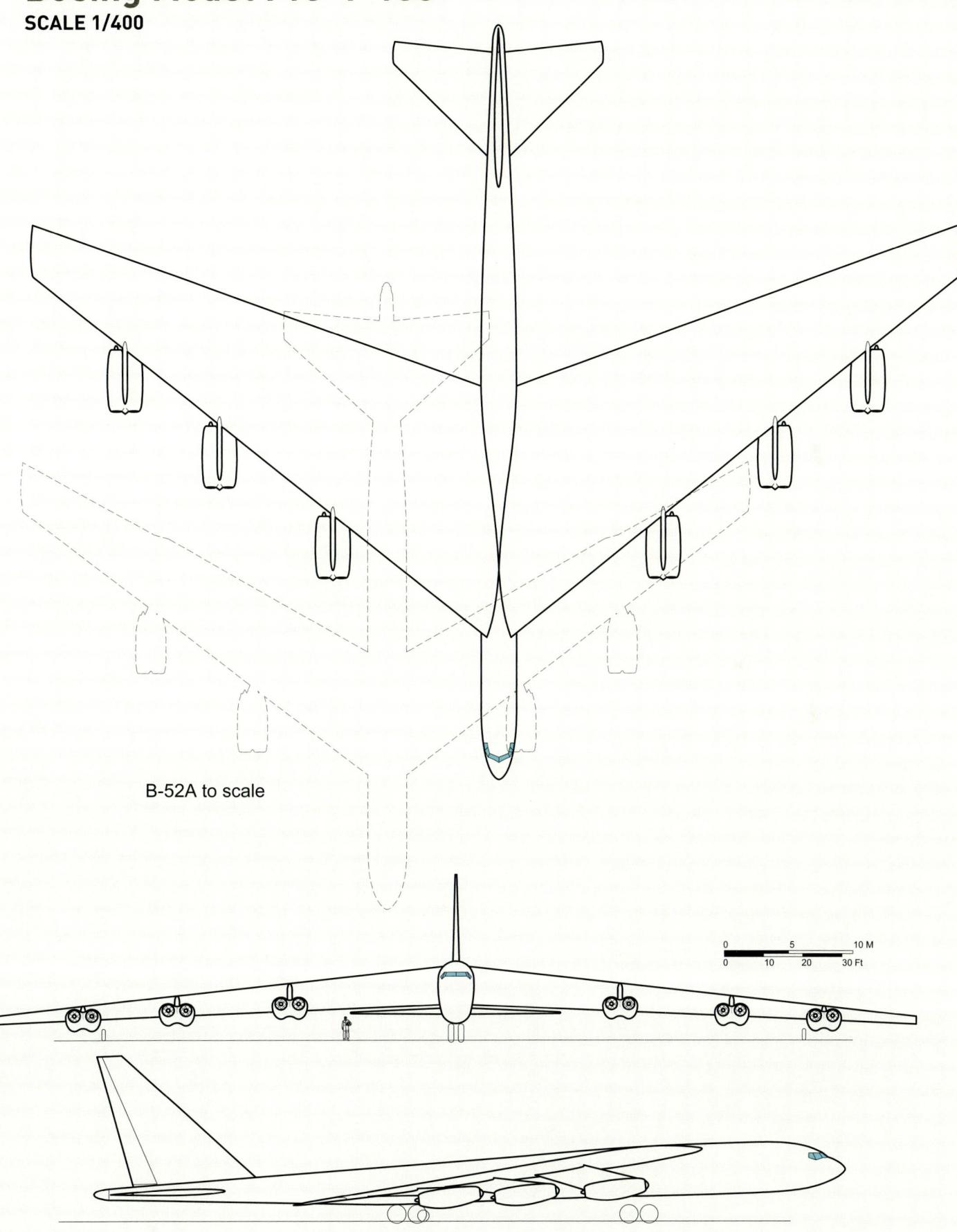

B-52A to scale

Boeing Model 713-1-133 Alternate
SCALE 1/400

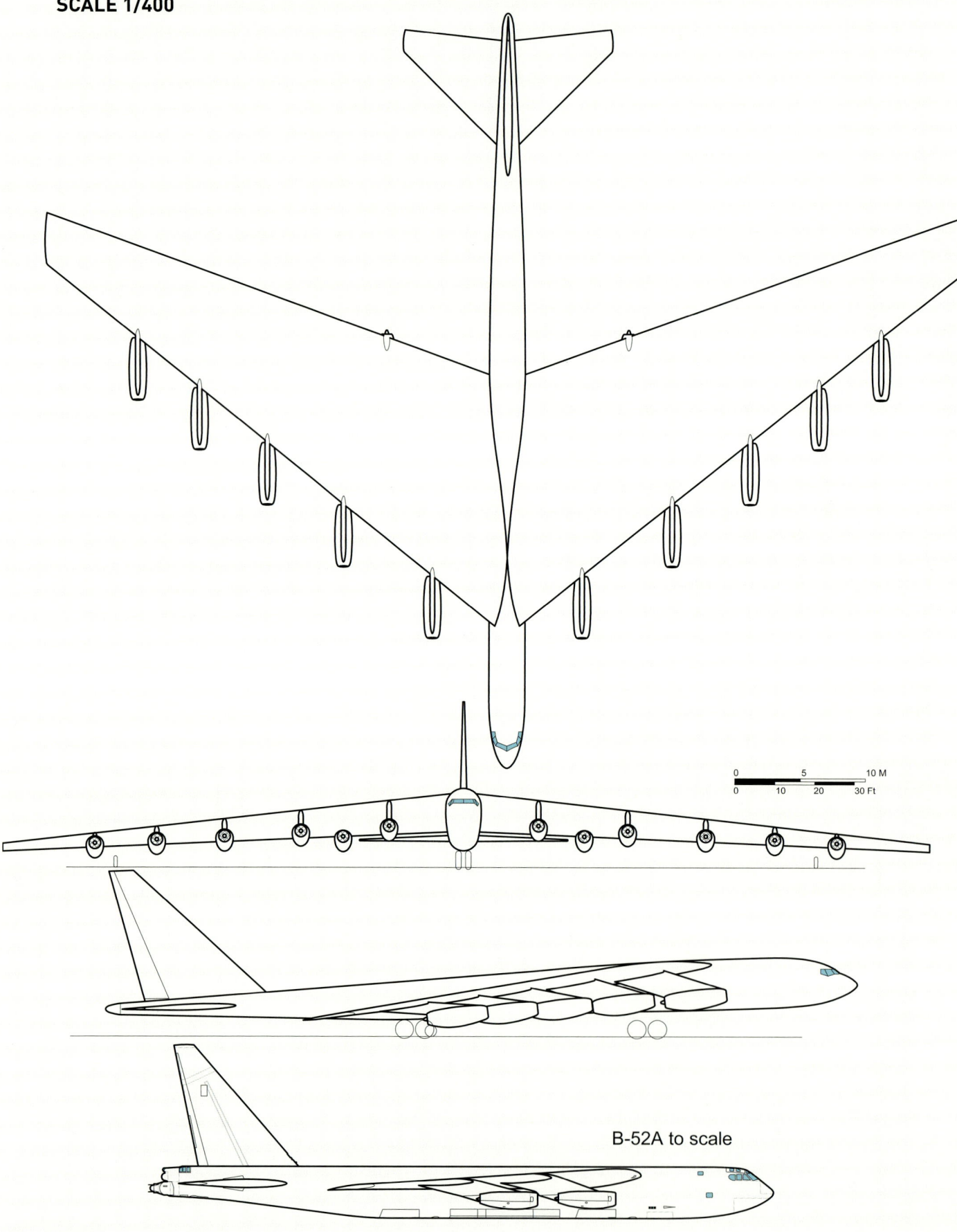

B-52A to scale

Boeing Model 713-1-133C
SCALE 1/400

Control surfaces speculative

B-52A to scale

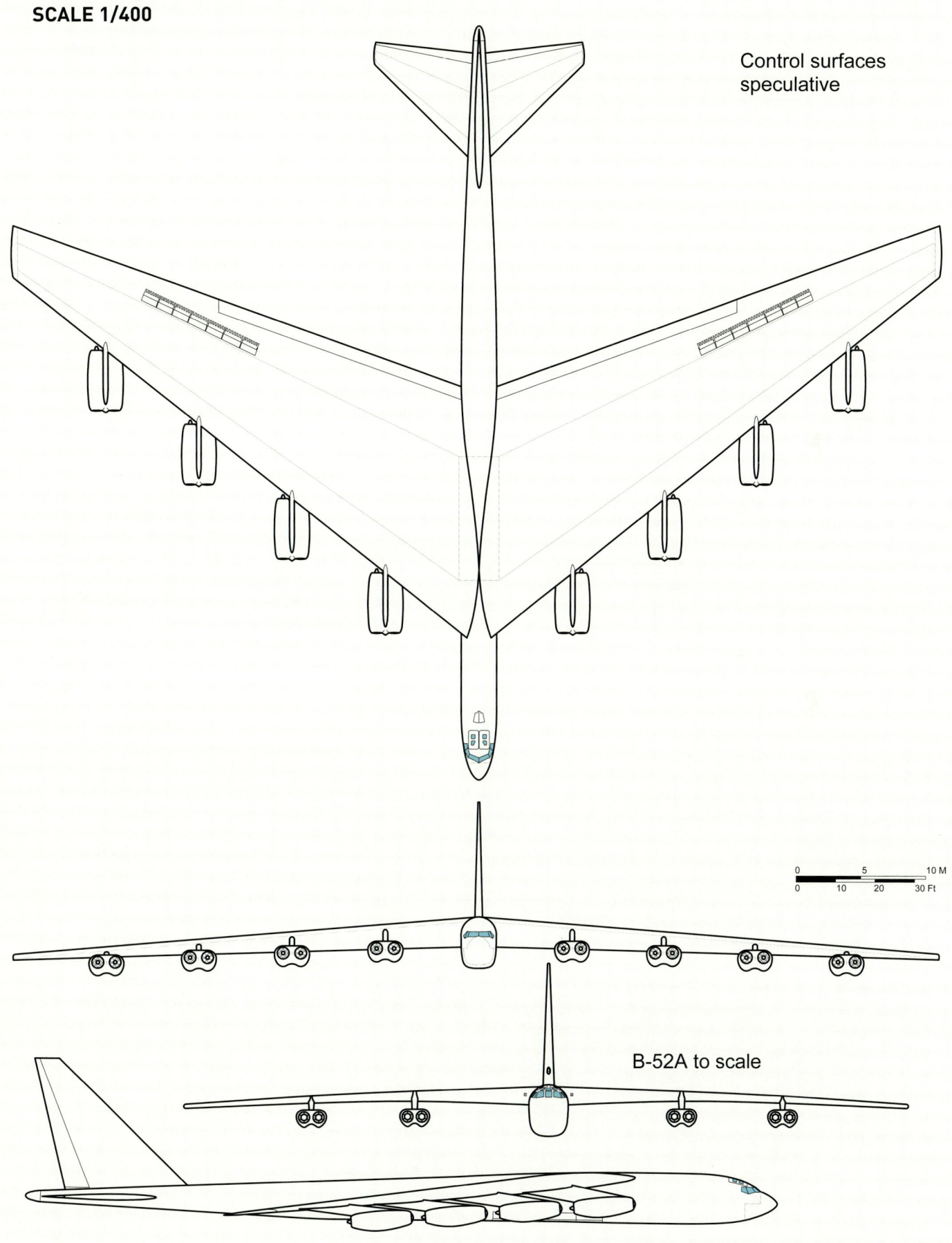

Boeing Model 713-1-138
SCALE 1/400

Control surfaces speculative

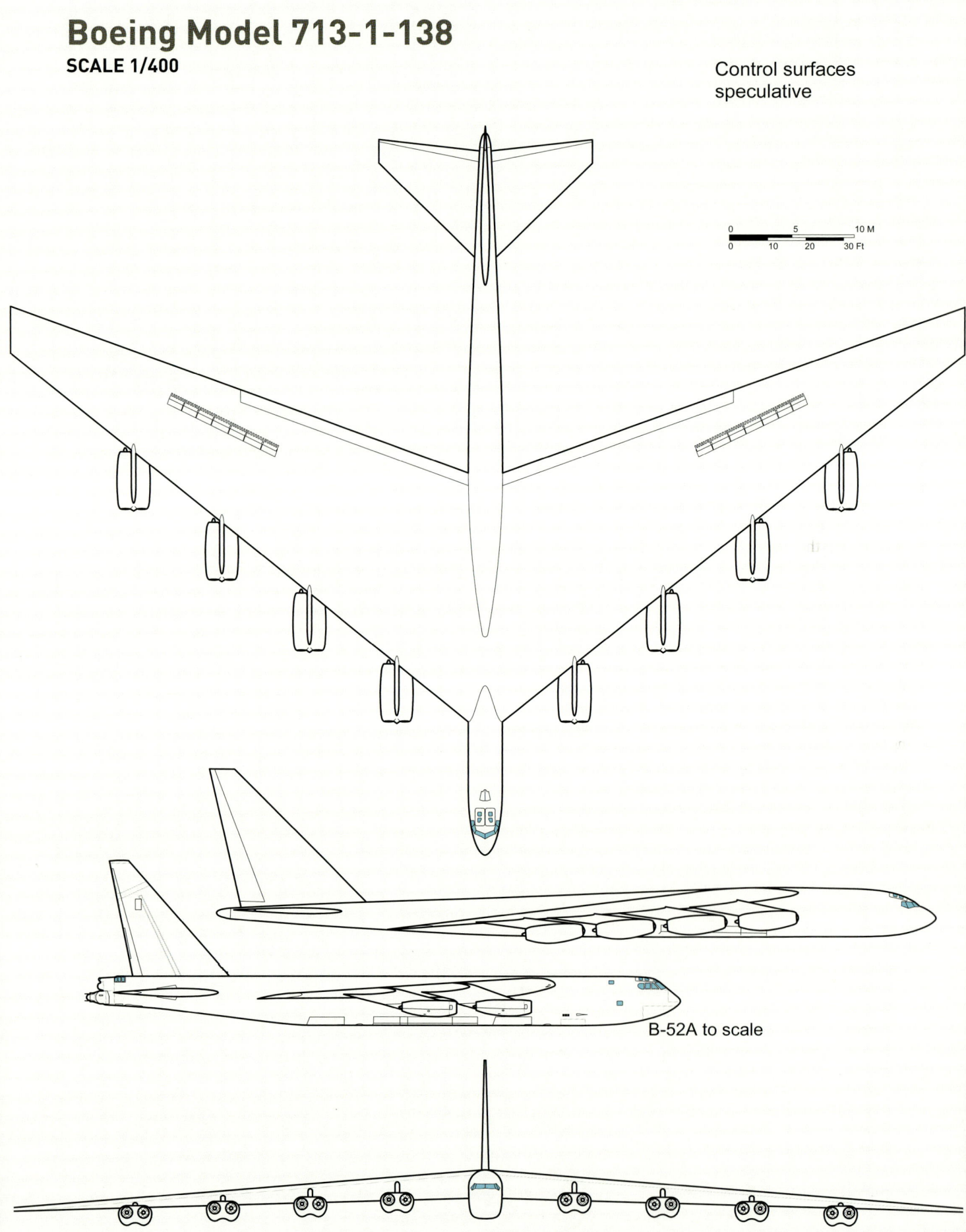

B-52A to scale

large bomb; by this time, the weapons load was almost certainly meant to be entirely nuclear, so a 50,000lb nuclear device sounds… enthusiastic).

Liquid hydrogen fuel

There was a mid-1950s fascination with liquid hydrogen as an aircraft fuel. The combustion energy contained within a kilogram of hydrogen is approximately three times what's available in jet fuel; this is the sort of advantage that will cause many aircraft designers to try to work around the fact that the density of liquid hydrogen is only one-tenth that of jet fuel, and that to keep hydrogen a liquid it must be stored at dangerously low temperatures. But the 1950s was the great 'can-do' era in American aerospace development, so studying hydrogen was inevitable.

The NACA Lewis Flight Propulsion Laboratory studied the theoretical side of hydrogen propulsion for aircraft and in 1955 showed sketches of several possible configurations. Hydrogen aircraft could in principle use conventional turbojet engines with modest modifications, but the airframes would be distinctive. The incredibly low density, cold temperature and high pressure of the propellant meant that the propellant tanks could no longer be just jammed into whatever space was available in wings and fuselage, but instead must be large to the point of being vast.

The aircraft that the NACA-Lewis looked at were mostly supersonic designs with voluminous spindle-shaped fuselages with small wings, but the notion of a subsonic bomber/reconnaissance plane clearly inspired by the B-52 was also kicked around. Powered by four engines, it was laid out like the B-52 with large cylindrical tanks filling the fuselage; but even then there was not enough fuel storage volume. Additional liquid hydrogen was stored in the wings in banks of long and relatively thin cylindrical tanks; lacking volumetric efficiency, these would have made the wings both fat and stiff.

Boeing Models 464-245 and 464-246

At the same time, the American aeronautical industry also looked at hydrogen fuel. The best known of these efforts was Lockheed's CL-400 'Suntan', a Mach 2+ high-altitude reconnaissance plane; a great deal of time and effort was spent on it, producing some important advances… but not a hydrogen powered aircraft. Boeing also studied hydrogen fuelled aircraft of numerous kinds including supersonic bombers and reconnaissance vehicles of entirely new configurations… but hydrogen conversions of the B-52 were also examined. Two such configuration are known: Models 464-245 and 464-246. Performance data is lacking but the configurations are well established.

Model 464-245 used a largely conventional B-52 wing and fuselage, but added two very large fixed external fuel tanks. These tanks, each roughly the size of the B-52's fuselage, were fitted to the wings in the positions formerly occupied by the inboard engine pylons. The outer pylons still held two engines each, but these were now non-afterburning Pratt & Whitney J75s with modifications to burn both jet fuel and hydrogen. The J75 could produce something like 70% more thrust than the contemporary J57, so the dropoff in thrust was not that great.

Each of the giant wing pods would contain 41,500lb of liquid hydrogen and there would be additional hydrogen tankage in the fuselage. The wing tanks were still to be filled with jet fuel, totalling 134,000lb. Exactly how the aircraft would use both fuels is unclear. It's interesting that the wing tanks still managed to hold a much greater mass of jet fuel than the giant hydrogen tanks.

Model 464-246 was obviously designed at the same time. This design dispensed with the large external tanks in favor of a redesigned and much larger fuselage. The new fuselage was circular in cross section, about 17 or 18ft

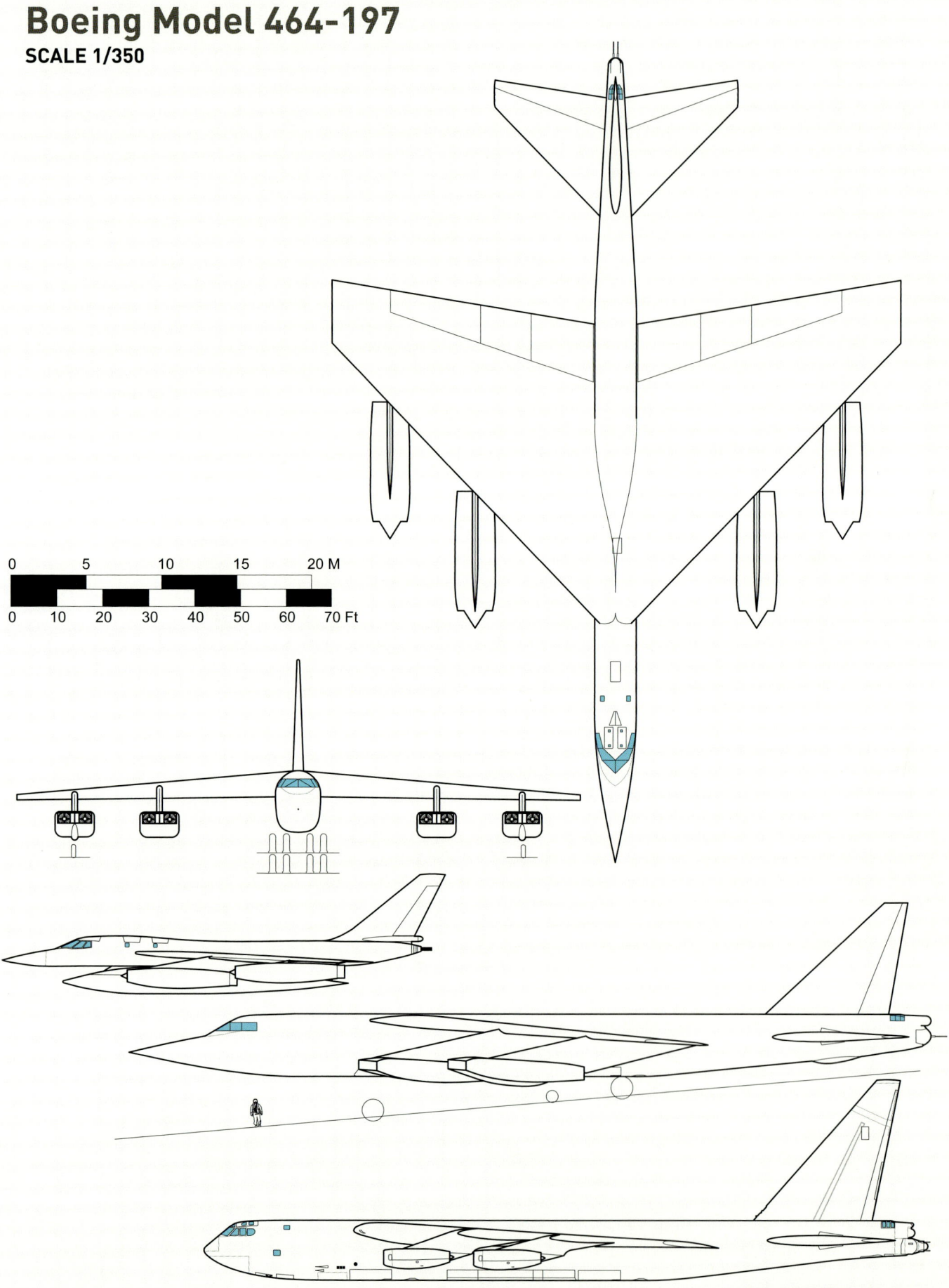

Boeing Model 464-197
SCALE 1/350

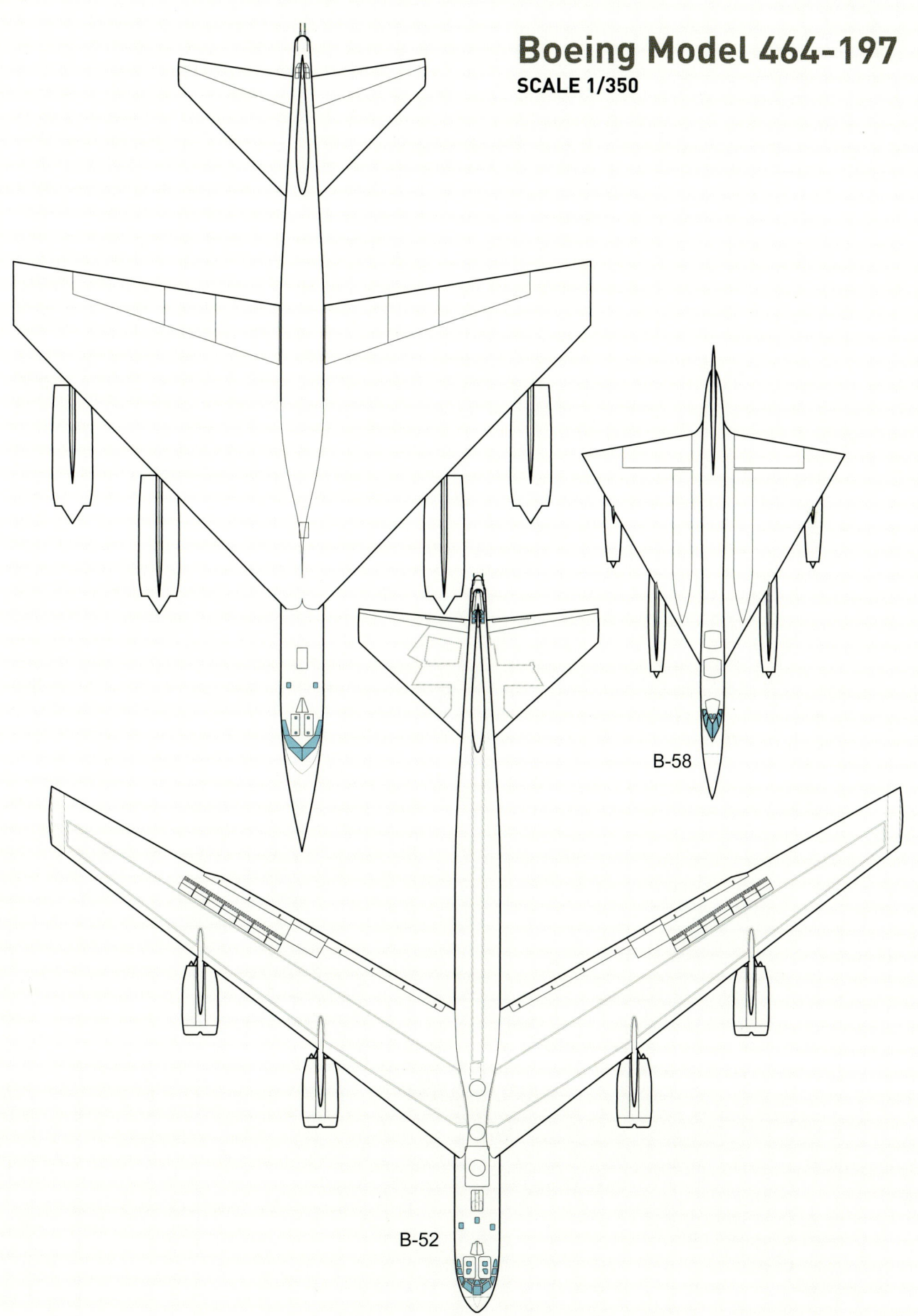

Boeing Model 464-197
SCALE 1/350

B-58

B-52

in diameter, with six separate fuel cells holding a total of 89,500lb of liquid hydrogen. The wing tanks would hold 152,000lb of JP-4 fuel. This aircraft had six J75 engines, two in inboard nacelles, and single engines in outboard nacelles. This aircraft would have required substantially more modification than Model 464-245, but it would have been a more aerodynamic vehicle, lacking the drag of the two giant external tanks.

Nothing seems to have come from Boeing's brief foray into modifying the B-52 for hydrogen fuel. The difficulties associated with liquid hydrogen have kept it from being an aviation fuel for nearly 70 years. Future issues with hydrocarbon fuels may lead to hydrogen fuelled aircraft which may resemble Model 464-245 or 464-246, but it is very unlikely that a B-52 will be retrofitted in this manner.

Inflight refuelling

This process was introduced to the B-52 design evolution with Model 464-33 in January 1948 and while it promised greatly increasing range, there was also a major problem: the tanker aircraft. The first tanker aircraft to enter US Air Force service, in 1951, was the capable but limited KC-97 Stratofreighter. Derived from the B-29, it was powered by piston engines and flew slowly – so slowly in fact that B-52s attempting to rendezvous with it generally had to not only deploy flaps but also lower their landing gear in order to match its speed. This of course was not optimal.

Boeing Model 491

The inflight refuelling problem was foreseen early on; a faster, higher-flying tanker was needed to let the B-52 become all it could be. This would eventually appear in the form of the KC-135, derived from the Boeing 367-80 prototype. The 367 project did not begin in earnest until 1952, with first flight in 1954, and the KC-135 did not fly until 1956.

Back in 1949, the KC-135 was well beyond the horizon. But the B-52 had already become a fully jet-propelled design by that point and a piston-engined tanker simply wasn't going to be adequate. So in December of 1949, Boeing designed Model 491.

Unlike the KC-97 and the later KC-135, Model 491 was an entirely new clean-sheet concept. It was a twin-boom design with a central truncated fuselage, not unlike the Lockheed P-38. Equipped with six Allison T40 turboprops, each with an eight-bladed contra-rotating propeller of 15ft diameter, it could reach 490mph at 25,000ft. Coupled with this speed, about 115mph faster than the KC-97G, Model 491 could carry 39,000 gallons (or 40,000 gallons – see below) of fuel… some 30,000 gallons more than the KC-97. Model 491 could fly high and fast enough to comfortably refuel the B-52.

Fuel would be contained within the central fuselage (19,500 gallons), in the wings (15,000 gallons) and boom (5,500 gallons). This unfortunately totals 40,000 gallons, a thousand gallons more than the total given in the data.

A single flying boom taken from the KC-97 was attached to the right-hand boom. The boom operator was located in the central fuselage pod, rather than the boom, which would have meant for a drastic change in practice for the operator.

Not much information is available on Model 491, indicating that it was probably not deeply studied or seriously proposed. But it does show that Boeing understood that the B-52 would need a new tanker aircraft even before it flew.

Boeing Model 464-245
SCALE 1/350

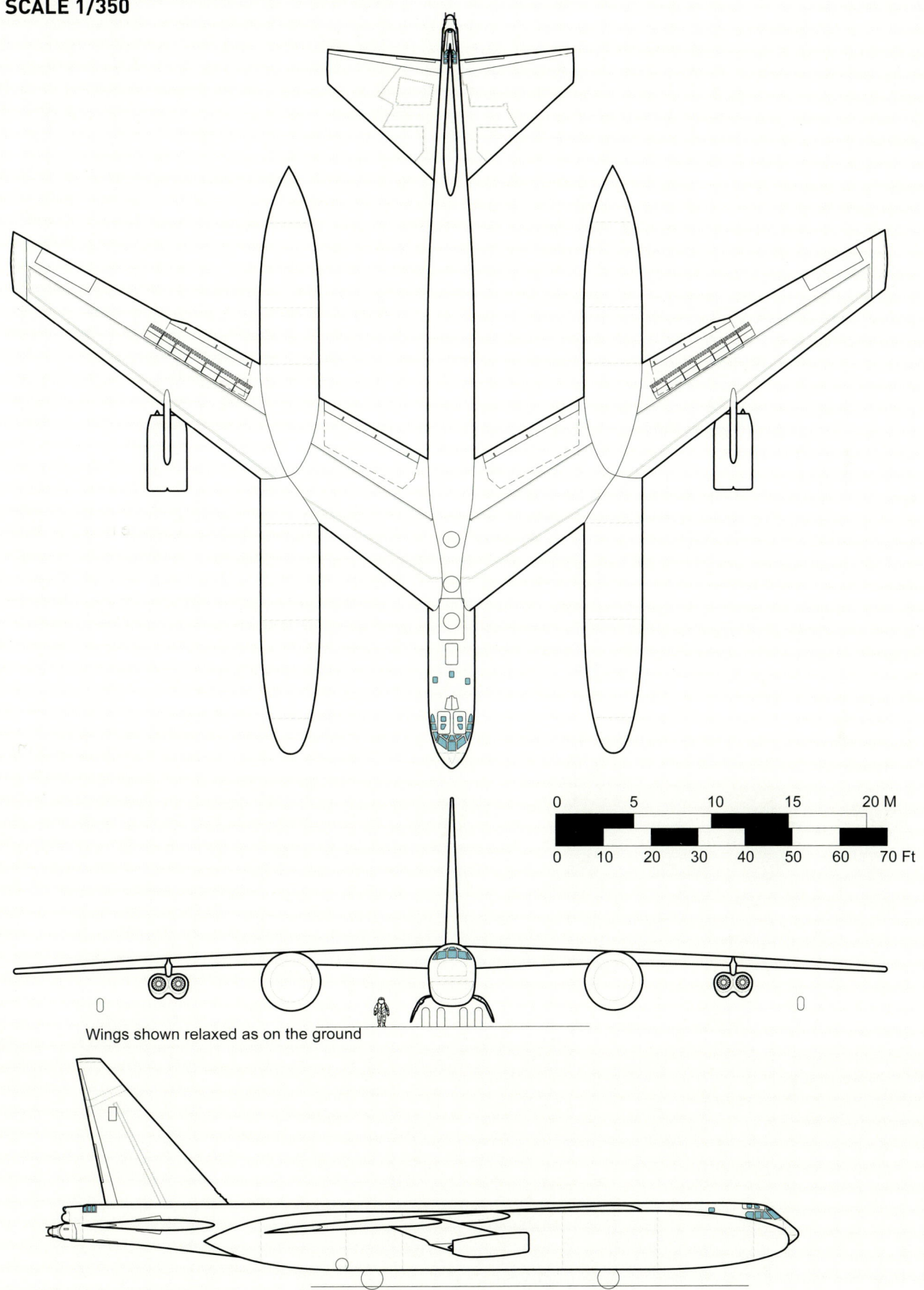

Wings shown relaxed as on the ground

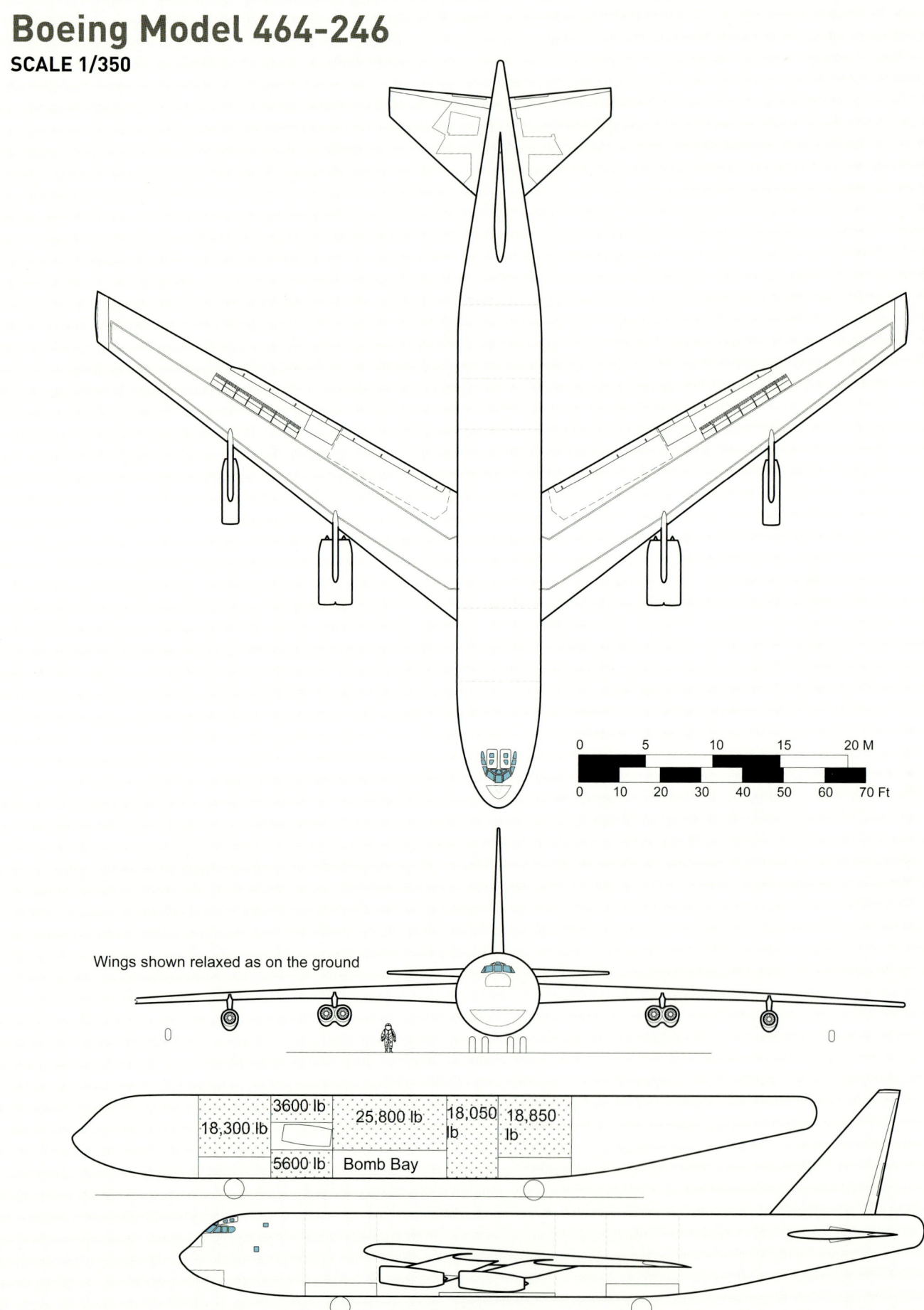

Boeing Model 491
SCALE 1/350

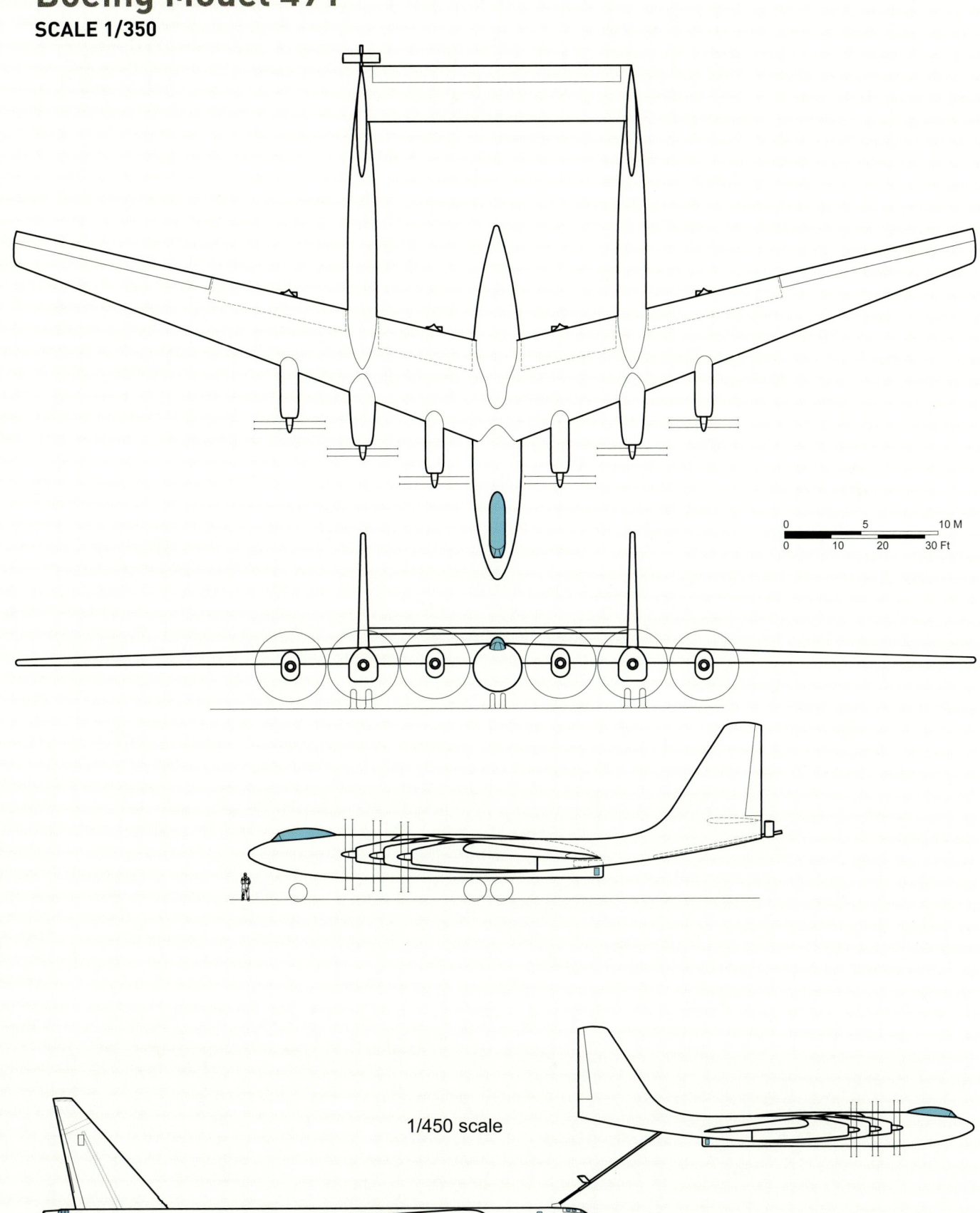

1/450 scale

CHAPTER 6: B-52 with Nuclear Power

As atomic power burst onto the scene in the mid-1940s, it brought with it the potential to power… well, everything. Nuclear powered ships, submarines, rockets, tanks, trains, airships, trucks and even cars were proposed at various levels of seriousness, with nuclear power proving itself practical for ships, submarines and space probes. One area that saw nearly as much effort, but nowhere near as much success, was the concept of nuclear power for the propulsion of aircraft.

There was apparently some interest in nuclear aircraft propulsion as far back as Second World War Germany, though they had no hope of making it work. It was not until the NEPA programme (Nuclear Energy for the Propulsion of Aircraft) was begun in 1946 by the United States Army Air Forces that a realistic and systematic effort was started to study the physics and engineering of nuclear aircraft. Until the last gasp of the programme in 1961, NEPA and its successor ANP (Aircraft Nuclear Propulsion) studied a wide range of reactor-based engines and a wider range of aircraft to be powered by them.

Most of the engines married a reactor with a turbojet, some a reactor with a turboprop, a few a reactor with a ramjet. In all cases the function of the reactor was the same: to provide a vast amount of heat. To recap: a conventional turbojet (or turboprop) works by compressing atmospheric air, forcing that into a combustion chamber, combusting it with a chemical fuel, and blowing the resulting really hot air past a turbine that is spun by the exhaust. The spinning turbine is used to power the compressor. Turbojets obtain thrust directly from the high velocity exhaust gas while turboprops extract as much mechanical power as they can from the exhaust and use it to turn a large propeller (or a smaller propeller at very high rotation rates).

A nuclear turbojet works almost exactly the same, except that it replaces the chemical combustion with nuclear heating of the compressed air. This can be done directly or indirectly. In a direct system, the compressed air flows through the reactor itself. This system has the advantages of being relatively lightweight with a high efficiency of transfer of heat from the reactor to the air. But it also is a safety and environmental nightmare: little bits of the reactor, sometimes as small as individual atoms, are constantly spraying out of the exhaust.

Anything the engine might swallow, from dust to rain to snow to insects to birds to shrapnel, could get lodged in the reactor and caused localized superheating, or could knock chunks out of the reactor, or could get irradiated enough to become fallout. Direct nuclear turbojet designs are thus often not highly thought of except for weapons systems to be used on Doomsday, when leaving a trail of radioactive carnage in the vehicle's wake is not a bug, it's a feature.

Indirect nuclear turbojets separate the reactor from the jet. The reactor does not heat the air directly, but heats a fluid. In this case the fluid could be a gas such as helium, or it could be a relatively low melting temperature metal such as sodium, potassium or bismuth. The fluid is passed through the reactor and heated, then passed down the length of a high-temperature metal pipe from the reactor – usually safely ensconced within the fuselage – to the turbojet.

The fluid then passes into a heat exchanger in the combustor section of the jet engine; the extremely hot fluid gives its heat to the compressed air, replacing the combustion of air and jet fuel. The cooled fluid then flows back to the reactor to be heated again. This is a complex, heavy, less efficient system – but it has the important benefit of separating the reactor from the atmosphere. Anything that comes into direct contact with the reactor is separated from the atmosphere by impermeable metal walls.

NEPA began on May 26, 1946, when a contract was signed with the Fairchild Engine and Airplane Corporation. While other companies would enter the world of nuclear aircraft and engine design in the coming years (Lockheed, Boeing, Convair and Northrop gave considerable study to nuclear powered aircraft, while General Electric and Pratt & Whitney did a great deal of work on the subject of nuclear turbojets), Fairchild had the lead.

For several years however, the designs Fairchild produced seem to have been mostly vague and speculative, doubtless a result of the technology being itself vague and speculative. Still, Fairchild designed a wide array of military aircraft that would use nuclear power for flight. Most were entirely new configurations but several were based on existing aircraft. Unsurprisingly Convair's B-36, with its vast payload capability, was a common choice of aircraft to base nuclear powered designs on; this would halfway come to fruition with the development and flight, starting in 1955, of the NB-36H. This aircraft housed a reactor capable of producing one megawatt. It used air for the reactor coolant, but did not use the resulting heated air to actually power anything. Instead, the NB-36H was merely a testbed to see how materials, systems and maintenance techs would react to a flying reactor. A planned follow-on, the Convair X-6, would have incorporated actual nuclear-powered turbojets.

While the B-36 had the lift capacity to carry a reactor, it was slow and not at all representative of the sort of high-speed combat aircraft that the Air Force might someday operate with pure nuclear power. A demonstrator for such an aircraft was needed, and even though the B-52 had not flown and the design wasn't even necessarily firmly established, it was still used as the basis for a number of early concepts.

A Fairchild report of 1950 illustrated a pair of concepts for B-52 based nuclear powered aircraft. The first utilized the wings, tail surfaces, landing gear and, it seems, the cockpit of the XB-52 configuration, with an entirely new fuselage and a very different propulsion system. This vehicle would use a direct (air cycle) nuclear turbojet system with a single heavily shielded reactor.

Air from wing leading edge inlets would be ducted to the compressors of the six jet engines (modified XJ-57s) embedded in the rear fuselage. The compressed air would then be ducted through the reactor; the air would cool the reactor and in the process be raised to incandescently hot temperatures, then fed into the combustor sections of turbojets to drive the turbines. The resulting exhaust would be directed down to a series of nozzles wrapping around the underside of the rear fuselage. It was found that an XJ-57 modified for nuclear heating would be 15% heavier than a conventional version of the engine. The convoluted paths the air would take from inlet to compressors to reactor to turbines to exhaust nozzles would make the ducting complex, space-consuming and heavy.

While the wings were based on those of the B-52, the reality was that they would be substantially redesigned. The B-52 does not, after all, have sizable leading edge air inlets nor internal ducting. The engine nacelles spread along the wing of the B-52, along with the fuel in the wing tanks, provided a spanwise load that reduced the moment imposed at the wing root… but the wing of the nuclear version did not have those loads, increasing the torque and thus necessitating structural strengthening. As a result, the wing of the nuclear aircraft would be about 45% heavier than the stock B-52 wing.

Fairchild believed that the crew could be kept safe from reactor radioactivity by putting a massive shield on the forward side of the reactor, and a separate circular shield at the aft end of the crew compartment.

A similar configuration was presented that was to use a helium cooled reactor. The low density of compressed, super-heated helium meant that the pipes required would have to be large in diameter. Also, the low molecular weight of helium coupled with its high temperature would make leaks inevitable; some would simply seep through the metal itself. While sitting on the ground, helium in the system would remain a gas with the reactor powered down.

As mentioned earlier, an alternative to helium would be a liquid metal such as sodium. Helium had the advantage that a substantial leak, such as might be expected due to the aircraft getting hit with shrapnel or buffeting, would result in a jet of superheated inert helium. Unfortunate, to be sure. But compared to a leak of superheated and highly flammable liquid sodium metal generating a white-hot flame akin to a magnesium fire, the jet of hot helium would be comparatively benign.

Additionally, sodium would not simply remain liquid with the reactor powered down on the ground – it would start to solidify. This would structurally stress the pipes and make re-melting the whole system difficult, so an operational nuclear aircraft would have to either offload the liquid metal or keep it molten during down times. The former process was the plan where liquid metal was specified, with the need to tap off hot, readily flammable (sodium and lithium will both spontaneously combust on contact with air at room temperature) liquid metal, filling the lines with an inert gas instead. This doubtless would have been the cause of many an incendiary incident.

These designs based on B-52 components were expected to be technology testbeds or at best interim tactical aircraft. To obtain the best performance and most optimized designs using nuclear propulsion, blank-sheet designs would be required.

Fairchild continued to study nuclear powered aircraft, including B-52 derivatives, for a several more years. In 1952 the company released a report discussing a range of aircraft designs, both clean-sheet and modifications of existing aircraft such as the B-36, the C-99 and the B-52. Here the designs used a liquid metal (sodium) coolant system, with the metal cycling through the reactor on a short loop, passing through a heat exchanger not within the turbojets but right next to the reactor. The thermal energy would be passed on to a second loop of liquid

Fairchild XB-52 NEPA 1
SCALE 1/320

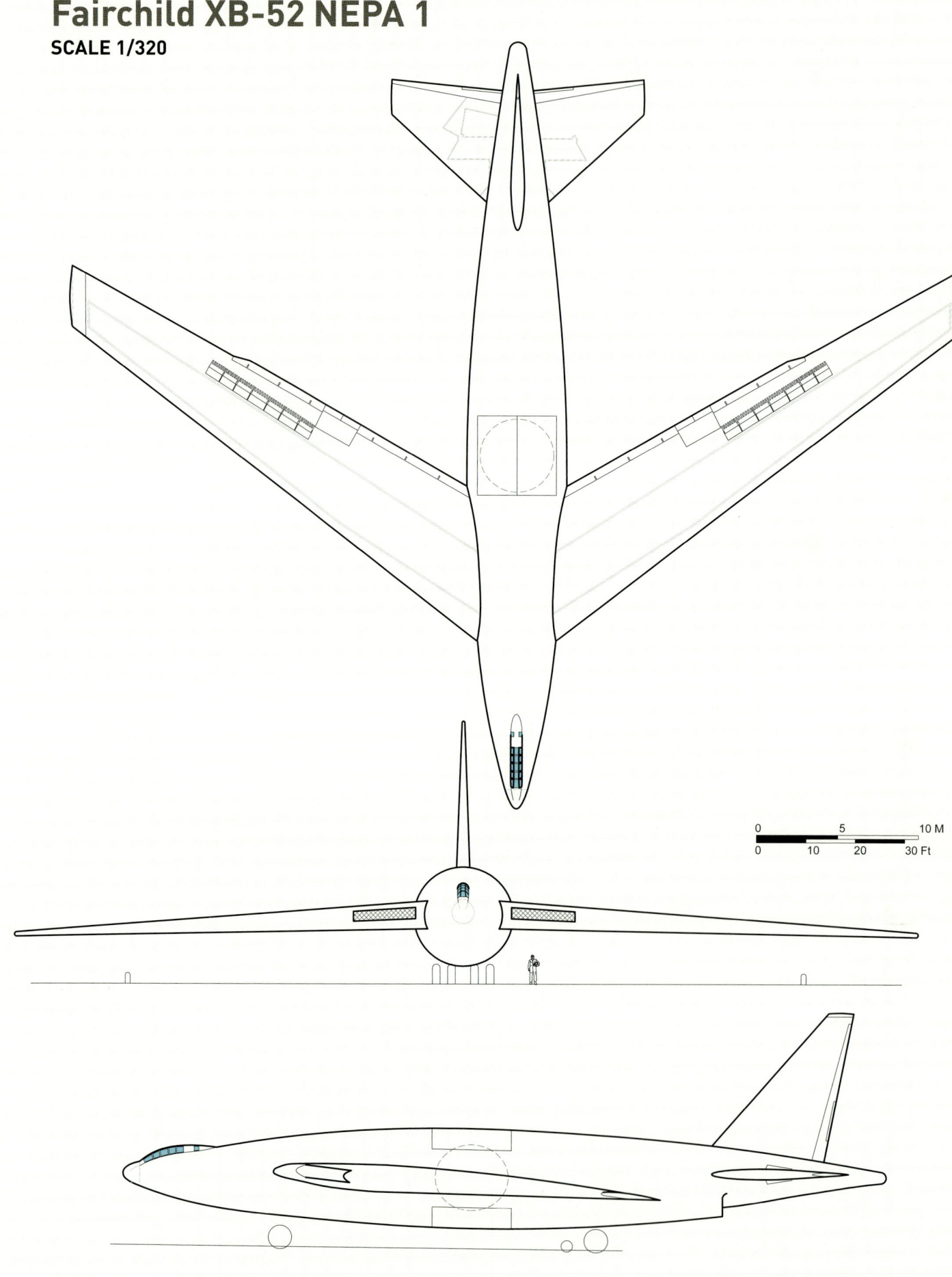

Fairchild XB-52 NEPA 2
SCALE 1/320

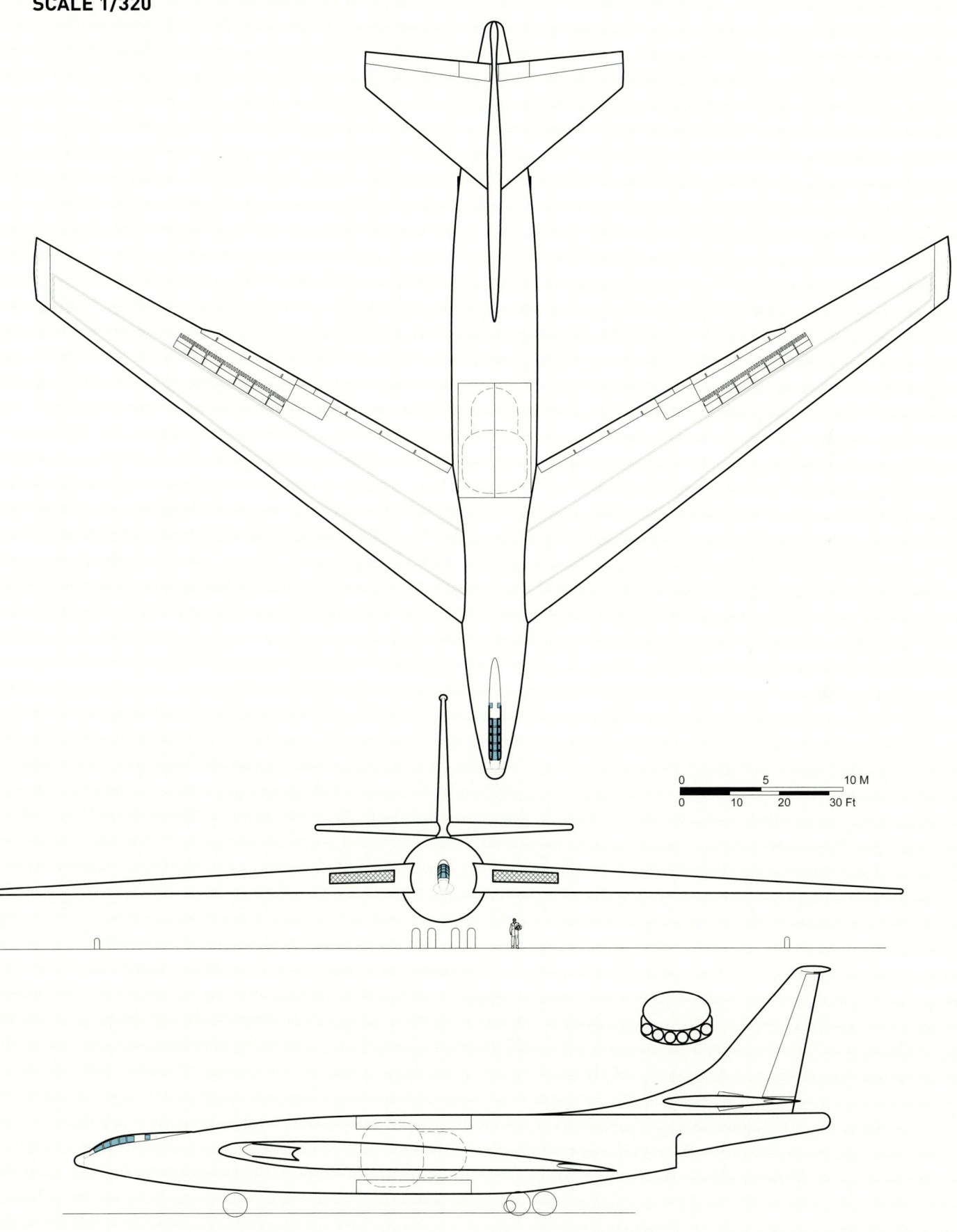

metal and that second much longer loop would pass through the turbojets.

The purpose of this was to aid in maintenance. As the metal passes through the reactor it becomes dangerously radioactive. So by keeping it to a short loop, the total mass of radioactive liquid metal would be minimized. The rest of the liquid metal in the system could be dealt with much more readily, the maintenance crews only needing to worry about being burned to ash by a small leak. There were of course inefficiencies in the system: the reactor had a maximum wall temperature of 1,400°F, but the turbine temperatures dropped to between 1,225° and 1,250°.

The reactors had multi-level shielding... an inner layer of borated water surrounded by a layer of borated gasoline to absorb neutrons; steel and lead plates on the forward side provided gamma ray protection. The crew compartments were to be shielded with lead and plastic plates, though the details on that are lean.

Fairchild NEPA N-14

The N-14 was a relatively straightforward modification of the XB-52 configuration and appears to have existed two years earlier as the 'IL-1' configuration. The N-14 was a liquid metal cooled design using long insulated pipes to carry hot liquid sodium from the L-14 reactor to turbojets in pods much like those of the XB-52. The engines were different than those of the XB-52, however, being modifications of the planned General Electric J53 turbojet. There would only have been six of these, with inboard two-engine nacelles similar to (but longer than) the inboard nacelles on the actual B-52 and with each outer nacelle holding a single engine.

The nuclear reactor, with a maximum power output of 151,000 BTU per second, would be positioned in the fuselage just aft of the wings. The reactor and its shield were so large (outer diameter of 58in) that the fuselage needed to be bulged outwards to accommodate them; in addition the rear fuselage was extended a further 12ft. The secondary liquid metal lines required valves on either end to minimize leakage. The lines averaged 123ft in length and had a total planform area of 520sq ft.

Pumps kept the coolant circulating at 190lb per second per engine with a pressure of 75psi. This sped the liquid through the lines with a transient time of 4.9 seconds. The total powerplant weight was a hefty 179,400lb; much of this was the reactor and especially the shielding (58,600lb for the reactor shielding, another 30,000lb for the crew compartment shield), but the pipes themselves would have been a major weight hit too... they would have had to be made of a material that remained strong and impermeable at the operating temperature of more than 1200°. This likely would have meant a high temperature alloy of steel and nickel, molybdenum, and/or tungsten.

All this led to aircraft performance that was somewhat lackluster: top speed on pure nuclear power was a respectable 480 knots at 30,000ft, but rate of climb was, putting it charitably, pathetic. At sea level the aircraft clawed its way skyward at 1.5ft per minute on pure nuclear power, though rate of climb improved as the plane gained altitude. Thus the aircraft carried a supply of conventional jet fuel for some chemical thrust augmentation; rate of climb at sea level would go to a more respectable 5,800ft per minute. Time to climb from sea level to 10,000ft was 28.4 minutes; sea level to 20,000ft was 42.2 minutes; sea level to service ceiling was 85 minutes. With chemical augmentation these times would decrease to 1.8, 4.0 and 25.3 minutes.

While it might seem that the N-14 would be a straightforward modification of the XB-52, integrating the liquid metal pipes into the airframe would be no small task. The end result would be an aircraft capable of flight, but not capable of carrying out meaningful tactical missions; payload would only be 2,000lb. The relatively simple diagrams of the N-14 and N-15 (see below) design seem to show the outline of the as-built XB-52 canopy, but in reality the crew compartment would probably have been built quite differently to accommodate the required crew shielding.

Fairchild NEPA N-15

The N-15 was another design that was meant to be a relatively straightforward modification of the XB-52 configuration. The same basic reactor would be increased in power to 183,000 BTU/second, but that increased power would be fed to a reduced number of engines. The lone outer engine on each wing would remain but each inboard pair of modified XJ-53 engines would be swapped for a single engine. The four engines would receive a higher input of thermal power and would see higher turbine temperatures. This would lead to improved performance per engine, and substantially improved performance of the aircraft at low altitude.

On nuclear power alone, the rate of climb at sea level would be 1,100ft per minute, a vast improvement over the N-14. However, high altitude performance would not be as good as that of the N-14, with a lower top speed and substantially reduced service ceiling. Time to climb on pure nuclear power from sea level to 10,000ft was 9.2 minutes; sea level to 20,000ft was 20.2 minutes; sea level to the service ceiling was 45 minutes. With chemical augmentation, these times would be 3.2, 7.6 and 32 minutes. This is noticeably

Fairchild NEPA N-14
SCALE 1/300

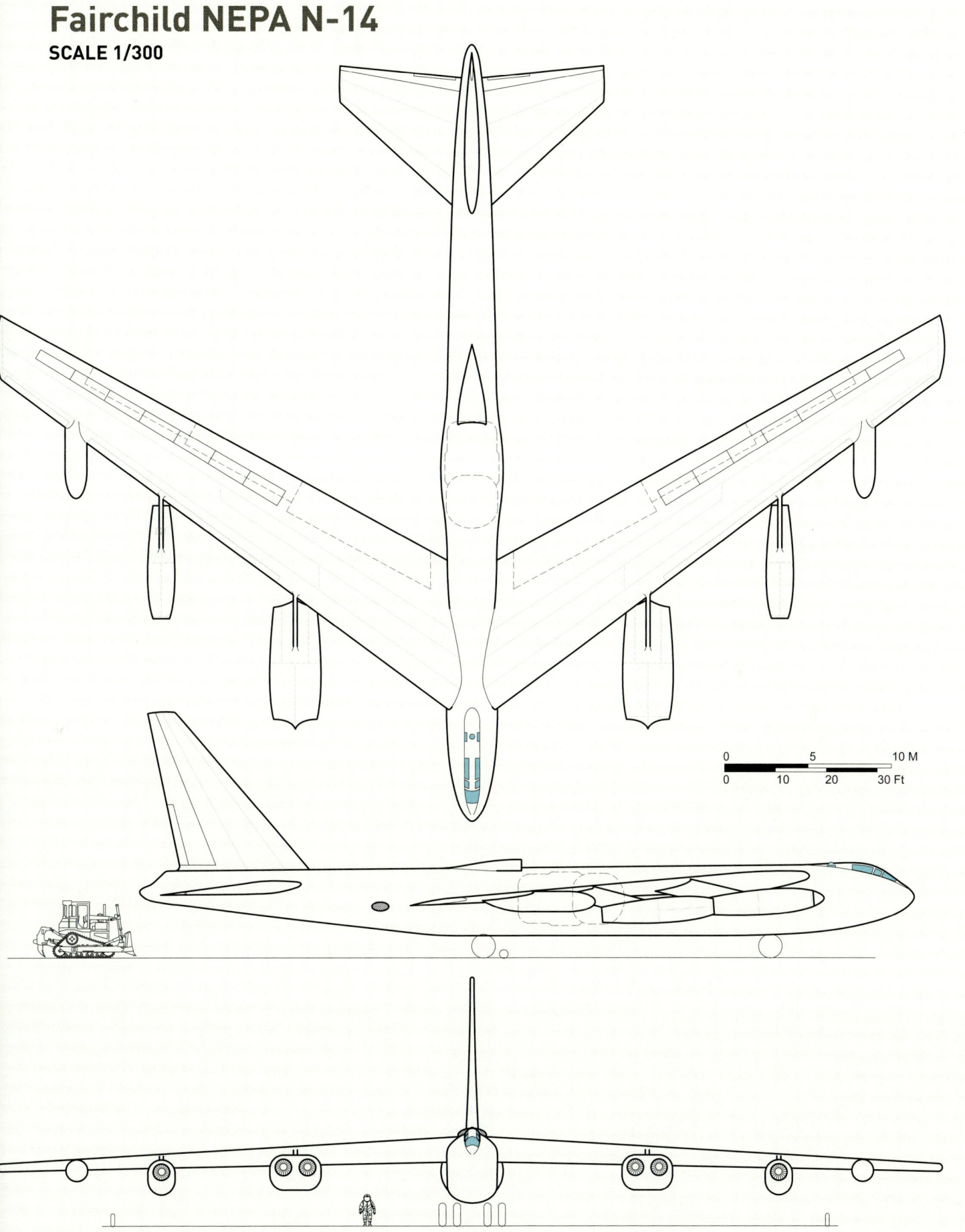

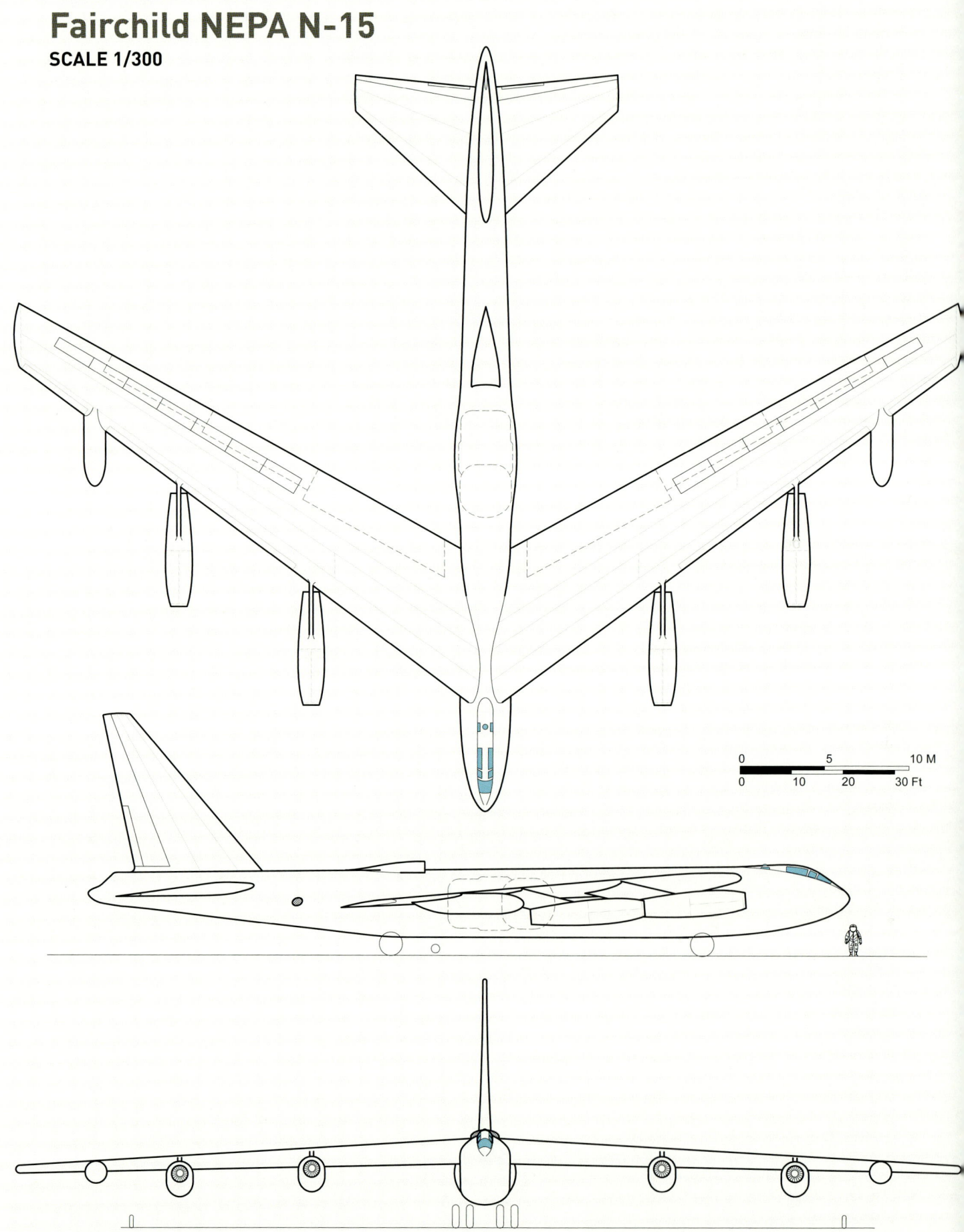

slower than the N-14, most likely due to the reduced number of engines.

Even with the increased reactor power output, shield weights would remain the same. The rear fuselage would not be stretched like the N-14's was. Payload for the N-15 was even worse than for the N-14 at a mere 400lb; clearly this was not the way to achieve an operational military vehicle. But for a test vehicle, it would have proved instructive.

Fairchild NEPA N-16

The N-16 was a modified B-52 in name only. The wings and tail surfaces would have been used but substantially modified; the fuselage would have been entirely new. The turbojets would have been relocated to within the fuselage in order to reduce the length, weight and losses of the liquid metal pipes. The B-52 landing gear was expected to remain largely unchanged.

Nuclear powered time to climb from sea level to 10,000ft was 16.2 minutes; sea level to 20,000ft was 26 minutes; sea level to the service ceiling was 54 minutes. With chemical augmentation, these times would be 1.5, 3.3 and 20 minutes. Nuclear climb performance was middling compared to the N-14 and N-15, but superior to both with chemical augmentation. But once again payload was essentially nonexistent at only 300lb.

Boeing B-52A/General Electric LF-2

General Electric proposed a modified B-52 with a liquid metal cycle nuclear powerplant, the LF-2, in 1953. This used a single reactor that provided heated liquid sodium/potassium (NaK) to eight J73-GE-3 turbojets modified to have radiators in the combustors. The reactor was embedded within the core of a spherical shield 112in in diameter; this was in turn implanted within the B-52 fuselage just ahead of the rear landing gear.

The turbojets were clustered around the sides and rear of the fuselage, aft of the reactor. These would replace the engines under the wings, necessitating structural strengthening. However, this would still result in lower mass than the pipes, liquid metal and insulation required to feed the engines had they been in their normal positions.

The diagram included here is a provisional reconstruction based on diagrams of the engine installation included in a much later report. While the engine diagram is clear and detailed, it is unfortunately rife with inconsistencies (including showing the wrong diameter for the reactor shield and a substantial reduction in the depth of the B-52's rear fuselage). The configuration shown here is a best attempt at reconciling those issues.

Boeing Model 702-138(1) MX-2145

Boeing reported on the results of its studies for Air Force Wright Air Development Center project MX-2145, high altitude heavy bombers, in January 1954. The basic aircraft was interpreted by Boeing to require an unrefuelled radius of 3,000 nautical miles, with a radius of 1,500 nautical miles from the target being flown at an altitude of at least 50,000ft. Payload would range from 10,000 to 40,000lb. For all aircraft, the bomb bay of the B-52 was chosen as the default since the size of the bombs that the Air Force might want to carry in the future was not available for the study.

Among the subsonic, supersonic, chemical and nuclear-powered designs was the Model 702-138(1). This was an all-nuclear design with a B-52-like configuration, including four pairs of engines in distinctly B-52-like nacelles. These were connected via high temperature pipes (filled with molten NaK) to a 124 megawatt reactor in the fuselage just behind the bomb bay. The crew were all contained in a heavily shielded cockpit compartment and were expected to be exposed to 0.1 rem per hour, for a mission lasting up to 23 hours. The technology of the aircraft, apart from the reactor, was expected to be available for a 1959 first flight.

Unlike many other available studies, here Boeing reported on some of the operational issues and characteristics of the nuclear aircraft. While grounded for maintenance or other reasons, the liquid metal reactor coolant would be drained and replaced with helium. This would allow the reactor to power down and be removed from the aircraft, being dropped down into pits for storage and maintenance in below-ground 'hotshops'.

It was estimated that the reactor could be powered down sufficiently to safely allow the flight crew to leave the aircraft within ten minutes of landing; the process of de-fuelling the aircraft – removing the liquid metal and nuclear fuel – would take ground crew up to three hours.

In order to fly again, either the chemical-burning engines or external auxiliary power systems would be used to get the liquid metal system up to temperature; the reactors would be loaded and activated at the last possible moment. However, it was found that it would take approximately three hours to prepare a bomber to fly, each bomber on a base requiring a dedicated 'filling station' during that time. A base with 12 such stations would only be able to prepare 12 aircraft every three hours – hardly the lightning fast response time needed to respond to a Soviet sneak attack.

Boeing designed hundred of nuclear powered bombers. The designs ran the gamut from surprisingly small to gigantic; subsonic to supersonic;

Fairchild NEPA N-16
SCALE 1/320

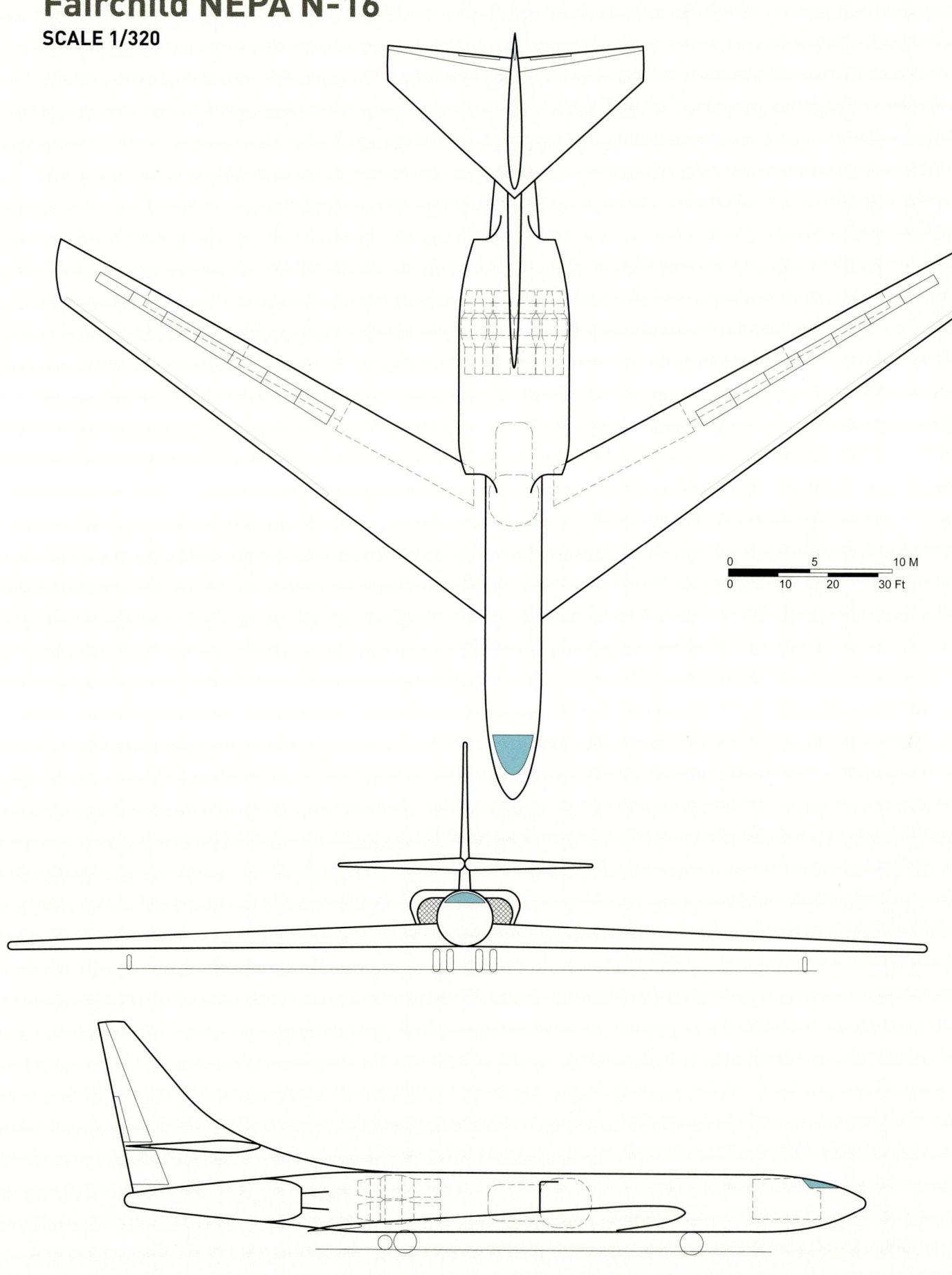

Boeing B-52A/G.E. LF-2
SCALE 1/300

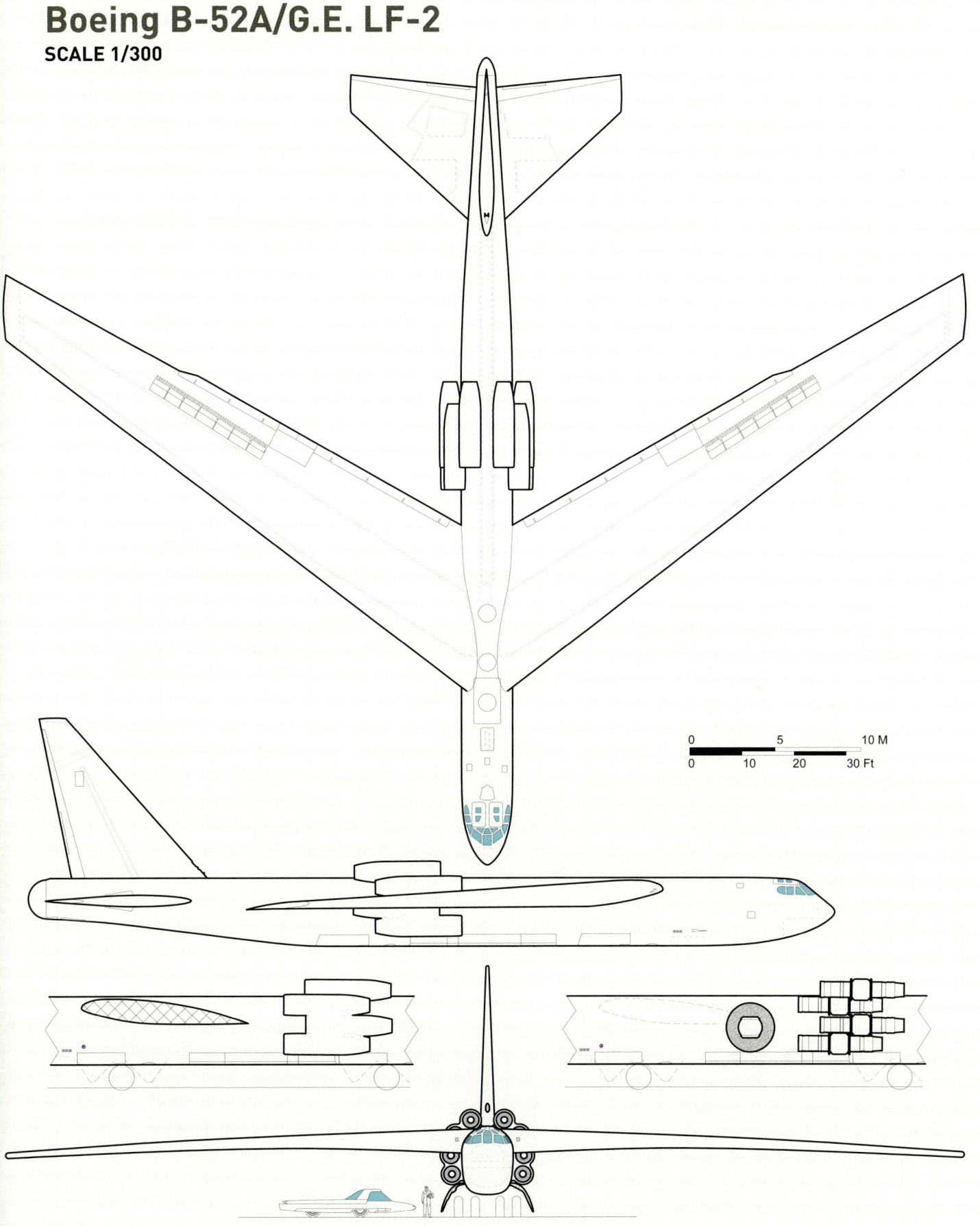

Wings shown relaxed as on the ground

Boeing Model 702-138 (1)-1
SCALE 1/400

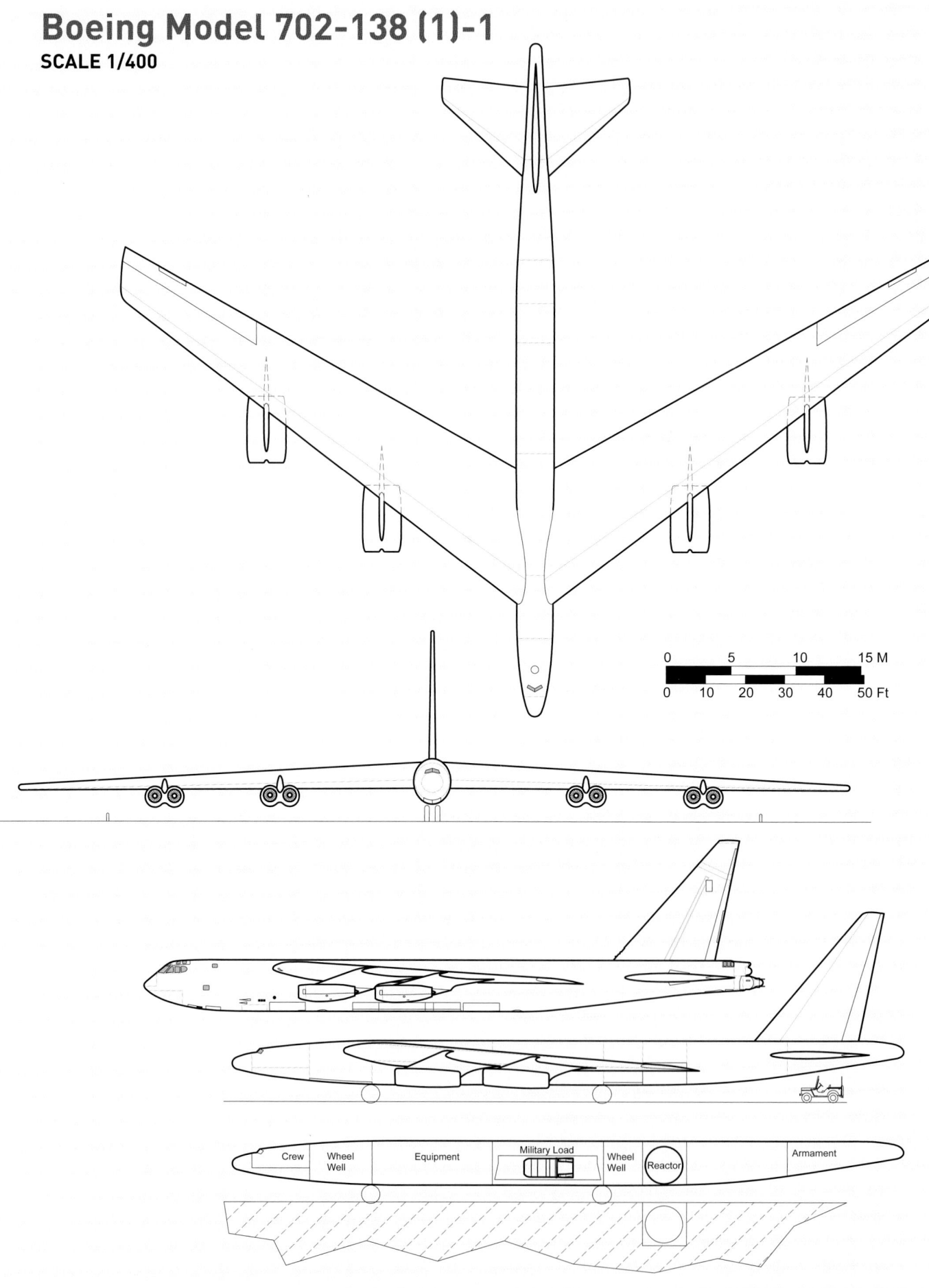

B-52 Flying Nuclear Testbed
SCALE 1/350

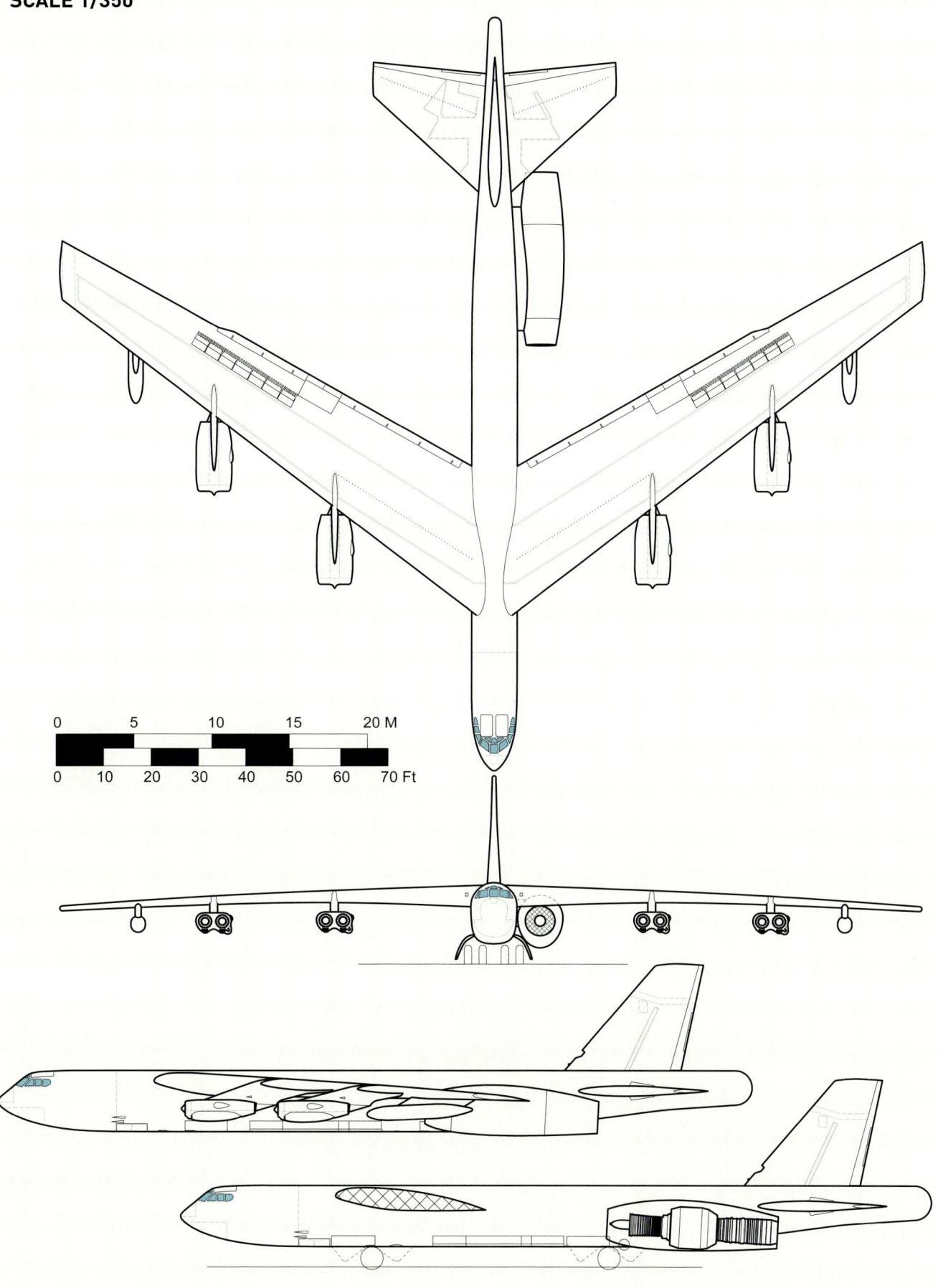

conventionally configured to extremely weird. But something Boeing did not seem to do as much as one might expect is design nuclear powered versions of the as-built B-52.

B-52 Flying Testbed

General Electric spent years working on nuclear engine designs. By the time President John F Kennedy cancelled the programme in 1961, the company had two nuclear jet engines in advanced development: the XMA-1 used a single reactor integrated with two X211 turbojets, and the XNJ140E married a reactor to a single jet engine. Both systems were direct air cycle designs; both attached the compressors as close as possible to the front of the reactor and the turbines as close as possible to the back. Consequently these looked like turbojets with very fat combustors, rather than complex assemblies of pipes and engines and reactors.

While GE had tested the X211 turbojet components at operational conditions for the XMA-1 engine, it had then cancelled that engine in favour of the XNJ140E which seemed to hold more promise. Unfortunately, by the time of overall programme cancellation, that engine had not yet made it to hardware stage. GE had, however, drawn up details of how to test the engine. The Air Force had planned on using the Convair NX2 aircraft as a multi-engine demonstrator, an all-new aircraft configuration meant to demonstrate a practical tactical weapons system configuration. But GE felt that the B-52G could serve as a single-engine demonstrator.

The company studied several configurations, placing the XNJ140E-1 engine inside the rear fuselage or on top of it, but eventually it was concluded that hanging it in a special nacelle off the left side of the rear fuselage was the most practical option. The result seemed to mirror the Canadair CL-52, with a similarly outsized nacelle seemingly clumsily nailed to the rear of the aircraft, making it appear ill-balanced. As the engine assembly weighed 60,000lb, it certainly seems remarkable that the B-52G did not end up tail heavy – though the additional 30,000lb of shielding required for the cockpit would have provided some counter balance. Additional studies suggested that the B-52G could take a second XNJ140E-1 nacelle on the other side; two of these nuclear engines would allow the aircraft to shut down all four of its conventional J57 jet engines (the other four having been removed) and fly on purely nuclear power. Illustrations of this two-engine configuration are not currently available.

An interesting detail was pointed out: the crew's oxygen supply would need to be located in a low-radiation area of the aircraft... not because gamma rays of neutrons from the reactor would 'activate' the oxygen and make it radioactive, but because the radiation would spark the creation of ozone (O_3) – chemically toxic to the crew. Other concerns were raised regarding some of the standard materials used in the B-52: elastomers in hoses, tyres, wires and seals, along with semiconductors and lubricants and hydraulic fluids would be negatively affected by radiation. These problems were not as well understood a decade earlier, but now they were known issues to be addressed.

The cockpit as shown looks unchanged but would have had to be completely new – a heavily shielded bunker like that of the Convair NB-36H.

CHAPTER 7

B-52 Miscellaneous Designs

Boeing Model 703

A massively modified B-52 minelayer aimed at the Navy, Model 703 was designed in December 1951, at the same time as Model 450-150-30 (aka Model 704), a B-47 modified for the same role. It was about the same size as the Model 450-150-30, but featured aerodynamic refinements in line with B-52 design principles. The wings were not as swept as those of the B-52, but they shared the B-52's variable thickness ratios. The wings differed from those of the B-47 and B-52 in that they had substantial ailerons. The tail surfaces resembled those of the B-47.

Unlike that of the Model 450-150-30, the cockpit was fully faired into the fuselage like that of the B-52 and indeed bore a distinct resemblance to the B-52 cockpit. The four turbojets were in two pylon-suspended nacelles like those of the B-52, though fitted with Westinghouse J40-WE-8 turbojets rather than J57s. Unlike either the B-47 or the B-52, Model 703 was equipped with reasonably conventional tricycle landing gear. The main gear was composed of two dual-wheel bogeys that folded up into the fuselage and a conventional nose gear stowed under the cockpit.

The bomb bays, one in the forward fuselage and the other in the rear, separated by the wing and main landing gear, would be filled with up to 30 x Mk 36 1,000lb mines, or 15 x Mk 25 2,000lb mines, or 36 x Mk 50 500lb mines. For defence, Model 703 had a radar-aimed tail gun turret with two 20mm cannon. The gunner was in a unique location – beneath a low streamlined dome atop the middle of the fuselage. Performance data is not available.

Boeing Model 809-1004

This was one of the odder B-52 derivatives. Designed in early May 1957, this bomber was meant specifically for low-altitude operations. The B-52 had originally

Boeing Model 703
SCALE 1/250

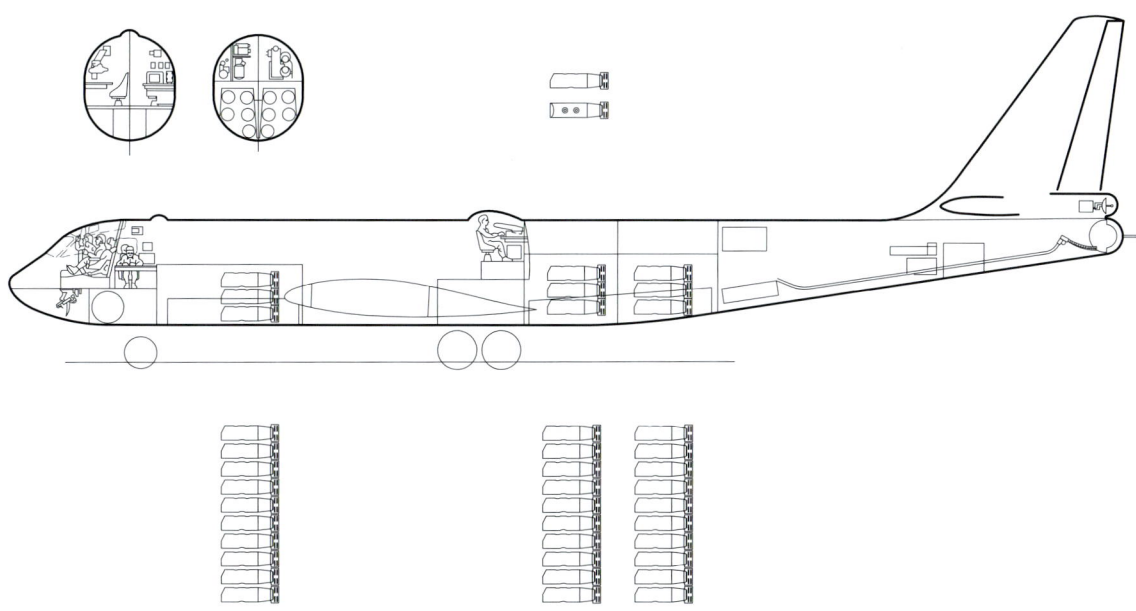

225

Boeing Model 703
SCALE 1/250

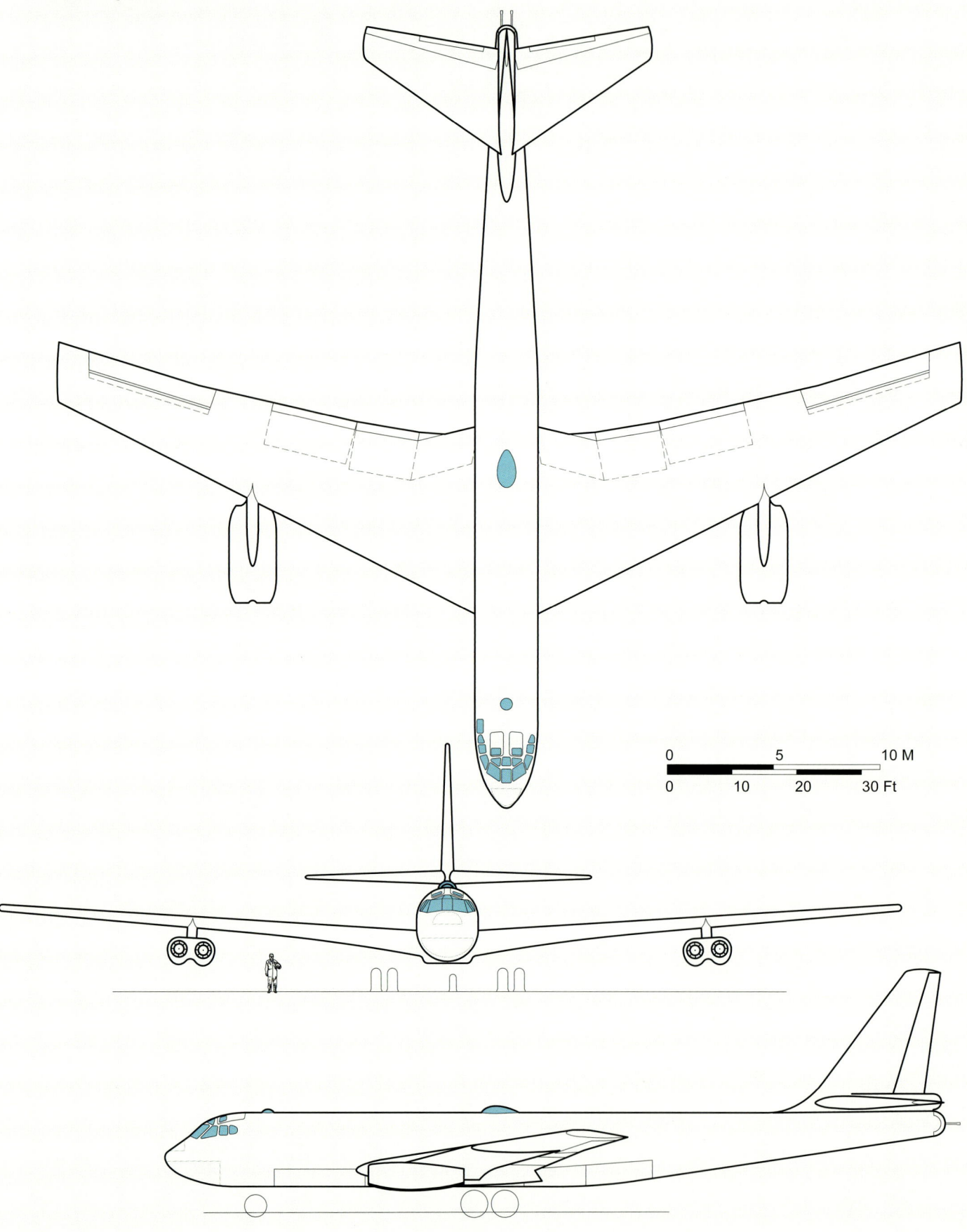

Boeing Model 809-1004-1
SCALE 1/250

B-52 Miscellaneous Designs

227

been designed for high altitude flight, to keep it out of the reach of enemy fighters, but as mentioned earlier advances in not just interceptor technology but also surface-to-air missiles made high-altitude subsonic bombing questionable at best.

Low-altitude bombers have different aerodynamic requirements than high-altitude designs; the thin air high up means lift is reduced, so wings need to be large. Low down, large wings are actually detrimental to an aircraft… making it susceptible to the turbulence and gusts that characterize the chaotic air closer to the ground. So Model 809-1004 was to be equipped with wings and tail surfaces that look shockingly small for a B-52.

But it wasn't the small wings that truly set the Model 809-1004 apart – it was the complete absence of podded engines below the wings. A casual glance might lead an observer to think that the aircraft had no engines. But in fact the aircraft not only had engines, it had a lot of them: 34 or 38 General Electric YJ-85-GE-1 turbojets were embedded within the wings.

The J85 was originally designed for the McDonnell ADM-20 'Quail' decoy. For this application it was made from lower performance materials; adequate to survive the single flight that the Quail would have. But it was soon improved with better materials and went on to successfully power aircraft such as the Northrop T-38 Talon trainer, the related F-5 fighter and the Cessna A-37 Dragonfly. All of these are quite small aircraft and the J85 itself is a small engine. The YJ-85-GE-1 planned for Model 809-1004 would produce a mere 2,450lb of thrust, and to do that it required a 20% water injection.

The small turbojets would be buried within the wings near the trailing edge. The wings would serve as long linear nacelles – the entire leading edge was a rectangular inlet and the entire trailing edge was an exhaust. A long narrow-chord flap lined the bottom edge of the exhaust; during normal flight this would block about half of the nozzle area; at slower speeds the flap would be deflected downwards and would open up the bottom half of the exhaust, which would be deflected directly downwards by the flap. The diagrams show that the wings could be fitted with 34 engines or extended slightly to fit in four more.

Model 809-1004 seems like the sort of thing that would be intended as a technology demonstrator, but it evidently it was aimed at becoming an operational bomber. The single 10,000lb payload (whether a bomb or a missile is not specified) it would be semi-submerged within one of the fuel tanks, rather than carried in the bomb bay. If it was a nuclear gravity bomb, that bomb would definitely have to be of the 'laydown' variety in order to give the bomber a chance of escaping the blast.

At least two other Model 809 designs are known, a STOL bomber and a tail-sitting VTOL bomber. Model 809 was clearly used for some of the more unconventional thinking in bomber design.

Standby Alert

A somewhat crude preliminary concept for a 'Standby Alert Bomber' was shown in September 1959. This married the fuselage and tail surfaces of a standard B-52 to the Pratt & Whitney T57 turboprop engines and wings from a Model 820-101 transport design (see Chapter 10) to create a new aircraft.

After the launch of Sputnik in October 1957, the US Air Force was keenly aware that the Soviet Union was rapidly developing the ability to strike American targets at a moment's notice. Having bombers sitting around inert, requiring hours to get into the air, was no longer acceptable. The Strategic Air Command resolved this problem with the 'Alert' system where bombers would be fuelled, armed and manned; essentially sitting at the end of runways, engines idling, ready to launch skywards within minutes of the command to do so.

It seems that the 'Standby Alert bomber' was intended to fulfill that very role, to either sit near a runway ready to go, or perhaps to be kept flying for extended periods, regularly refuelled. A turboprop bomber might not be as capable as a turbojet bomber, but a less capable bomber that survives a Soviet first strike because it is already in the air is infinitely better than the best jet bomber in the world after it has been evaporated by a Soviet ICBM.

Unfortunately the fragmentary information available to the author does not include performance data.

Engine Testbeds

Numerous attempts were made to replace the aging and inefficient engines of the B-52 over the decades, all of which have so far come to nothing since the introduction of the B-52H. Ironically, the B-52 has played host to more powerful and more efficient engines several times… but only as a flying testbed for the engine itself, not to demonstrate the potential of a re-engined B-52.

Boeing XB-52/J75

The virtually handmade XB-52 was substantially different from the operational B-52, so converting it into an operation bomber was not a practical possibility. Nevertheless it continued to provide some service even after its flight testing phase had ended. It arrived in Wichita for modifications in March 1957, the outboard J57 engine nacelles being removed and replaced with single afterburning Pratt & Whitney J75s.

Boeing Standby Alert
SCALE 1/350

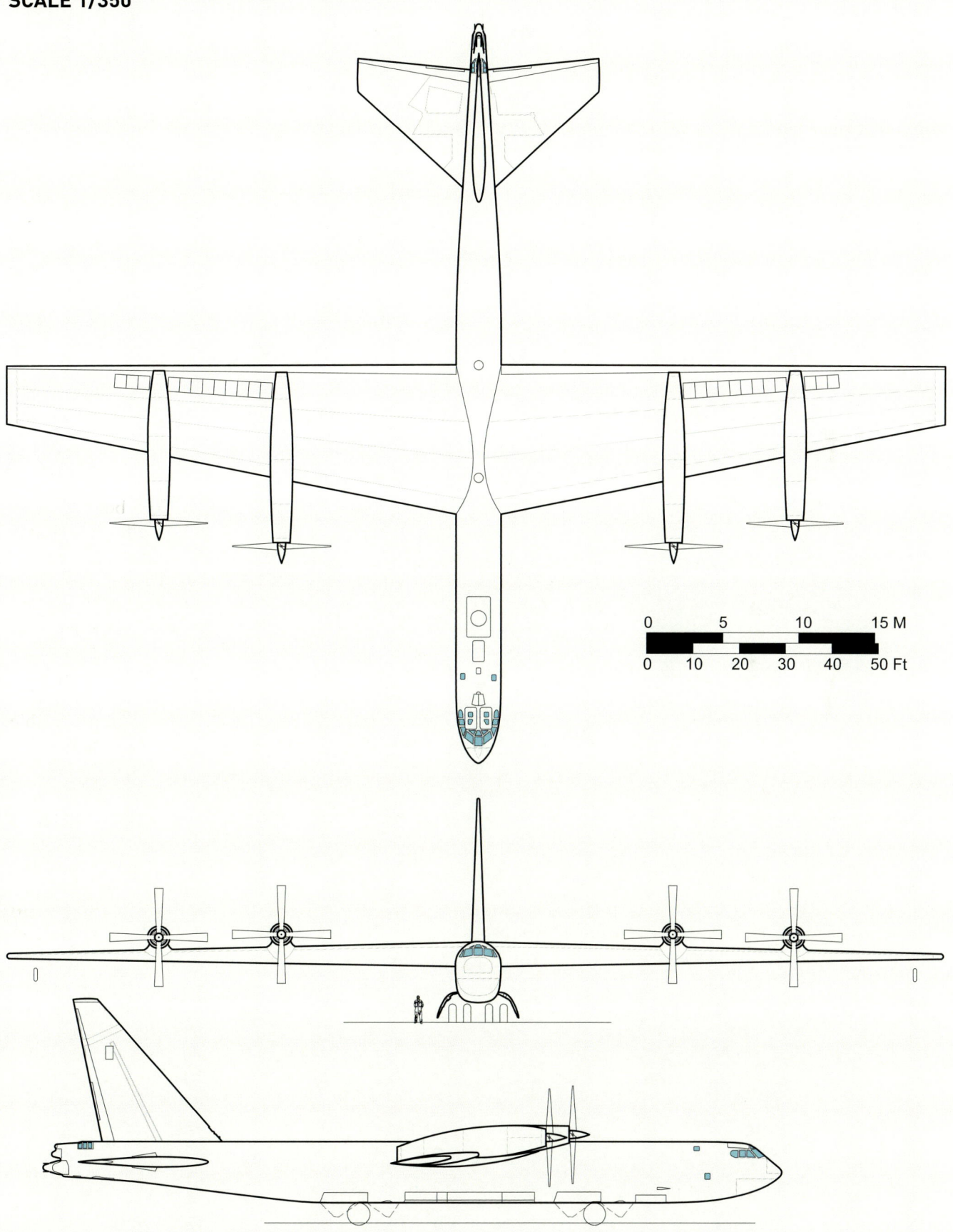

Boeing XB-52 + J75
SCALE 1/300

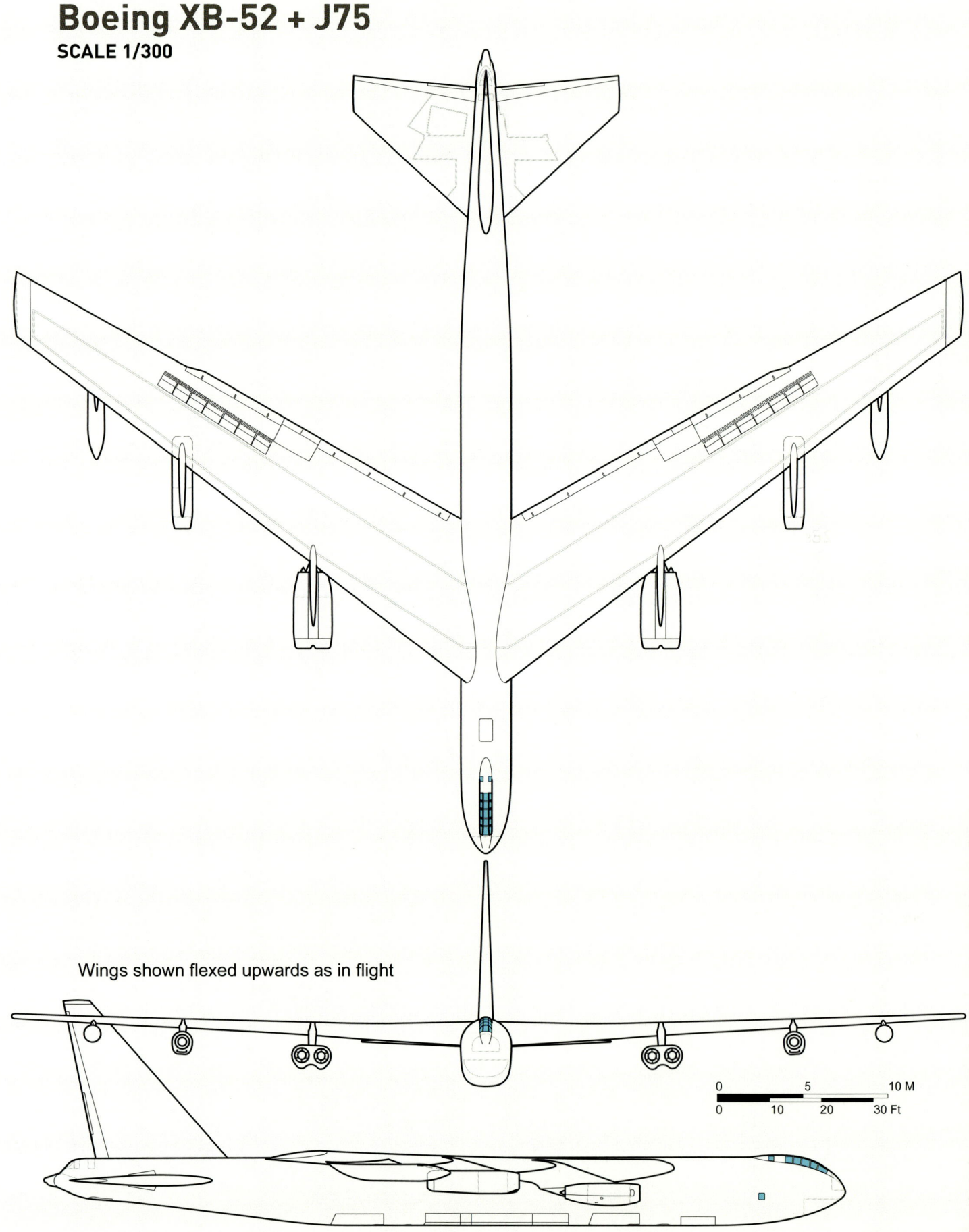

Wings shown flexed upwards as in flight

The J75 in its military and afterburning form powered the Convair F-106 and Republic F-105; without the afterburner it propelled the Lockheed U-2. The civilian non-afterburning version, the JT4A, provided thrust for the Boeing 707 and Douglas DC-8 jetliners.

Some structural modifications were made to the wings to allow them to withstand the much higher thrust levels that the new engines would produce. The XB-52's YJ57-P-3 engines produced a maximum of around 8,700lb of thrust, but the J75 with afterburning could generate 24,000lb. A single J75 could thus produce around 1.4 times as much thrust as the two-engine J57 nacelle that it replaced.

The re-engined XB-52 was delivered to Wright-Patterson Air Force Base for testing in November 1957 but the result, as well as the extent of the testing, is unclear. Since the XB-52 survived to be unceremoniously cut apart for scrap in the mid-1960s, clearly the engines did not rip themselves off the wings.

Boeing JB-52E

Several B-52Es became notable in the late 1960s as they each had the inboard engines under their right-hand wings replaced with large turbofan engines. This was done to test out the new generation of turbofans intended to power the 'jumbo jets': Boeing's own 747, the McDonnell Douglas DC-10 and the Lockheed C-5 Galaxy. These test flights provided valuable information but did not lead to serious efforts to re-engine the B-52s themselves.

All of the engines were high-bypass turbofans of much greater diameter than the TF33-P-3 turbofans used on the H-model of B-52, never mind the J57 turbojets used on the E-model. The large diameters of these new engines made fitting them under the wings of the B-52 a bit tricky; the end result being ground clearance comfortably measurable in inches. B-52E number 56-636 was given a Pratt & Whitney JT90-7R4, intended for the 747. B-52E number 57-0119 got a GE TF39 military turbofan meant for the C-5 programme; that type of engine evolved into the civilian GE CF6 which ended up on the McDonnell Douglas DC-10 jetliner. The aircraft was retired in 1981.

Flashback

One truly effective way to be intriguing is to be mysterious. And the 'Flashback' modification of a B-52C in the mid-1960s fits the bill perfectly. In this case, the modifications made to the B-52 are not the interesting part, but the 'Flashback Test Vehicle' (FBTV). This was a payload built and carried by a B-52, a payload intended to be dropped like a bomb. But was it a bomb? Was it a pure science experiment, or was it meant to lead to a practical device? These and other questions remain unanswered, at least as far as this author is aware.

What is known is that in January 1965, the Flashback Test Vehicle was tested at Kirtland Air Force Base in New Mexico. These tests did not involve the dropping of a nuclear bomb which then detonated; such atmospheric tests were banned under the Partial Test Ban Treaty of 1963. Instead, the susceptibility of the Flashback device to electromagnetic emissions from the ARC-58 transmitter in the B-52 carrier aircraft and the device's own telemetry transmitters was the point of the test.

The Flashback device itself was large: large enough that in order to fit into the B-52 the bomb bay doors had to be removed – even then the device protruded from the belly of the aircraft. It was about 8ft in diameter and 24ft 9in long, not counting the protruding parachute pack or antennae.

The configuration of the Flashback Test Vehicle was distinctly bomb-like. At the front was a round-nosed cone that in configuration and dimension closely – though not precisely – match those of the Mark 6 re-entry vehicle from the Titan II ICBM. A short cylindrical section followed, with a tapering tail section with three wedge-like tail fins. In the rear of the tail was a large parachute package. The tail fins, it should be noted, are quite similar to the fins on large hydrogen bombs of the time. A number of antennae protruded from the rear of the device, presumably transmitters of onboard telemetry generated during a drop.

The Flashback Test Vehicle and modified B-52C were shown in some poor-quality photos and some decent-quality wind tunnel model diagrams. These diagrams were used to create the diagrams here and provide a good guide to shape and dimensions. The wind tunnel testing was conducted at the Cornell Aeronautical Laboratory in 1966; the testing was conducted for the Sandia Corporation, the designers of the Flashback Test Vehicle. Sandia Corporation operates Sandia National Laboratories, which is a premier nuclear weapons ordnance engineering laboratory… i.e. they designed nuclear bombs. The wind tunnel models of the FBTV and the B-52 were built from aluminium and supplied to Cornell by Sandia. The tests involved determining the aerodynamics not only of the FBTV within the B-52's bomb bay but also during a drop. It seems that the FBTV was planned for an actual drop from the B-52 in September 1967 over Johnston Atoll as part of Operation Paddlewheel; whether that drop occurred is unknown.

Intriguingly, during the test "all HE (high explosive) and nuclear components were deleted," and "a simulator was used to replace the warhead".

Boeing JB-52E
SCALE 1/300

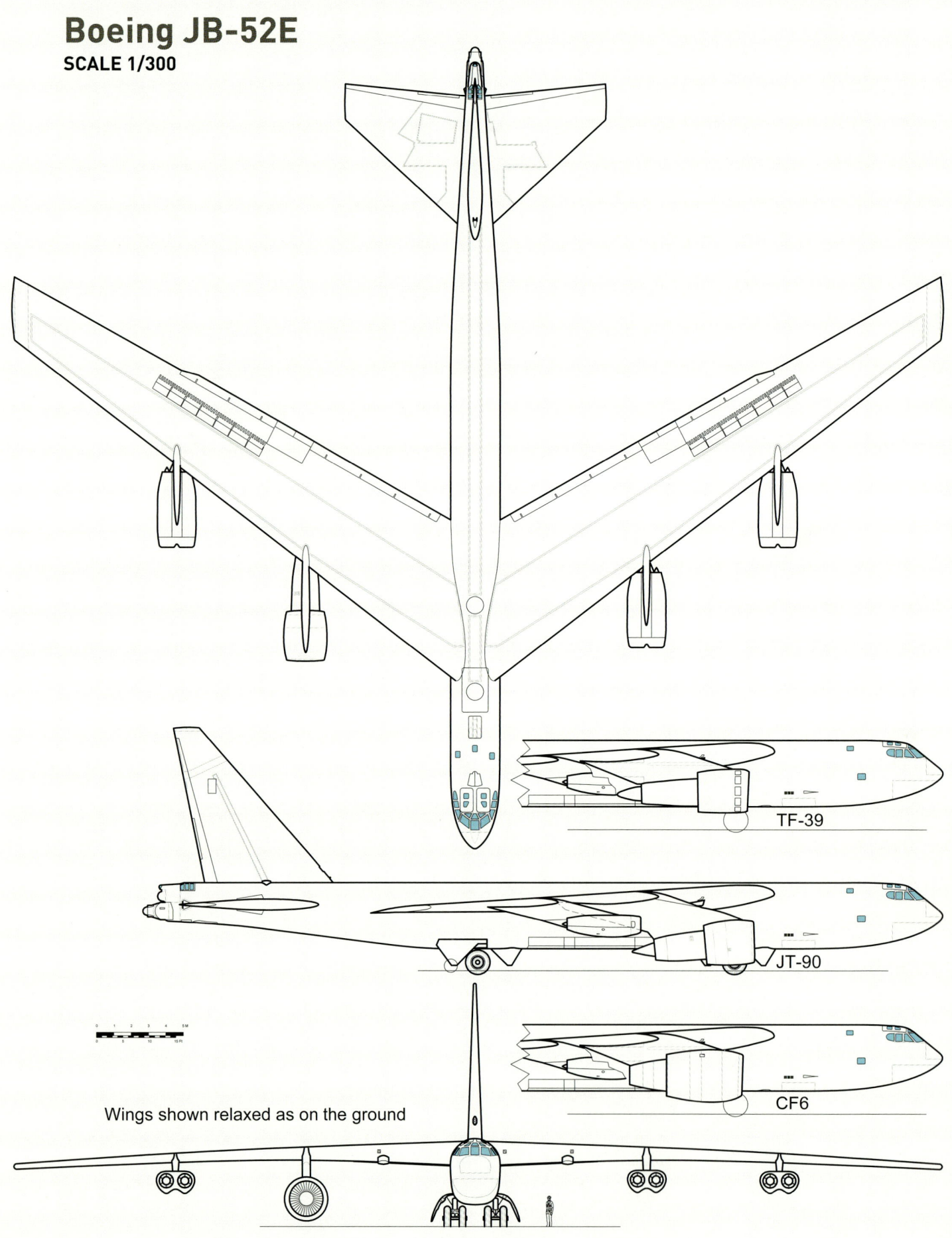

Wings shown relaxed as on the ground

TF-39

JT-90

CF6

Flashback Test Vehicle/B-52C
SCALE 1/300

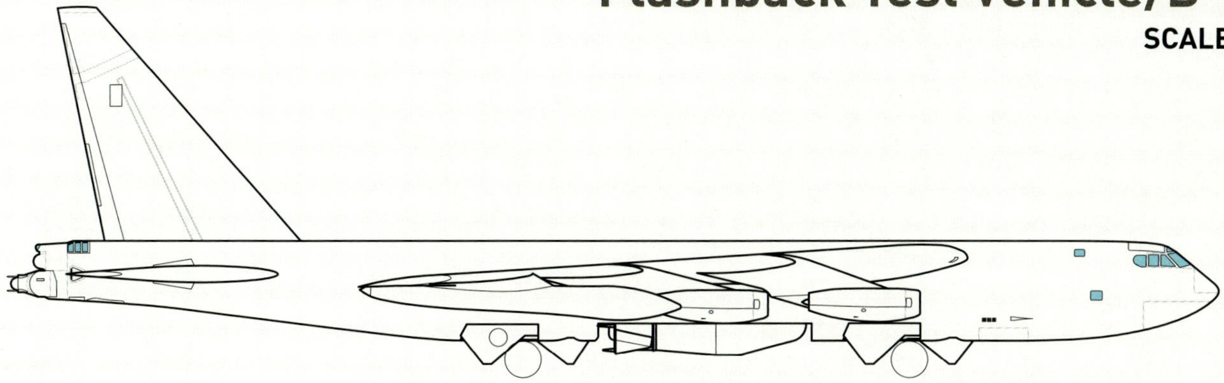

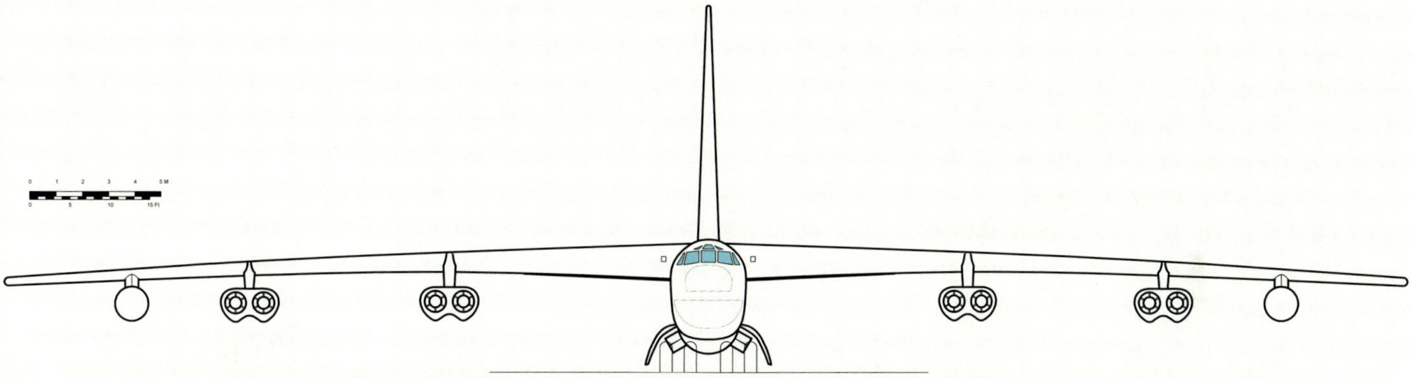

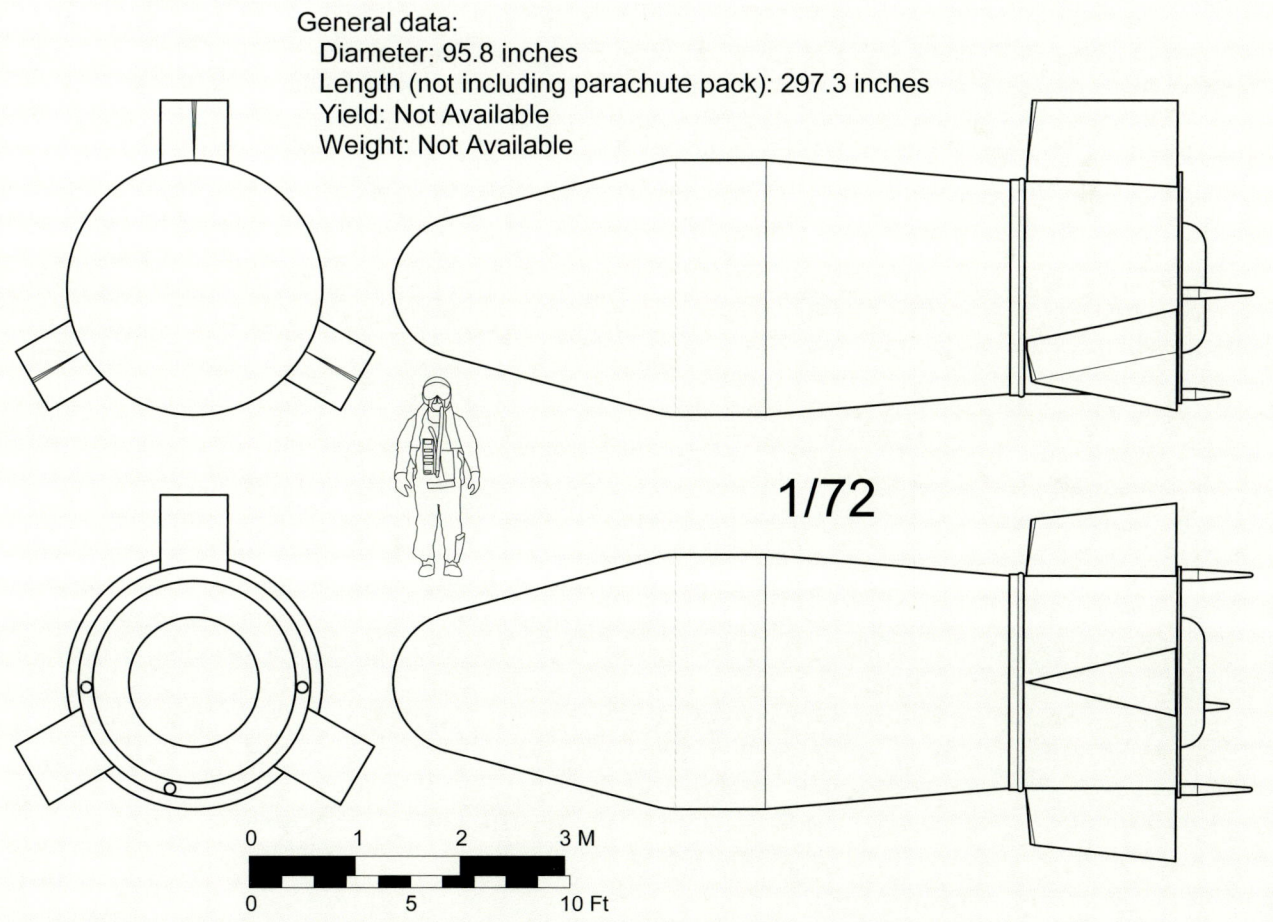

General data:
 Diameter: 95.8 inches
 Length (not including parachute pack): 297.3 inches
 Yield: Not Available
 Weight: Not Available

1/72

So… what we have is a very large device that was intended to be dropped from a B-52, and which was supposed to have a nuclear warhead. It seems unlikely to be a stand-in for an existing bomb; any bomb then in existence could have been easier to use than something entirely new.

The Flashback Test Vehicle was larger by far than any other nuclear bomb in the US inventory at the time. It is a little bit bigger than the Soviet Tsar Bomba – the 50 megaton monster detonated in 1960 and designed to be 100 megatons in yield. Was Flashback a response to the Tsar? If so, it certainly seems like it could have been designed for a yield in the hundred-megaton range. But, after all that, why use the Mark 6 RV from the Titan? The Mark 6 RV contained a W-53 warhead with a yield of nine megatons.

There was no need to use Titan components in order to match the W-53 to the B-52; the B-52 was perfectly capable of carrying the B53 bomb, which used the W-53 and also generated a nine megaton blast. Dropped from a B-52, it was certainly not going to encounter re-entry conditions. The inclusion of a parachute pack shows that it was not going to be allowed to break the sound barrier on the way down; most likely it would begin deployment of the chutes almost immediately after being dropped.

Sandia was the original promoter of the 'laydown' bomb, one which used a parachute to lower the bomb relatively gently to the ground; upon landing the bomb would not necessarily immediately detonate but would instead wait a set period of time. This, coupled with the retarded descent, would give the bomber aircraft the time needed to escape; for a bomb with a yield potentially measuring in the hundreds of megatons, the B-52 would need every second it could get to get away.

Hopefully the Flashback story will come out one of these days.

Boeing NB-52E Control Configured Vehicle

The advancement of computers in the 1960s and 1970s allowed new ideas to be incorporated and tested. One such was 'Control Configured Vehicle' (CCV) technologies. CCV marries sensors with fast-acting computer control of aerosurfaces. When it works, this permits manoeuvring and ride control that human pilots are unable to provide. Ride control is especially important for aircraft that fly low to the ground, where air is disturbed by downdrafts, thermals and air flowing around and over terrain. An active control systems computer is able to detect the beginnings of a 'bump' and command aerosurfaces to create counter-forces, smoothing out the ride without pilot involvement.

The earliest computers that might have been able to handle these duties were too big, heavy, expensive and power-hungry. But by the 1970s, computers had reached the stage where aircraft designers could seriously consider adding active controls to military aircraft. Like any new technology, it needed to be tested. And so Boeing B-52E number 56-632 was selected for modification by the Air Force Flight Dynamics Laboratory and Boeing to test CCV technologies.

This aircraft had been previously modified in 1969 for the Load Alleviation and Mode Stabilization (LAMS) programme, a similar project to demonstrate gust load alleviation. LAMS included adding a sensor boom to the nose of AF56-632; hydraulic controls were added to the control surfaces and a fly-by-wire system was incorporated. The LAMS flight tests showed that using automated systems to tinker with the existing control surfaces in real time could indeed reduce loads on the aircraft imparted by gusts.

And so the CVV programme began in 1971, Boeing again working with the Flight Dynamics Laboratory. AF56-632 was given two horizontal nose canards for pitch control plus a vertical canard for yaw control below the nose. Ailerons were added outboard of the outboard engine pylons; three-segment longer-chord flaperons replaced the inboard flaps. Each external fuel tank had 2,000lb of ballast added to its nose and two Electronic Associates Inc. Model TR-48 analog computers were installed in the lower section of the fuselage nose. The flight control system remained fly-by-wire and the long nose boom previous added to the aircraft was retained. The changes were sufficiently drastic that the Air Force rebranded this aircraft as the lone NB-52E. It was painted in a distinctive red color scheme with patterns on its wing and fuselage to aid in measuring responses when caught on film.

Flight testing CVV began in January 1973 and ended in November, helping to develop several systems in the process, such as Ride Control, Flutter Model Control, Manoeuvre Load Control and Augmented Stability. The systems demonstrated a 30% reduction in accelerations from turbulence, a 10 knot increase in flight speed above what would normally wreck it due to flutter, reduced wing root bending moments during manoeuvres, satisfactory flight characteristics when flown at neutral static stability and reduced fatigue.

After the successful test series, the aircraft was retired to Davis-Monthan Air Force Base in 1974 where it sat in the Arizona desert until it was scrapped in 1993.

Boeing NB-52E
SCALE 1/300

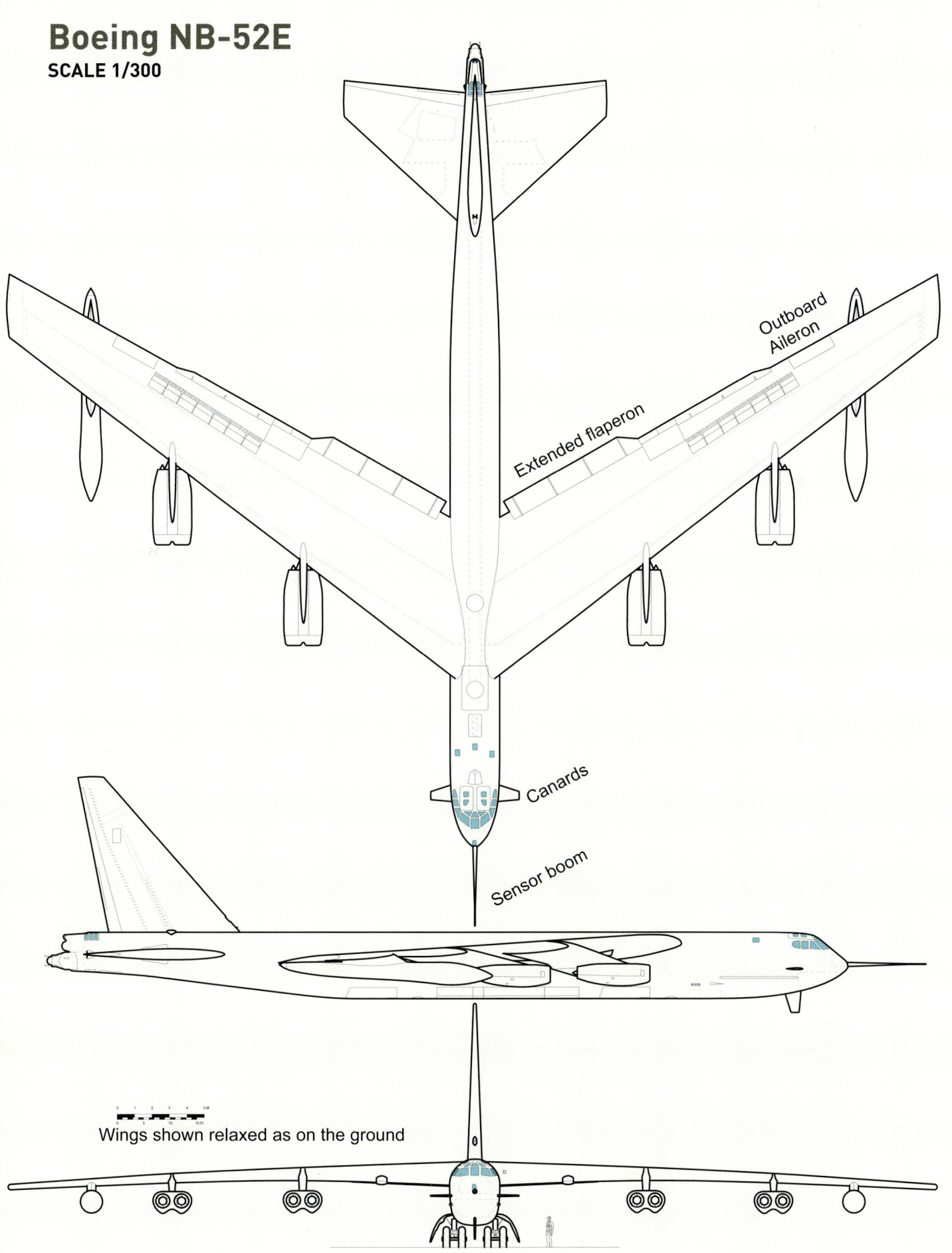

Outboard Aileron
Extended flaperon
Canards
Sensor boom

Wings shown relaxed as on the ground

B-52 Miscellaneous Designs

CHAPTER 8: B-52 as a Carrier

During the last years of the Second World War, the Boeing B-29 was proposed as a carrier aircraft for high-speed rocket powered aircraft. These proposals became a reality with B-29s (and the improved B-50) used to carry the Bell XS-1, X-2 and Douglas D-558-2 rocket planes, as well as the GAM-63 RASCAL. As time passed and rocket planes grew bigger, faster, heavier and more ambitious, the B-29's carrying capacity quickly became insufficient.

Rocket plane designs in the 1950s had ambitions far beyond simply breaking the sound barrier. As the 1950s dawned two aircraft, the B-36 and the B-52, were seen as good options. Indeed the B-36 was often the baseline – it could lift a massive payload and had a giant bomb bay that could be readily modified to fit rocketplanes. But while the B-52 was not as voluminous, it could fly higher and faster.

In the end, the improved flight performance of the B-52 not only meant that the X-15, originally intended for a B-36 launch, would fly from the newer jet bomber; it also meant that anyone who had an idea for an aircraft that needed to be carried to high subsonic speed at reasonably high altitude would design their system for a B-52 launch. The following collection of designs put forward for B-52 carry and/or launch is far from comprehensive.

B-52 and the Dyna Soar

The X-20 Dyna Soar was America's first serious attempt to build a manned, reusable spacecraft. Through the life of the project its design and contractor changed drastically but it always remained a manned lifting vehicle, generally somewhat dart-shaped, relying on advanced high temperature metals to survive the aerothermal heating nightmare of re-entry.

The concept can be said to have originated in wartime Germany with Austrian rocket scientist Dr Eugen Sänger's 'Silverbird', a theoretical design for a long-range rocket powered bomber capable of leaving the atmosphere and able to achieve a once-around global range by 'skipping' off the upper atmosphere. Silverbird enthralled the planners and intelligence services of the victorious allies in the years immediately after the war, though in actuality the vehicle had many design features that would have been impractical or unworkable. The realities of re-entry heating were not adequately understood until many years later; the airframe of the Silverbird would have likely been reduced to tumbling molten debris on the first skip.

But the idea of a rocket powered intercontinental bomber held merit. From the late 1940s through the 1950s there were many who felt that the only way for a long range rocket to effectively strike targets deep within the Soviet Union would be if that rocket was manned, the pilot able to spot the target and adjust the course of the vehicle in order to put a single thermonuclear weapon within a few thousand feet of the target point.

Bell Aircraft under Dr Walter Dornberger led the effort with multiple studies for BOmber MIssiles (BOMI) starting with MX-2276 in 1955. At first the design was a rather imaginative – and unlikely – three-stage system with a manned winged first stage, an unmanned and expended second stage and a manned winged third stage. There was a definite family resemblance with the old Silverbird. But as engineering studies continued, the design became geometrically simpler and more restrained, eventually becoming a cone-cylinder with faceted delta wings atop a conventional expendable rocket booster.

This was all well and good for its time but the march of technology killed BOMI. The Atlas ICBM eventually showed that an unmanned rocket, far simpler than any possible variation of a manned rocket bomber, could be accurate enough. Even though Atlas was still in development, the writing was on the wall. Multi-megaton warheads meant pinpoint accuracy wasn't exactly a necessity, and yet Atlas looked like it could get reasonably close. Suddenly the need for a recoverable manned vehicle, with all the extra weight, complexity and cost, just didn't make sense. And yet, the engineering and design work did not go entirely to waste.

Three programmes were spun off from BOMI in 1956: Brass Bell, a manned, rocket-based reconnaissance system; a direct continuation of BOMI known as ROBO from 'ROcket Bomber' and HYWARDS: Hypersonic Weapons Research and Development System.

236

These programmes would have likely continued on their merry and separate ways if Sputnik had not gone into orbit in October 1957. Later that same month, the Air Research and Development Command (ADRC) took the three programmes, jammed them back together (although some aspects of HYWARDS would continue separately – see below) and called it Weapons System 464L, better known as Dyna Soar (for DYNAmic SOARing, the same sort of skip-glide that Silverbird had been intended to employ).

The ARDC wanted Dyna Soar to produce a pure research vehicle (Dyna Soar Step I) which would in turn be followed by an operational reconnaissance vehicle (Dyna Soar Step II) and then an orbital bomber (Dyna Soar Step III). Nine aerospace companies tendered submissions to the WS 464L competition in March 1958. These varied widely in detail but all were some form of a smallish spaceplane atop an expendable rocket booster. Many of those boosters were uncomfortable kludge of stages, liquid or solid propellant rockets, squat clusters of existing motors and engines in order to save development costs.

HYWARDS

The HYWARDS programme began in late 1956 and the design study period was brief, running from January to April 1957. The concept was to develop a manned research aircraft that, in its ultimate form, could reach a speed of 12,000ft per second and an altitude of 360,000ft... both far in excess of what the X-15 could achieve. To do this, the HYWARDS vehicle would need to be boosted to speed and altitude atop rocket boosters derived from then-ongoing ICBM work. Before this could happen though, the vehicle would need to be tested lower and slower. This would require it to be carried aloft under the wing of a B-52 and dropped. Initial drop tests would be pure glide tests, but the vehicle's onboard rocket propellant tanks would eventually be filled and the HYWARDS craft would expand its flight envelope.

What design work was done by the USAF and the NACA for HYWARDS has been obscured by time and security classification. However, a few designs that were repeatedly tested by the NACA have been associated with the HYWARDS effort, and are very likely the HYWARDS craft or descended from them. Two configurations (A and B) were reported on in 1957 and subsequently over the next few years would receive considerable examination in wind tunnels both subsonic and supersonic. These craft were intended to reach a maximum of 18,000ft per second, or around Mach 16. Both would carry a single pilot and 1,200lb of research instrumentation, using an XLR99-RM-1 engine for thrust. Both would require a cluster of booster rockets to achieve their maximum potential.

Configuration A used a highly swept delta wing with the fuselage entirely above. The wing thus served as a heat shield for the fuselage. The aerodynamic control system was unconventional: wingtip cones. These served the same role as conventional aerosurfaces, but were thought to be more efficacious at hypersonic speeds. The wings also had conventional flaps, primarily useful at low speed. Small dorsal and ventral fins at the rear could potentially be employed for low-speed control.

Configuration B was almost Configuration A flipped upside down, with the wing on top and the fuselage fully exposed. The concept here was that the vehicle would, using downward-canted wingtip fins, generate the same sort of compression lift at supersonic speeds as the North American XB-70, thus reducing drag and weight and increasing hypersonic glide range. It used a deployable ventral fin at low speed to improve stability, and conventional control surfaces.

Both configurations would be carried by the B-52, though as can be seen in the diagrams, it would be a tight fit.

Wind tunnel testing of these configurations would continue into the early 1960s. They were not developed into actual aircraft, nor do there seem to have been detailed engineering designs; they appear to have remained largely in the realm of the theoretical. That said, the basic shape of Configuration A has come back again and again right up to the present day as the basis for a long string of hypersonic glide vehicles. Launched by ballistic missiles, lobbed by aircraft-launched rockets or dropped from orbit, arrowhead-shaped vehicles such as the Hypersonic Glide Vehicle from the 80s and the Common Aero Vehicle from the 2010s have kept that configuration alive.

Martin-Bell 464L CTV

Martin teamed with Bell for the initial WS464L studies, forming the BoMI Division at Martin. This produced a General Management Proposal in March 1958, calling for two entirely separate Dyna Soar designs – the Dyna Soar I 'Conceptual Test Vehicle' (CTV), and 'Proposed Aircraft Configuration' (PAC). The CTV was a minor modification of prior BOMI/ROBO/Brass Bell designs created at Bell and was a secondary vehicle designed for a pure research role. It was also the last obvious major Bell design in the X-20 Dyna Soar story.

It was along the same lines as the prior BOMI concepts – a cylindrical fuselage, a long conical nose, simple facetted clipped-delta wings and dorsal and ventral fins, with an expendable booster fuelled with hydrazine and fluorine. The PAC was geometrically

Ames Mach 10 Demonstrator
HYWARDS Configuration B
SCALE 1/115

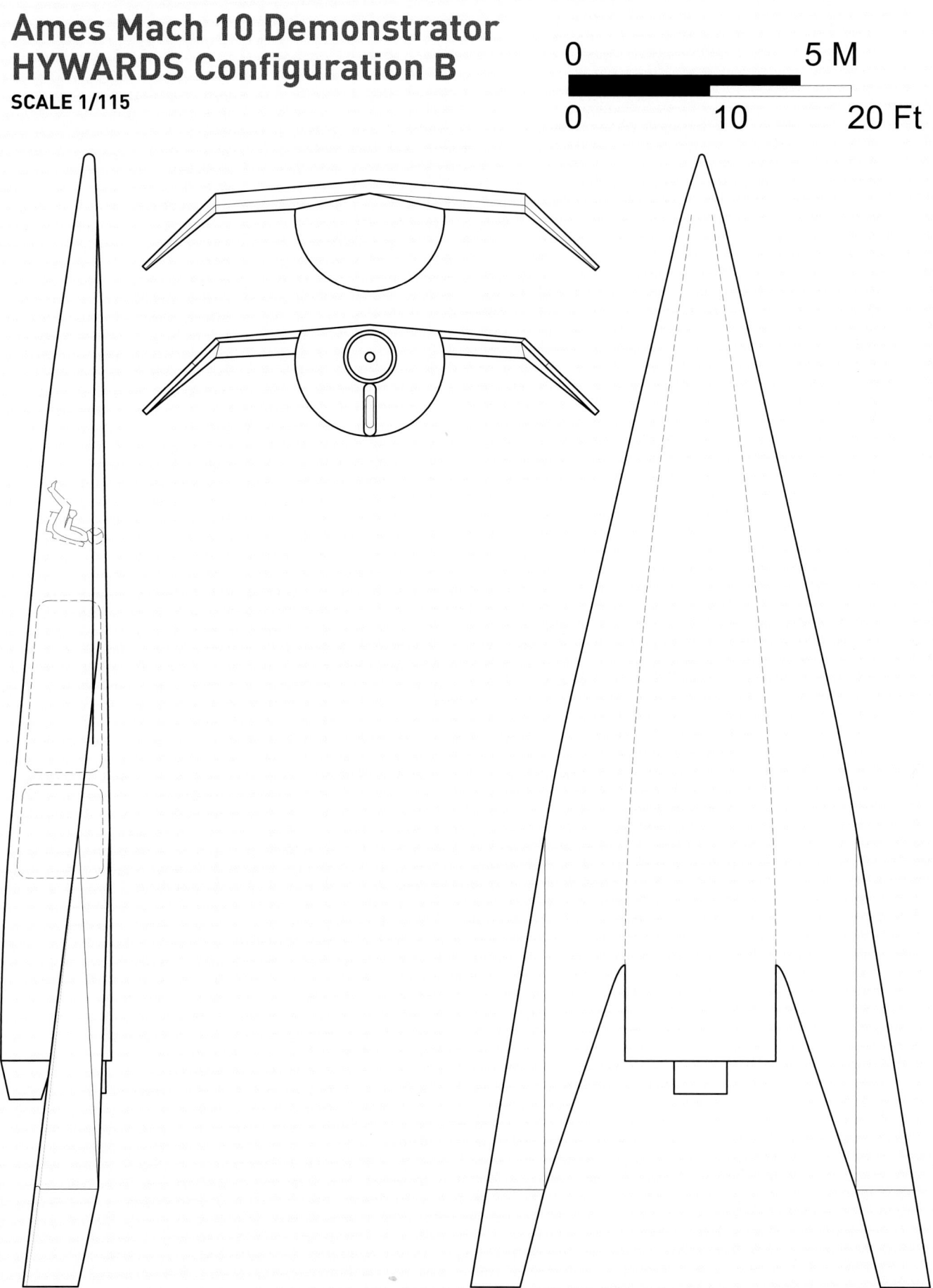

B-52A Carrier + HYWARDS Configuration B
SCALE 1/288

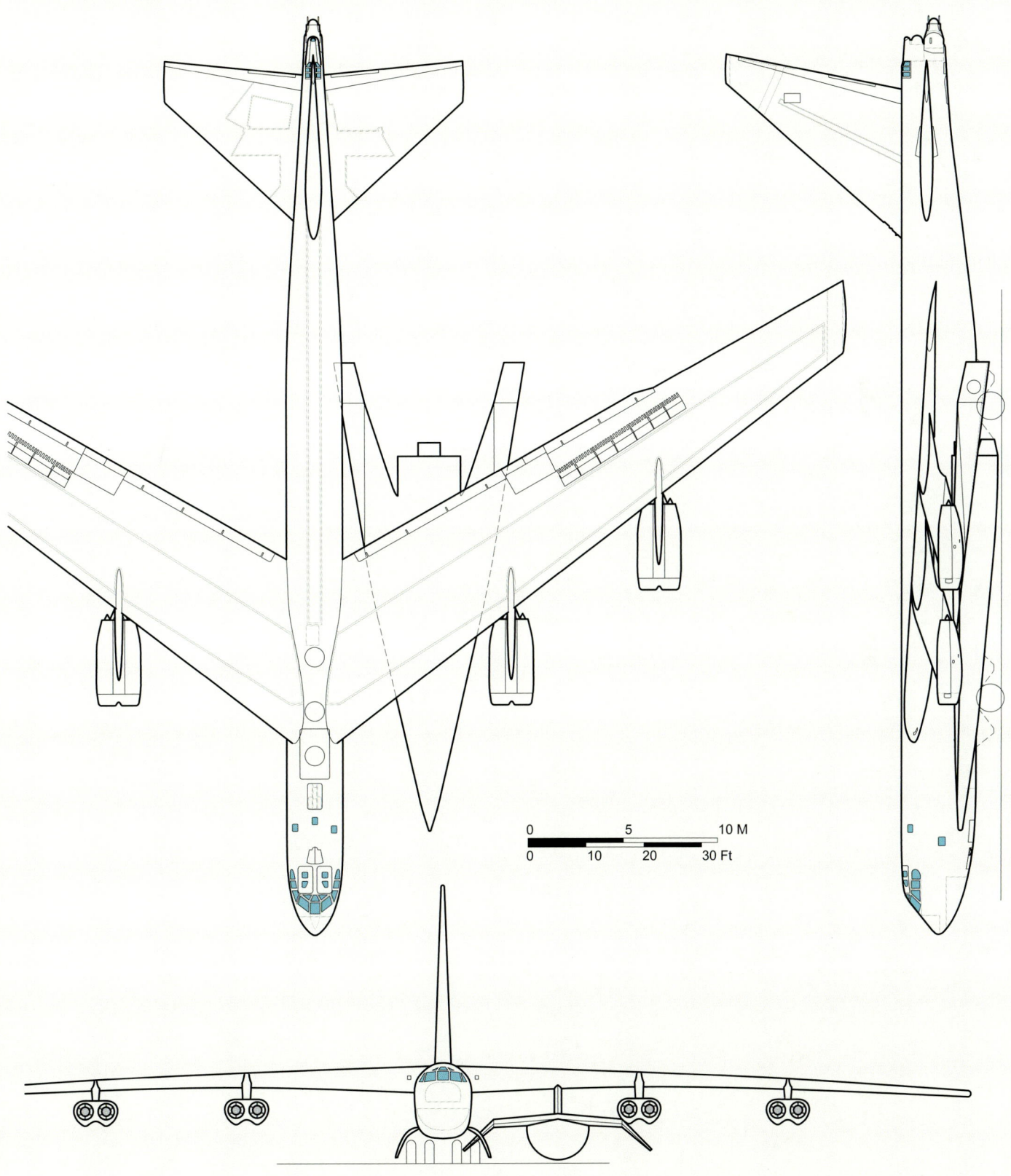

Ames Mach 10 Demonstrator
HYWARDS Configuration A
SCALE 1/115

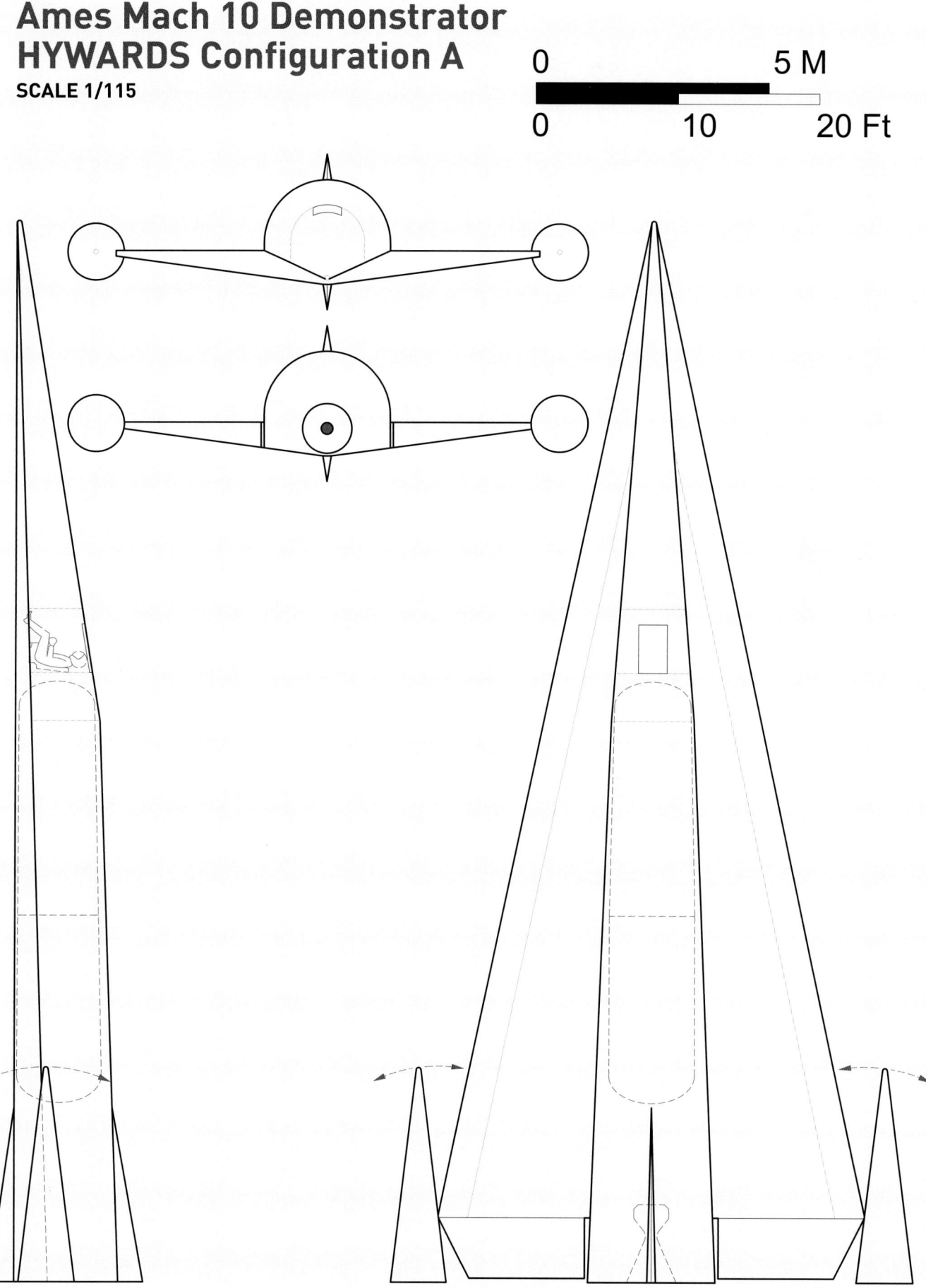

B-52A Carrier + HYWARDS Configuration A
SCALE 1/115

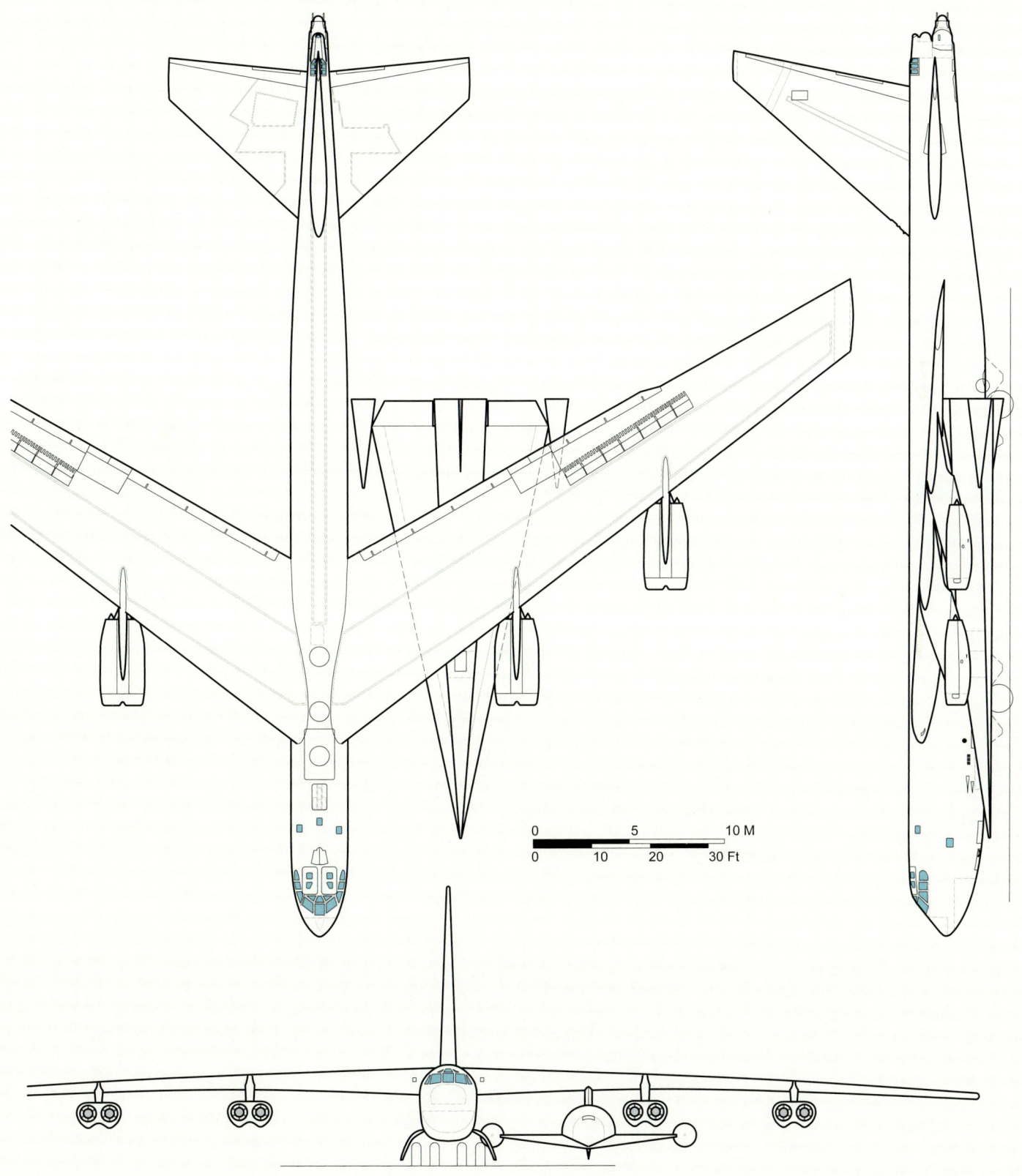

Boeing B-52B/Bell CTV
SCALE 1/300

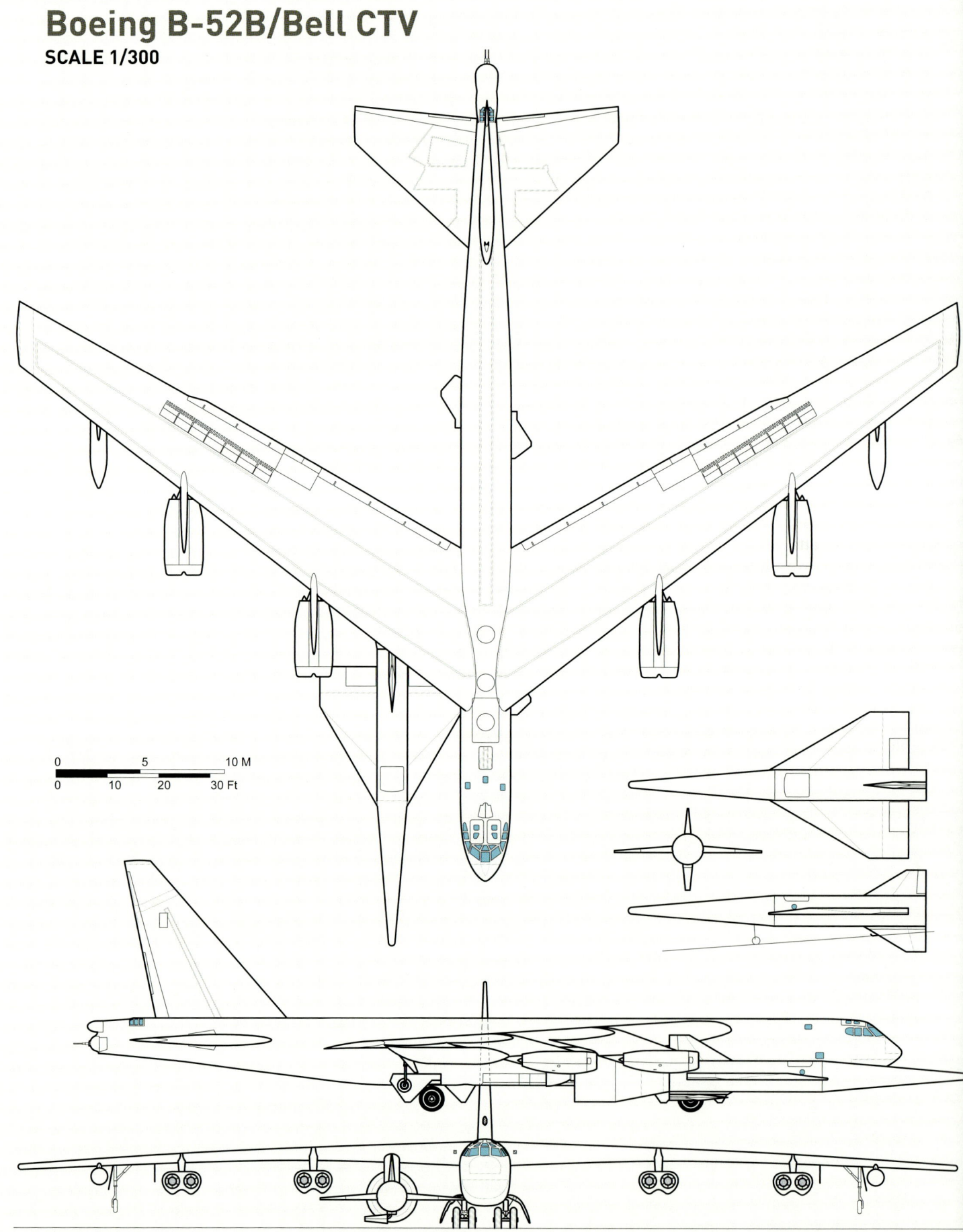

and aerodynamically a wholly different vehicle, one intended to be an operational system.

The CTV was a glider and, as with the BOMI and Brass Bell designs that preceded it, featured a large payload volume in the rear of the fuselage. Along with basic science equipment needed to analyze the performance and responses of the vehicle, various reconnaissance subsystems could be added, as well as bomb simulators. These would test not only bomb separation (a not-inconsiderable issue given the airspeeds involved), but also accuracy and bomb aiming techniques. The CTV would not be an operational bombardment or recon vehicle, but would be useful in developing the techniques and systems needed by an operational system.

Compared to some of the other WS464L designs, and to some of the later Dyna Soar concepts, the CTV was geometrically quite simple. Even so, it would need to be tested incrementally. It was proposed by Martin-Bell that it be carried by a B-52. This would have been an incredibly tight fit using the standard pylon attachment point under the right-hand wing. The pylon would not touch the CTV itself, but would instead be attached to the expendable upper stage behind the CTV. Most of the glider – whose wings seem to have actually touched the B-52 fuselage on the one side, and reached underneath the inboard jet engines on the other – would have projected well forward of the wing leading edge, the nose of the CTV reaching beyond the nose of the B-52 carrier. The ventral fin of the CTV would have to fold to the side to provide clearance and even then it would be a close-run thing.

Despite the years of effort they had put into the enterprise, the Martin-Bell team lost the WS464L contract to Boeing. Ironic then that the glider eventually known as Dyna Soar would look more than a little like the Martin-Bell BOMI Division design for the PAC. In fact, it looked a great deal more like the Martin-Bell design than the relatively bizarre Boeing entry, which looked like the offspring of an arrowhead and a 1950s Cadillac with giant tailfins.

As Boeing refined their concept, it lost everything that made it unique and turned into what Martin-Bell had wanted to build in the first place. The PAC was to be launched by boosters derived from Martin's Titan I ICBM. The PAC was the end of the road for the BOMI, and the beginning of the road for Dyna Soar. And the B-52 was not done with the Dyna Soar.

Republic 464L CTV

Most of the WS464L designs are known only fragmentarily. But like the Martin-Bell proposal, the Republic Aviation proposal is known from documentation which provides decent technical information. Republic's entry was unique in not having a sharply pointed dart-like shape. Instead it was a surprisingly blunt lifting body. But it was roughly the same size and weight as the other entrants and, like at least the Martin-Bell submission, included plans for a CTV that would be dropped from a B-52.

The Republic diagrams depict the CTV fitting under the left-hand wing of the B-52. The test phase would consist of eight unpowered glide flights and four powered flights. The unpowered flights would take off and land at Edwards Air Force Base, while the powered flights would originate there before flying out over the ocean to the Point Mugu Sea Test Range for the drop. The CTV would have a solid propellant third stage rocket motor attached to it and this would push it to high altitude and supersonic speeds. After burnout, the motor would be jettisoned into the ocean; the CTV would glide to a landing at Edwards.

Dyna Soar Drop Test

Once Boeing had the Dyna Soar contract, the company went to work designing and redesigning its concept. While the Boeing Dyna Soar soon had its final basic shape – a tailless glider with a low-set delta wing, flat underside and wingtip fins – several years were spent nailing down the precise configuration, materials and equipment.

Boeing's Model 814-1050 configuration of March 1959 was one of the major design iterations. This was planned for launch into space atop an Atlas-Centaur booster, but for basic flight testing something less aggressive was required. The B-52 was the proposed carrier and studies were made of carrying the Dyna Soar either in the bomb bay (with the bay doors removed to provide clearance for the spaceplane's wings) or under the wing. In the end the underwing position was preferred. There was experience of external carriage via the X-15, and this position also gave room for growth – though Dyna Soar itself was not expected to grow.

In the event it was, compared to some of the other vehicles considered for B-52 underwing carriage, relatively small. But then Dyna Soar itself was a glider. Some design iterations gave it a small turbojet for landing, some installed a pair of Rocketdyne AR-2-1 liquid fueled rocket engines in the rear for flight testing. But such a system would not provide a great deal of performance, given the limited volume in the small fuselage for propellant tanks.

A pair of AR-2-1 engines would get the spaceplane up to about Mach 1.9 – but at least one design would have given the air-dropped Model 814 Dyna Soar considerable performance by adding an XLR-99 rocket engine, sizable external tanks (with 6,000lb

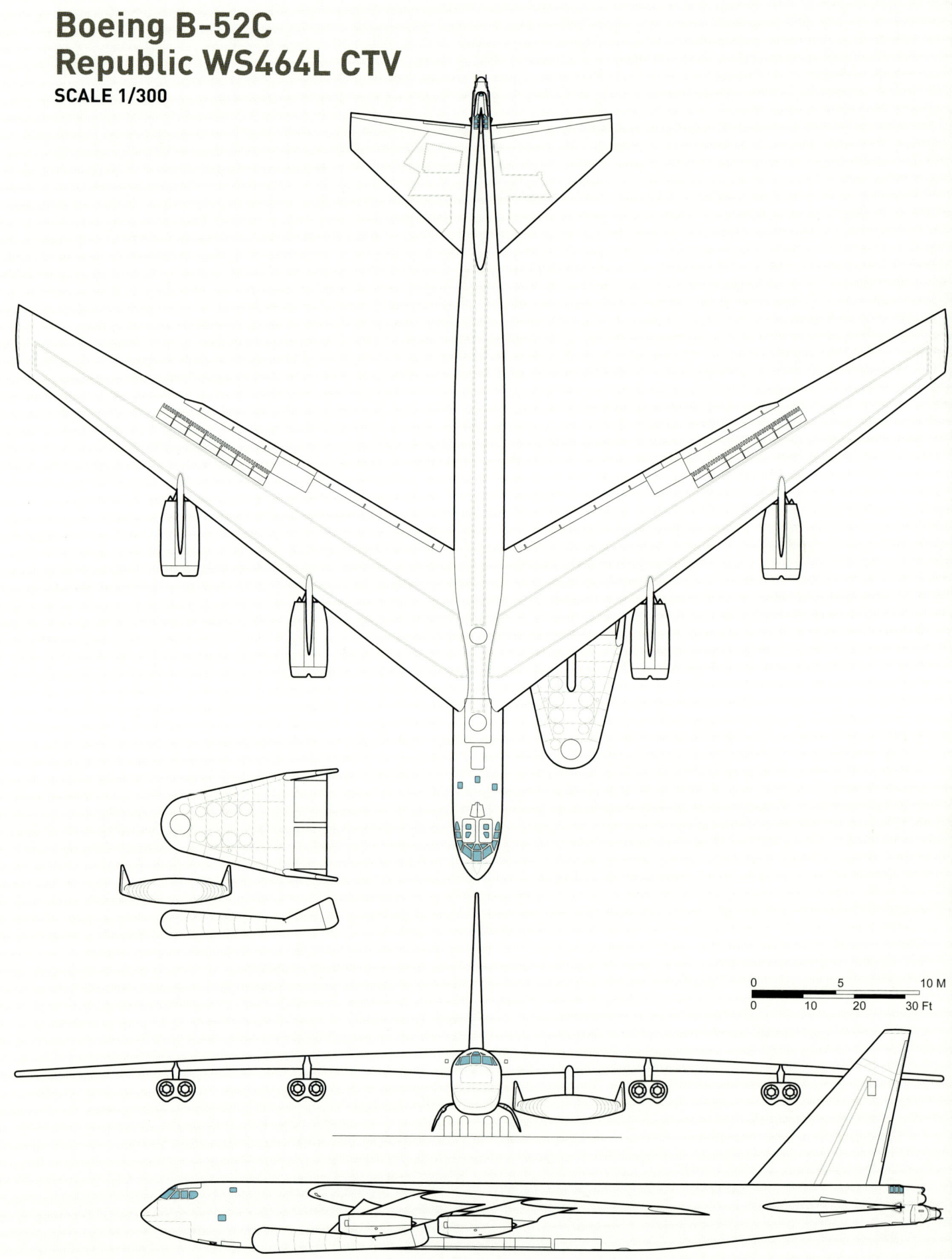

Boeing B-52C
Republic WS464L CTV
SCALE 1/300

Boeing B-52G + Model 844 Drop Test
SCALE 1/300

of ammonia fuel and 7,500lb of liquid oxygen) and stabilizer fins. Complete performance data is not currently available for this configuration, but it would have had performance akin to that of the X-15. Likely able to climb above the sensible atmosphere, it would have had at least a few seconds where the pilot could have used small reaction control thrusters to orient the craft and demonstrate the Dyna Soar's ability to pitch, roll and yaw while in space. Top speed would have been Mach 6; not enough to truly replicate the inferno of re-entry, but enough to make the thermal control systems and thin sheets of refractory metal skin stand up and take notice.

The Model 814 series of Dyna Soar would give way to the Model 844 series. The basic size and configuration of the Dyna Soar was now established, as was the use of the B-52 as a carrier aircraft; but the large booster rocket for drop testing was apparently a short-lived notion.

Dyna Soar Orbital Launcher

Even more so than the Space Shuttle Orbiter, the Dyna Soar spaceplane was in essence a reusable payload rather than a reusable launch system. It needed a substantial launch system to make it into orbit and provided no propulsion of its own to that enterprise. From the beginning of the programme, reusable boosters were proposed for both orbital and suborbital missions as a way to reduce costs for what was hoped to be a constantly-flying operational system. Most were winged rocket powered vertical launch vehicles; they would likely be cheaper than a Titan IIIC in the long run, but would represent a serious development expense and effort as well as suffering similar operational issues to any other launch vehicle.

A compromise concept was theoretically available however: use an existing strategic bomber aircraft to carry a rocket boosted Dyna Soar to altitude. The rocket booster would still be expended, but at lower cost than a full Titan IIIC and with added operational flexibility.

The obvious choice, and initially really the only choice, for a carrier aircraft was the Boeing B-52. In early 1962, NASA researchers at the Flight Research Center in California released the results of a preliminary study on air-launching Dyna Soar (using the Model 814 Dyna Soar) with both subsonic and supersonic aircraft. So long as the expendable booster rockets could generate 325 seconds Isp, a 200,000lb booster system could be carried by a conventional bomber and would get a 10,000lb winged glider to orbit. Several design variations were examined but all shared a nitrogen tetroxide/mixed hydrazine propellant system.

Solid rockets were rejected as being too massive due to their lower Isp; hydrogen and other cryogens were rejected for operational reasons (primarily boiloff and the need to have top-off tanks in the carrier aircraft); liquid hydrogen was explicitly excluded due to its bulk. Four booster layouts were therefore considered: Types A to D.

Type A used a single-stage booster with three 100,000lb thrust rocket engines. Large fins at the aft of the booster gave lift and offset the pitching moment produced by the winged payload. Type B used a two-stage booster vehicle; the first stage with two engines and the second with one engine. Each engine had 150,000lb thrust. There would be large fins at the rear of the first stage booster

Types C and D both used the same system: three separate boosters strapped side by side, each with a single 150,000lb thrust engine. The outer two boosters served as the first stage; as they burned out and fell away, the central core ignited. Swept fins provided stability and lift. Types C and D differed in their placement of Dyna Soar on the booster; Type C had it on its nose, while Type D placed the vehicle on the 'back' of the booster cluster.

It was found that suborbital missions (between about 20,000 and 22,000ft per second) could be carried out fairly easily with available structural and propulsion technology, specifically a vacuum Isp of 320 seconds. An improvement to 335 seconds vacuum Isp was needed to attain orbit when using a subsonic carrier aircraft. It was found for most configurations that a 1% change in vacuum specific impulse led to a 1% change in burnout velocity. Launch for the subsonic aircraft was to be 35,000ft, while the supersonic aircraft would launch from 70,000ft.

It was determined that the Type D configuration was best suited for supersonic launch, while the others were best suited for subsonic carry. All were laid out so that they would fit within a minimally modified carrier aircraft. However, 'minimal' was a somewhat subjective term; for the B-52 subsonic carrier, the bomb bay would be extensively modified and forward landing gear would be fixed into the extended position, since the booster and spaceplane would prevent it from retracting.

For the supersonic carrier aircraft (presumed to be the XB-70; this was not explicitly said, but it was the only possible aircraft to carry out the mission), considerably greater effort towards aerodynamically fairing in the booster and spaceplane would be needed. However, when drag considerations were entered into the computer modeling that was done, the added drag imposed reduced the value of supersonic launching to no greater than subsonic launching. And as it turned out, the B-70 did not enter production; even if it

B-52/Dyna Soar Launcher
SCALE 1/144

Alternate arrangement
1/200 scale

B-52/Dyna Soar Launcher
SCALE 1/350

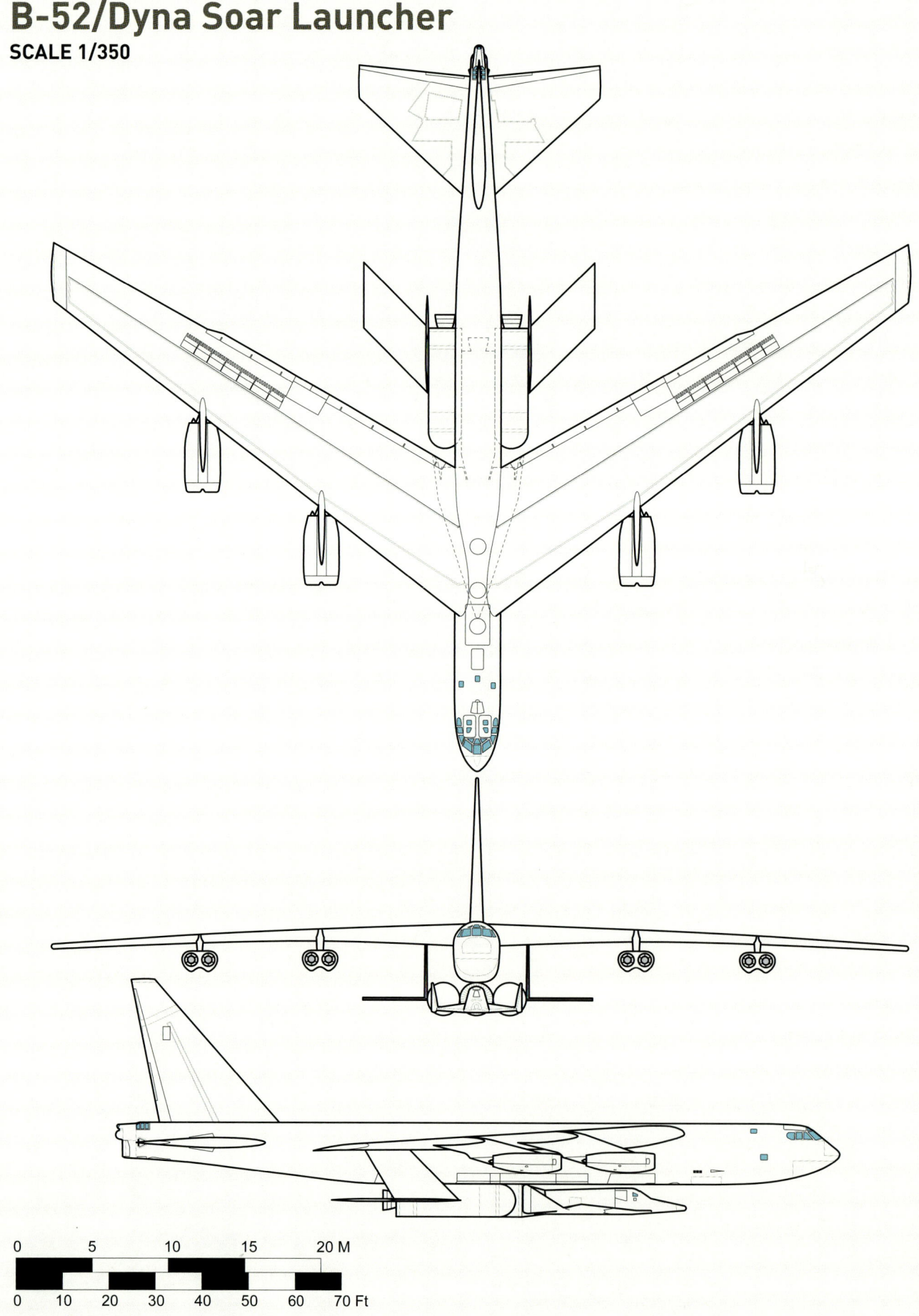

had, it likely would have been produced in strictly limited numbers. Modifying even one B-70 to serve as a Dyna Soar launcher would have been a hard sell, while B-52s were available in abundance. Yet as it turned out, no Dyna Soar would ever be air-launched at any speed.

Dyna Soar Model 844-2050E

While the Dyna Soar design evolved, so did the mission. The original concept of a manned rocket bomber decreased in priority as ICBMs began to prove themselves, but never fully faded. Boeing also proposed to use it as a satellite inspector, a satellite interceptor (using rockets and a modified M-16), a reconnaissance platform, an electronics intelligence platform, a satellite-deploying mini-space shuttle, a passenger-carrying space taxi and as a tiny little space station of its own. But it was officially a pure research vehicle, the obvious next step beyond the X-15. Hence the official 'X-20' designation.

The final Dyna Soar design, Boeing Model 844-2050E produced in December of 1961, was designed down to the last detail by late 1963. This included not just the relatively small spaceplane itself, but all the associated auxiliary equipment and vehicles to go along with it.

After considerable wrangling about using a Titan I or Titan II ICBM to loft the Dyna Soar onto a sub-orbital flight, by the end of the programme the plan was to go straight to the Titan IIIC for a fully orbital vehicle. In fact, the development of the Titan IIIC was largely a result of the needs of the Dyna Soar. Photos of early Titan IIICs show circular features on the nose cones of the UA-1205 solid rocket motors strapped to the side of the Titan core: these circles are the blow-out ports for the thrust termination system.

In the event that a disaster were to occur early in the flight of a Dyna Soar/Titan IIIC, linear shaped charges would blow out a pair of circular holes from the forward domes of the booster rockets; these holes would allow for combustion gases to escape forward and would counter the thrust from the aft nozzle. They were positioned specifically so that the exhaust gases would miss the projecting wings of the Dyna Soar vehicle, and were features deleted from the UA-1205 once it was determined that Dyna Soar would not be a payload of the Titan IIIC.

As with the earlier iterations, Boeing and the Air Force intended to carry out an incremental series of low-speed flight tests. Some consideration had been given to using the Lockheed C-130 to carry the Dyna Soar for both transportation and drop testing, but unsurprisingly the B-52 was ultimately tapped for that role. A specific B-52C, number 53-399, was set aside for this role and work began on necessary modifications at Wichita. As per normal, the Dyna Soar would be suspended beneath the right-hand wing under a pylon connecting to the existing underwing hardpoints. The pylon itself was to be made largely of aluminium and would weigh 2,000lb. It was not a modification of an existing X-15, Skybolt or Hound Dog pylon, but was entirely new.

Drop tests would have included both powered and unpowered flights. Powered flights would have seen the Dyna Soar carried aloft to 50,000ft with a special 'transition section', a semi-conical aerodynamic adapter that faired the round-topped, rectangular aft face of the Dyna Soar's rear fuselage to the circular cross section of the Titan IIIc Transstage. Contained within the transition section was a single Thiokol XM-92 solid rocket motor.

In an operational Dyna Soar flight, the XM-92 would be used as either

Boeing Model 2050E
Dyna Soar
SCALE 1/72

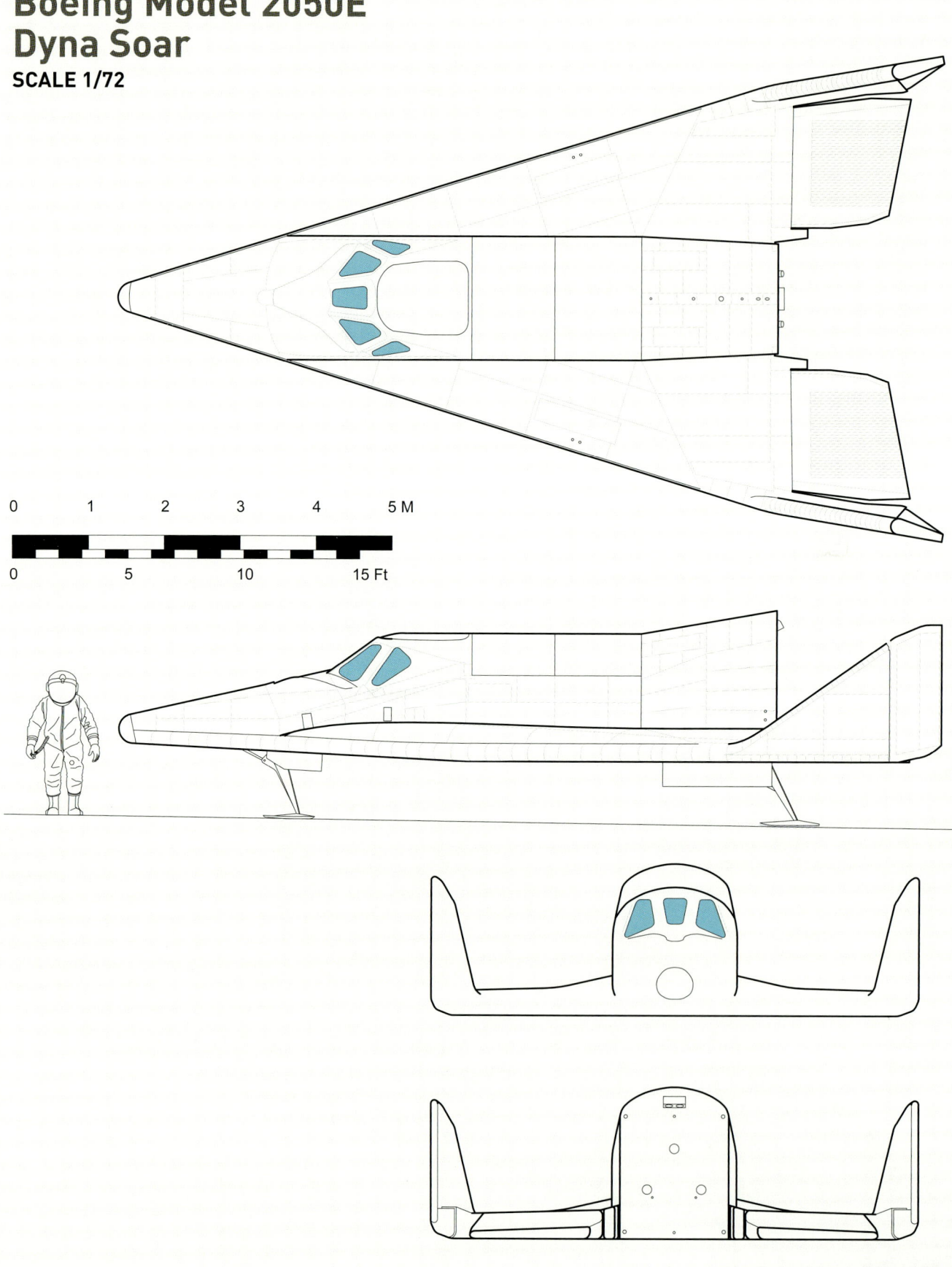

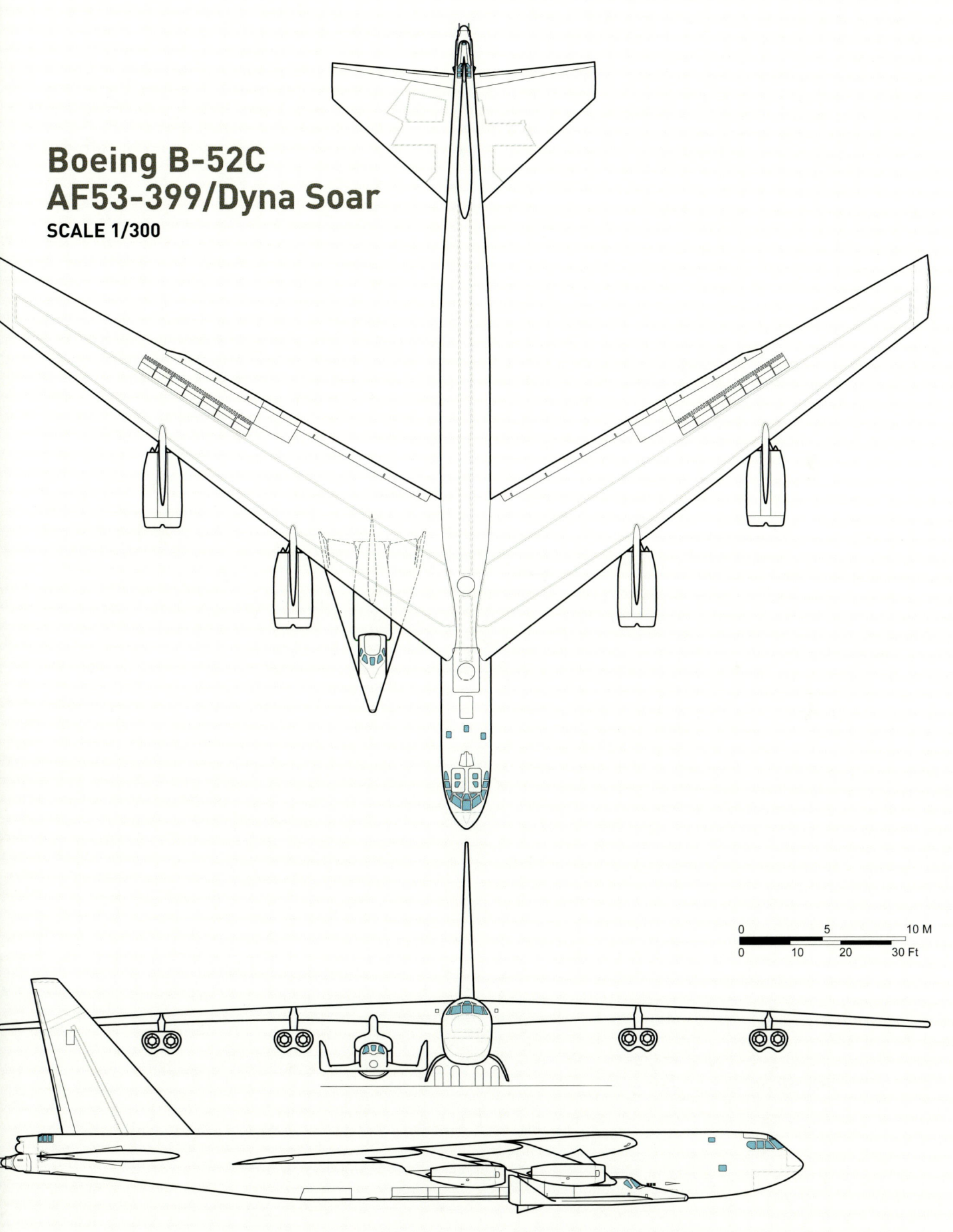

Boeing B-52C
AF53-399/Dyna Soar
SCALE 1/300

Boeing Model 2050E
Dyna Soar w/transition
SCALE 1/72

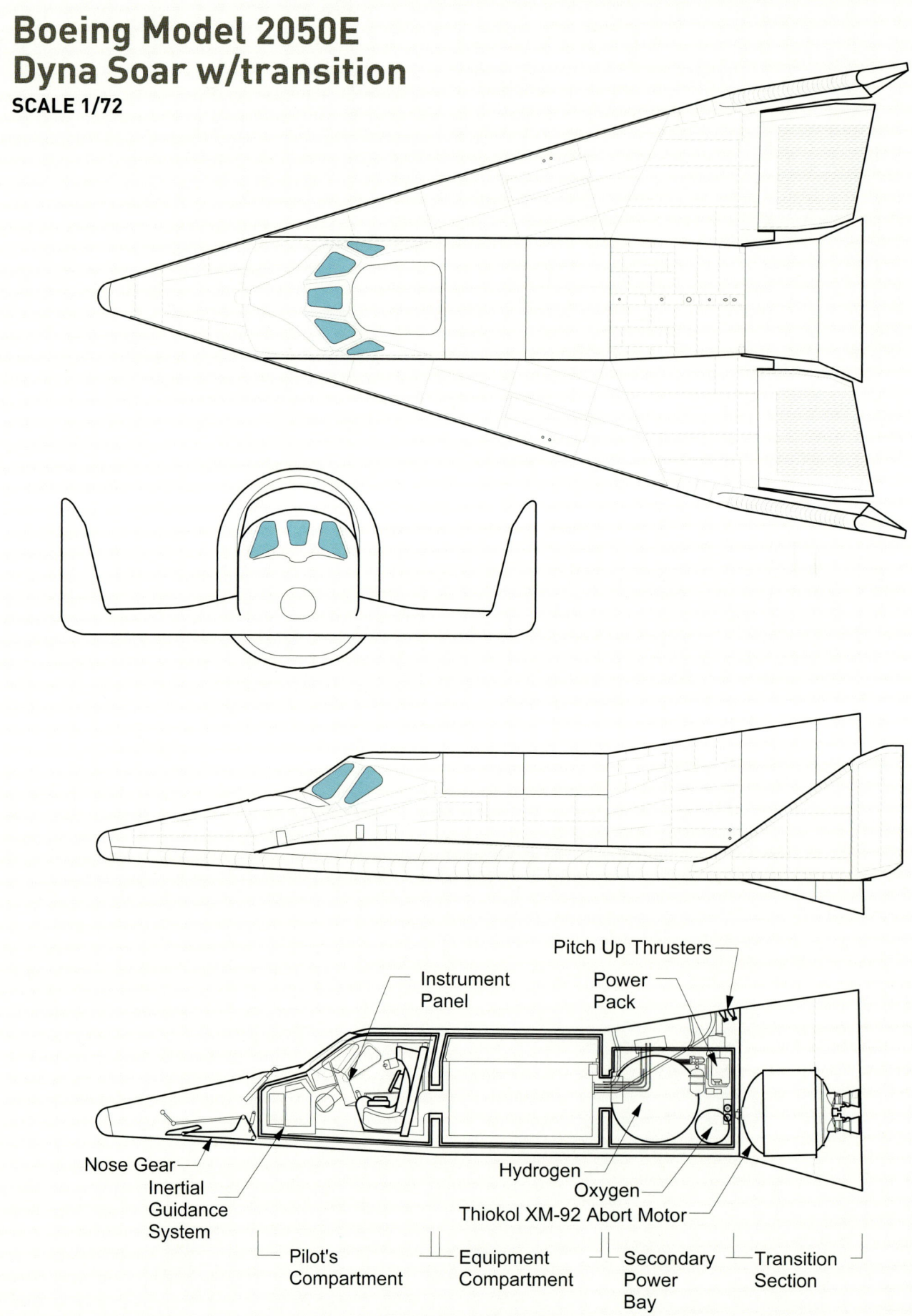

Boeing Model 2050E Dyna Soar w/pylon
SCALE 1/72

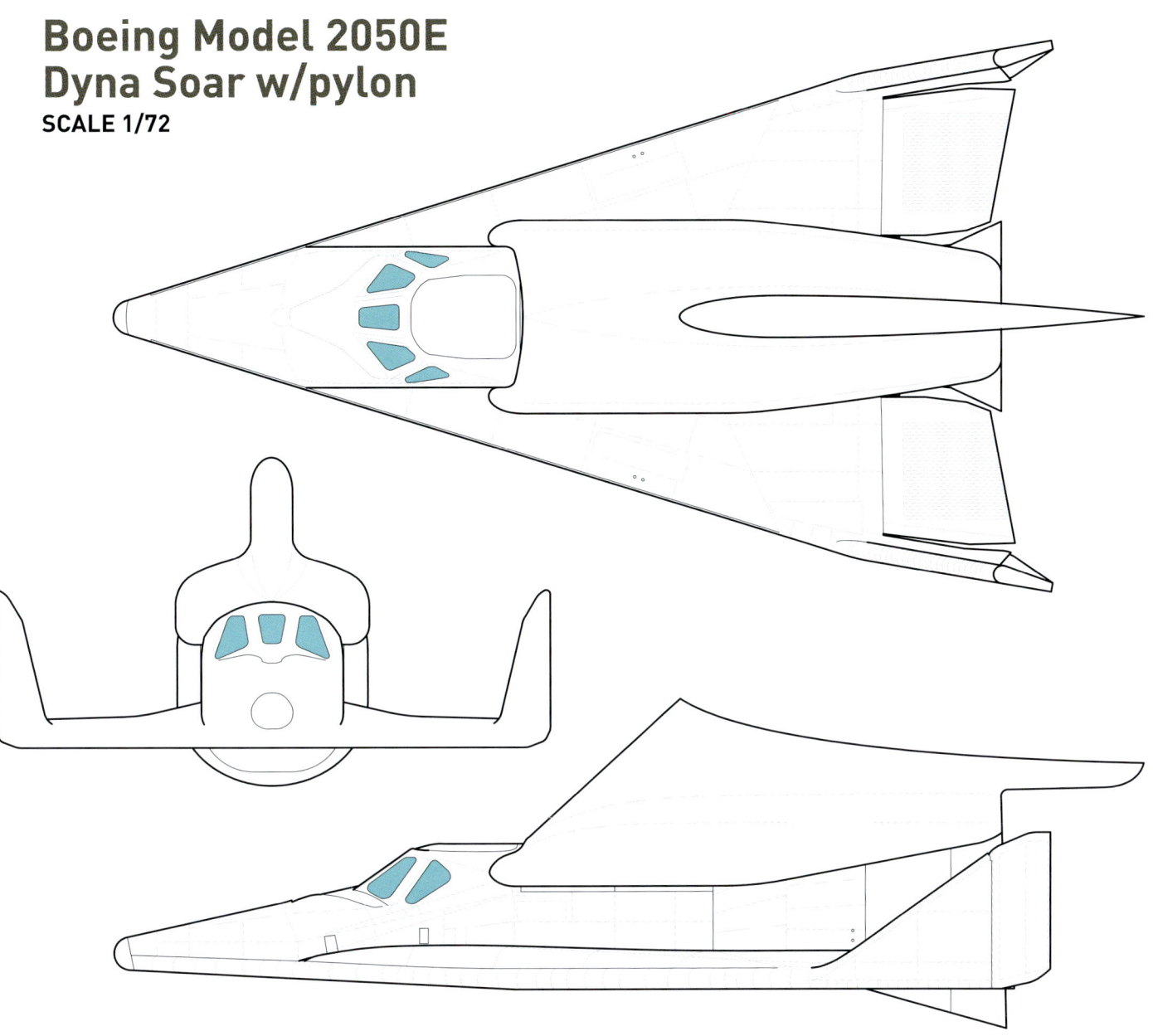

an abort motor, jettisoning the spaceplane away from a failing Titan IIIC booster, or as a de-orbit motor at the end of a successful orbital flight. For the drop tests, the motor provided power enough (40,000lb thrust for 13 seconds) to push the Dyna Soar upwards and to accelerate it to supersonic speeds. Dropped from the B-52 at 50,000ft and Mach 0.8, the XM-92-boosted Dyna Soar would climb to 70,000ft and Mach 1.4. In the summer of 1963, Boeing expected that these drop tests would begin in March of 1965. Unmanned once-around orbital flights would begin in January 1966, with manned once-around orbital flights in July of that year. In August 1967, the Dyna Soar would begin multi-orbit manned flights. But… construction on the first prototype was about 40% complete when the whole programme was cancelled in December 1963. The costs had skyrocketed and the mission had contracted. Long range bombardment could be done far more cheaply by ICBMs; space-based reconnaissance could be done better by the then-planned – and in due course also cancelled – Manned Orbiting Laboratory. The need for a space taxi was at best uncertain, and the research aspects, which the Dyna Soar would have been doubtless spectacular for, did not seem to be worth the expense.

Convair 'HAZEL' MC-10

There were many proposals for carrying smaller craft either submerged within the B-52's bomb bay or suspended beneath one of its wings – but very few called for one to be carried above the fuselage. The Convair MC-10 was one of those designs and

Convair 'HAZEL' on B-52C
SCALE 1/350

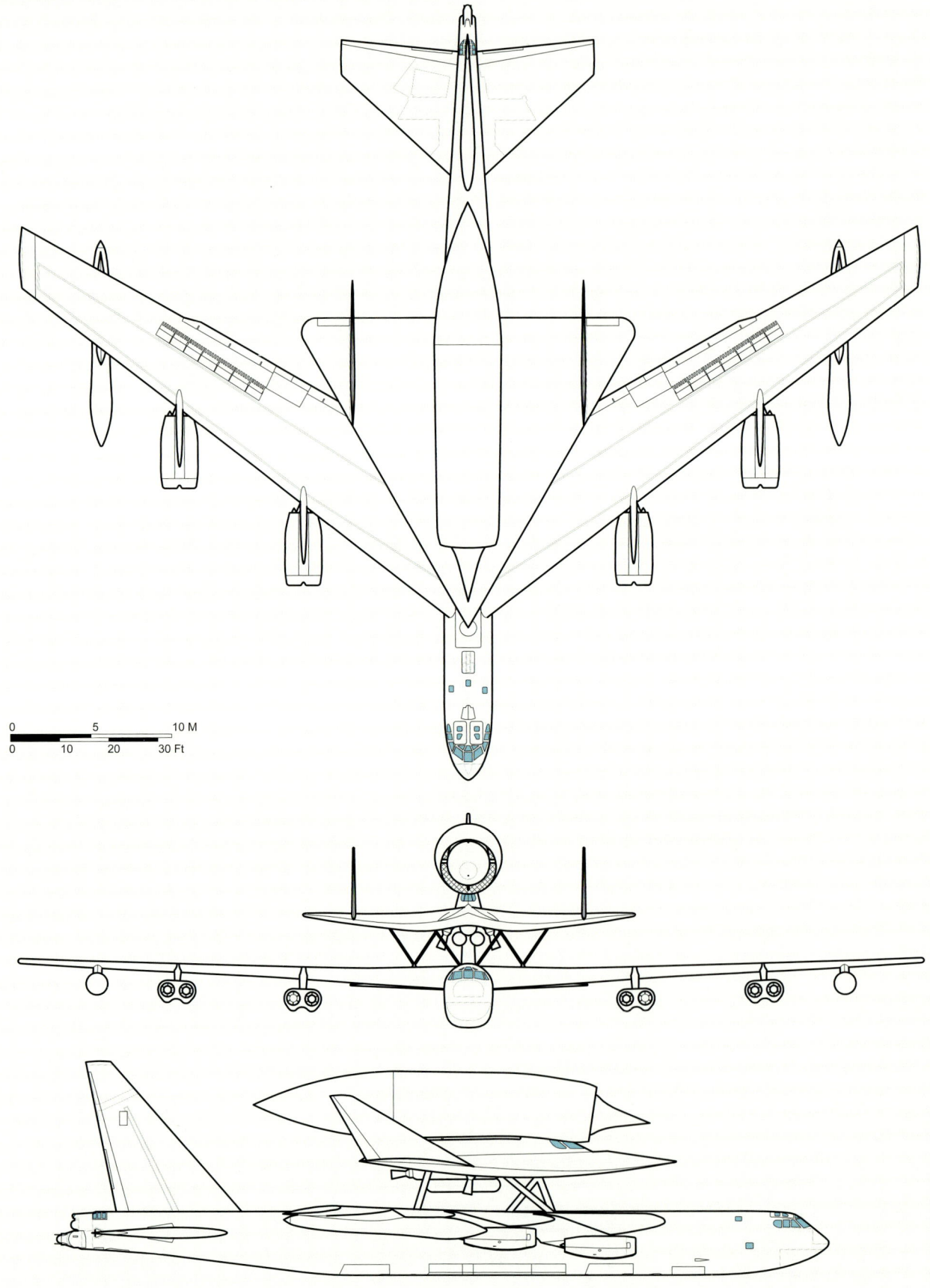

Convair 'HAZEL' MC-10
SCALE 1/144

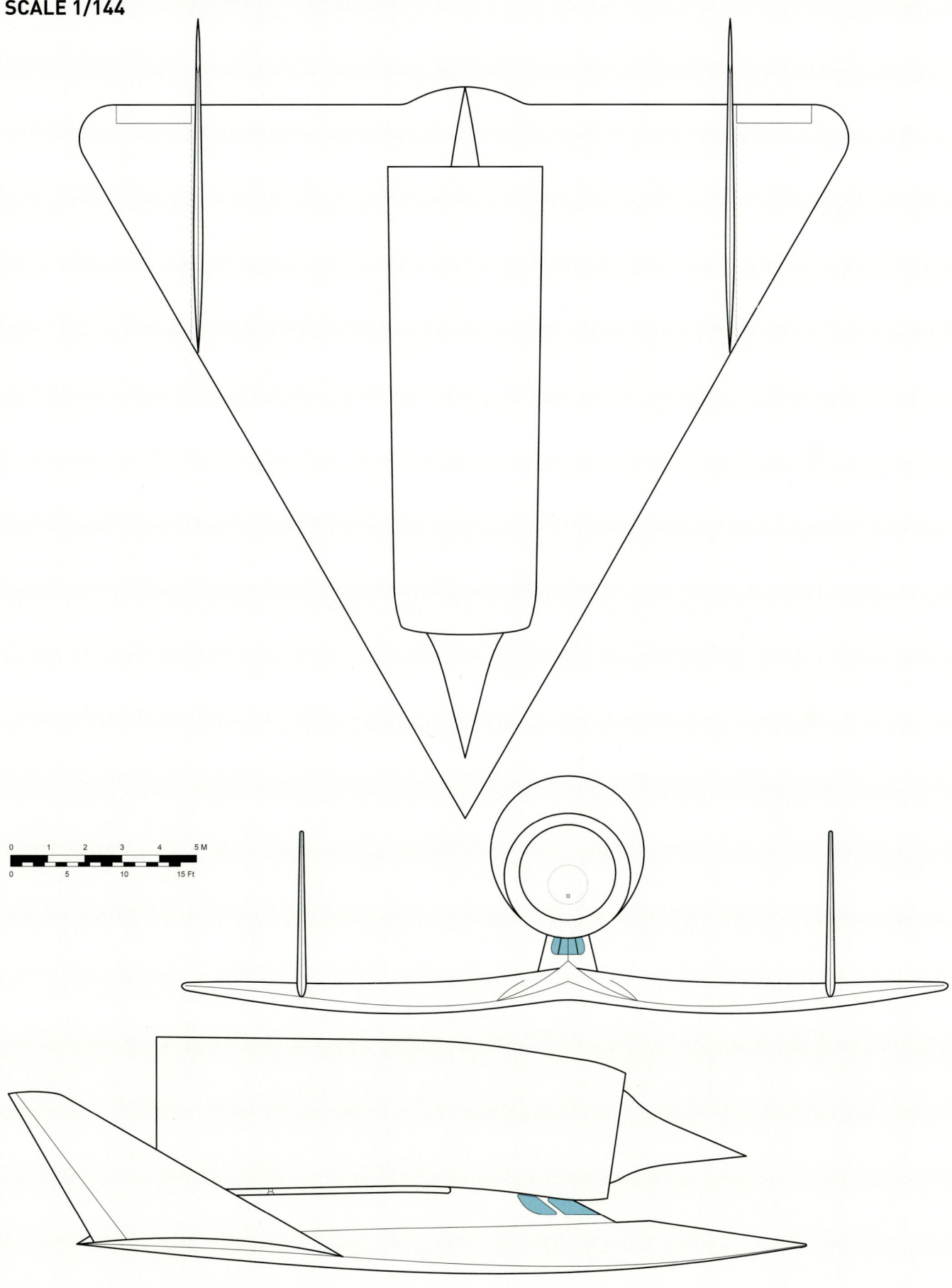

it would have made a remarkable sight, both on the tarmac and in the air.

During the late 1950s, Lockheed was working on Mach 3 reconnaissance aircraft for the Air Force, a process that would eventually lead to the SR-71. At the same time, Convair had high speed recon projects of its own, both for the USAF and the United States Navy.

In late 1958, Convair reported on concepts for high-speed, high-altitude manned reconnaissance platforms for the US Navy under Project 'Hazel'. A range of designs were proposed; rocket powered, turbojet powered, ramjet powered. Many used pentaborane fuel, which at the time seemed highly promising. Even with the very energetic green-burning fuel, though, the aircraft were not capable of fulfilling their mission objectives in a single stage. They needed to be launched by some other system in order to save fuel.

One of the designs put forward by Convair was the MC-10, a Mach 3 ramjet powered aircraft. It used fibreglass pressure-stabilized delta wings (silicone impregnated fibreglass fabric), basing its structural rigidity on internal air pressure. The company was driven to use this radical construction technique by the need for low wing loading at high altitude; the fact that it was radar-transparent was also a useful feature.

The fuselage contained a rigid pilot capsule, topped with a single gigantic Marquardt ramjet engine. The latter was made largely out of a type of plastic honeycomb, a high temperature ceramic fibre impregnated with phenolic resin similar to the commercially available Refrasil. The fuel system, flame holder, cooling shroud and engine mounts were to be made of metal. The central section of the aircraft was a rigid structure made of laminated fibreglass. Aft of the pilot's capsule was a bay containing the recon equipment. Aft of that was the fuel tank, containing the pentaborane fuel.

As a pure ramjet aircraft, the MC-10 was incapable of launching itself to high speed and altitude. Ramjet engines are capable of incredible performance including great range when they are flown at their optimum design point, but they are also very sensitive and small divergences from the optimum altitude and airspeed quickly degrade performance. So the MC-10 needed to be a 'staged' aircraft. Convair studied, and rejected, several approaches, firstly: high-altitude balloons. Considered practical for a test programme, using vast helium balloons to haul the aircraft to high altitude prior to release was considered unworkable from an operational or tactical standpoint.

Secondly there was rocket launch but this would impart unwanted stresses on the aircraft, forcing it to endure intolerably high speed and acceleration while still in the lower atmosphere. The third approach, using a specially designed high-speed airbreathing carrier aircraft, was rejected by the Navy – probably for being simply too much of a development effort.

In the end, the baseline launch system was to carry the MC-10 to altitude atop a B-52. When properly positioned, the MC-10 would separate and be boosted higher and further yet by rocket boosters. It was not expected that the MC-10 would be mechanically latched to the B-52 though. Instead, the B-52 would be fitted with a launch platform that had the same planform as the MC-10, and with a concave upper surface that closely matched the convex outer geometry of the MC-10's lower surface. The MC-10 would fit snugly into this and then be held in place via suction.

The mating surface of the launch platform would be perforated and a powerful vacuum pump would attempt to draw air down through the skin, like an air hockey table in reverse. This approach would allow the recon aircraft to be held securely without the need for penetrations into its inflatable skin, as well as providing a strong hold that could be released by simply turning off a switch. Most importantly, this system eliminated aerodynamic loads on the inflatable wing structure during the turbulent, bumpy low-and-slow portion of the initial flight.

The upper surface of the support platform would be made of a honeycomb sandwich with a perforated outer skin to get a vacuum grip on the lower surface of the MC-10's wing. The ramjet engine was so vast that it was to be fitted with an ogival fairing for the exhaust in order to reduce base drag.

Just before separation, the B-52 would slow to minimum safe airspeed – 200 knots at 45,000ft. A rocket booster system attached to the underside of the MC-10 – projecting through the launch platform via a cutout in the structure – would then boost the aircraft to 125,000ft and around 3,150ft per second (more than 2,100mph). At that point the ramjet powered cruise would begin.

It should be noted that the B-52 is, of course, a US Air Force aircraft, not US Navy. Left unclear was whether Convair was suggesting that the Navy rely on co-operation with the Air Force for launches, or if the Navy would buy its own B-52. In that case, it's interesting to ponder how those Navy B-52 carrier aircraft might have been painted. Like the SR-71 and its relatives the A-12 and the YF-12, the MC-10 doubtless would have been painted black – or very nearly so – as a way to radiate away heat. The B-52, though, can only be speculated at. An all-black B-52 with Navy markings, carrying on its back the bizarre all-black MC-10 with its supremely oversized ramjet engine, would have been a startling sight.

For a more complete description of the MC-10 and its stablemates, see Lockheed SR-71: Origins and Evolution.

McDonnell Model 192 ISINGLASS

Still frustratingly classified is the 'ISINGLASS' reconnaissance system. Meant to be the next great thing to follow the SR-71, McDonnell Aircraft Model 192, code-named ISINGLASS (and sometimes also referred to as RHEINBERRY, indicating either some overlap in programmes or some confusion in the documentation), was a 1965 project to develop a rocket-powered globe-spanning boost-glide vehicle that would fly far too high and far too fast for even the most wildly optimistic Soviet air defence system to intercept.

Only scattered scraps of information are available on this CIA programme, but what is known is that Model 192 was designed for the role. It was a semi-conical long lifting body with a single Pratt & Whitney XLR-129 rocket engine and a single crewman. The reconnaissance cameras peered downwards through windows in the forward fuselage, protected from the Mach 20+ airstream by a small ramp.

Model 192 was meant to have a range of some 7,000 miles, reach a top speed of Mach 21 and top out at 205,000ft. These are not easy goals even more than half a century later, and in the mid-1960s it was beyond the state of the art for a single stage vehicle of manageable size. So Model 192 needed to be hauled to altitude and closer to the target. It was quickly determined that the B-52 would be the best option for that, and that Model 192 would be carried under the wing in the normal position for external payloads. However, the XLR-129 burned hydrogen and oxygen, leading to voluminous fuel tanks and a large vehicle – Model 192 being one of the largest vehicles proposed for underwing B-52 carriage, barely fitting in the available space.

There is little more that can be said about the Model 192 at this time. The project lasted to at least 1967 and seems to have perhaps made it as far as 1970, but nothing came of it other than the XLR-129, which became the basis of Pratt & Whitney's proposal for the Space Shuttle Main Engine.

Junkers RT 8

Dr Eugen Sänger, designer of the legendary 1944-vintage Antipodal Bomber (aka Silverbird), helped produce a somewhat similar design for the revived Junkers company in the 1960s. But instead of a sub-orbital spaceplane meant to bomb New York, the RT 8 was a design for putting small payloads into orbit. The RT 8 (from 'Raum Transporter', literally 'Space Transporter', with the '8' likely referring to the 8th design iteration) was, like Sänger's earlier Silverbird design, launched from a horizontal track using a booster rocket to push the vehicle to high speed before liftoff. There were many major differences, however, between the RT 8 and the Silverbird.

The RT 8 was a two-stage spaceplane, unlike Sänger's earlier single-stage Silverbird. And more, both stages were powered by hydrogen/oxygen rocket engines, much more advanced that the LOX/fuel oil engine proposed for the Silverbird. The first stage had three main engines and the second stage had one. Both stages had a single crewman. Combined thrust was 200,000kg (441,000lb), with a vacuum specific impulse of 430 seconds. The first stage had a dry weight of 10 tonnes and a propellant mass of 69 tonnes. The second stage had a dry mass of 18.5 tonnes, a propellant mass of 13 tonnes and a payload of 2.5 tonnes.

The track booster was a hot-water rocket. This is one of the conceptually simpler forms of rocket; the propellant tank is filled with water, which is then heated to high temperature. The water boils and creates high pressure before the resulting steam is allowed to escape through a rocket nozzle, generating thrust. Performance is relatively poor compared to most forms of liquid fuelled rockets… but for a non-flying booster stage, this performance loss is offset by the cheapness of the propellant (not many things come cheaper than water) and the ease of handling.

Track length was about 3km. Liftoff would occur at 900km/h; the booster remaining on the track. This would require the construction of a sizable and expensive track, consuming not only the land the track sits upon but likely some distance downrange to account for the possibility of a failed launch catapulting flaming wreckage into the surroundings.

An alternative launch system replaced the track and hot-water rocket booster with a carrier aircraft, such as a highly modified Boeing B-52. How Junkers thought they could lay hands on a B-52 is anyone's guess; perhaps they imagined that it would be possible to buy or lease one. After all, the Canadians had managed to obtain the use and modification of a B-47 to serve as a testbed for the Orenda Iroquis jet engine. But if the Raum Transporter could be launched from a conventional aircraft, then there would be no need for the fixed track and the associated infrastructure; the carrier aircraft could in principle haul the vehicle south of Germany to launch from some more equatorial airport and obtain improved orbital performance.

While diagrams exist of the RT 8, the B-52 carrier aircraft, so far as this author is aware, exists solely in the form of some concept art. The reconstruction diagrams here should therefore be taken with a

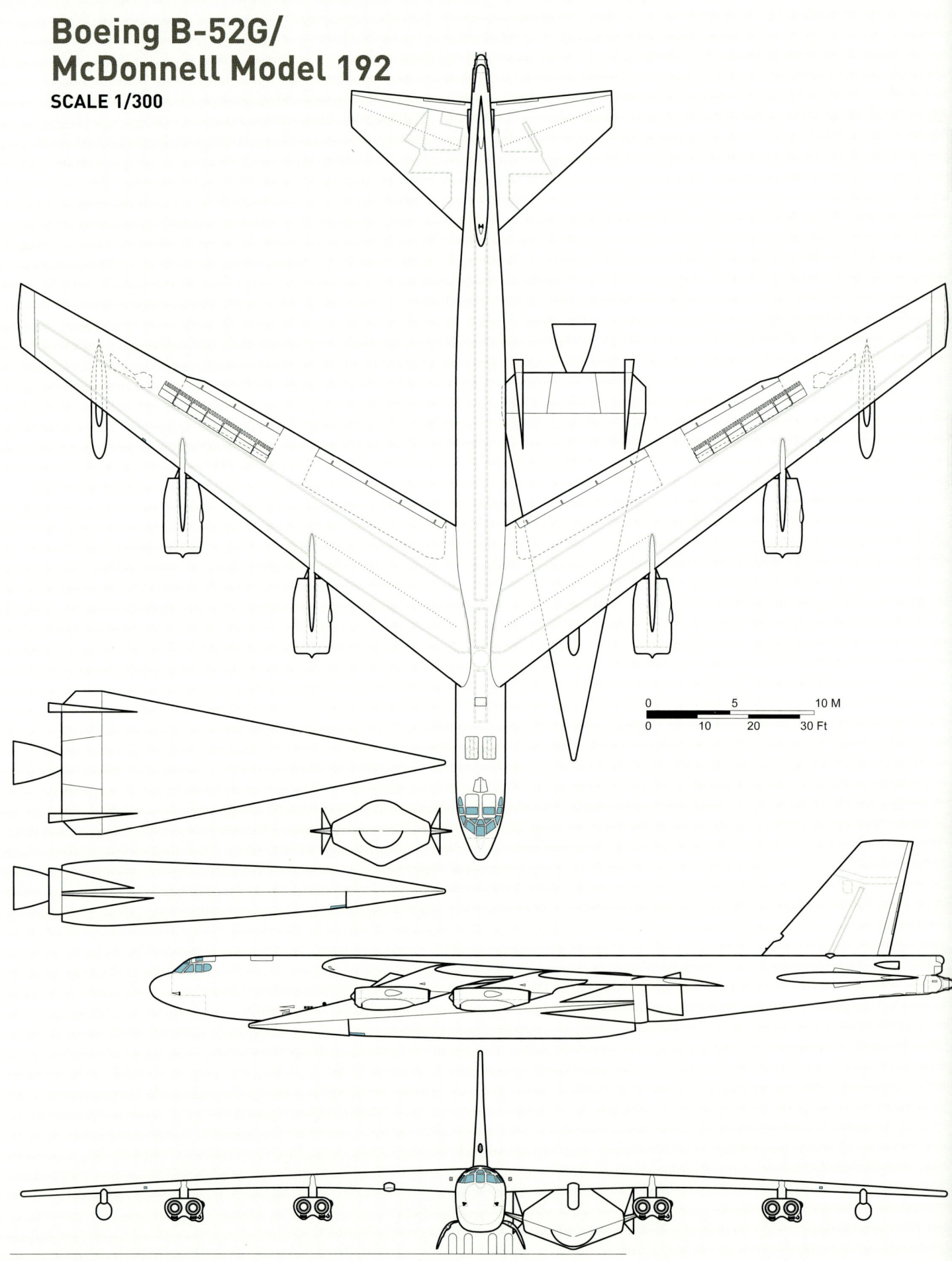

Boeing B-52G/ McDonnell Model 192
SCALE 1/300

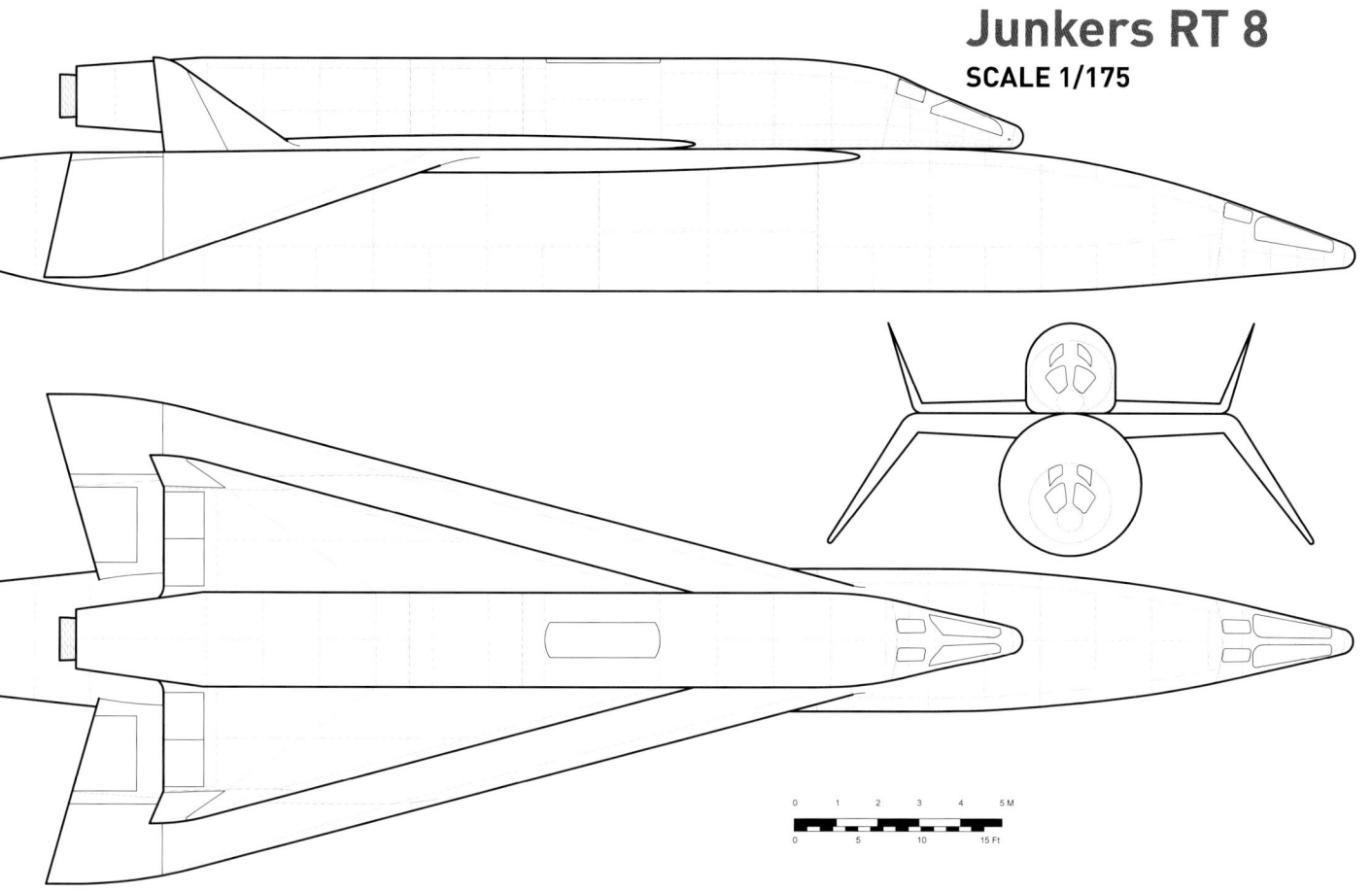

Junkers RT 8
SCALE 1/175

grain of salt; not only are there potential vagaries in translating the painting to diagrams, but the painting itself might not have been wholly reliable. What was shown was a B-52 with a G-model short tail and substantial modifications to the underside. This included gutting the bomb and landing gear bays in order to fit the upper-stage spaceplane within the body of the aircraft.

The forward fuselage of the B-52 had a bizarre-looking downward extension, apparently containing new nose landing gear that would provide ground clearance for the Raum Transporter. The aft landing gear, though, is a bit of a mystery. A new gear bay seems to be in evidence, but it would need to include very long legs to keep the Raum Transporter from scraping on the runway.

It should also be noted that the artwork has not RT 8, but RT 7 on the side of the spaceplane. This seems to indicate that the B-52 launch option came along before the tracked steam rocket in the design process.

Other than the launch method, though, the designs of the spaceplane and its booster seem identical between the RT 7 and RT 8.

One hundred and fifty seconds after ignition of the first stage engine, the second stage would stage off of the first at an altitude of 60km (this is for the track-launched concept; the altitude should be higher for a B-52-launched version). The second stage would then continue to orbit at 300km. Illustrations show the upper stage rendezvousing with a space station of NASA design. After re-entry, the upper stage would land on tricycle skids similar to those planned for the Boeing X-20 Dyna Soar.

The Junkers RT 8-I-01 was a curiously small vehicle, one of the smallest manned payload-carrying orbital vehicles of its kind so far envisioned. With a liftoff mass of 113 tonnes, it was only nominally more massive than the original Silverbird, and with much greater orbital payload potential.

It used stages that were geometrically similar, although not exactly so. While the two stages were nearly mirror images of each other, with the lower stage looking like an upside-down version of the upper (which in turn resembled a stretched X-20), the operational environments of the two vehicles determined that inverted geometries actually worked out for the best. The shoulder mounted wings and drooped wingtips of the first stage served to increase lift through compression lift effects (similar to the principle used on the XB-70 and waveriders, where shockwaves from the nose are trapped under the wings). The flat underside and upswept wingtips of the upper stage helped smooth out hot spots from orbital re-entry.

Boeing B-47 Stratojet and B-52 Stratofortress: Origins and Evolution

Junkers RT 8
B-52 Carrier/Launcher
SCALE 1/350

Junkers Raumtransporter

Rather less well defined than the Junkers RT 8 is another concept, believed to have been produced at around the same time. This was another Junkers Raumtransporter concept with two rocket-powered stages carried underneath a B-52 derivative, but the resemblance ends there. The launch vehicle was quite similar to the Lockheed STAR Clipper concept from a few years later, featuring a lifting body nestled within a V-shaped drop tank.

The lifting body had sizable wingtip fins hinged at the roots; prior to separation from the drop tanks the wings would droop downwards, but raise upwards at some point after separation and before re-entry. The drop tanks had rocket engines of their own and were intended to be recovered. To facilitate recovery, a membrane was stretched between the tanks, over the lifting body payload; this would serve as enough of a wing to give the drop tanks some measure of gliding ability.

This appears to have been an alternative to the RT 8-I-01. Several generally crude illustrations of it were published; they agreed on the general nature of the design, but differed on the specifics. The diagram here is thus rather provisional, an attempt to rationalize the variations.

Boeing B-52 as an aircraft carrier

Given how popular the idea of a 'flying aircraft carrier' has been over the decades, both before and after the advent of the B-52, there is a surprising lack of proposals for a B-52 modified into such a thing. So far as this author is aware, there is not a single serious proposal for adding features to a B-52 which would allow it to carry fighters for defence, as had been proposed for the B-36.

However, while there were few to no proposals for carrying manned aircraft to aid the B-52, as has been seen there were many proposals for using the B-52 as a carrier of aircraft to help those other aircraft fulfill their missions. That said, the B-52 has served as an operational aircraft carrier... of sorts. Almost from the beginning there have been proposals and operational systems that added jet-propelled 'parasite' aircraft to the B-52; thousands of such aircraft have flown attached to B-52s. It's just that these aircraft were unmanned and unrecoverable.

McDonnell GAM-72/ADM-20 Quail

The B-52 was never going to be a stealthy aircraft. Flying at more than 40,000ft high, it was a giant slab-sided radar reflector and jet-powered infrared source, detectable by Soviet sensors at a great distance. There was little in the 1950s that anybody could do to hide the B-52 at altitude, though at first it was expected to survive by simply outflying Soviet interceptor aircraft.

But as Soviet interceptors flew higher, what the B-52 could in principle do was fool them with decoys. Studies for a jet-powered B-52 decoy began in the USAF in 1955, finally formalized in January 1956 as General Operations Requirement (GOR-139). Shortly afterwards, in February 1956, the Air Force selected McDonnell Aircraft as prime contractor for the GAM-72 Green Quail, soon shortened to simply 'Quail'.

This was an unassuming little aircraft powered by a single diminutive General Electric J85 turbojet. The wings were clipped deltas with sizable downward hanging tip fins and a pair of vertical stabilizers, the fins perpendicular to the wings. The nose was blunt, the fuselage trapezoidal with large flat sides angled outwards. Normally this would be a disastrous layout; everything about it served to make the radar cross section much larger than would normally be desired. But the purpose of the Quail wasn't to sneak past Soviet defences. The purpose of the radar-reflecting design features was to make the tiny airplane look as big as a B-52. To further the illusion, the Quail would fly at the same altitude as a B-52, at the same speed and for a substantial distance. While on the ground it could be programmed to make up to two turns in flight and a change in altitude in hopes of drawing off Soviet forces.

The vast expanse of flat planes that formed the sides of the fuselage hid radar reflectors. Much of the structure of the aircraft was made of radar-transparent fibreglass; this made the components designed specifically for radar reflection easier to configure correctly to replicate the radar cross section of the B-52. The wings were of conventional aluminium construction, with some aluminium in the dorsal fins; the wingtip fins were primarily fibreglass, as were the fuselage and inlet ducts.

Captive tests with prototypes carried by B-52s began in July 1957, with glide tests following in November. Successful powered flights began in August 1958 and the Strategic Air Command began accepting production Quails in September 1960. The first squadron of B-52s to be operational with Quails was in February 1961, with 11 such squadrons operational by December and the 14th and final squadron going operational in April 1962. A total of 616 Quails were built, with 492 in operation at the peak.

A 'clip' with two Quails, one above the other, was installed inside the B-52. Four such clips could be carried for a total of eight Quails, but the usual load was two, installed within the rear of the bomb bay to one side. When two clips were carried they could fit side by side and to conserve space the

Boeing B-52A +Junkers System B
SCALE 1/300

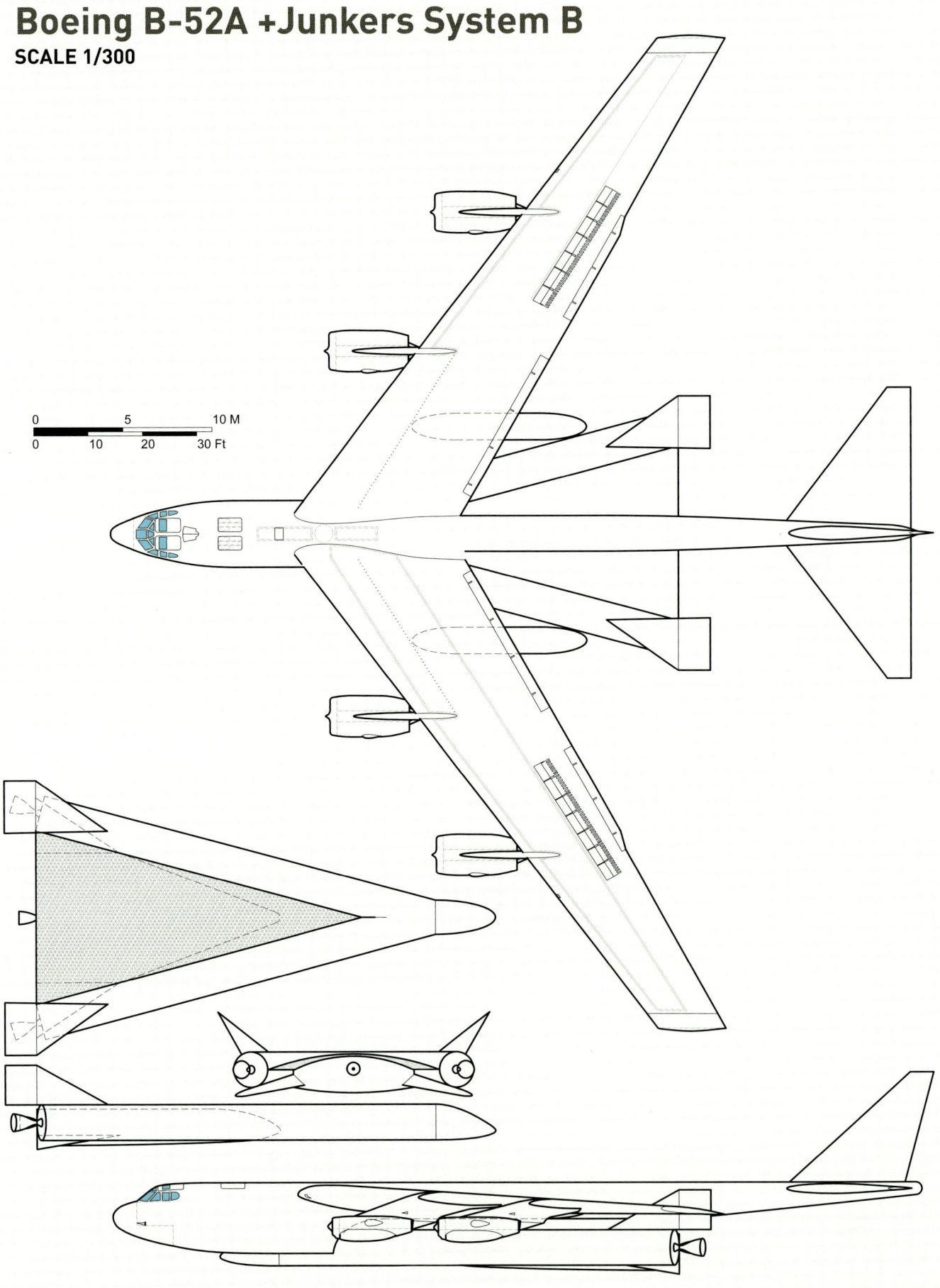

GAM-72
SCALE 1/20

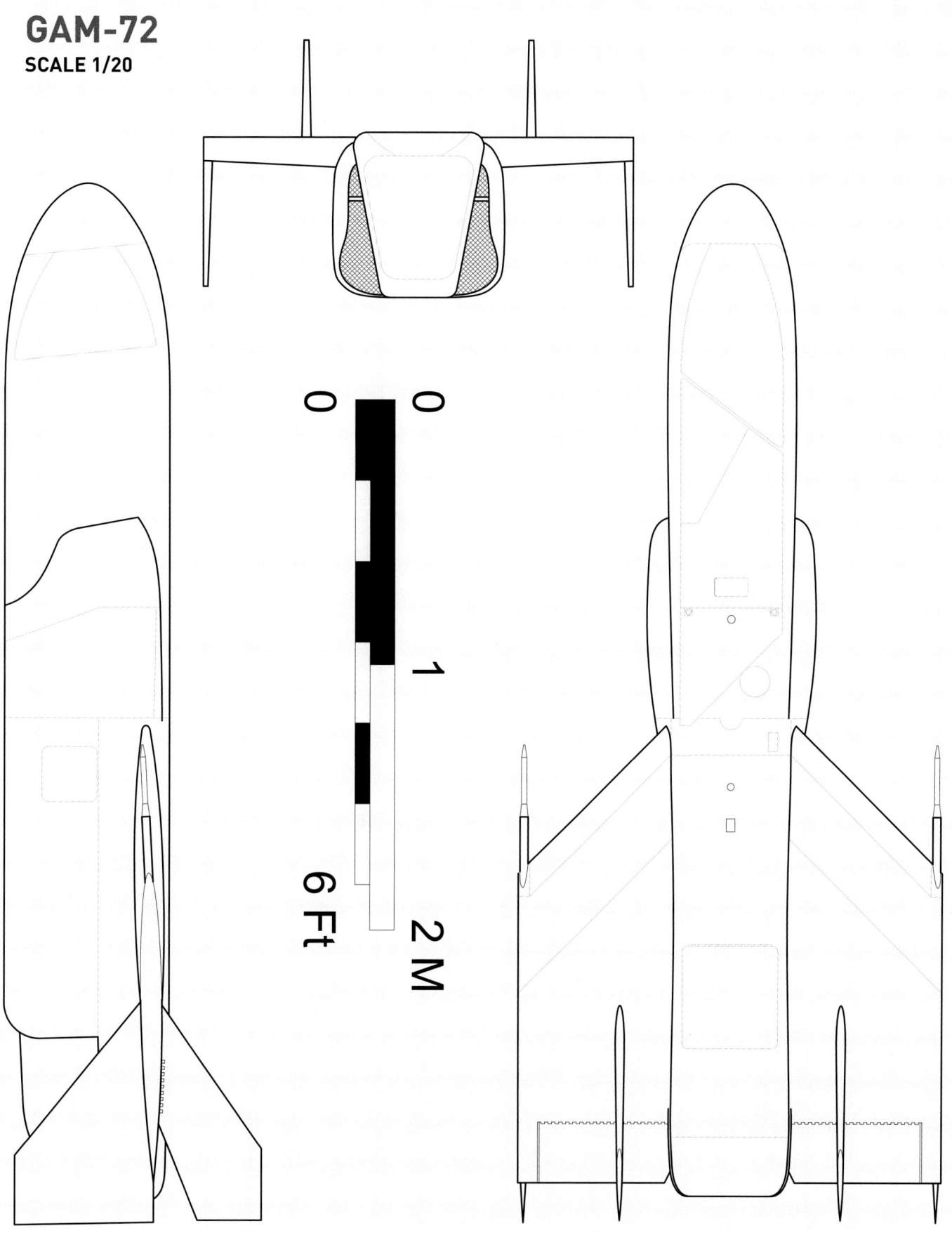

wings and vertical tails of the Quail would fold down approximately 90° while stowed.

For launch, a quail would be mechanically extended from the bomb bay. While held below aircraft in the relatively undisturbed slipstream, the wings would unfold and the engine would be run up and only then would the Quail be released, safely flying away from the B-52 and starting on its journey.

The GAM-72's original J85-GE-3 engine proved unreliable and was substantially redesigned. The Quail itself then had to be redesigned to accept the larger and heavier J85-GE-7 – becoming the GAM-72A in the process. All but the first 24 Quails were built as GAM-72As. In 1963 a barometric switch was added which allowed the Quail to fly at lower altitudes, becoming the GAM-72B, and in June 1963 the GAM-72 was redesignated the ADM-20. The Quail served into the 1970s, when it was found to be ineffective in its role. Even with radar repeaters, chaff dispensers and infra-red flares, it could no longer convincingly pretend to be a B-52 to the aircraft detection systems of the day. It finally ended its service in June 1978.

GAM-77/AGM-28 Hound Dog

The Air Force issued General Operations Requirement 148 (GOR 148) in March 1965, calling for a standoff missile for the B-52. North American Aviation, Northrop and Vought are all known to have turned in proposals – all being single-turbojet designs capable of Mach 2 at high altitude. North American's design was chosen to become the GAM-77 'Hound Dog' in October 1956.

Stories vary about where this unusual name came from; one of the most popular, and not altogether unlikely, explanations points out that Elvis Presley released his cover of the song 'Hound Dog' in July 1956. While naming a missile after a song does not seem to be entirely in keeping with Air Force tradition, the fact was that at the time 'Hound Dog' seems to have been an omnipresent cultural phenomenon.

The desire for the Hound Dog grew from the increasing dread of Soviet bomber interception capabilities. The B-52 had started off as a bomber that flew so high and so fast that it could not be reached by either fighters or missiles but by the late 1950s it began to look like the Soviets could soon swat it from the sky at will if it crossed into their territory. A turbojet-powered missile would extend the reach of the B-52, allowing it to launch weapons while still far from the target… and far from the ring of interceptors sure to circle it.

The Hound Dog benefited greatly from North American's work on the Navaho cruise missile. Many of the design features carried over from the earlier programme, with a similar forebody, canards and main wing. But where the Navaho had two ramjet engines on the sides of the fuselage, sitting atop low-mounted wings, the Hound Dog had a single Pratt & Whitney J52-P-3 turbojet in an underslung nacelle. The engine had an inlet spike and an unconventional annular plug nozzle, this unusual appearance occasionally leading some to believe that the engine was a ramjet. An interesting feature of the Hound Dog is that the port and starboard wings are seemingly identical and interchangeable units. This was doubtless a cost saving feature.

The Hound Dog was much too large for the B-52 to carry internally – the only option being external carriage using underwing pylons. This required modification of the wing skin and structure, with new wiring and plumbing. While the Hound Dog did not serve anywhere near as long as the B-52, it did make an important impact on the aircraft going forward as the Hound Dog pylons and their wing connections survived and have served multiple roles up to the current day.

The non-afterburning J52 engine was designed originally for the Navy to power aircraft such as the A-4 Skyhawk. It was a scaled-down version of the J57 which powered the B-52 itself, an interesting circular development. Unsurprisingly it burned the same jet fuel that the B-52's own engines did, a fact that was employed on takeoff. The Hound Dog's jet engines could be run at full thrust to increase the takeoff performance of the B-52, and then shut down once the bomber was airborne, working like the JATO system used by the B-47. The missile's fuel tanks were then topped off from the B-52 using lines piped through the pylons.

The Hound Dog was guided by an inertial navigation system. Prior to launch, the missile was updated with readings from star trackers built into the pylons. It could then fly a pre-programmed course at either high or low altitude, following a dog-leg path to throw off air defence systems. The basic mission involved a high-altitude launch followed by a high-altitude cruise, terminating in a high-speed dive, with a range of 683 nautical miles.

There were three alternate missions. The first had a high-altitude launch followed by a low-altitude (selectable between 200 and 1,500ft) cruise using a radar altimeter to maintain height. This would greatly reduce range to 378 nautical miles, and cruise speed would be only Mach 0.83. The second alternate mission involved a low-altitude launch followed by a pop-up to a high-altitude cruise (range: 638 nautical miles); alternate mission three used a low-altitude launch and a low-altitude cruise (range: 366 nautical

AGM-28
SCALE 1/60

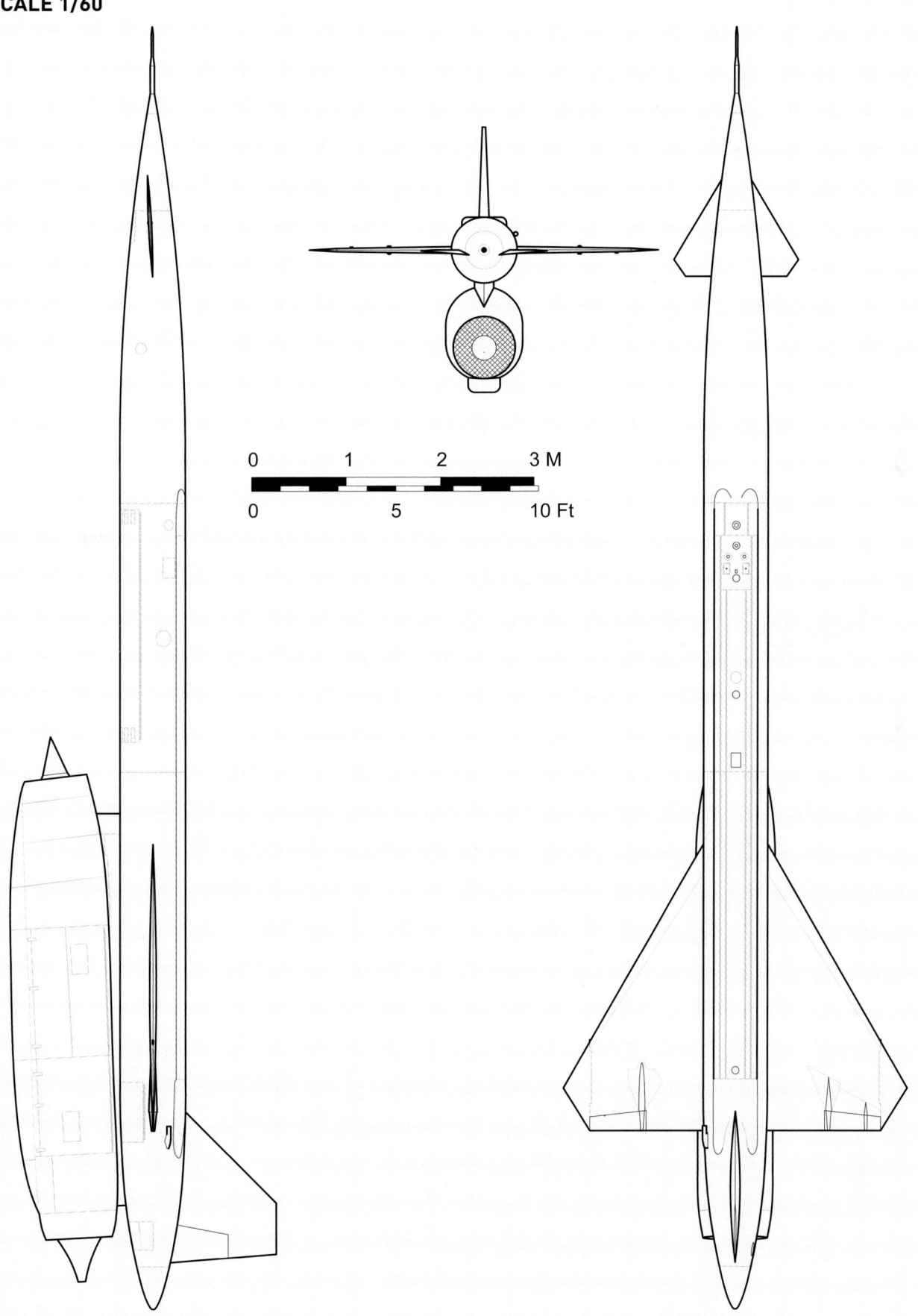

B-52G w/AGM-28
SCALE 1/300

miles). At high altitude – 50,000ft or so – the Hound Dog would be clearly visible to defensive radar systems, but would have maximum range. At low altitude, under 5,000ft, it would be much less visible, but its range would be greatly reduced.

A primary role of the Hound Dog was to clear a path for the B-52 by taking out known fixed surface-to-air missile sites ahead of it. The missile could fly much faster than the bomber and had a range great enough that it could, even with a somewhat meandering course, strike a missile site well before the B-52 got within range of the SAMs. The passage of the Hound Dog, along with the inevitable Earth-shattering kaboom at the end of its mission, would sow enough confusion and chaos to allow the B-52 to sneak by and drop further nuclear destruction with a measure of security and precision.

The Hound Dog itself was apparently not especially accurate, with a circular error probability of 3,600ft when flying a short-range low altitude mission, and 5,800ft when flying a high-altitude long-range mission. But given its mission and its W28 warhead – capable of nearly one and a half megatons – the destruction of relatively soft targets like above-ground SAM sites was virtually assured.

By modern standards, development of the Hound Dog happened at a blistering pace. Following selection of North American Aviation as the prime contractor in October 1956, the first launch of a GAM-77 from a B-52 occurred less than three years later in April 1959. Delivery of the first production unit occurred in December 1959, and the first wing of GAM-77-equipped B-52s was operational in July 1960. An improved navigation system led to the GAM-77A which first flew in June 1961; this version was operational in September. At the peak of its deployment some 600 Hound Dogs were operational in 1963.

The USAF redesignated the GAM-77 and GAM-77A as the AGM-28A and AGM-28B in June 1963. It had been planned that the Hound Dog would be a relatively short-lived programme, to be replaced with the faster and longer-ranged Skybolt missile… but Skybolt was cancelled in December 1962, leaving the Hound Dog to soldier on for years longer.

Tests using a terrain contour matching (TERCOM) system occurred in 1971, a system that would have allowed the missile to safely fly very low altitude missions following the contours of the Earth as well as precise targeting. This would have led to the AGM-28C, but it was not adopted. However, TERCOM would appear on another missile that the B-52 carried, the Air Launched Cruise Missile. The Hound Dog was finally retired from operational service in June 1975.

D-21B Senior Bowl

As the Lockheed Skunk Works A-12 spyplane was developed, the CIA became concerned about the risks of sending a manned vehicle – even one as advanced as the A-12 – over enemy territory. Kelly Johnson and his team were asked to study unmanned drones to be launched from the back of the A-12 as an alternative to manned overflight of enemy territory. The programme was codenamed 'TAGBOARD' by the CIA in September 1962.

Lockheed's feasibility study was completed in January 1963, showing that such a drone was possible; in the following February Lockheed was given the go-ahead to produce 20 of them. The project was to be carried out with the highest of security classification.

The design of the drone itself was a blended single-ramjet that clearly bore a familial resemblance to the A-12, but was not merely a reduced-scale version of the aircraft. This design had a metallic central structure with the chines and large portions of the wings and vertical tail to be made from composites.

The fuselage was a semi-monocoque structure of rib and skin construction, the bulk of which was titanium with a few parts made of stainless steel. The wings were of a modified delta planform with sizable rounded chines that served a canard-like aerodynamic function. The wings had substantial anhedral. Control surfaces were few: a rudder on the vertical stabilizer and elevons on the wing trailing edges. The wing structure was again of titanium, sheets covering ribs that were integral with the fuselage structure. The chines, wing leading edges and elevons were made from a honeycomb covered in a silicone-asbestos laminate. 'Plastic' elevon fences – small vertical fins – were located just inboard of the elevons on the upper surfaces of the wings. The control surfaces were hydraulically actuated.

To prevent confusion, in October 1963 the A-12 'mothership' was designated 'M-21', M for 'Mother', the '21' being simply a reversal of the '12' to prevent confusion. The drone was designated D-21, D for 'Daughter'.

The D-21 was powered by a single Marquardt RJ-73 Model MA20S-4 ramjet. This was derived but greatly modified from the Marquardt RJ43-MA-3 ramjet engine used on the IM-99 BOMARC surface-to-air missile. The engine was given a larger exhaust nozzle with a greater expansion ratio for improved high-altitude performance. A number of other design changes were made to account for the higher cruise altitude and much longer flight duration.

The engine itself was in the extreme rear of the fuselage, fed by a long and somewhat sinuous duct from a circular inlet – equipped with a fixed inlet

Lockheed D-21 w/Booster and Pylon
SCALE 1/100

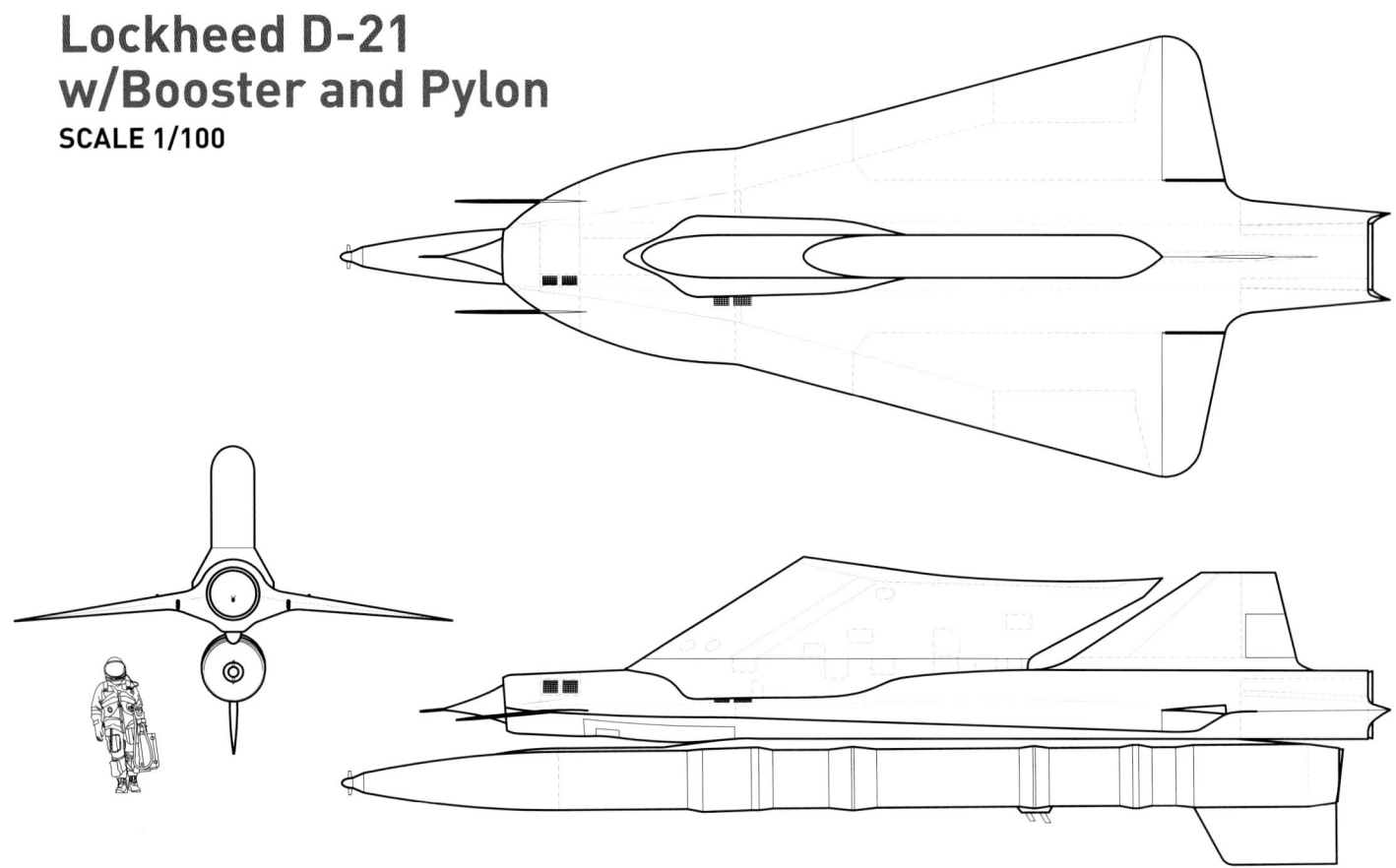

spike. The D-21 used triethylborane to ignite the ramjet, both during the initial engine startup and in the event of a flameout. The fuel was designated PSJ-100B, a low aromatics, low volatility concoction.

The duct bulged upwards near the front in part to provide room beneath for the actual surveillance system. The D-21 used the 275lb Aerial Reconnaissance Camera System Model HR-335 with a 24in focal length f/5.6 camera. This created 9x9in images on one 4,500ft-long roll of film, enough for 5,600 frames, with a resolution of 2ft at an altitude of 85,000ft.

It could operate in one of two modes – known as Mode 3 and Mode 5. In Mode 3, assuming a flight altitude of 80,000-95,000ft the width of the images would be from 15 to 18 nautical miles for a coverage of 3,700 nautical miles. Mode 3 would take images at three pointing angles… 19° left, vertical and 19° right. In Mode 5, the width of the images would be 26 to 31 nautical miles for a coverage of 2,780 nautical miles. Overlap between sequential frames would be 60% for both modes. Mode 5 would take images at three pointing angles… 36° left, 19° left, vertical, 19° right and 36° right.

The D-21 included a Honeywell MH-390(D) inertial navigation system capable of guiding the D-21 to up to 16 destinations. Position could be updated while on the M-21 up to the moment of launch, with a maximum initial position error of +/- 1.7 nautical miles using the M-21's stellar tracker. While the D-21 was self-navigating, it was capable of receiving signals for the purpose of fuel shutoff, self destruct and jettison of the camera system for recovery.

The D-21 was entirely expendable with no recovery system… no landing gear, no parachute. At the end of each mission it was to plunge into the sea for complete destruction. The camera system was attached to a hatch on the lower side of the vehicle along with the inertial navigation system, the automatic flight control system and a 'Sarah' radio beacon, an X-band transponder and an automatically deployed parachute. Jettisoning the hatch, which would descend beneath the 'chute, meant all these systems could then be recovered.

In an ideal situation the hatch would be jettisoned at 60,000ft over a pre-determined spot of the sea. There a specially modified C-130 would be waiting to air-snatch the parachute and its payload; the hatch would be brought on board and promptly flown to a base for film processing. In the event the C-130 failed to capture the hatch and parachute while in flight, it was designed for floatation, with all of the equipment in a water-tight package. The hatch and associated structures were largely made of titanium.

The first launch of a D-21 occurred in March 1966. The drone successfully separated and flew 120

Boeing B-52H
Lockheed D-21
SCALE 1/300

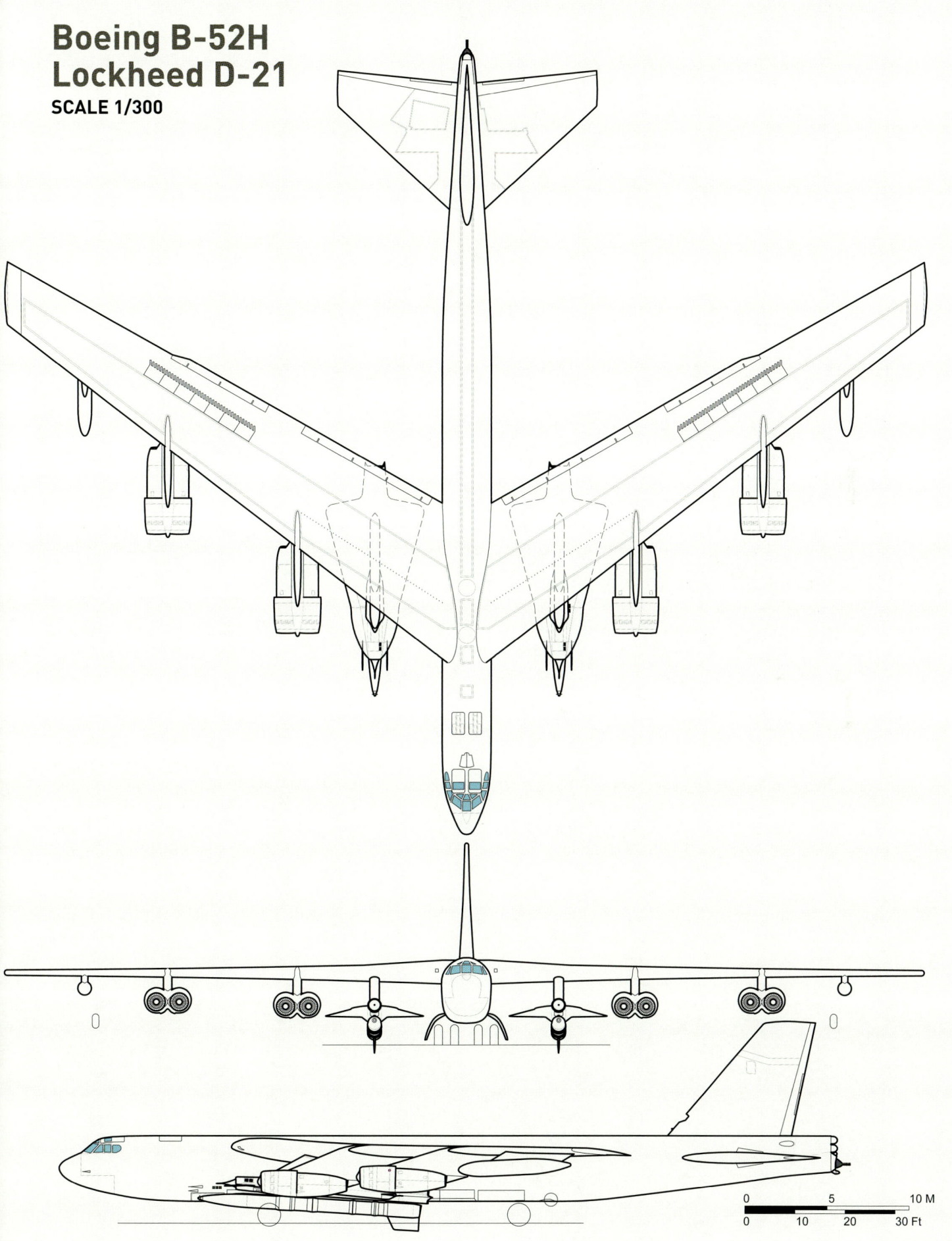

miles. Two more successful D-21 flights followed, in April and June, flying 1,200 and 1,600 nautical miles respectively. But in July 1966 the fourth launch failed catastrophically. Shortly after separation, the D-21 bounced off the shock wave shed by the M-21, came back down and stuck the aircraft, causing both to break apart in flight. The crew ejected safely and parachuted to splashdowns in the ocean, but the launch systems officer drowned. With the loss of one of only two M-21s, the decision was made to abandon the idea of launching the D-21 from a Mach 3 manned aircraft.

This was not the end of the D-21, however. In early 1966 Kelly Johnson suggested using the B-52 as a carrier aircraft for the D-21. In order for that to work, the drone would be equipped with a large solid rocket booster to get it up to speed and altitude. The idea was approved under the code name 'Senior Bowl', and two B-52H bombers were provided for modification.

Lockheed built four pylons to carry the drone, redesignated D-21B (the D-21s already built were modified to have new attachment points in their upper surface to mate with the pylons). The pylons were similar in appearance to the pylon used to carry the X-15, and would attach to the inboard hardpoint already built into the B-52H for the AGM-28 Hound Dog and other payloads.

The B-52H could, and occasionally did, carry two D-21Bs and their boosters, but only launched from the left pylon; the right pylon was used solely for transportation of the D-21B. The B-52H flight deck was modified to include two launch control officer stations, taking the stations formerly occupied by the electronic warfare officer and gunner. A stellar inertial navigation system was added along with telemetry and redundant command communications systems. A periscope above the flight deck could be used to observe both D-21B stations in flight.

The launch control officer received telemetry data regarding the position and speed of the D-21B for the first ten minute of flight, during which time he could order the D-21B to self-destruct if necessary. After that the drone would be out of range of the B-52 and it would automatically self-destruct if its altitude fell too low.

The booster for the D-21B consisted of a Lockheed Propulsion Company A-92 'Avanti' solid rocket motor, a fairing assembly (including a nose, a pylon to connect to the D-21B and a tail unit with a deployable ventral fin) and a ram air turbine in the nose to generate power. The overall length of the rocket motor itself was 422in with a main tube diameter of 30.16in and a diameter over the flanges of 32.62in. With an overall weight of 12,500lb, it could produce an average thrust of 27,300lb for about 87 seconds. The weight of the booster including the rocket motor and external structures (but not ballast) was 13,000lb. The propellant was low burn rate Polybucarbutene R, a high energy propellant with 68% ammonium perchlorate and 17% aluminium powder.

The booster would burn out at an altitude of about 80,000ft and above Mach 3.3. After that the drone would continue on its mission, initially climbing to a little short of 90,000ft before settling down to around 82,000ft. Through the course of the mission, as fuel was burned off, the drone would maintain a speed of Mach 3.3 but climb to around 94,000ft. At the end of the mission it would quickly descend, jettison the camera hatch and then self-destruct. The circular error probability of the D-21B at end of mission was 4.7 nautical miles.

The first launch of a D-21B from a B-52H was carried out in November 1967. The booster took the drone to altitude, but the drone failed, coming down about 150 miles away. A number of aborted launches and unsuccessful test flights followed; it was not until June 1968 that a truly successful flight occurred. An altitude of 90,000ft, a speed of Mach 3.3 and a range of 3,000 miles were achieved, along with camera recovery. The ramjet flamed out during programmed course turns but the TEB system re-ignited the engine and the aircraft continued.

More test flights, successful and unsuccessful, followed. Finally an operational mission over China was launched in November 1969; on this flight the D-21B simply vanished, likely due to navigation system problems. The second operational mission was more than a year later in December 1970. The drone flew the programmed route successfully, but the hatch was not recovered. The third operational mission was in March 1971. Again the flight itself was successful, but the hatch was damaged during recovery and the photos were ruined. The fourth and final operational flight was also in March 1971; the drone was lost over heavily defended enemy territory.

The D-21 programme consumed many years, many man-hours and many dollars without a single useful reconnaissance photo to show for it. And while the D-21 was struggling on, spy satellites were becoming ever more capable; the usefulness of the programme fell into serious doubt. In July 1971, the D-21B programme was cancelled, the surviving airframes ordered into storage.

For more on the M-21/D-21, see Lockheed SR-71 Blackbird: Origins and Evolution.

AGM-86 Air Launched Cruise Missile

As the ADM-20 Quail began to age and show its inadequacies, the Air Force sought a modern replacement. This led to the Subsonic Cruise Aircraft

Decoy (SCAD) programme. Early (1966) studies at the RAND Corporation envisioned decoys that the B-52 could carry in substantial numbers. More, the decoys could be armed. Instead of merely flying about and looking like a B-52, they would overwhelm Soviet air defences with numbers and nukes, though they would not be true attack missiles as their range would be low.

Even if the Soviets could tell these were not B-52s, they would not be able to ignore them. The idea of a new advanced decoy was popular in the Air Force; the idea of an armed decoy, however, had opposition. For several years the idea over whether to arm the SCAD or not raged back and forth. The coming of the B-1 supersonic bomber added complications: missiles that the subsonic B-52 could happily carry externally might create too much drag for the Mach 2+ B-1.

SCAD was approved in 1970 as a low-cost decoy that could potentially be armed and was designated ZAGM-86. The notion of a true cruise missile capable of deep penetration into enemy territory appealed to many… and many saw the development of such a missile as a risk to the B-1. Why develop this new expensive aircraft when a cheap missile could be built by the thousand and carried by existing B-52s? Thus the Air Force, the branch of the military that could most use a cruise missile of this kind, was unenthusiastic about it since it threatened their new bomber project. A decoy could have a simple guidance system, such as that of the Quail, but a cruise missile had to be much more accurate, including TERCOM radar and control systems for flying down low near terrain.

Progress on SCAD was slow. Finally a Request For Proposals was released in February 1972; in June Boeing won the contract to design the airframe. Previously, in May, Teledyne and Williams won parallel contracts to develop the small turbojet engine that the missile would require. But all did not go well. Design work proceeded on the decoy, with various notions about how to extend the range to turn it into a true attack missile capable of meaningful penetration. These options posed problems.

It was required that SCAD could be launched from the rotary launcher developed for the Short Range Attack Missile and this limited both diameter and length. An extended missile might not fit the length requirement, while a short missile with a belly tank would most assuredly not fit within the available diameter. Squabbles within the Air Force and between the Air Force and Congress, compounded by ballooning costs of the supposedly cheap and austere decoy, led to the ZAGM-86 SCAD programme being cancelled in June 1973.

In December of 1973, however, the Air Launched Cruise Missile (ALCM) programme was born. The AGM-86 designation was retained (now the AGM-86A), as was Boeing's basic configuration. It was to be a true cruise missile, even though Air Force interest in it remained soft.

Opinions began to change, however, as the United States Navy continued to successfully develop the BGM-109 Tomahawk cruise missile. The two missiles were similar in role and capability, and Congressional leaders began to wonder if it might not be better to cancel the ALCM and require the Air Force to adopt the Navy missile instead. As it turns out, few things motivate the Air Force quite like being told that they will have to use Navy hardware, and interest and enthusiasm for the ALCM suddenly spiked.

The SCAD and then the ALCM had an unconventional trapezoidal fuselage. This was not unprecedented: the ADM-20 Quail had a very similar cross-section. However, the section was flipped upside-down for the SCAD, with the widest part at the bottom. The fuselage looked, from the front, like a pie-wedge, and for good reasons. Eight of the cruise missiles would be wrapped around the launcher, the trapezoidal fuselages allowing them to pack as efficiently as possible into the limited space.

The initial ALCM design shared the SCAD's short fuselage and 'duck-bill' nose. These features allowed the missile to fit onto the unmodified SRAM launcher. The tail surfaces folded flat against the fuselage, the wings (fibreglass on the SCAD, metal on the ALCM) were slim and swept aft far enough to stow against the underside of the fuselage. The inlet for the turbojet was a scoop above the rear fuselage; it too folded down, burying itself partly within the rear fuselage to lower the height of the missile. The main change from SCAD to ALCM was the removal of the ECM package the drone carried and the addition of a thermonuclear warhead.

As an attack missile, the AGM-86A's range was still insufficient. A drop-tank option remained unworkable, as it would simply not fit on a rotary launcher and would add considerably to the drag of an externally carried missile. The idea of a belly tank was finally dispensed with in 1977. An 'Extended Range Vehicle' design simply stretched the AGM-86A configuration (and added more than a foot to the wingspan), adding range by adding fuel. The trouble was fitting the notably longer missile onto the SRAM launcher… it was no longer possible. Nevertheless, development went forward on both the AGM-86A and the extended AGM-86B. The duck-bill nose of the A-model was dispensed with, substituting a blunter, rounder, more volumetrically efficient nose. The B-model was visually very distinct from the A, as well as being substantially longer.

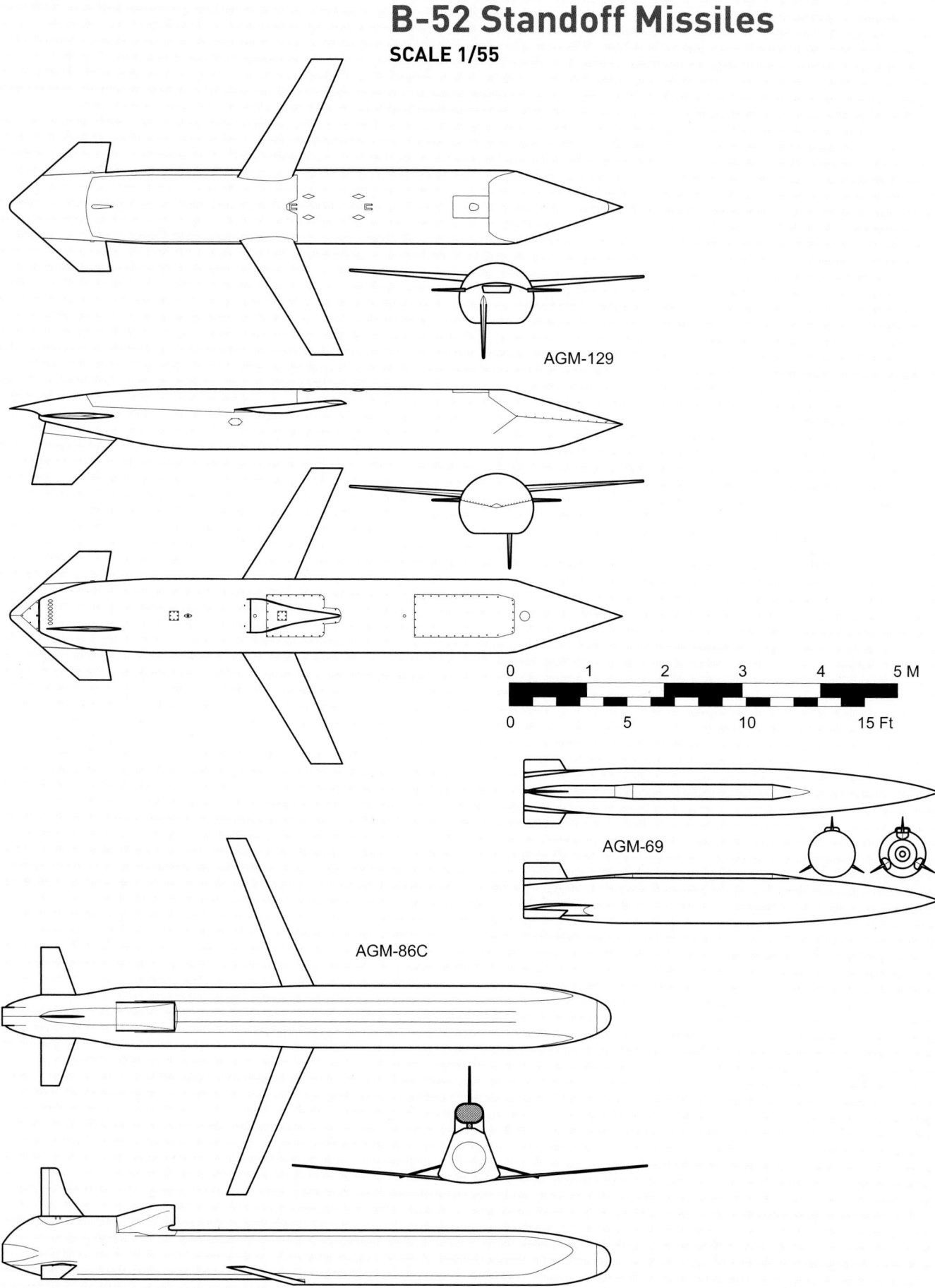

The first powered flight of an AGM-86A occurred on March 5, 1975; the first powered flight of an AGM-86B was on September 6, 1979. In March 1980 the ALCM won a competition against the AGM-109 Tomahawk to be the new Air Force cruise missile. Late that month, Boeing received a contract to build 3,418 of the missiles. The first production missile was delivered in November 1981 and a total of 1,833 were actually built by the time production ended in October 1986.

The first modified B-52G squadron became operational with AGM-86Bs in December 1982. These B-52Gs were initially only able to carry AGM-86Bs on external pylons. The pylons, attached to the existing Hound Dog pylon fittings, could carry six missiles under each wing. To comply with the SALT II treaty, 'strakelets', smoothly-blended fillets in the wing leading edge roots, were added to B-52s equipped to carry the AGM-86B. These were treaty-stipulated "functionally related observable differences" – aircraft modifications that could be seen by reconnaissance satellites for treaty compliance verification.

However, the SALT II treaty was never fully ratified by either nation; as it expired in 1985, the addition of strakelets became unnecessary. While the ALCM did not fit onto the existing SRAM rotary launcher, the Common Strategic Rotary Launcher was developed that fits eight missiles and plugs relatively easily into the bomb bays of the B-52 and the B-1B.

The AGM-86B had a single W80 nuclear warhead. In contrast to the ADM-20 Quail, the drone it was originally intended to replace, the ALCM does not make a big production out of being jettisoned from the carrier aircraft. It is simply ejected, wings and control surfaces still folded, inlet tucked down, engine not yet running. The missile drops about 450ft before it is under full power and flying correctly.

The Quail needed to be fairly gently placed into the air because its control system was not at all intelligent; if it got a bad start, it would simply fly off in the wrong direction. But the ALCM had TERCOM systems, and later GPS, that would let the missile know where it was. It would correct its own course as needed.

The Air Launched Cruise Missile never fulfilled its nuclear mission, or at least it has yet to do so. However, it has been used in combat. The AGM-86C Conventional ALCM (CALCM) programme started 1986, carried out by removing the nuclear warhead from AGM-86Bs and giving them 1,360lb pound conventional warheads and GPS guidance (rather than TERCOM). These missiles have been used in numerous Middle East conflicts starting with Operation Desert Storm. The AGM-86D was a further change to the C-model which replaced the blast warhead with a penetrating one for taking out underground bunkers and the like.

The Air Launched Cruise Missile is 1970s technology and while efforts were made to reduce its radar cross section (ironic, given that it was originally supposed to mimic the reflectivity of the B-52), but it was not a truly stealthy missile. That role was supposed to go to the AGM-129 Advanced Cruise Missile, designed from the ground up to be as invisible to radar and infrared as possible. But the AGM-129, like many stealth aircraft, was a maintenance headache.

It was expensive to develop and expensive to produce; rather than being made in bulk to replace all AGM-86Bs, only 460 AGM-129s were built. So it became fairly easy to kill the AGM-129 early; it was withdrawn from service in 2012, leaving the aging AGM-86 series to soldier on. At the time of writing, the best hope for an ALCM replacement was the Long-Range Stand-Off weapon being developed by Raytheon. But even if that comes to fruition, the AGM-86 was expected to continue on into the 2030s.

Skybolt

The Bold Orion launched by the B-47 proved that an air-launched ballistic missile was feasible. The B-52 was intended to carry the missile that would benefit from Bold Orion's teachings: the GAM-87 Skybolt.

The USAF began feasibility studies of air launched ballistic missiles in January 1958. An early part of this effort was directed to the Martin Company for the design and development of the Bold Orion test vehicle, previously described in the B-47 section. By November of that year Bold Orion had successfully flown, demonstrating that the idea was at least feasible.

In January 1959, therefore, General Operational Requirement Number 177 was established. This formalized the requirement for Weapon System 138A, a "Rocket Powered Strategic Air-to-Surface Missile System". This was not to be a true intercontinental ballistic missile, but more of a stand-off missile, extending the range of bombers or clearing the road for them by delivering a nuclear beatdown to targets in their path.

This new missile was to be compatible with the B-58 Hustler (specifically the proposed B-58B version) and compatibility with B-52s already modified to carry the GAM-77 Hound Dog was greatly desired. The Air Force was flexible in how the missile was to work… a boost-glide missile was as acceptable as a ballistic missile. The ability to be retargeted while still attached to the carrier aircraft and to be at least 85% reliable after extended periods of alert were a must.

Skybolt Missile on Pylon
SCALE 1/72

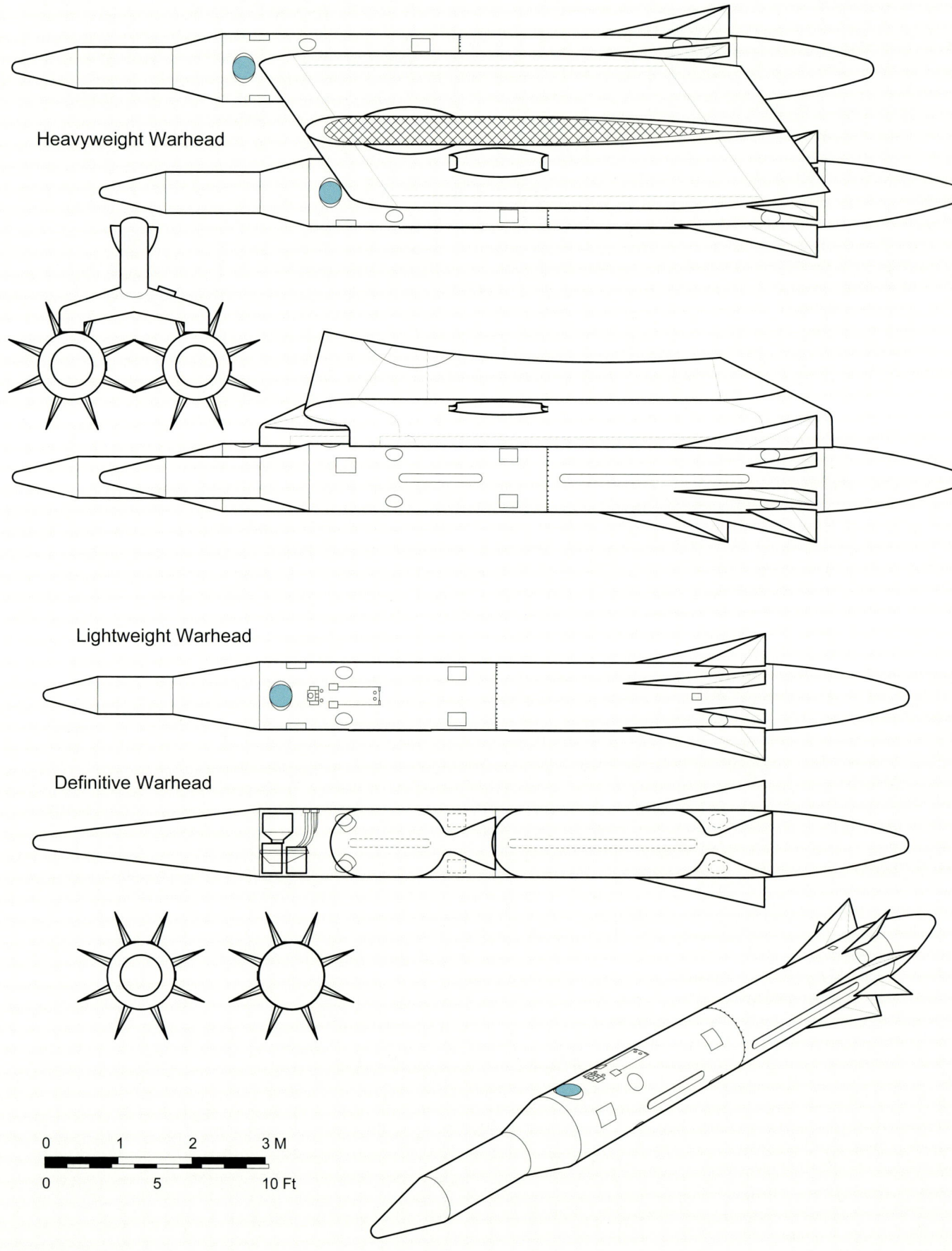

Heavyweight Warhead

Lightweight Warhead

Definitive Warhead

Boeing B-52H - Skybolt
Airborne Missile Launcher
SCALE 1/300

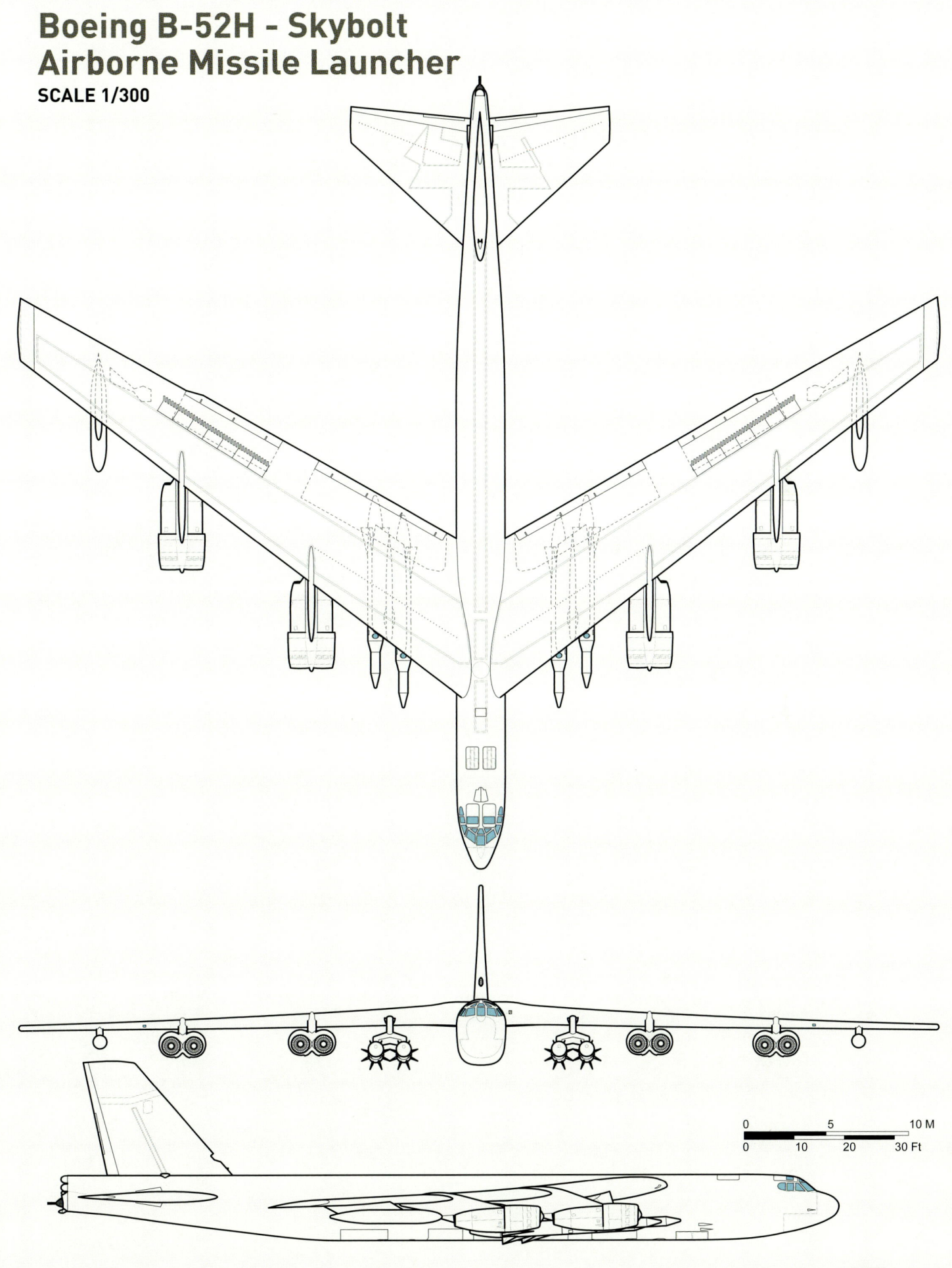

Almost from the beginning, the programme was intended to be multinational. The Royal Air Force met with the US Air Force in Washington, D.C. in April 1958 as a task group to consider strategic air-to-surface missiles of the kind represented by WS138A. In March 1960 the United Kingdom formalized an agreement to buy Skybolt missiles from the United States, while at the same time granting US Navy Polaris missile submarines the right to berth in the submarine base in Holy Loch, Scotland.

A month later, Britain announced the cancellation of its home-grown Blue Streak intermediate range ballistic missile, then under development. Blue Streak was, like the Atlas ICBM, liquid-fuelled, requiring kerosene and liquid oxygen to be loaded on board prior to launch. This made it slow to respond. Coupled with the expense incurred in developing it and the expenses yet to be incurred in deploying it, it rather quickly fell out of favour when Skybolt came on the scene. Skybolt would be able to respond just as fast as the British V-bombers could get off the ground and would be relatively cheap. Skybolt would, it was expected, be an important part of the British nuclear deterrent force in the years to come.

Twenty-three prospective contractors attended a meeting at Wright Field to be briefed on the new project in February 1959. Fifteen of them had turned in proposals by late March and the proposal from the Douglas Aircraft Company was selected in May. Douglas initially had five distinct configurations (dubbed Able, Charlie, Delta and presumably also Baker and Echo), varying considerably in layout. The Douglas design as built – the Delta 2 Phase H/J configuration – was a straightforward two-stage design using 36in diameter motors by Aerojet General.

Initially, the second stage was to have four small nozzles but the design was refined to just one; this lengthened the body but made it lighter and more straightforward. The first stage had a fixed nozzle with yaw, pitch and roll control provided by all-moving tailfins. The second stage nozzle was vectorable to provide pitch and yaw control; roll control was via small separate thrusters.

General Electric was the contractor for the re-entry vehicle. The original plan was to have one heavyweight and one lightweight RV, providing differences in warhead yield and range. Both of the original RVs were cone-cylinder-cone configurations similar to the RVs used on the original Minuteman ICBM, with an 18in diameter cylindrical section of different lengths.

By March 1961 General Electric had determined that a geometrically simpler round-tipped cone would make a more efficient RV, extending range a few dozen miles. Conical RVs became the standard for Skybolt from that point on (early mockups and test articles used the prior configuration), and became the standard that is still used in ICBMs today.

The B-58B was cancelled in July 1959, leaving the B-52 as the baseline carrier aircraft. Some consideration was given to equipping the B-70 supersonic bomber, the KC-135 or some future hoped-for nuclear powered bomber with Skybolt, but little effort was made on working those options out. The B-52 received the lion's share of attention.

Skybolt used a star tracker built into the 'upper' side of the missile as it was carried by the B-52. This could, at the high altitudes where the B-52 would normally carry and launch the missile, see stars in the daytime and would be able to determine which stars were which. By comparing those sightings with a database of where they would be in the sky at any time, the missile would know where it was and where it was pointed just before launch. The B-52's onboard navigation system would tell the missile where it was too and also where to go, targeting being a process that could be done or changed relatively quickly (unlike the hours it could take to retarget an early Minuteman).

After launch, the Nortronics-built onboard inertial navigation system would keep it on course. For range control the second stage featured thrust termination ports. The warhead was not manoeuvrable, though some thought was given to that capability early in the programme.

The first stage weighed 5,500lb in total, its motor generating 36,800lb of thrust for 40.7 seconds with a specific impulse of 245 seconds. The second stage weighed 2,714lb and generated 18,400lb of thrust for 41.3 seconds, at a specific impulse of 245 seconds. Gross weight of the missile was 11,353lb; length was 39.4ft. Maximum range was 950 nautical miles with an apogee of 210 nautical miles.

Maximum velocity was 12,380ft per second and time from launch to impact was 710 seconds. The warhead planned for use with the Skybolt was the J-21, later designated the W-59. This was a thermonuclear device weighing 550lb and yielding 1 megaton.

Skybolt was revealed to the public in early January 1961, with captive-carry flights in the middle of the month. A dummy missile was dropped from a B-52G modified with operational Skybolt pylons in February 1961. This and a subsequent drop test in March demonstrated not only clean separation but the good aerodynamic stability of the missile. Drop tests of instrumented but unpowered missiles began in July, showing proper flight control systems. At this time the Air Force expected to buy 1,122 Skybolts (equipping 23 B-52 squadrons), while the RAF would buy 192.

Boeing B-52H - AML II
SCALE 1/300

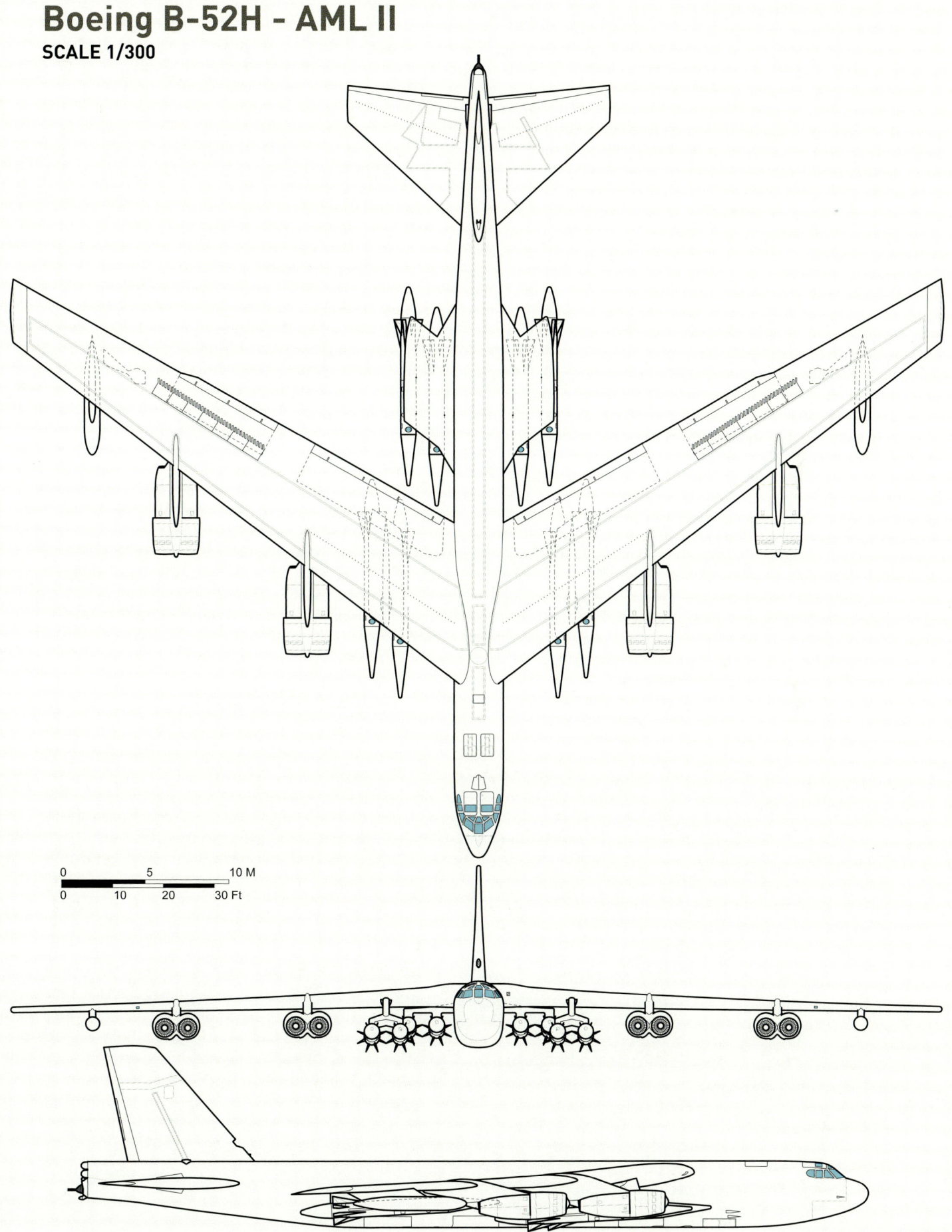

April 1962 saw the first powered flight. The first stage performed normally but the second stage failed to ignite. Upon the second launch in June, the first stage failed to ignite. As a result, the igniters for both stages were revised to be more energetic. In Britain, the Vulcan Mk II bomber began captive flights with mechanically complete missiles in July 1962. September saw two launches, both failing. The fifth launch in November failed due to a fault in the gas generator. In December, the first successful launch was made, succeeding in all objectives. Unfortunately this flight came too late: the programme had already been cancelled.

Skybolt was done in by the usual dreary problems of rising costs and slow development – the repeated test flight failures only added to this. By the summer of 1962 the writing was on the wall and the British were becoming nervous. The United States had fallback options in the event that Skybolt didn't work out – the Minuteman ICBM, Polaris sub-launched missiles and Hound Dog missiles carried by the B-52. Britain had none of those. So the British government was understandably annoyed when the rug was pulled out from under its deterrent plans. The British were somewhat mollified, though, when the United States agreed to provide Polaris missiles to the Royal Navy for inclusion in its submarines.

Boeing B-52H AML II

There exists a diagram of an evolved version of the B-52H/Skybolt weapons system. Along with the four missiles carried under the wing, the Airborne Missile Launcher (AML) II would have carried a further four Skybolt missiles. Rather oddly, these would not have been carried under the wings, but instead under pylons that were attached to the sides of the rear fuselage.

This would seem to have been an unfortunate position from the standpoint of centre of gravity; presumably some fuel tankage would have been deleted from the rear fuselage to offset it. This would have greatly reduced overall fuel capacity and thus the range and duration of the aircraft, making in-flight refuelling essential. Once in flight, the rear fuselage pylons would have been able to generate some measure of lift to offset their own weight and the weight of the missiles. No further information on the configuration is available at this time, however. The missiles are depicted with the final conical GE re-entry vehicle, so the design likely dates from later than March 1961.

Longbow

The Air Force System Command began a study of a strategic air-launched ballistic missile dubbed 'Longbow' in 1978. DARPA added funds to the effort with the aim of making the missile dual-use. It could be armed with a single nuclear warhead for the ground-attack role or it could be equipped with a seeker for long-range air-to-air missions.

Longbow came about during a rather dark period in American strategic weapons development; the Carter administration had just cancelled the B-1 in favour of a rather theoretical stealthy bomber. The Peacekeeper ICBM and Pershing II medium range ballistic missile were in development, but it was not smooth sailing as both had substantial opposition from both many politicians and public protesters at home and abroad.

Two cruise missiles were under development, but there was no certainty that these would work as expected or survive political hurdles, and many of the proposed launch options for them involved cheap options such as modified jetliners. Longbow provided the possibility of a different capability altogether, harkening back to the Skybolt.

As a strategic missile, Longbow would have carried a single nuclear warhead and would have had a range of about 2,000 nautical miles. This put it ahead of Skybolt and the Pershing II battlefield support missile in terms of range (each could go about 1,100 miles). In order to achieve this, it would have been a two-stage solid rocket weighing 4,000-5,000lb. Launch trajectory for both roles would involve the missile lofting above the atmosphere; much of the distance travelled would be in space with no meaningful drag. The anti-air version would turn some of the re-entry energy into manoeuvre energy, using an undefined seeker system to hunt down its targets.

The air-to-air version was to have a manoeuvring entry vehicle and was meant to take out multiple aircraft. Available documentation does not detail how it was to do this; presumably it would have been equipped with a nuclear warhead in the kiloton range, the only really effective approach to defeating large numbers of aircraft with a single missile.

It's unclear how the B-52, carrying 12-16 Longbows on external pylons and within the bomb bay, would know where to send them for the anti-aircraft mission. The B-52's onboard radar systems were in no way capable of spotting aircraft thousands of miles away. Presumably the bomber would receive targeting information from other aircraft or even satellites.

The details available on Longbow are few, but it was about half the size of either Pershing II or Skybolt, yet could surpass them in terms of range while carrying an RV of about the same size as the Mk 12. That's impressive, if not downright questionable.

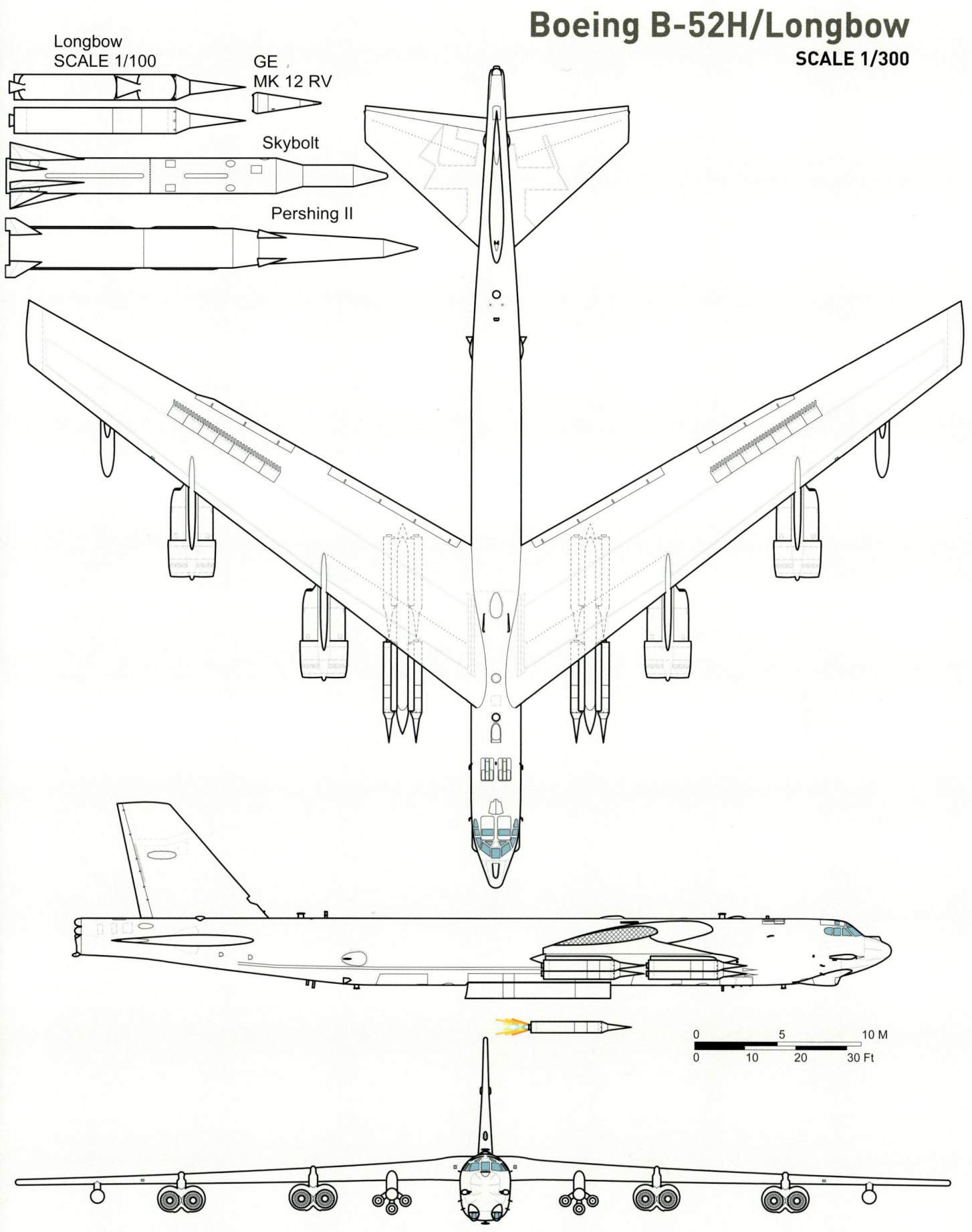

Boeing NB-52B

As previously mentioned, proposals for high-speed rocket-powered demonstrator aircraft during the 1950s culminated in the X-15 programme. This was a high-prestige, high-priority programme; virtually the entire United States aircraft industry wanted to be a part of it and numerous proposals were tendered for the contract to design and build the aircraft.

Part of those proposals involved the carrier aircraft. The B-29 and B-50 were clearly not up to the task; Convair's B-36 was, at least initially, the obvious solution. It had the lift capability and the fuselage/bomb bay were large enough that the proposed rocketplanes could be semi-submerged within it. This would allow the X-15's pilot to ride within the spacious and comfortable pressurized main cabin of the bomber during takeoff and climb, boarding the rocketplane just before drop.

Indeed, the B-36 was the carrier aircraft of choice for North American Aircraft when it won the X-15 contract in August 1955. The winning X-15 design was not quite what would actually be built, but it was very similar. North American projected that the B-36 would be able to carry the rocketplane to 35,000ft and Mach 0.7; an adequate, if not spectacular, performance.

But as with the bombardment role, speed and altitude were important capabilities for the carrier aircraft and the B-36 simply could not compete against the B-52 in either area. Additionally, the B-36 was nearing the end of its operational life while the B-52 was looking young and fit… the expertise to fly, maintain and update a B-52 carrier aircraft would remain for some years, while the B-36 and its many, many components would soon be museum pieces at best. So in 1957 the NACA directed North American to ditch the B-36 carrier concept and focus on the B-52. This would raise the drop altitude to 45,000ft and the airspeed to Mach 0.9.

There was some early thinking that the two prototypes, XB-52 and YB-52, would be transferred to the X-15 programme. This made sense insofar as these aircraft were never going to be operational bombers, thus the Strategic Air Command would not lose strike capability by turning them into carriers. But the prototypes were non-standard in nearly every way they could be; maintaining them would be almost the same logistical nightmare that maintaining a last-of-its-kind B-36 would be.

Fortunately, it was decided instead that a B-52A (52-0003) and a B-52B (52-0008) would be sent to North American for the necessary modifications, creating aircraft officially dubbed the NB-52A and NB-52B.

The B-52 fuselage was not large enough to accommodate the X-15; it would need to be carried under a wing pylon. This was a bit of a disappointment, but hardly a true stumbling block. Unlike a bombardment mission, test launching a rocketplane did not involve many long and uncomfortable hours flying great distances, so the pilot having to board the X-15 prior to liftoff was not a major issue.

The pylon did need to be designed so that the pilot could eject if need be though. This was accomplished by positioning the cockpit well ahead of the wing leading edge, the leading edge of the pylon being behind the X-15's canopy. The geometry of this arrangement meant that a portion of the inboard flap on the starboard wing needed to be cut out to provide clearance for the X-15's vertical stabilizer. The inboard flaps on both sides were to be locked in place, and were therefore not used for either takeoff or landing. This made those phases rather faster than normal for a B-52.

A 1,500 gallon liquid oxygen tank (roughly 15,000lb of LOX) was added to the bomb bay in order to top off the X-15 just before launch, with lines being plumbed through the wings and down through a new 1,170lb pylon. Vents were added to the bomb bay to prevent fume accumulation; after launch of the X-15, the remaining liquid oxygen would be slowly jettisoned prior to landing. This process would take up to 15 minutes, depending on the amount of LOX remaining. In the event that the X-15 was not launched, the LOX and rocket fuel would be jettisoned before landing.

Three shackles in the underside of the pylon supported the X-15. These would normally release using hydraulic activation; in the event of a failure there was a pneumatic backup.

Structural stiffeners were added to the right wing while a fuel tank located over the pylon was removed. Both film and video cameras were added to the starboard fuselage to keep an eye on the X-15. Nitrogen and helium bottles were added to the fuselage, the helium to pressurize the liquid oxygen tank and the nitrogen to operate the liquid oxygen system valves. The tip tanks were removed; the tail turret was removed and a truncated fairing was added. A Plexiglas dome was added for the use of the launch panel operator. Bombardment system instruments were removed or deactivated and instruments to monitor the aircraft and the X-15 were added.

The NB-52A was retired in 1969, after launching the various iterations of the X-15 a total of 72 times, the M2-F2 four times, the HL-10 11 times and the X-24A twice. It was sent to the Military Aircraft Storage and Disposition Center at Davis-Monthan Air Force Base in Arizona. Unlike many of the unfortunate airframes that languished in the dry desert air for a while before being turned into scrap

Boeing NB-52
SCALE 1/300

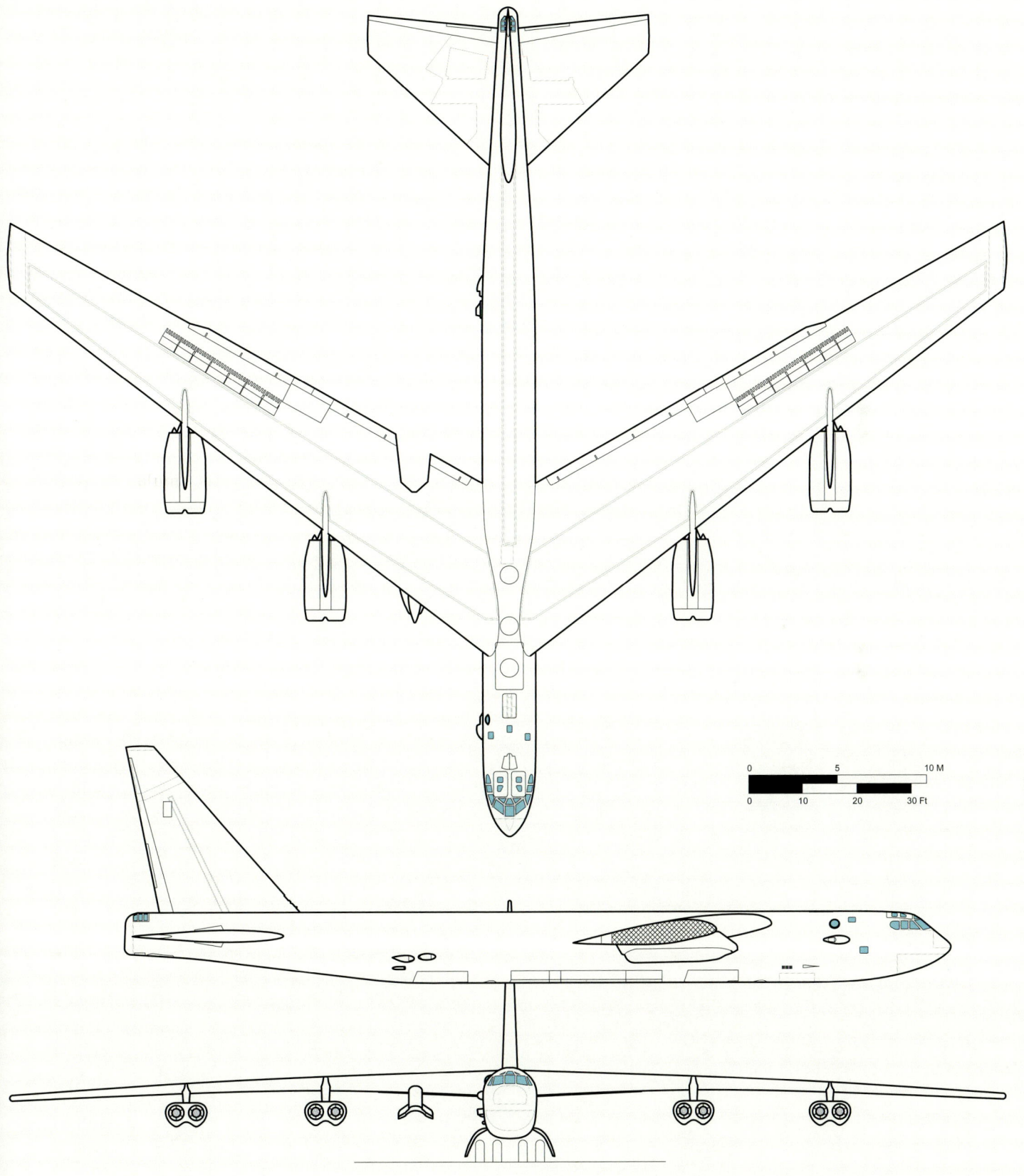

metal, the NB-52A eventually wound up on display at the Pima Air Museum near Tuscon, Arizona, where it still exists today.

The NB-52B soldiered on until 2004. At that point it was the oldest B-52 in service and the only non-B-52H model, and was replaced in NASA services with an NB-52H (aircraft 61-0025). The NB-52B is currently on display at Edwards Air Force Base. The NB-52H has not had as noteworthy of a career; in the years since the retirement of the NB-52B NASA has not produced the sort of flight vehicles needing the services of a large and capable carrier aircraft.

During its decades of service, the NB-52B carried many test aircraft, each requiring its own modifications to the NB-52B or the pylon. To fully describe all the numerous vehicles and test objects that the NB-52s carried over the years would require a substantial book all its own; what follows are brief descriptions of some of the more interesting and important NB-52 payloads. Not all were manned and not all were aircraft by any rational definition.

North American X-15

The aircraft that the NB-52B was created to carry was the North American Aviation X-15, intended to fly to the edge of space and reach speeds well beyond anything a manned aircraft had previously hoped to attain. It was a single-seat rocket aircraft with relatively small trapezoidal wings that had only modest sweepback.

The horizontal stabilizers were sharply swept and angled downwards; the vertical stabilizer was a fat wedge with a blunt trailing edge. It was equipped, ultimately, with an XLR-99 rocket engine that burned a mix of liquid oxygen and ammonia. However, this mighty engine took longer to develop than originally expected, so the first two of the three X-15s were initially equipped with two XLR11 rocket engines that burned LOX with alcohol. The XLR11, even paired, did not provide the thrust that the XLR-99 would but had the virtue of being a reliable and venerable system. It was, after all, the engine that first shoved the Bell X-1 past the sound barrier.

The first flight of an X-15 occurred on June 8, 1959, the one and only glide flight of the X-15 programme. This and the other four of the first five X-15 flights were carried by the NB-52A; the first flight using the NB-52B occurred on January 23, 1960. The first powered flight was on September 17, 1959, reaching Mach 2 and 52,341ft.

The XLR-99 was finally fired up in flight on November 15, 1960. The highest altitude reached was 354,200ft on August 22, 1963; the maximum Mach on that flight was 5.58. The maximum Mach reached was 6.7 on October 3, 1967; the peak altitude then was a mere 102,100ft. The final flight in the X-15 programme was on October 24, 1968, with a total of 199 flights. Two X-15s survived; one is at the National Air and Space Museum in Washington, D.C. and the other at the National Museum of the United States Air Force in Dayton, Ohio. The X-15 has of course been described fully in many other books.

Lifting bodies

The 1960s saw growing interest in a new category of aircraft: the lifting body. Lifting bodies are just as described: instead of relying on sizable wings to generate lift, lifting bodies are typically partially or wholly wingless – generating lift through the shape of the fuselage instead.

In general these configurations are moderately terrible for low speed flight and not all that great for high speed cruise. They generate relatively little lift at low speed and a great deal of drag at high speed. But interest in lifting bodies was based not on them being great aircraft, but rather on them being the best aircraft that could serve as spacecraft.

The wings of a conventionally-configured aircraft would have a tendency to melt during re-entry. 'Orbital aircraft' designed over the decades have generally relied upon variable geometry or otherwise stowable wings; while this can theoretically work and result in an aircraft with some fair aerodynamic properties when fully deployed, it's added mass and complexity.

Additionally, the fuselage of a conventional aircraft tends to be long and slim, with poor volumetric efficiency. At the far end of the scale, space capsules such as the Soyuz and Mercury and Gemini have fair volumetric efficiency and are simple with reasonably low cost… but they do not 'fly', they drop. That is fine for spacecraft which can be aimed at a wide expanse of steppe or ocean, with a sizable military operation in place to recover them. But if you want to fly often and cheaply, using capsules for crew return is not a great option. With minimal cross-range, crews cannot simply decide on the spur of the moment that it's time to go back to Earth; they need to wait until the orbit is just right so that the de-orbit burn will drop the capsule close enough to recovery forces.

Lifting bodies, starting in the late 1950s, looked like the answer to this problem. The US Air Force and NASA both, by the early/mid-1960s, believed that they would soon have a series of space stations in orbit, ranging from tiny orbital laboratories sized for a crew of just two to great assemblages of modules lofted by a series of Saturn V boosters. These would require the shuttling of crews up and down on a regular basis, as well as the possibility of 'life boats' to recover crew who were sick or injured.

B-52 as a Carrier

X-15A-3 + NB-52A
SCALE 1/300

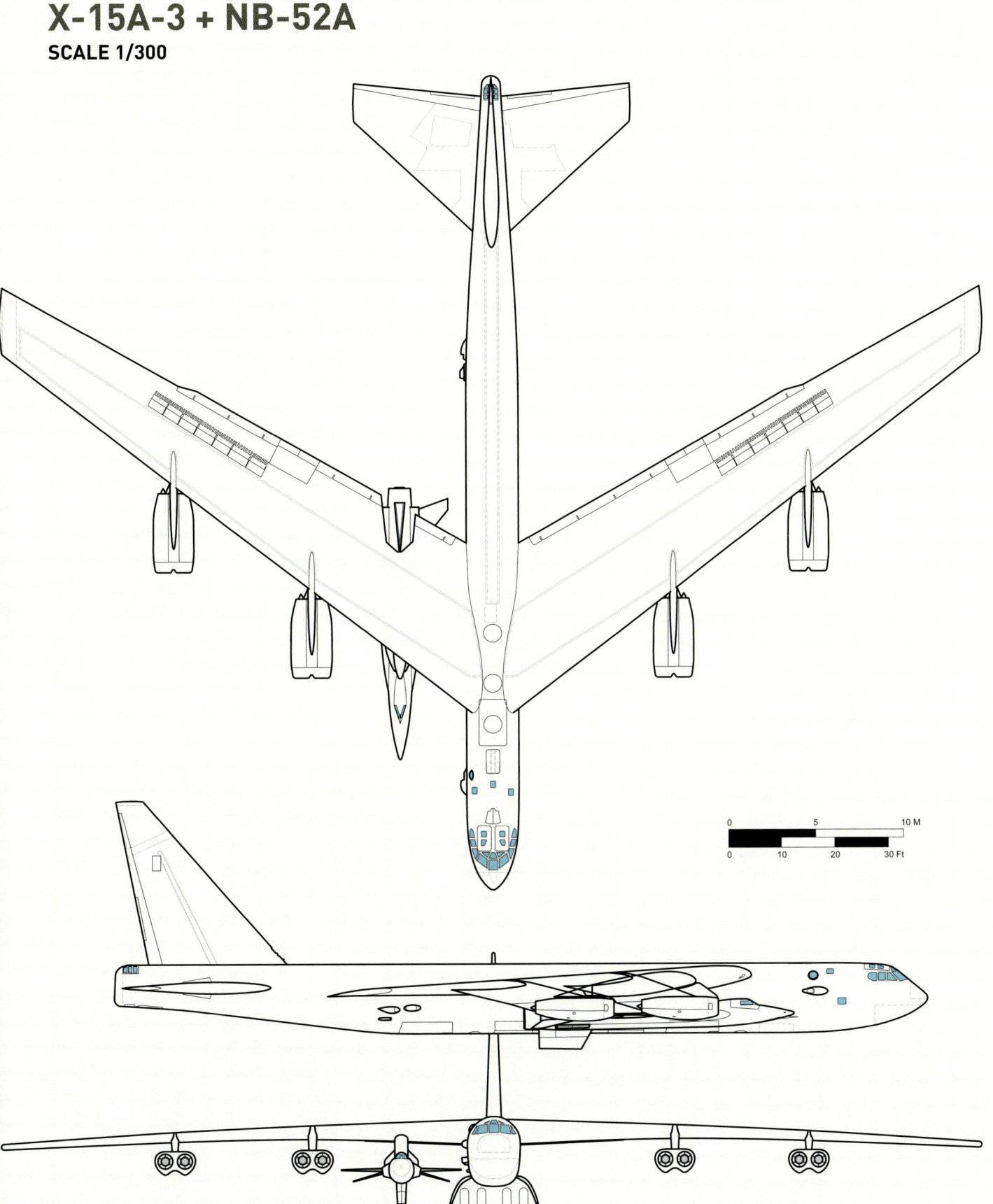

The ability to land at certain Air Force bases or NASA sites was vital, as was the ability to do so with minimal fuss. Lifting bodies could, it was proposed early on, do this. With the configuration being generally a rounded single form with few to no projecting surfaces to present themselves to the heartache of being burned off on re-entry, lifting bodies seemed a good alternative to capsules.

In general they have good volumetric efficiency, enabling them to easily accommodate crews. While they are not great fliers, they are good enough to cover hundreds of miles of cross-range, letting them re-enter from a much wider expanse of the sky. They fly well enough to target and land on decent sized runways, which Air Force bases and NASA launch sites tend to have in abundance.

Lifting bodies were dynamite in wind tunnels. But a wind tunnel test of a subscale model does not teach you everything you need to know, especially when dealing with an entirely new kind of aircraft. Full-scale, or near full-scale, lifting body aircraft are needed.

While not the first lifting body aircraft, the NACA-Ames M1 ('M' for 'Manned') concept was presented in 1958 as a potential manned space vehicle. It was a stubby 30° cone split in half, with a rounded underside and a flat top, with flaps for aerodynamic control. The M1 was considered early on as a potential competitor in the Mercury programme against the simpler conical capsule. It would be able to land as far as 230 miles to left or right, with a landing footprint some 700 miles long.

As if was a lifting vehicle, it drew out the re-entry process, compared to the purely ballistic Mercury capsule. So where the capsule hit the atmosphere hard, generating a dangerous 8+ gees peak deceleration, the M1 would top out at around two and a half gees… not only more comfortable, but also safer for space station crews who might be weakened from long periods in space, sick or injured. Unfortunately, a lifting body, even one as stubby as the M1, not only weighs more than a simple capsule, it make integration onto a launch vehicle more difficult. Thus the M1 was not chosen for the Mercury programme.

As the NACA became NASA, capsules became all the rage. But interest in lifting bodies remained, particularly at NASA-Langley and NASA-Dryden. It was at Dryden in 1962 that several years of efforts using hand-made balsa models of M1-like lifting bodies, towed behind remote controlled model aircraft, finally paid off with the decision to fund the construction of the M2-F1 ('F' for 'Flight').

The M2-F1 was a low speed demonstrator – a plywood skin wrapped around a welded tubular steel structure, with room for a pilot. It was built entirely in-house at Dryden on a very low budget of $30,000. It was completed in four months (compare with modern aircraft development programmes which cost billions and take years). A half-cone vehicle with, again, a flat top, it had a raised canopy along with Plexiglas glazing in the round nose for pilot visibility. Stability was provided by a pair of slab-sided vertical stabilizers; control was provided by rudders and 'elephant ears' that projected horizontally from the stabilizers.

Towed into the air, the M2-F1 would demonstrate that a lifting body could indeed fly in a stable and safe manner. Although it was technically a glider, small solid rocket motors would later be installed for burn at landing ('instant L/D').

The first flight occurred on April 5, 1963, behind a 1963 Pontiac Catalina convertible (modified into a tow vehicle by a hot-rod shop, with adjustments to the engine, the addition of roll bars, the passenger seat was turned 180°, a new transmission). On August 16, 1963, the M2-F1 flew behind a C-47. The M2-F1 was towed to an altitude of around 12,000ft and 100mph. Upon release of the cable, it would begin to descend, landing about two minutes later. The final flight was on August 16, 1966.

This was all well and good for demonstrating the low speed potential of the craft, but more information would be needed. Thought was given to hauling the M2-F1 to higher speeds and altitudes under larger aircraft, including the B-52. But the M2-F1 was, after all, a lightly constructed plywood vehicle; it would never survive unless greatly modified. A heavyweight version was needed.

M2-F2/M2-F3

An aluminum M2-F1 with a gross weight of around 10,000lb was thought capable of reaching about Mach 2 if carried to altitude by a B-52 and using a single XLR11 rocket engine, now available due to the X-15 moving on to the XLR99. The new vehicle, it was soon decided, would not be merely an M2-F1 made from metal, but an all-new vehicle designed to replicate an orbit-capable re-entry vehicle.

Some configuration changes were necessary for the M2-F2 design. The 'elephant ear' elevons used to good effect on the low-speed M2-F1 had to be deleted; these stark projections would form very definite hot spots during entry and would burn off. Stability and control would have to be provided by split flaps on both the top and bottom of the rear fuselage.

The XLR11 would be installed directly in the centre of the trailing edge of the M2-F2. Retractable landing gear replaced the simple fixed gear of the M2-F1. This was a much more complicated undertaking than the rather simple M2-F1, so a request for proposals went out to the industry to find a contractor to build one M2-F2 and one HL-10. In June of 1964,

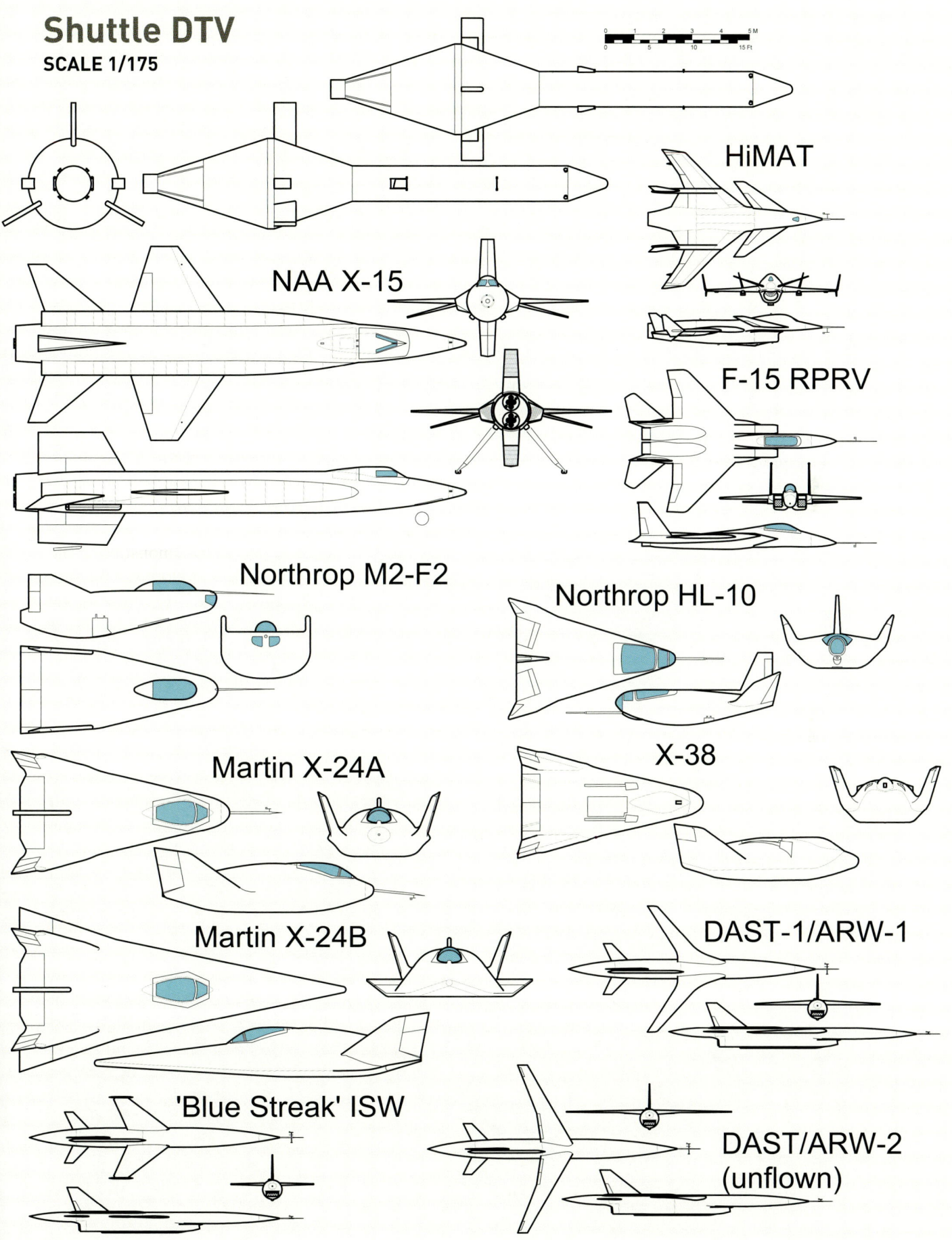

NB-52/'Blue Streak' ISW

NB-52/DAST-1/ARW-1

NB-52/DAST/ARW-2

NB-52/X-24A

NB-52/X-24B

NB-52/X-38

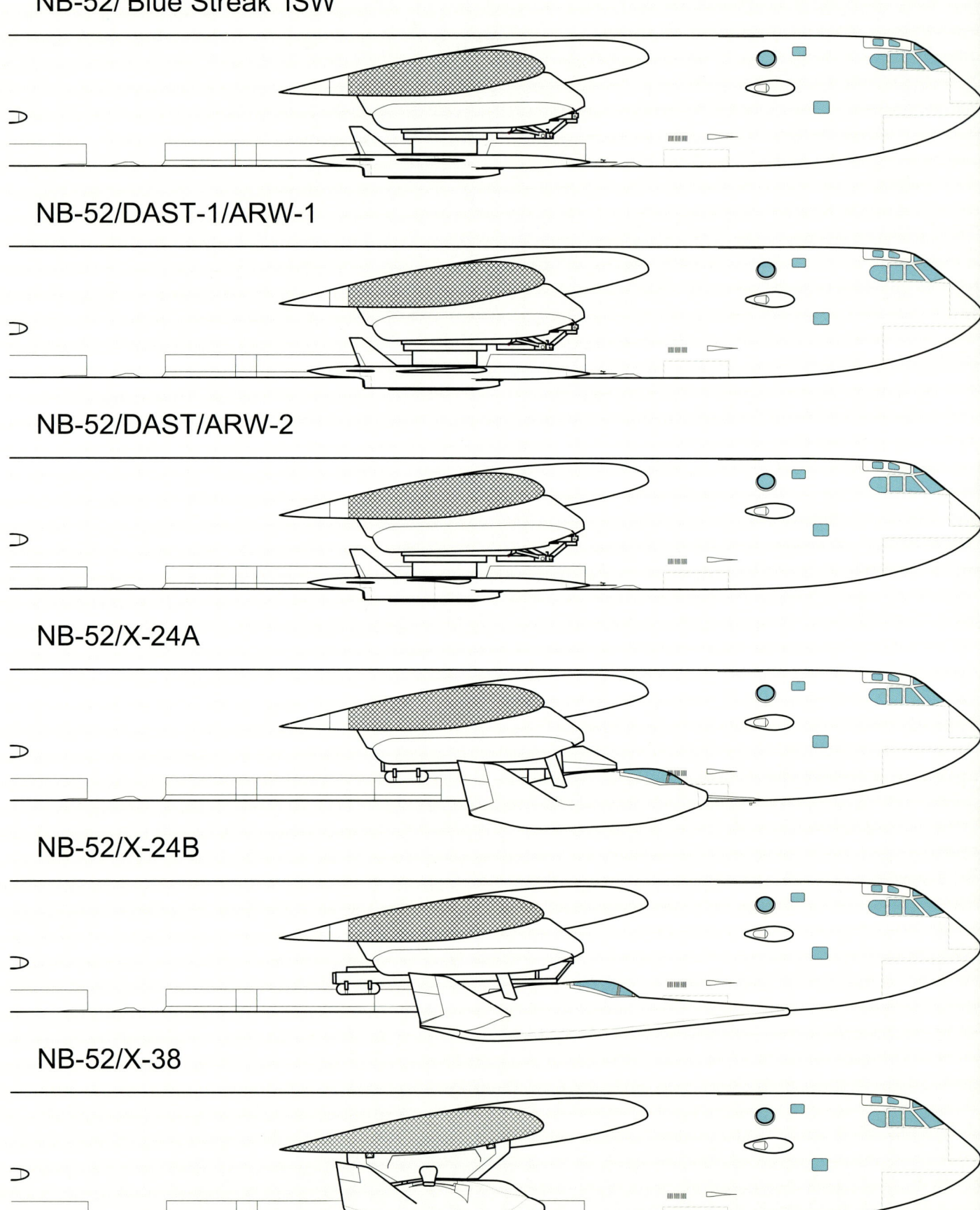

B-52 as a Carrier

NB-52/DTV

NB-52/X-15

NB-52/F-15 RPRV

NB-52/HiMAT

NB-52/M2-F2

NB-52/HL-10

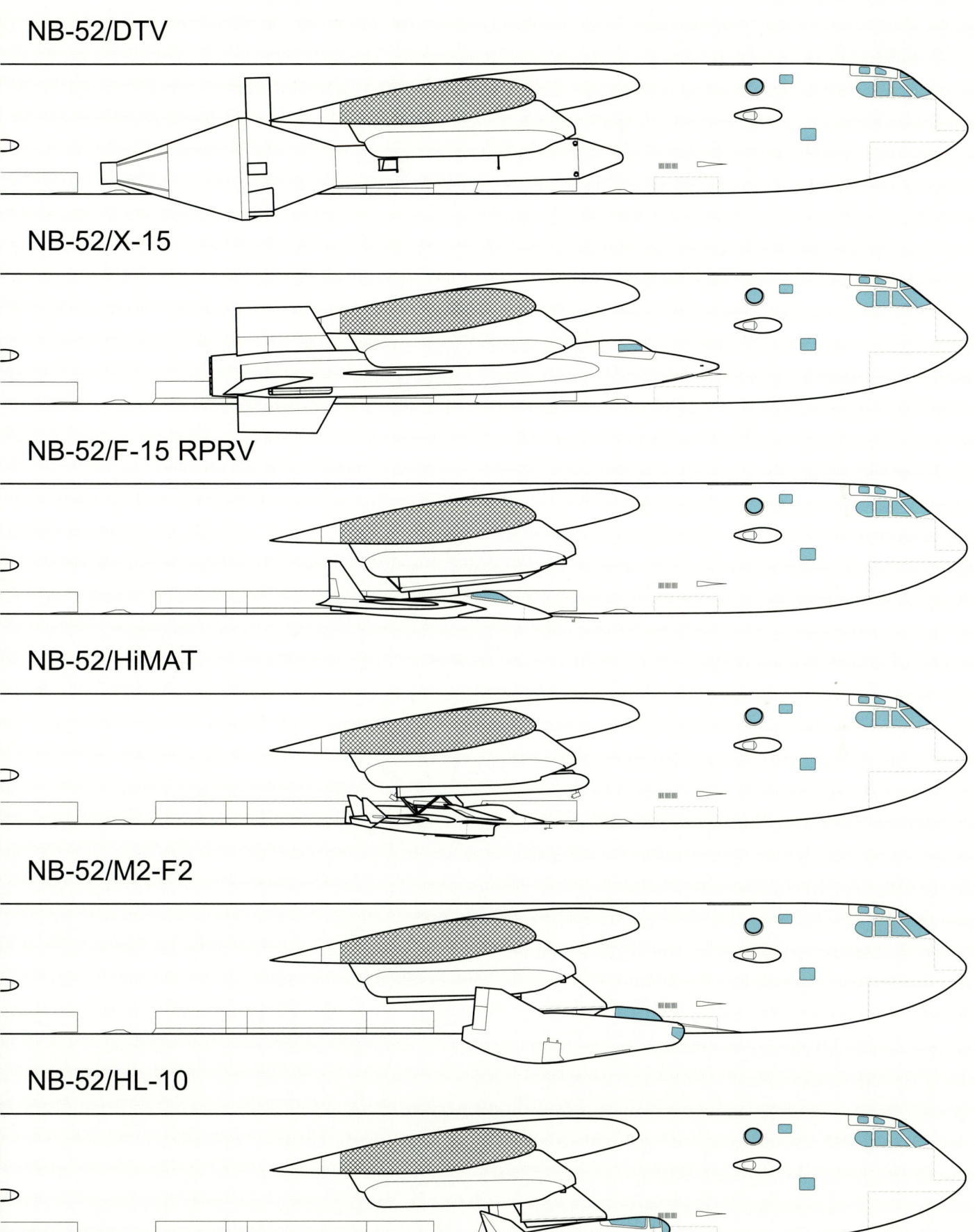

287

the Northrop Corporation won the contract to build both vehicles.

In order to keep costs down, the M2-F2 used off the shelf components where possible – including parts from the T-37, T-38, T-39, X-21 and F-5. And prices were kept low, at least by aerospace standards, with both vehicles delivered for $1.2 million each (1964 dollars). The M2-F2 was rolled out on June 15, 1965, from the Northrop plant in Hawthorne, California. Though similar in shape and size to the M2-F1, it weighed about ten times as much when fully loaded with propellant.

While the airframe was being built, and subsequently tested in the large wind tunnel at NASA-Ames, another series of wind tunnel tests of a subscale model of the M2-F2 and the B-52 carrier aircraft were carried out at NASA-Langley. These indicated that the flow field produced by the nose of the B-52 in the vicinity of the lifting body payload on its pylon would cause the little vehicle to roll immediately upon release, striking the pylon with it tail. As a result, the adapter that would connect the lifting body to the existing X-15 pylon was modified. It was lowered and pushed forward a substantial distance.

The M2-F2's first flight was on July 12, 1966, carried aloft by the NB-52B from Edwards Air Force Base to an altitude of 45,000ft. Once released, the M2-F2 dropped and successfully glided to a landing. A series of further glide flights followed (16 total flights, 12 lifted by the NB-52B, the other four by the NB-52A), none surpassing the speed of sound. Even though the airspeed was slow, the flights were over quicker than X-15 flights were; the lifting body may have been controllable, but it dropped like a stone. An aerodynamic stone under qualified control, but a stone nonetheless.

All of the flights were successful (though not without various issues) until the 16th, made on May 10, 1967. During that flight, a series of issues conspired to create a disaster. After an otherwise standard glide flight, as the M2-F2 came in for a landing it began a series of Dutch rolls, rapidly rolling from side to side. Pilot Bruce Peterson corrected these by raising the angle of attack, but the lifting body was now off course. The dry lake bed made a perfectly serviceable landing area, but without the runway markings pilots used to orient themselves and judge altitude.

As the craft was off course, it found itself heading towards a Piasecki CH-21 rescue helicopter. This, following the disorienting Dutch rolls, threw Peterson off. The landing rockets were fired and the lifting body flared correctly for landing; the landing gear was lowered. But it was lowered very slightly too late, hitting the ground before the gear was down and locked. The round-bottomed M2-F2 was an especially unfortunate configuration for a belly landing and it ended up rolling six times, seriously injuring Peterson.

Rolling six times, bouncing 80ft into the air and landing upside down is typically the sort of thing that is remarked on in an aircraft's history just before the entry on how the wreckage was sold for scrap. But the M2-F2 was a stoutly built little machine... and a round one. It suffered serious damage to be sure (the vertical fins were destroyed, the landing gear mashed or torn off, the outer skins and transparencies wrecked, etc.), but the damage was repairable. The basic structure held, the interior equipment and instruments basically intact. The decision was made to restore it to flight status.

While it was in the shop getting fixed, Northrop and Dryden took the opportunity to add a third vertical stabilizer, this time smack-dab in the middle of the rear fuselage. The roll problems experienced by the M2-F2 had been driven by unanticipated flow interactions resulting from the upper elevons and rudders: as an elevon deflected upwards, it increased air pressure on the vertical fin next to it, pushing the fin sideways and contributing to an adverse yaw. But a fixed central fin saw the same increased pressure in the direction opposite. Thus the yaw force was cancelled out. The improved result was dubbed M2-F3.

The wind tunnel tests, redesigns, checks, repairs and modifications were not done quickly and the first flight of the M2-F3 was not until June 2, 1970. This and the following two drops were glide flights. But on November 25, 1970, the M2-F3 finally fired its XLR-11 engine briefly, reaching Mach 0.809. The first supersonic flight was on August 25, 1971, reaching Mach 1.095. The fastest flight was the 26th, on December 13, 1972, reaching Mach 1.6313; the highest flight was the next – and last – flight on December 20, 1972, reaching 71,500ft.

The M2-F3 flew successfully and safely, demonstrating the concept. It did not go through its life entirely without incident: in June 1970, while hanging under the NB-52B being fuelled on the ground, a valve failed and allowed the liquid oxygen to flow into the alcohol/water fuel tank. This was a nightmare scenario: liquid oxygen will happily mix with many hydrocarbons... and happily detonate with the slightest provocation.

Liquid oxygen mixed with liquid or solid fuels often goes by the name 'oxyliquit' and has been used in various industries as a very powerful explosive. In its way it is considered safe: a bucket of liquid oxygen will not detonate, a jar of carbon black will not detonate; the two can be safely transported down into, say, a mine and mixed on site. But once mixed, the explosive formed is very sensitive to shock. Liquid oxygen poured onto a charcoal briquette will soak in and, if set off by heat, shock, harsh language or simple spite, detonate with the force of half a stick of dynamite.

Thus the crew and support personnel had on their hands a lifting body with a fuel tank full of a high explosive on par with nitroglycerin, sitting out in the desert sun. They did the only thing they could do: carefully open all the valves, and then evacuate. Supersonic flights over Edwards Air Force Base were banned for several days as the M2-F3 and the NB-52B sat alone, the liquid oxygen slowly and calmly boiling away. In the end the aircraft was recovered and the offending valve replaced.

By the time the M2-F3 flew, the idea of the Air Force or NASA needing a fleet of small 'space taxis' to service a flotilla of space stations was passe. It was now the time of the Space Shuttle. The M2-F3 dutifully flew its missions, learned its lessons, and then was sent to the National Air and Space Museum in Washington, D.C. for display.

HL-10

The stablemate of the M-2 series was the NASA-Langley designed HL-10 ('Horizontal Landing', tenth in a series of designs). The HL-10 configuration originated in 1962, coming from studies of space logistic vehicles capable of cramming up to twelve astronauts onboard in order to keep the soon to come space stations properly manned.

The HL-10 was, like the M2, a lifting body... but the designs were unrelated. The HL-10 was, unlike the half-cone M2 series, a flat-bottomed 74° delta with negative camber (i.e. the airfoil cross section appeared upside down compared to the sections of normal wings), with a large central vertical stabilizer at the rear and smallish outward-canted wingtip fins. Unlike the M2, the HL-10 did not have a raised canopy; the cockpit was flush with the contours of the nose of the craft.

Where NASA-Ames evolved their lifting body design by building and flying low-cost models and then a low cost manned flight vehicle, the NASA-Langley team lavished wind tunnel runs and design teams on developing the HL-10. During this time the original wingtip fins, originally little more than small stubs, grew in size. Many variant concepts were investigated; not just differences in shape to determine what would fly best, but differences in scale to fulfill different missions, with designs exceeding a 100ft in length.

Later in the 1960s the HL-10 design would be used by several aerospace contractors as part of their early Space Shuttle concepts. But in early 1964, the main goal remained to simply fly a manned test vehicle to prove the concept. As mentioned previously, an RFP went out, and Northrop won the contract to build both the M2-F2 and the HL-10.

Just as it had with the M2-F2, Northrop used components from the T-38, T-39 and F-5A to save cost. It also had a system involved a pair of hydrogen peroxide monopropellant thrusters, each able to produce 500lb thrust for 30 seconds. These were to be used for the landing flare... 'instant L/D' as with the M2-F1. However, they were not used in flight.

The HL-10 was delivered on January 18, 1966, and the first flight occurred on December 22, 1966. A glide flight, the aircraft demonstrated aerodynamic problems at high angles of attack. The flow separated over the rear fuselage, causing the control surfaces to lose effectiveness; the problem was severe enough that the pilot – and the rest of the HL-10 team – wanted it fixed before the aircraft flew again.

During much of 1967, wind tunnel tests were carried out on models with a range of modifications, looking for a fix for the problem. In autumn of that year, a pair of 'gloves' were made from fibreglass that fit over the existing structure, adjusting the leading edge and inboard airfoil shape of the wingtip fins. It was a subtle change (far less obvious than the new central fin on the M2-F3), but it solved the problem. On March 15, 1968, the HL-10 flew again, launching at the standard altitude of 45,000ft and the standard speed of Mach 0.65. This flight, decked out in tufts to look for flow separation, was a complete success. The problem had been solved and the HL-10 was cleared for further testing.

The first powered flight was on October 23, 1968, but a premature shutdown of the rocket engine meant that it only attained Mach 0.67. The rocket ran correctly on November 13 and the first supersonic flight was on May 9, 1969, reaching Mach 1.13 and an altitude of 53,300ft. The fastest flight was on February 18, 1970, reaching Mach 1.86 and 67,310ft; the highest flight was on February 27, 1970 – reaching Mach 1.31 and 90,303ft. The final flight was on July 17, 1970, reaching Mach 0.73 and 45,000ft. The HL-10 survived its testing to become a 'gate guard' at the Dryden Flight Research Center.

X-24A

The United States Air Force was not to be left out of lifting body research. After all, in the early 1960s a good fraction of those many future space stations would have the USAF insignia on the side. Starting in 1960, the USAF had contracted with Martin Aircraft to study lifting bodies for space logistics; for several years these studies ran alongside other programmes for winged Air Force vehicles, such as the Dyna Soar. But by 1963, the M2-F1 had demonstrated that the lifting body would fly, at least at low speed. And the greater internal volume afforded by lifting bodies compared to configurations such as the Dyna Soar began to sway Air Force opinion in that direction.

In November 1964, the Martin SV-5 lifting body configuration was chosen for the Air Force START (Spacecraft Technology and Advanced Reentry Test) programme, which was to feature two scales of vehicle. The smaller would be the PRIME (Precision Recovery Including Maneuvering Entry) vehicle, leading to the X-23, while the larger would be the PILOT (Piloted Lowspeed Test), leading to the X-24. The PILOT was essentially the same shape as the PRIME, simply scaled up by about a factor of four.

SV-5 was unlike either the M2 or the HL-10. It was a somewhat difficult to describe shape, sort of a squashed football. The underside was flattened, the rear sloped downwards and spread out; from above it looked like a blunt-nosed delta configuration. At the rear two quite sizable fins projected up and outwards; on the PILOT vehicle a central vertical fin sat at the rear of the fuselage, separating the body-flap elevons in much the same fashion as the new central fin on the M2-F3.

Four of the PRIME vehicles (designated SV-5D) were built, and three were launched on suborbital flights by Atlas boosters, demonstrating the hypersonic (Mach 24.5, just short of orbital velocity) flight characteristics of the configuration. The first flight was in December 1966, the second in March 1967. Both were successful but the recovery systems failed and both were lost as sea. The third fight was in April, 1967 and this time the vehicle was successfully recovered and is now on display at the National Museum of the US Air Force in Dayton, Ohio. PRIME did not demonstrate how the configuration flew at low speed however. That was the job of PILOT.

The PILOT branch of the START programme was approved for construction in March 1966. Unlike the NASA lifting bodies, Martin built more than one. A single SV-5P vehicle was paid for by the Air Force as a rocket-powered manned test vehicle, but Martin built two more on their own. These two were SV-5Js and would be powered by turbojet engines rather than a rocket. They would have been slower, but capable of longer duration flights. But it appears that neither SV-5J ever had a turbojet installed, and neither ever flew.

Like the M2-F2 and the HL-10, the SV-5P was built largely from aluminium and powered by a single XLR-11 rocket engine. Like the HL-10, it had two 500lb thrust hydrogen peroxide monopropellant rocket engines for landing thrust. While under construction in June 1967 the SV-5P was officially designated X-24A by the Air Force and it was completed in August. A C-130 delivered it to Edwards Air Force Base but it was then sent on to Ames Research Center for wind tunnel testing, this commencing in February 1967. It was returned to Edwards in March.

As with the Northrop lifting bodies, it would attach to the X-15 pylon of the NB-52B using a special adapter of its own. While similar in appearance to the Northrop adapter, this did not project downwards as far. It also connected to the X-24A with only a single shackle; the aircraft was stabilized on the adapter with snubbers.

Captive flights finally led to a glide flight on April 17, 1969. The first powered flight was on March 19, 1970, reaching Mach 0.865 and 44,384ft. Problems were encountered with the controls, requiring a series of fixes over several flights. But the aerodynamics of the vehicle were good and did not need revision. The highest flight was on October 27, 1970, reaching Mach 1.357 and 71,407ft; the fastest flight was on March 29, 1971, reaching Mach 1.6 and 70,500ft. The last of the X-24A's 28 flights occurred on June 4, 1971, reaching Mach 0.817 and 54,373ft. Eighteen of those flights had been under power; all had been under the wing of the NB-52B.

X-24B

The hypersonic performance of the M2, HL-10 and X-24 lifting bodies was rather uninspiring, with a L/D little better than 1 at speeds from re-entry down to the lower hypersonic. This was adequate for basic peacetime space logistics use, and the lifting bodies studied certainly were volumetrically efficient. But there were other approaches to lifting re-entry vehicles that could do quite a bit better at higher speeds.

The USAF Flight Dynamics Laboratory studied its own range of lifting entry vehicles, producing designs such as the FDL-5, FDL-6 and FDL-7 by the mid-1960s. These would have far greater cross-range potential – up to 1,500 miles. NASA-Langley produced a similar class of high performance lifting body concepts. But the tradeoff was that these designs, which were pointier and less volumetrically efficient, tended to have much lower L/D than the other lifting bodies at low speeds. Without tricks such as deployable wings, these craft would have difficulty in landing at sane speeds. Perhaps even more than earlier lifting bodies, the FDL series needed a low speed landing demonstrator.

Hearkening back to the development of the M2-F1, NASA-Langley engineers built a number of low cost subscale radio control models of their designs, launching them from a specially made radio control 'mothership'. These tests demonstrated that the new sleeker lifting bodies could work, could be flown and could be landed. But the speeds achievable were low and the weights small; something grander was called for.

The FDL-7 configuration, it was found, could be built using the SV-5 shape as a nucleus. Just as the HL-10 had added 'glove' components to adjust its aerodynamics, gloves could be added to the SV-5 in the form of a new longer nose and wing-like fuselage extensions. This would produce the 'FDL-8X' configuration. Unlike some of the other FDL configurations, the FDL-8X was thought to have an adequate L/D at low subsonic speeds and did not need deployable wing surfaces to land safely.

Martin still had the shells of its two unflown SV-5J jet-propelled lifting bodies and staff from the FDL were interested in acquiring them to modify for their purposes. But the SV-5Js had never flown for good reason: they would have been, at best, marginal aircraft. The lifting body is an adequate way to control a descent, but it's a terrible way to lift off from a runway. Simulations showed that the SV-5J would barely be able to lurch itself into the sky. Launched from a B-52, it would not be able to break the sound barrier. It soon became clear that rocket propulsion was the way to go. In April 1971 the project was given the go-ahead to add the new components to the rocket-powered X-24A after the completion of its current test series.

The aircraft was taken to the Martin Marietta plant in Denver for modification. The new metal components were grafted to it – with only the cockpit canopy and control surfaces seemingly left unaltered. A new nose, slim with a 78° sweep, was grafted on, adding 14½ft to the length of the craft. Span was increased 10ft by adding new horizontal wings, flush with the underside. The underside itself was now quite flat. This was useful for more than just the aerodynamics of the lifting body – the flat underside of the nose could, it was thought, serve as the compression inlet ramp for an underslung scramjet package.

Now dubbed X-24B, the vehicle was delivered to Edwards Air Force Base in October 1972. Many of the X-24A's systems remained; some being upgraded. The XLR-11 engines were increased in thrust by boosting the chamber pressure and adding nozzle extensions. The modifications and additions greatly increased weight and shifted the centre of gravity. To keep from over-stressing the X-15 pylon, the X-24B had to be held further aft than the X-24A had been. This meant that the X-24A adapter could not be used, and this meant that the X-24B pilot could not eject while his vehicle was still attached to the NB-52B. In the event of an in-flight emergency, the X-24B pilot would need to separate his craft from the B-52 and either glide it to a landing or eject while in flight. As it happened, the need never arose.

The X-24B flew 28 times, all from the NB-52B. The first flight, a glide flight, occurred on August 1, 1973. Dropped from 40,000ft, the X-24B reached Mach 0.652. The power was used for the first time during sixth flight, with Mach 0.917 and 52,764ft achieved.

The fastest flight was the 16th, reaching Mach 1.76 and 72,150ft; the highest flight was the 23rd on May 22, 1975, reaching Mach 1.63 and 74,130ft. The final flight was a glide flight on November 26, 1975. Dropping from 45,000ft and reaching Mach 0.7, this was not only the final flight of the X-24B but also the final flight of the American manned lifting body programme… and the very last time that the NB-52 would carry a manned payload.

F-15 RPRV

In the wake of the manned lifting body research programme it would be smaller remotely piloted research aircraft that kept the NB-52B flying – when it might otherwise have been destined for the scrap heap. The first of this new generation of research vehicles was the F-15 Remotely Piloted Research Vehicle (RPRV).

It was noted in 1971 that research on stall and spins was low. These are problems which tend to be fairly specific to individual designs, so testing these issues for a particular aircraft would require putting one of those aircraft – as opposed to a differently configured cheaper aircraft – at risk. Yet a subscale version of the aircraft would have similar stall and spin characteristics, so in 1973 NASA authorized the construction of 3/8 scale unpowered glide models of the McDonnell Douglas F-15, then under advanced development.

McDonnell Douglas built three out of fibreglass for a quarter million dollars each; expensive to be sure, but nowhere near the $6.8 million price for a real F-15. The F-15 RPRVs would not have jet engines, but they would have a television camera in the 'cockpit', providing live video to a pilot sitting in a simulated cockpit on the ground. After being carried to altitude and dropped, the pilot would put the aircraft through a series of manoeuvres. These would be risky for a pilot and an aircraft, but the RPRVs were expendable.

The first glide flight for the F-15 RPRV took place on October 12, 1973. The vehicle was carried under the NB-52B's X-15 pylon, a simple framework adapter connecting to the top of the model. The first 16 flights were to be recovered in air: the RPRV popping a chute and descending slowly; an Air Force helicopter would then snag the chute in flight and carefully lowered the model to the ground.

This did not always work, however. On one flight, the RPRV descended all the way to the desert floor under the parachute and suffered substantial damage in the landing. In the 16th flight, the helicopter was unable to reach the RPRV, so the pilot landed the small plane on its belly. There was damage, but the landing was good enough that it was decided to modify the RPRVs with simple deployable skids for landings. This proved quite successful and the helicopter was no longer needed.

The original F-15 RPRV configuration flew 27 times, the last time being on December 17, 1975. Modifications then created the Spin Research Vehicle (SRV). The basic F-15 configuration was retained, but the nose could be replaced and aerodynamic features could be added, changed or taken away; these included strakes, helical trip strips, localized surface textures, vortex control nozzles and deployable anti-spin parachutes.

The first flight of the F-15 SRV was on November 29, 1977, again from the NB-52B. The final flight was on July 15, 1981. The F-15 SRVs flew 26 times. One of them was restored and is on display at the Dryden Flight Research Center.

The pilot had only a single TV screen by which to fly the vehicle. The angle the view provided was narrow and the resolution – by today's standards – poor. This made flying difficult yet it proved important experience for future remotely piloted vehicles.

Shuttle Booster DTV

Significantly unlike the other vehicles dropped from the X-15 pylon by the NB-52B was the Space Shuttle Solid Rocket Booster Drop Test Vehicle (DTV). This was a 'vehicle' in only the vaguest sense. With no wings or propulsion systems, it was essentially just a big heavy thing that dropped from the sky, travelling towards the Earth as fast as gravity could drag it.

The Space Shuttle Solid Rocket Boosters were essentially large steel tubes filled with rocket propellant, with a nozzle assembly at the rear and a nose cone at the front. The intention from the beginning was that they would be recoverable: splashing down in the ocean under parachutes before being recovered, refurbished, refuelled and relaunched. But nothing quite like that on this scale had ever been done before, so every aspect of their design needed to be tested to ensure that it would work in real-world conditions.

The parachute system in particular, including the drogue, needed to be checked out. An obvious approach would be to air-drop an actual SRB, but the only aircraft that came close to being able to carry such a thing, the Lockheed C-5A, would require several million dollars worth of modifications, including removal of the rear doors and the development of a sled extraction system (similar to the sled used to extract Minuteman ICBMs from the rear of C-5s in flight). A cheaper approach would be to drop a lower-mass test item.

Martin Marietta-Denver therefore built a Decelerator System Drop Test Vehicle (DTV). This was a test mass of about one-third the weight of an actual SRB casing (48,000lb compared to 170,000lb) fitted with the parachute pack of an SRB. The only aircraft that could conveniently carry it was the NB-52B, so it was designed to fit within the packaging limitations created by the NB-52B's X-15 pylon and surroundings, including the need for ground clearance.

It was an unusual shape: a long narrow cylinder (48in diameter) with a very large conical flare at the tail (maximum 122in diameter). The cylindrical section included 22,000lb of ballast; the forward conical frustum contained the instrumentation and the rear conical frustum contained the drogue chute and either one or three main chutes. The SRB contains three main chutes. Consequently, on drops with only a single main chute the DTV, with one-third the weight and one-third the chutes, generated approximately the same loads on the chute and lines, albeit for a shorter period.

Instrumentation included rate gyros, chute load sensors, accelerometers, extensometers, a pressure probe and cameras. The nose of the forward ballast section was a single ductile iron casting, filled with lead. The cylindrical section was made from rolled and welded half-inch T-1 steel. The aft conical frustums and internal structures were also made of steel; this was a case where weight reduction was not important but strength and the ability to be repaired in the field were. Two DTVs were built.

The X-15 pylon required no modification to support the DTV as some of the later X-15 variants had been heavier than the DTV. But as the NB-52B sat on the runway, the DTV was quite close to the ground.

The typical DTV flight saw the NB-52B release it at an altitude of about 21,700ft and 200 knots. The process of parachute deployment would begin almost immediately, a slug-gun firing to deploy a vane chute. This caused the release of an 11.5ft diameter drogue chute that pulled the 'nose' cap off which, as it drew away, extracted the DTV pilot chute, which extracted the drogue chute, which opened its first reefed stage.

The second stage then disreefed to its second stage and finally pulled the conical frustum away from the DTV, exposing and deploying the main chute. The frustum would descend to the ground under the drogue chute; the main chute would lower the DTV to an impact at 45ft per second. The ballast would be separated from the DTV prior to impact to reduce the weight and thus impact velocity.

For the first two flights, the DTV had no fins, relying on the ballasted nose and the wide conical tail for aerodynamic stability. Unfortunately, this was not as stable as hoped; the DTV 'coned' on its way down, generating more drag than planned and slowing decent. So a trio of simple rectangular fins were added which provided the desired stability.

The first captive flight was in June 1977, with the first drop a few days later. The third drop in December 1977 suffered a drogue chute failure and consequently lawn-darted into the desert floor at high speed and was destroyed. The remaining DTV was dropped three more times in that configuration, the final time in September 1978.

The DTV returned to flight in 1983 to test a new, larger parachute for the SRB. The first planned drop flight in February 1983 was cancelled due to the weather but as the NB-52B taxied back to the hangar, the rear hooks holding the DTV to the X-15 pylon failed and the rear of the DTV dropped to the ground. It was found that the hooks had been overstressed and a redesign, not to mention repair, was needed. The first flight of the improved system occurred in September 1983 over the US Navy's China Lake Test Range. The eighth and final drop occurred in March 1985.

DAST

Aircraft efficiency became increasingly important as the cost of petroleum – and thus jet fuel – began to spike in the early 1970s. Aircraft could be made more fuel efficient through improved engines, of course, but also through improved aerodynamics and lighter weight structures.

A structure that was both lighter in weight and lower in drag would be ideal, but a lighter, slimmer structure is also one that is naturally less stiff. The B-52 was an early pioneer in the field of flexible wings, but wing design in the world post-oil crisis could take flexibility to new extremes. Lighter weight wings could easily mean increased flutter, decreased service life through fatigue and a decrease in stability.

New wing designs and new materials could not be sanely tested at full scale and reasonable cost, but they could be cheaply and safely tested subscale. The Drones for Aerodynamic and Structural Testing (DAST) was a joint project of Dryden Flight Research Center and Langley Research Center to demonstrate the ability of active controls to reduce the down sides of lightly constructed highly flexible wings.

To achieve this, a Ryan BQM-34E/F Firebee II supersonic aerial target drone was fitted with new wings and an active Flutter Suppression System.

The wing incorporated into the DAST was the supercritical Aeroelastic Research Wing (ARW-1). The modification changed the external appearance of the Firebee little, except for replacing the existing highly swept 2.72m span wing (aspect ratio 2.5) with a 4.343m span wing with an aspect ratio of 6.8.

The new wing also incorporated a substantial leading edge glove that smoothly faired it into the fuselage. The control surfaces, turbojet engine, main structure and recovery parachute were not changed. The equipment compartment, however, saw substantial changes from the stock target drone. A new autopilot, radio receivers and other instruments necessary for remote piloting were added; computers to analyze flutter in real time and to adjust the controls accordingly were also necessary.

The wing was similar to, though smaller than, the wing added to the TF-8A Supercritical Wing research plane in the early 1970s. Such wings have an unusual airfoil section, flat on top with a concavity on the underside near the trailing edge. This delays the onset of wave drag as the wing approaches the transonic, greatly reducing drag at high subsonic speeds.

Prior to the modification and flight of the ARW-1 equipped DAST, NASA flew a Ryan Firebee II under the X-15 pylon on the NB-52B. The programme began a few years prior, with the same Firebee having been captive-carried under the wing of a Navy DC-130A in 1975 and 1976. In July 1977, the Firebee was fitted to an adapter attached to the underside of the X-15 pylon and carried to an altitude of 20,000ft.

Once dropped, it was piloted remotely (with a ground station being the primary site for the pilot, and with a two-seat TF-104G flying nearby so that the back-seater could take over piloting duties in the event the Firebee lost contact with the ground site), climbing to 40,000ft under the power of its single turbojet.

At the end of its mission the engine shut down and the vehicle descended under a parachute deployed from its tail. The Firebee was recovered and the wing was removed and replaced with a NASA-Langley provided Instrumented Standard Wing. Dubbed the 'Blue Streak' due to a painted triangular stripe on the wings, the drone had modified flight control instrumentation and was flown once on March 9, 1979, reaching a top speed of Mach 0.97 before being recovered by helicopter while descending under parachutes.

The ARW-1 wing was then added to the craft, along with the active flutter suppression instrumentation. It was flown on October 2, 1979, reaching Mach 0.75 and again recovered in flight. While the vehicle was successfully recovered, the flight itself was troubled with numerous mechanical faults and telemetry problems.

No data was successfully received from the onboard flight data sensors, resulting in the flight being ended early. The second flight, on March 2, 1980, was considered very successful. It reached Mach 0.926 and was again recovered via helicopter. The onboard flutter suppression systems worked, providing promising data.

During the third flight on June 12, 1980, the right wing experienced the flutter that the whole programme had been about, but experienced too much of it. The wing came off at the root and the Firebee crashed into the desert floor.

A second wing was then manufactured and attached to another Firebee. This one differed in having a belly pod with additional fuel and instrumentation. The DAST-II only flew once from the NB-52B, on November 3, 1982. Several further flights were attempted but aborted before release due to mechanical problems, though the DAST-II did fly several more times (two captive carry flights, one powered flight) from the DC-130A.

On that flight the drone crashed into a farm field. A different wing, the ARW-2 with lower sweep and longer span, was built, but never flown. With that, the DAST programme was over.

HiMAT

The same advances in computer technology that made the NB-52E a practical test vehicle could, it was thought by the early 1970s, carry through to the next generation of tactical fighters. Advanced controls, together with innovative designs and new materials, could make jet fighters of unprecedented manoeuvrability – with complete control through the difficult transonic stage of flight. But integrating everything into a single functional package was a challenge.

A full-scale fighter would be extremely expensive and, coupled with the uncertainty surrounding the many new technologies, potentially quite dangerous. There would be risk not only to the pilot but to the programme: the loss of an expensive aircraft and pilot could set the adoption of those technologies back years.

But there was a cheaper, safer option: a subscale unmanned demonstrator. The Highly Maneuverable Aircraft Technology (HiMAT) programme was born in 1974 at the Flight Research Center and NASA-Ames, with Grumman, McDonnell Douglas and Rockwell International competing. Requirements included a sustained 8 g manoeuvring capability at Mach 0.9 and 30,000ft, plus the ability to reach at least Mach 1.4. Early on, there was to be a full-scale manned aircraft and a scaled-down drone would be part of the development process.

Rockwell International had a design for an advanced dogfighting aircraft supposed to be capable of manoeuvring better than anything else in the air. Modern composite materials would be used to create a lightweight and strong structure with 2D vectoring nozzles, a close-coupled canard and a modestly swept main wing with wingtip fins.

The Air Force felt that this would be a good design to test, so Rockwell scaled it down to 44% and built two as remotely piloted craft. Like the F-15 RPRV, the HiMAT drones had a TV camera positioned where the pilot's head would be in the full scale vehicle; unlike the F-15 RPRV, these drones each had an afterburning General Electric J85-21 turbojet.

The HiMAT drones were among the first flight vehicles to take full advantage of carbon composites in their construction. Graphite-epoxy composites made up about 25% of the weight of the structure and about 95% of the external skins. The skins were of variable thickness, with the fibre tapes placed and oriented to tailor the aeroelastic properties of the structures to be rigid in some directions and flexible in others.

Aluminium made up another 25% of the structural weight, titanium 19%, fibreglass 4%, steel 9% and miscellaneous other materials 18%. The initially-planned 2D vectorable nozzles were excluded due to budget constraints. The wings featured supercritical airfoils and variable leading edge camber. For control it had two boom-mounted all-moving rudders, smaller than they would have been had it not been for the wingtip fins.

The large close-coupled canards had flaps, but these were fixed rather than mechanized. Unlike the prior F-15 RPRV and Firebee drones, the HiMAT drones were designed from the outset to land horizontally; for simplicity they had a tricycle landing gear composed of skids rather than wheels.

The HiMAT drones had advanced controls for the time including digital fly-by-wire systems and digital engine controls. The small size of the craft and limited internal volume necessitated miniturization of the electronics; this, along with the materials the HiMAT drones were built from, put them at the leading edge of aerospace technology of the time. The result was an aircraft that could outmanoeuvre any manned fighter in the sky, with a turning radius half that of contemporary dogfighters.

The airframe was built to withstand 12 g positive and six negative, and was designed to reach Mach 1.6. These g-forces were, of course, not felt by the pilots as they were seated in remote piloting stations on the ground, with a backup pilot in the back seat of a TF-104G. The lack of physical feedback, along with the limited field of view, did prove to be something of a hindrance to pilot performance.

The HiMAT drones were attached to the X-15 pylons using the M2-F3 adapters. The forward aerodynamic fairing was removed and the large tubular structural elements were converted into fuel tanks (230 gallon capacity) to replenish the fuel HiMAT burned off prior to release (the engine was started while the NB-52B was still on the ground). The HiMAT was attached in a 2° nose-down attitude; aerodynamic forces would thus push the little drone away from the adapter cleanly. The HiMAT was attached with two pneumatically-actuated hooks, with sway braces stabilizing it.

Twenty-six HiMAT drone flights were made for a total flight time of 11 hours. A HiMAT was typically launched at 45,000ft and Mach 0.68, with a mission lasting around 30 minutes. The first captive-carry flight was on July 11, 1979, with the first free flight on July 27. The first supersonic flight of a HiMAT was on May 11, 1982. The last flight was on January 12, 1983.

The first 12 flights were ballasted to be stable; the last 14 were in a relaxed static stability configuration. In other words, in the first configuration the HiMAT would glide stably if left alone, in the second it would have a tendency to tumble. This was by design: a lack of inherent stability meant that the active controls had to work constantly but it also meant that the drone was exceptionally nimble.

HiMAT was never really able to demonstrate its potential to become a world-class dogfighter due to the lack of pilot feedback. A full-scale manned vehicle would be needed for that… and such an aircraft was not forthcoming.

One HiMAT was restored and is currently on display at the National Air and Space Museum in Washington, D.C.

OSC Pegasus

The Orbital Sciences Corporation (now Northrop Grumman) 'Pegasus' is the only NB-52B-carried payload that has actually gone to orbit. Designed from the outset as a satellite launcher, design work on Pegasus began in April 1987, with preliminary design completed that December. It is a relatively simple and straightforward vehicle: a rocket booster composed of three solid rocket stages of equal diameter by decreasing length, the first stage having a shoulder-mounted delta wing and a trio of tailfins.

The idea with Pegasus was that launching from an aircraft would not only serve as a reusable first stage, it would also allow the booster to set off to orbit form virtually anywhere on Earth. The ability to shift launch latitude north and south is valuable as it allows a launch system to reach particular orbital inclinations much more easily.

The Pegasus payload is, by the standards of most launch vehicles, small at under 1,000lb. The original goal was that it would be inexpensive and fast-reacting. Launching often, it would fill the skies with a variety of small satellites, both commercial and military. Early on DARPA was quite interested in the vehicle for its perceived ability to quickly launch military satellites, though OSC largely self-funded the project by selling itself launches for the OrbComm and OrbView satellites. The decades it has been in service have not exactly born out those initial hopes.

The three motors were designed and produced by Hercules Aerospace (later ATK, now also Northrop Grumman). The first stage, somewhat unusually, does not have thrust vector control; all control of the vehicle during first stage powered flight is done by way of the three all-moving tailfins. The wings and tail surfaces were built from carbon composites by Scaled Composites.

Pegasus was designed to be carried to altitude and launched from a large carrier aircraft. The NB-52B was used for early development flights, carrying the Pegasus with the X-15 pylon. The top of the X-15 was basically a cylinder, while the top of the Pegasus – the relatively vast expanse of the simple delta wing – was flat. Rather than modifying the X-15 pylon, an adapter was made that replicated the top of the X-15 to fit within the pylon, with a flat underside to mate to the Pegasus.

For six early flights (on April 5, 1990; July 17, 1991; February 9, 1993; April 25, 1993; May 19, 1994; and August 3, 1994) Pegasus was launched from the NB-52B. Five of these flights originated from Edwards Air Force Base, with the February 1993 flight setting out from Kennedy Space Center instead. Four of the launches fully succeeded in putting their payloads into orbit.

Modifying the B-52 was fairly easy, but it was not considered economical; a Lockheed L-1011 'Tri Star' jetliner was purchased and modified into a carrier aircraft. The L-1011 has been used for 39 of the 45 total launches to date.

The Pegasus launched by the L-1011 is the slightly stretched 'Pegasus XL' variant, where the B-52 had only launched the initial baseline version. While Pegasus is technically still in service, its future looks bleak. The SpaceX Falcon 9 has flown about two and a half times more than Pegasus (123 to 45 times), has about 50 times the payload (50,000lb to LEO compared to 977lb) and actually costs less (a reused Falcon 9 costs about $50 million to the Pegasus XL's $56 million). There have been plans to launch Pegasus from the Stratolaunch Systems 'Roc' aircraft; renderings have depicted the Roc carrying three Pegasus vehicles at once. Why anyone would want to launch three satellites from one aircraft on the same flight is anyone's guess.

X-38

The 1960s and 1970s failed to produce the flock of space stations that both the Air Force and NASA

OSC Pegasus
SCALE 1/65

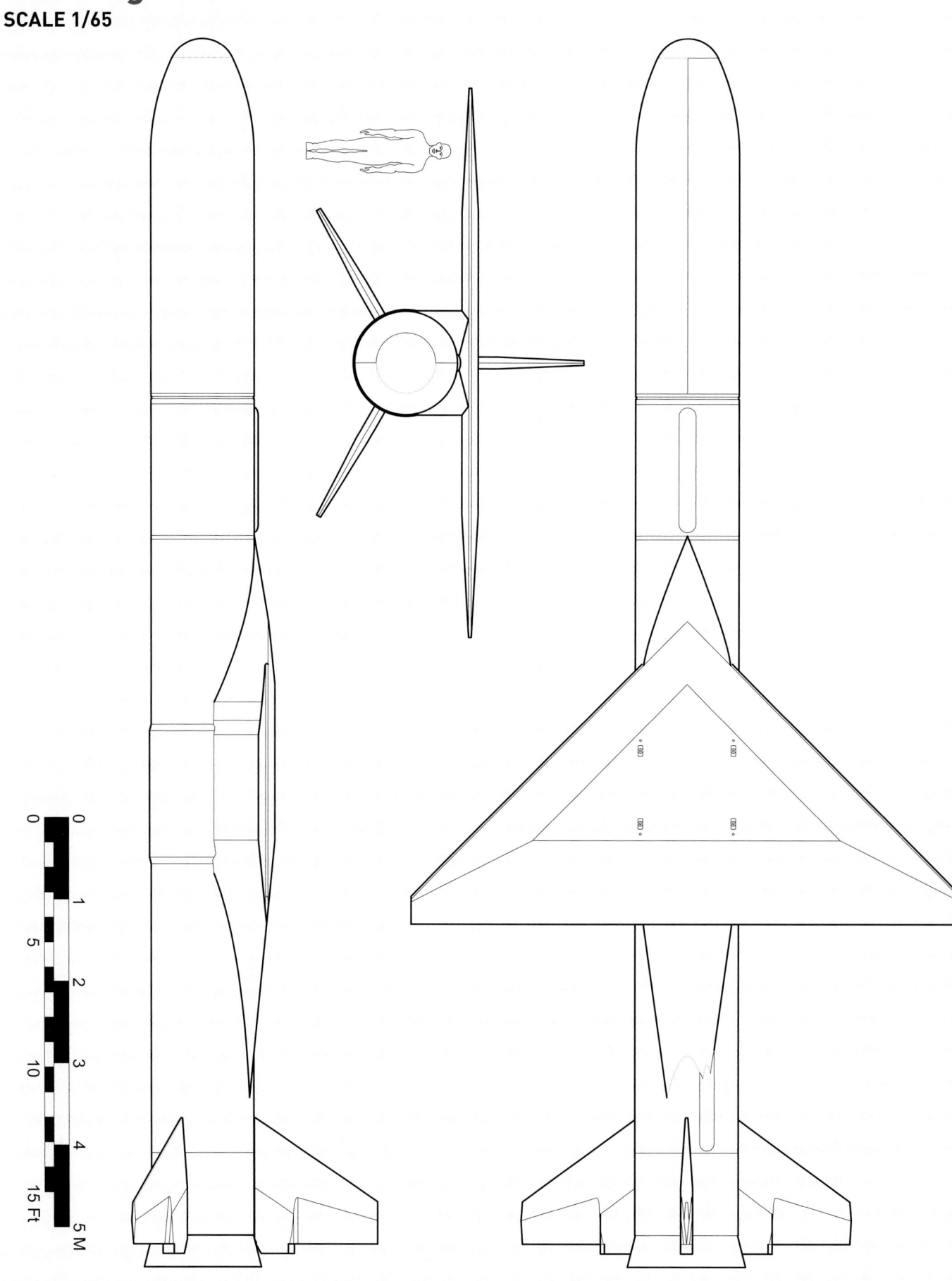

Pegasus + NB-52B
SCALE 1/300

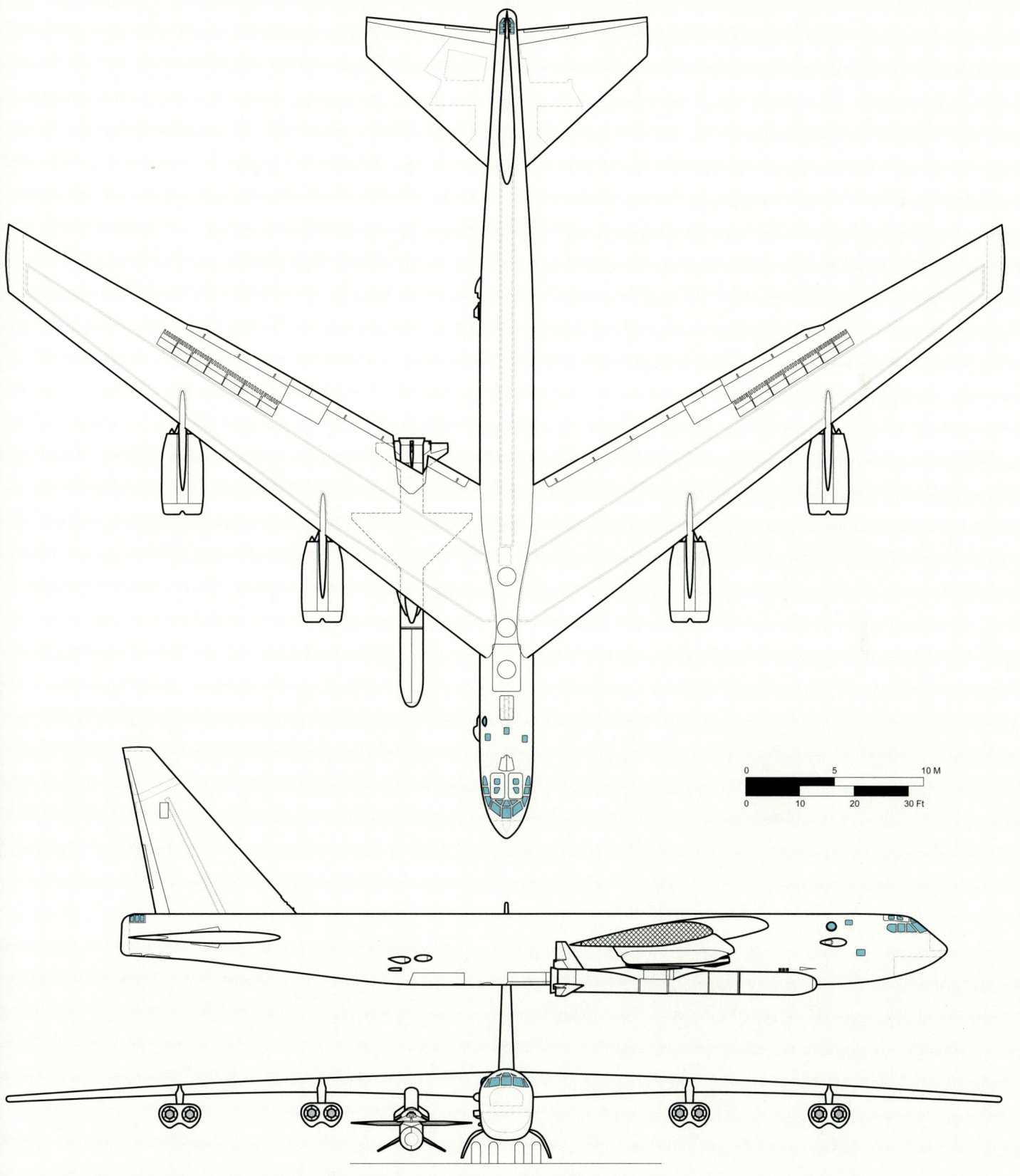

expected. Without them, there was nowhere for teams of astronauts to go, and no need to have simple space taxis to get them there. But the 1990s did finally see the arrival of a space station of sorts, the International Space Station (ISS). It could be reached by ether the Russian Soyuz craft or by the US Space Shuttle. Those were adequate systems for the low rate of crew replacement that the ISS was projected to have, but neither was good as an emergency 'lifeboat' in case crew needed to return home quickly.

The Space Shuttle was a great vehicle to bring home an injured crewman: spacious, relatively low-G re-entry, high cross range so that it could return home more often. But it was hideously expensive. It was not permanently docked at the ISS and getting one ready for launch was a process taking months. It could not be made ready to go at a moment's notice.

The Soyuz, in contrast, was permanently stationed at the ISS. Chances were excellent that there would be one waiting for you if you needed it. But it was small, seating just three (while the station could have seven or more), and it was ill-suited to fitting someone in a stretcher. Re-entry was rough and ended with a car crash of a landing in the back end of beyond on the Russian steppe, hell and gone away from hospitals. And with low cross range, the crew might need to wait a dangerous amount of time to come home.

Before the ISS was assembled, before the ISS was even the ISS, the Space Station Freedom of the 1980s needed a space taxi. Something small that could be used not only to inexpensively transport crew up and down, but also something that could be docked at the station for long periods and then be turned into a lifeboat at a moment's notice. NASA and the US aerospace industry studied a wide range of vehicles for the space taxi/lifeboat role, from capsules to spaceplanes to lifting bodies. In general these vehicles were designed to fit within the Space Shuttle orbiter cargo bay, or to fit atop a conventional launch vehicle as a more or less standard payload.

Starting in the late 1980s, one of the more developed concepts was the HL-20 Personnel Launch System, a NASA-Langley designed lifting body with aerodynamics based in no small part on photographs of the Soviet BOR-4 subscale test vehicle from 1983. It was to be launched by a Titan IV, the Space Shuttle or a variant of the National Launch System (a cancelled 1980s concept for a modular launch vehicle that would run the gamut from medium launch capability to Saturn V class) and could carry up to ten passengers. A great deal of wind tunnel testing, engineering design and even full scale crew integration in a mockup were put into the HL-20 effort, and it seemed to be a reasonable approach. Therefore, in the early 1990s, it was cancelled.

The Johnson Space Center began a study of a Crew Rescue Vehicle in 1995. Variously called the CRV, X-CRV, CRV-X and X-35 (that designation was never official, and went to the Lockheed prototype for the F-35 fighter), this concept was based on X-23 and X-24A aerodynamics. It even had the raised canopy of the X-24A, even though as designed the CRV would not have an actual cockpit; the crew would all be essentially passengers in an autonomous vehicle. The CRV survived the initial studies; after the X-35 designation was granted to Lockheed, the lifting body crew return vehicle was designated X-38.

The X-38 was to be a series of vehicles, with the first two being 80% scale mockups of the functional space vehicle. These were not mockups that would just sit on a factory floor and look pretty, but would be flying vehicles. They would be dropped from aircraft to demonstrate unmanned controllable flight and safe landing. But unlike the X-24A that it was largely based on, the X-38 would not swoop in to a high-speed horizontal landing. Instead, it would deploy a large parafoil; vastly increasing lift and drag, the parafoil would also greatly reduce landing velocity, thus making the craft safer for hypothetically injured crew. A further X-38 would be built full scale. This would be delivered to orbit by a Space Shuttle and would, it was proposed, return to Earth unmanned to demonstrate the full range of flight capabilities.

Early drops were conducted to test the parafoil using even more subscale models – 1/6 scale (4ft long) X-38s with subscale parafoils, as well as using simple weighted pallets. Two of the 80% scale X-38s were built by Scaled Composites under a contract awarded in 1996. The first, designated V-131, was originally delivered in September 1996 with fixed aerosurfaces. The craft was meant only to glide stably for a short time, then descend under the parafoil.

It was carried by the NB-52B on a captive flight in late July 1997. Like all of the other lifting bodies carried by the NB-52, it was suspended under the right wing with a pylon. But for the first time this was not the X-15 pylon with an adapter. Instead, an entirely new pylon weighing 3,000lb was built specifically for the X-38. It was a seemingly simpler structure, without the need to contain propellants for the payload since the X-38 was strictly a glider.

The first drop was conducted in mid-March 1998. This was not an unqualified success: the shroud lines twisted during deployment, becoming entangled with the drogue chute. The X-38 swayed back and forth, but it did sort itself out prior to landing. The 5,000sq ft parafoil underwent further study, including Army drop tests of nine-ton pallets, before the X-38 flew again.

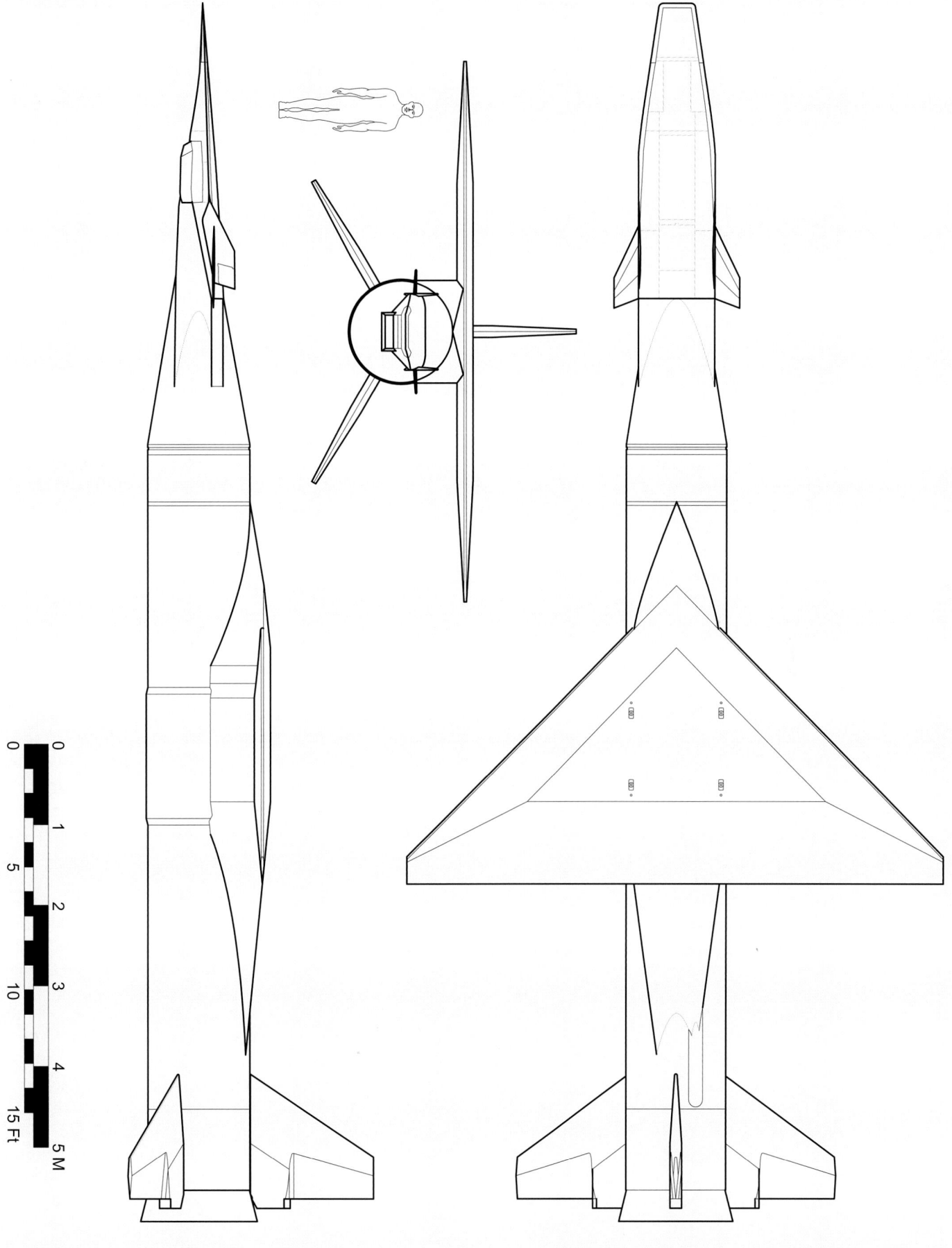

The second X-38, V-132, was delivered in September 1998. This vehicle had complete flight controls, allowing controllable flight after drop and before parafoil deployment. Captive flights occurred in October; the first drop flight occurred in December, demonstrating successful parafoil deployment and landing.

After a second (and successful) drop flight in February 1999, the V-131 was sent back to Scaled Composites for modifications. These were extensive: virtually the entire upper surface was covered in a new outer shell, greatly increasing internal volume and, in particular, raising the aft boat-tail substantially. This brought the aerodynamics of the V-131, now re-designated V-131R, into line with the full-scale orbital V-201 X-38. The control surfaces were also made fully functional. It was also given the 7,500sq ft parafoil meant for the V-201. The V-131R first flew in November 2000, successfully demonstrating the revised aerodynamics as well as the new parafoil.

The final X-38 flight was in mid-December, 2001. All in all the flights successfully demonstrated the flight characteristics of the configuration, the deployment and controllability of the parafoil, and the autonomous landing. The V-201 orbital vehicle was 90% complete at that time, with an orbital flight planned in the not too distant future. So, of course, the programme was cancelled in December 2001 as a cost saving measure.

V-131R is at the Evergreen Aviation Museum in McMinnville, Oregon; V-132 is at the Strategic Air and Space Museum near Ashland, Nebraska. The still incomplete V-201 sits outside at the Johnson Space Center in Houston.

X-43 Hyper-X

During the 1980s and early 1990s, NASA and the USAF spent a large amount of time and trouble working on the X-30 National AeroSpace Plane (NASP). This was touted as the solution to the problem of expensive space launch – instead of every satellite and astronaut being lunched atop a rocket that would be thrown away after each flight, they would instead ride to orbit within an aircraft that would take off and land from conventional runways and fly to orbit powered by fuel efficient scramjet (Supersonic Combustion Ramjet) engines.

It was a lovely dream with only one snag: scramjets have not been made to function at anywhere near the speeds the NASP would need to fly at. But with nearly 1950s-level optimism, NASP was begun with the assumption that this engineering issue would eventually get ironed out. And then hypersonic airliners would whisk passengers across the globe and space rockets would be a thing of the expensive past.

And in 1994, NASP was cancelled. Those in charge of the purse strings decided that it just wasn't going to happen, and as a result the dream of a near-term orbital aircraft died. But the dream of getting scramjets to work did not die. While an airbreathing engine that could run at speeds up to Mach 25 had faded from immediate need, the idea that an airbreather could function at, say, Mach 7 or even Mach 10 was not so ridiculous.

Such a propulsion system would not take a vehicle to orbit, but it could take a vehicle quickly across an ocean, or invulnerably over enemy territory, or high and fast enough to launch an upper rocket stage that could get to orbit. The scramjet was NASP's single most vital component, the technology without which the vehicle could never function. And yet despite the time and expense lavished on NASP, no scramjet had actually flown or performed anything beyond wind tunnel tests. This was an obvious technological gap.

During the NASP programme, American aerospace firms had designed a huge number of NASP-adjacent vehicles, the vast majority of which remain obscured by classification status, locked in file cabinets or simply lost to time (all too common, sadly). These vehicles were designed to fill any niche that might benefit from the ability to go insanely fast at unprecedentedly high altitudes: passenger transports, recon platforms, strike platforms, space launch systems.

In the mid-1990s, McDonnell Douglas reported on a series of vehicles intended to fly at or near Mach 10; vehicles derived from NASA work. One such design, the DF-9 (DF for 'Dual Fuel' as it burned both hydrogen and hydrocarbon, and Mach 9), originated in 1995 from a study performed for NASA. This was to be a fully functional operational vehicle, used for satellite launch. It was a waverider design with a flat wedge of a fuselage, small wings and control surfaces at the extreme rear, an almost square nose and an underslung scramjet package.

For propulsion at low speed it had turboramjets; for thrust when the vehicle shot itself above the sensible atmosphere, it had a linear aerospike rocket engine at the extreme rear. A 10ft diameter, 30ft long payload bay, largely to be filled with a rocket-powered upper stage and 10,000lb of payload, sat in the middle of the vaguely surfboard-like fuselage. It was an impressive design, but after the recent cancellation of the X-30 NASP it had approximately zero chance of getting built.

Still, the basic configuration was applicable to a wide range of sizes, including small research vehicles. By October 1996, McDonnell Douglas had taken the basic shape of the DF-9 and scaled it down to around 12ft in length (from the original's 208.5ft). This was just big enough to make a serviceable scramjet testbed. The programme was soon dubbed Hyper-X (rather obviously, 'HYPERsonic eXperiment'). This

Pegasus + NB-52B
SCALE 1/300

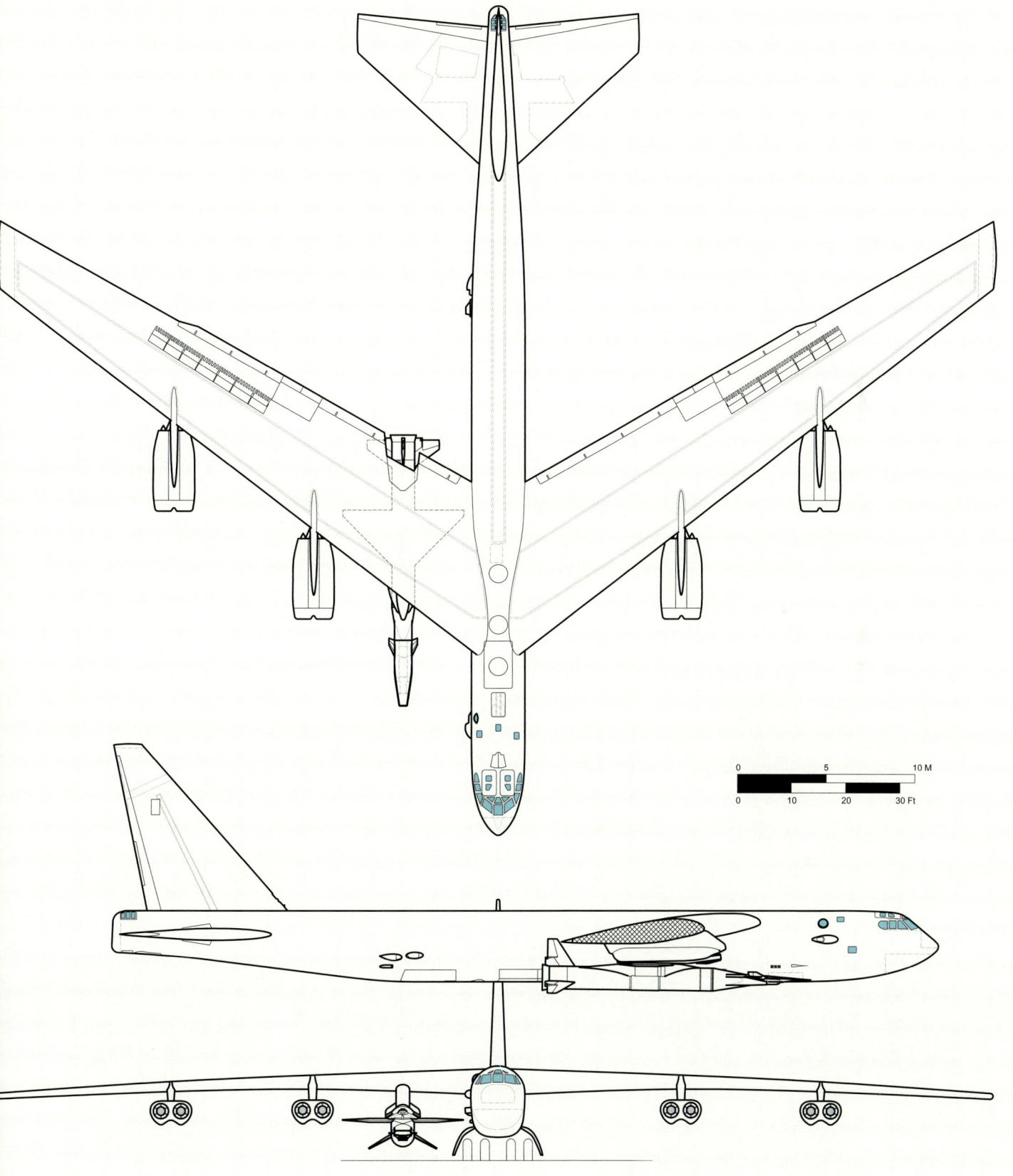

NB-52/X-43
SCALE 1/160

X-43
SCALE 1/30

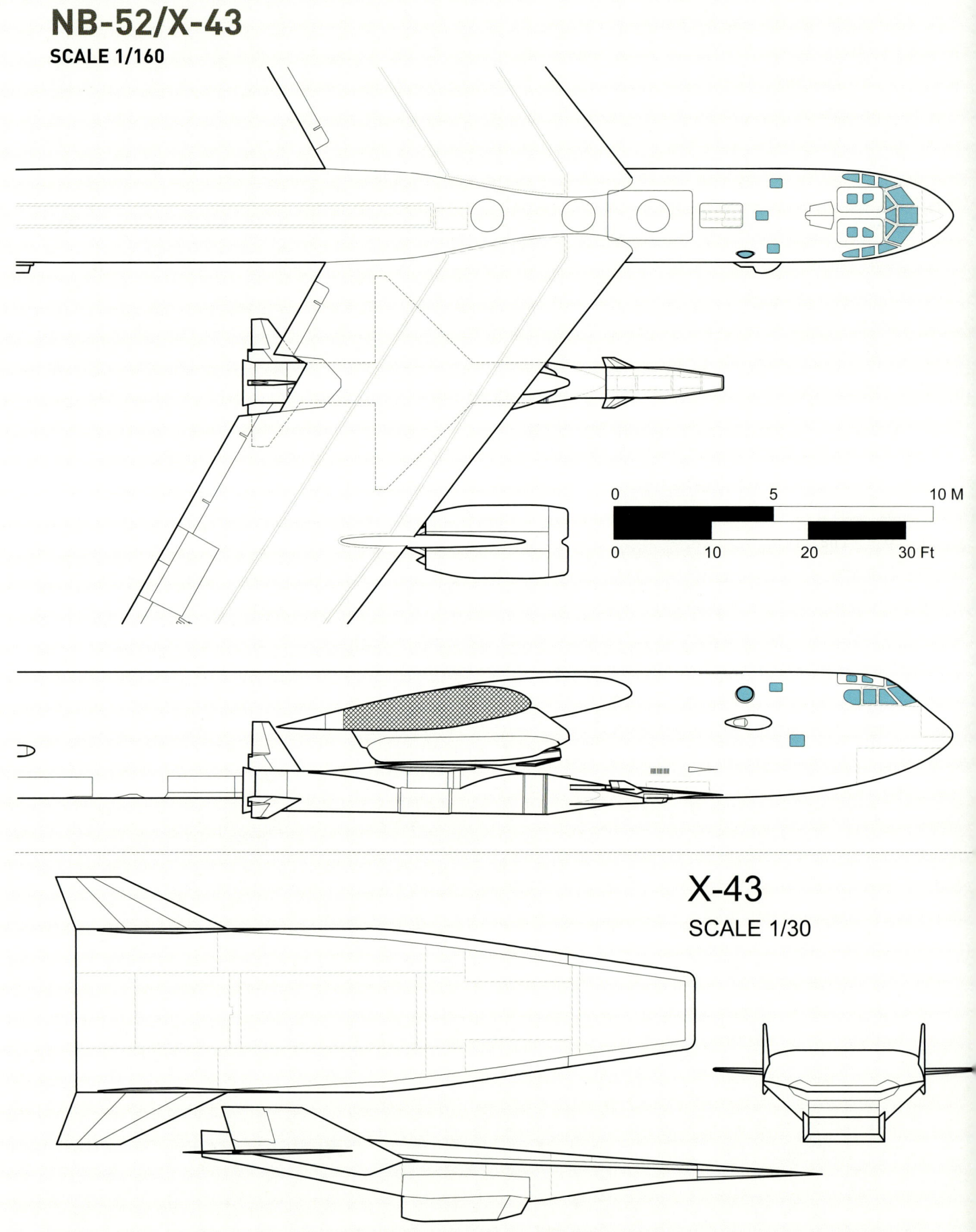

was a joint project combining the skills of NASA-Langley, NASA-Dryden, Alliant Techsystems, Orbital Sciences and Boeing (McDonnell Douglas having been absorbed by Boeing in 1997).

The NASA facilities were responsible for the basic vehicle design, wind tunnel testing, computer modelling, simulation and fluid dynamics. Alliant Techsystems and Boeing were responsible for building the engine and airframe; Orbital Sciences was responsible for the Pegasus first stage boosters. Manufacturing began in March 1997, with the first of three vehicles delivered in October 1999. On August 27, 1998, when a static test version of the scramjet was delivered by GASL, Inc for wind tunnel testing, the Hyper-X was officially dubbed X-43A by NASA.

The airframe was a combination of steel and aluminium, with tiles of alumina enhanced thermal barrier protecting it from the ravages of hypersonic flight. The sharp edges of the nose, wings and stabilizers were made of reinforced carbon-carbon. The large flat bulk of the nose was fabricated from tungsten, providing both thermal resistance and ballast. The scramjet was mostly copper with a water-cooled leading edge.

The X-43A had two relatively small liquid hydrogen fuel tanks; a full-scale vehicle meant to accomplish full missions would have had its fuselage essentially stuffed with liquid hydrogen. The duration of the powered test was so short that no chances could be taken with a delay in getting the scramjet engine running. Consequently, a tank of silane (SiH_4) was carried, injected at engine startup to assure combustion. Silane is a dangerous chemical, but the danger is wrapped up in its utility: it bursts into flames upon contact with air with great reliability and enthusiasm.

The Hyper-X booster rocket was the first stage from an Orbital Sciences Pegasus, complete with carbon composite delta wing and tailfins. Some thought had been given to the use of ground-launched boosters such as the Minuteman, but these would have to fly trajectories that such rockets were generally ill designed for. Conventional rockets are best at going straight up, getting above the atmosphere just as fast as possible; the Pegasus, in contrast, flies a lifting trajectory almost perfect for placing a hypersonic airbreathing testbed where it needs to be.

The Orbital Sciences L-1011 carrier aircraft would not have worked well for these tests; it would have required structural modification to provide clearance for the Hyper-X's fins, and it would not have had the plethora of instruments and recording systems that the NB-52B had. So once again the NB-52B was tapped to serve as a carrier. The Pegasus launch adapter was again used to mate the top of the Pegasus wing to the X-15 pylon.

The adapter holding the X-43A to the Pegasus provided an aerodynamic blending between the complex and distinctly non-cylindrical shape of the Hyper-X and the circular front face of the Orion 50S motor of the Pegasus first stage. It was built of steel and aluminium and also provided space for the water and nitrogen used for cooling and purging during the boost phase. Piston-based ejectors were built into the adapter that would bear on the titanium aft bulkhead of the X-43A and push it away with a 9 g acceleration.

Three X-43As were built by Micro Craft Inc. The first launch, in June 2001, was a failure; the Pegasus booster suffered a loss of control as it went transonic and the craft broke apart. The second X-43A was successfully launched in March 2004, reaching Mach 6.83 and 95,000ft. The engine ran for its full duration: 11 seconds. During that time the vehicle decelerated, indicating that thrust did not equal drag.

The third X-43A was launched in November 2004 and aimed for Mach 10 flight. At an altitude of 40,000ft and an airspeed of Mach 0.8, the B-52 released the Pegasus booster and its X-43A payload. The booster carried the X-43A upwards to 109,440ft and accelerated it to Mach 9.736, a bit faster than the targeted Mach 9.6. The scramjet ran for approximately ten seconds and maintained airspeed, indicating that thrust equaled drag. After burnout, the cowl closed over the inlet and the vehicle continued to glide, gradually losing altitude and speed. The telemetry data being transmitted from the aircraft was lost 721 seconds after launch when it had dropped to an altitude of 918ft and a speed of Mach 0.72. At that point it was some 850 nautical miles from the launch point.

There were proposals for an X-43B and an X-43C, but nothing came of them.

Unbuilt test vehicles

The NB-52B was also meant to carry a number of payloads that did not get built. The number of test aircraft that have been designed to fit under the NB-52's wing is vast and would make this a ponderous tome indeed, so what follows is an incomplete list of some of the more interesting and important projected aircraft that did not get as far as the wing of the NB-52.

X-15A-3

The most radical modification for the X-15 given serious study was the X-15A-3 variant. The fuselage would have been substantially stretched, giving an extra 91in of propellant tankage; the propulsion system would have been replaced; the nose would be lengthened and, most importantly, the small straight wings would have been replaced with large delta wings. The purpose of these modifications was not to make the vehicle go higher, but to go faster for longer.

A delta-winged X-15 was studied by North American Aviation, NASA and even Lockheed (in the form of the CL-839-28) over a span of some years, but the best documented design (at least, best documented by this author) is the North American D435-1-4 submitted to NASA Flight Research Center in May 1967. The intent was to modify the X-15 to serve as an experimental vehicle and technology demonstrator for future hypersonic cruise aircraft such as hypersonic passenger transports. An important consideration was the effect of extended hypersonic flight on delta wing skin panels and primary structures.

The D435-1-4 was intended to be launched from beneath the wing of the standard NB-52, using a slightly modified X-15 pylon. The basic mission profile called for release at an altitude of 43,000ft, promptly followed by engine ignition. The vehicle would climb to 100,000ft and accelerate to Mach 5 at about 120 seconds; it would then slightly dip to about 90,000ft and continue to accelerate to Mach 8.0. Burnout would then occur at about 160 seconds. It would maintain altitude for another 100 seconds while slowing to below Mach 4; it would then begin descent in preparation for landing. Total distance covered in the roughly 350 second mission would exceed 200 nautical miles.

While the X-15A-3 would not come anywhere near close to the maximum altitude attained by the X-15A-2, in many ways the mission profile was much harder on the airframe. It would 'cruise' at roughly the same altitude as the SR-71, but at a top speed nearly three times faster. As a result aerothermal heating would be immense, with leading edge temperatures expected to reach about 2,200° F. Due to the lower maximum altitude, the craft would never leave the sensible atmosphere, and thus had no need of the small reaction control thrusters that the earlier X-15 had in the nose and wingtips. Instead, all control was via conventional aerosurfaces.

The YLR-99 rocket engine would be modified with a 22.1:1 expansion ratio nozzle. Burning liquid oxygen with liquid ammonia, it had a thrust of 61,750lb at 100,000ft. Alternate engines were also examined, including the YLR91-AJ-9 used on the Titan. The YLR91 used an ablative nozzle extension which would be explosively separated from the rest of the engine prior to landing. With 33,500lb of propellant, this engine would permit the X-15A-3 to attain Mach 8 without the need for the external drop tank. This engine used storable propellants – nitrogen tetroxide oxidizer and Aerozine 50 fuel. As these are room-temperature storable propellants, the B-52 would not need to carry top-off tanks and systems to replace propellants that boiled off from the X-15A-3's tanks prior to separation. These were, however, impressively toxic chemicals.

Studies – which remain unavailable to this author – examined the use of a B-70 as a carrier and launch aircraft for the X-15A-3. By launching from a higher altitude and much higher speed than the B-52 could provide, the X-15A-3 could get up to altitude faster and either attain higher speed or maintain it for longer. However, this would have been a major engineering effort compared to the relatively minor changes needed for B-52 launch, and as the M-12/D-21 launch disaster showed, launching off the back of an aircraft travelling at Mach 3 is a non-trivial task with substantial risk.

One role that the X-15A-3 was expected to perform was as a high velocity testbed for subscale airbreathing propulsion systems. Diagrams show a Hypersonic Research Engine attached to the ventral fin. A dummy of this engine was flown on the X-15 but here it would be a functional system, briefly burning hydrogen. The purpose would be to simply demonstrate that the scramjet engine actually worked, not that it would work long enough to be truly useful. Other vehicles would be required for that role.

Landing approach speed was 300 knots, touchdown speed 199 knots at a 10° angle of attack. This was only slightly faster than for the standard X-15.

X-24C

For a fair stretch of time in the 1970s there was an expectation that the X-24B would be followed up by an improved design. The X-24B configuration was such that it lent itself to modification for testing air breathing hypersonic engines; it had a fair amount of internal volume, and the underside was shaped appropriately to serve as the inlet ramp for a scramjet system.

In early thinking at the Flight Dynamics Laboratory, the X-24C was expected to be a major modification to the X-24B airframe – just as the X-24B built upon the existing X-24A. At that point, 1973, the X-24C would be configured much like the X-24B and would use some components from that craft; but it would be better aerodynamically optimized for the planned Mach 5 top speed and would have an XLR-99 rocket engine. In order to protect the aluminium structure of the X-24B from aerothermal heating, thermal shielding would be bonded to the outer surface.

Other ideas included adding a rectangular scramjet package to the underside, as well as an internal air breathing engine fed by new 'cheek' inlets to the otherwise stock X-24B. The X-24C soon became an all-new airframe. Through the lifetime of the programme, the X-24C concept underwent many revisions not only in design but in concept.

North American X-15A-3 D435-1-4
SCALE 1/96

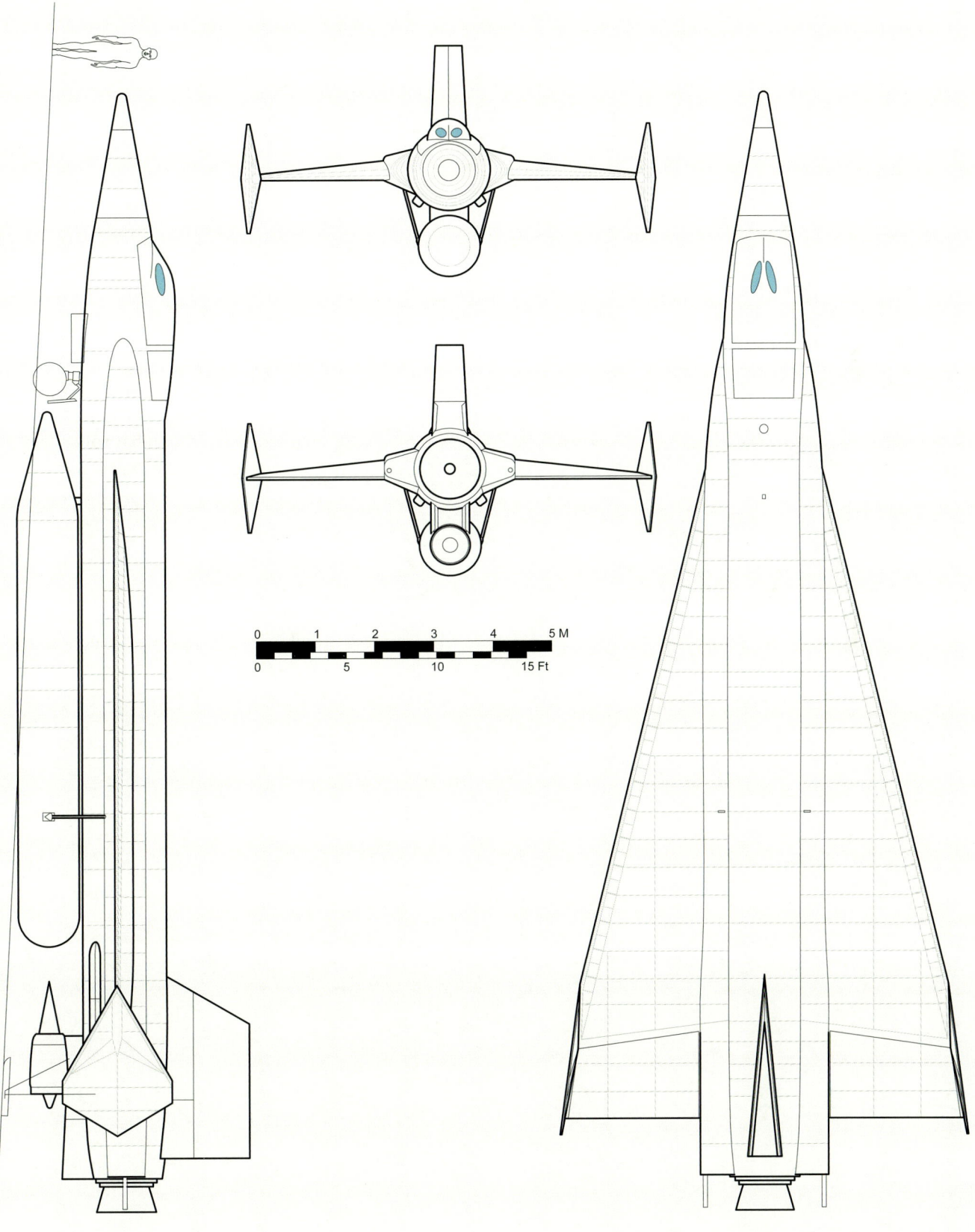

X-15A-3 + NB-52B
SCALE 1/300

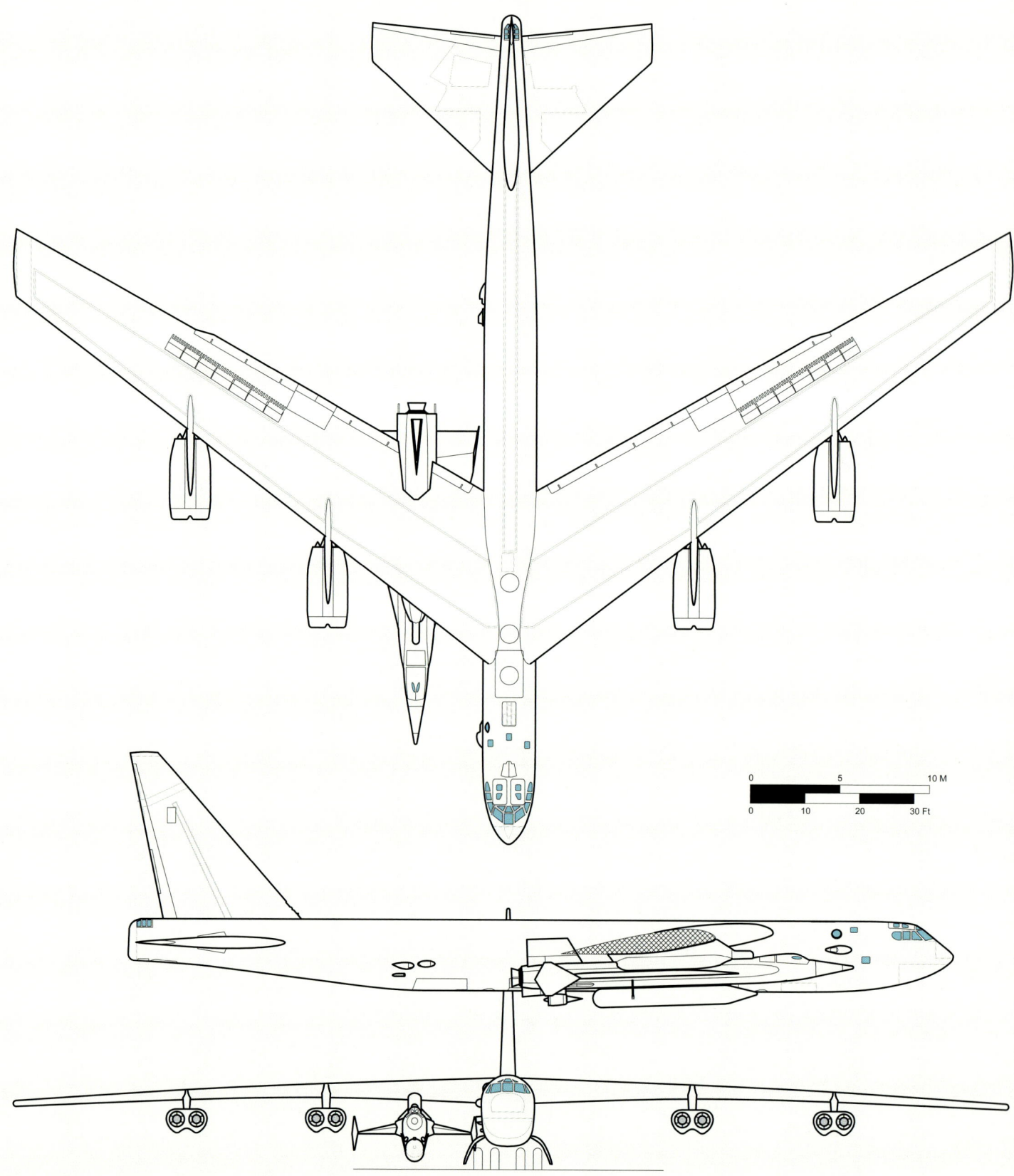

Both Martin and Lockheed put forward designs, as well as the Air Force and NASA. The X-24C was a joint venture between the Air Force and NASA, who both wanted a hypersonic research aircraft though they wanted it to demonstrate different technologies and capabilities.

The Air Force wanted a vehicle that would support its goals of hypersonic reconnaissance, interception and strike; NASA wanted hypersonic passenger transports and satellite launch first stages.

There is definitely overlap in some of the technologies these aircraft would require… but these concepts have wide divergences. NASA, after all, would not care so much about the separation dynamics and aerodynamic interactions between the aircraft and a small missile launched from the underside; the Air Force would be largely uninterested in passenger comfort or making long range high-speed flights economical. Producing a single design that could accommodate the full range of desired testing capabilities would prove too difficult and expensive.

One of the first X-24C configurations to receive substantial study was the FDL's X-24C-9. This was the 1973 configuration, basically a refined X-24B, including the its planform, wing and fin geometry and long nose with aft-set cockpit. It was intended to be a low cost approach; while greatly exceeding the X-24B in speed, reaching Mach 5, it was generally similar in size and outline and would use some of the same components.

However, its utility for anything beyond demonstrating the configurations' flight characteristics at higher speed was minimal. It had a small experiment bay on the underside that could be filled with instruments or deployable weapons, but it did not have the space for a meaningful liquid hydrogen tank. Without that, it could not be used for demonstration of useful high speed airbreathing propulsion systems such as scramjets.

The joint NASA/USAF study team went back to the drawing board and repeatedly tinkered with the design. A series of modifications followed, eventually leading to the X-24C-12I in 1975. This became, for a time, the baseline design of the X-24C, to be passed on to major contractors for revisions and refinements. The X-24C-12I differed from the original concept in some important ways: the fuselage was stretched and the cockpit moved forward, the nose greatly shortened. The result of these changes was to greatly increase the useful volume in the fuselage, allowing for increased propellant tankage. The fuselage was substantially less 'curvy' than that of the X-24B, losing the hump-backed appearance in favour of a straight, flat back.

The design included a 10ft payload section directly behind the cockpit. This could be used to contain deployable experiments, instruments or fuel tanks for externally mounted airbreathing engines. The payload section could be removed and swapped out, the outer skin being replaced with various experimental structures for dealing with the heat and drag of Mach 6 flight including ablatives, high temperature metals, low conductivity tiles or active cooling.

As the main structure was to be aluminium, a metal wholly incapable of withstanding the temperatures that Mach 6 can deliver, the bulk of the aircraft's outer surface would need to provide thermal protection for the inner structure. In the end, the general design used an aluminium main structure with a complex outer skin using Lockalloy (62% beryllium, 38% aluminium) as a heat sink, reaching thicknesses of up to 0.43in.

The propulsion system, using an XLR-99 engine from the X-15 programme but with a nozzle extension for improved performance, could not only accelerate the aircraft to Mach 6, it could cruise at that speed for 40 seconds. The main propellant tanks were multi-lobe with common bulkheads between the fuel and oxidizer.

They were non-integral with the structure of the X-24C so they could be relatively easily replaced with different tanks. This would permit the replacement of the XLR-99 with a different rocket engine (or engines) that used different propellants. A fly-by-wire system would be used; this would make wholesale replacement of the payload section much easier. Instead of mechanical linkages running through the bay, here only wires would pass through, easily unplugged and plugged back in.

Both Martin-Marietta and Lockheed took cracks at coming up with a complete design based on the X-24C-12I, initially producing designs virtually indistinguishable from the USAF/NASA original. But Lockheed at least went rather further and produced a very different derivative concept, greatly stretched, more elegant and clearly the result of the Skunk Works. This too was to be carried by and launched from the NB-52B. But the USAF and NASA were constantly changing what they wanted and the X-24C concept as a whole passed over into other programmes: the High Speed Research Airplane, the National Hypersonic Flight Research Facility. As the concepts evolved, the lifting body aspects faded in favour of a more traditional sort of aircraft layout.

Rockwell Hypersonic Research Airplane

The X-24C effort led to a series of manned hypersonic research aircraft designs meant to be launched from the NB-52B. Representative of this class of craft is

X-24C-9
SCALE 1/80

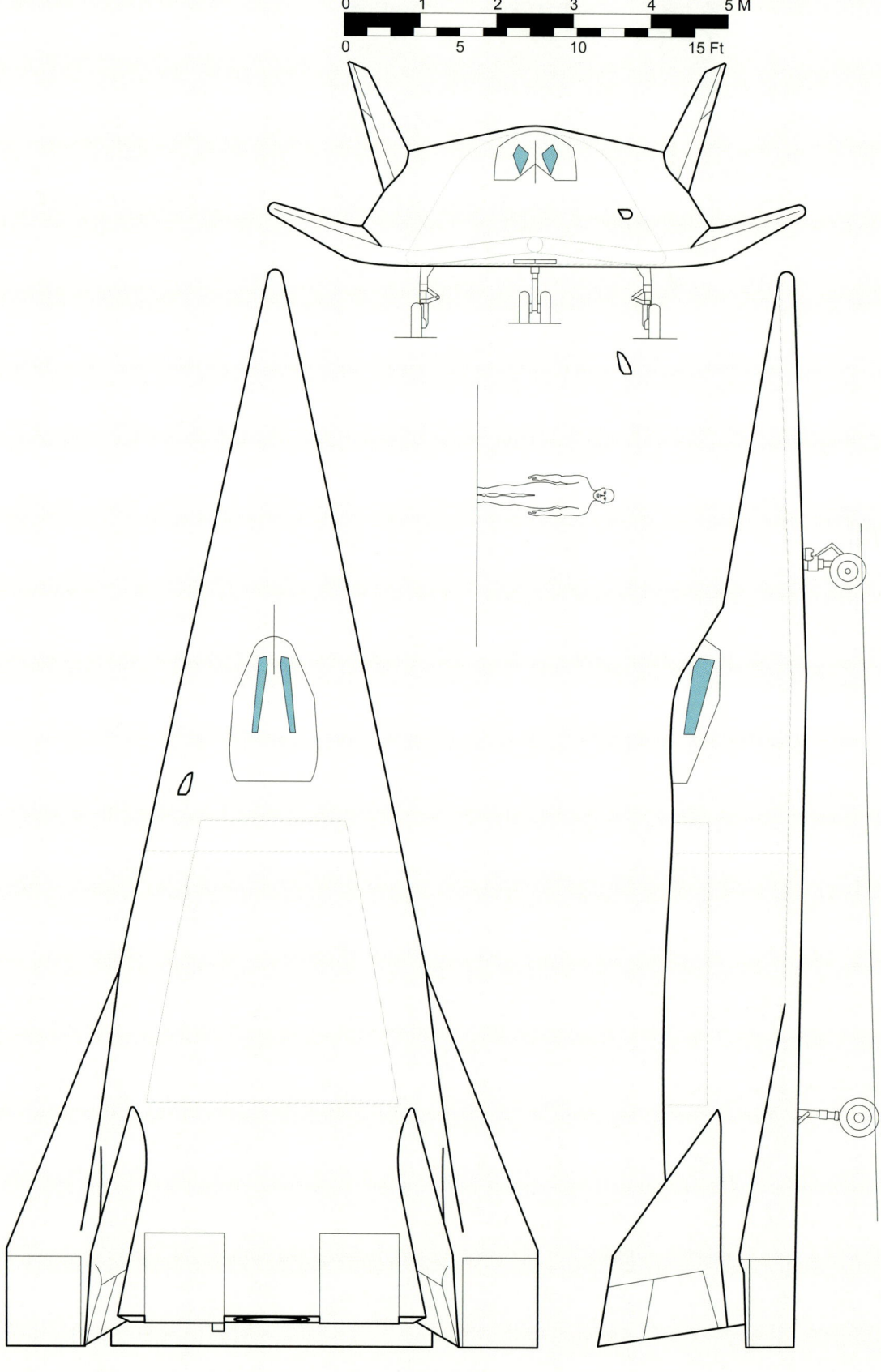

X-24C-9 + NB-52B
SCALE 1/300

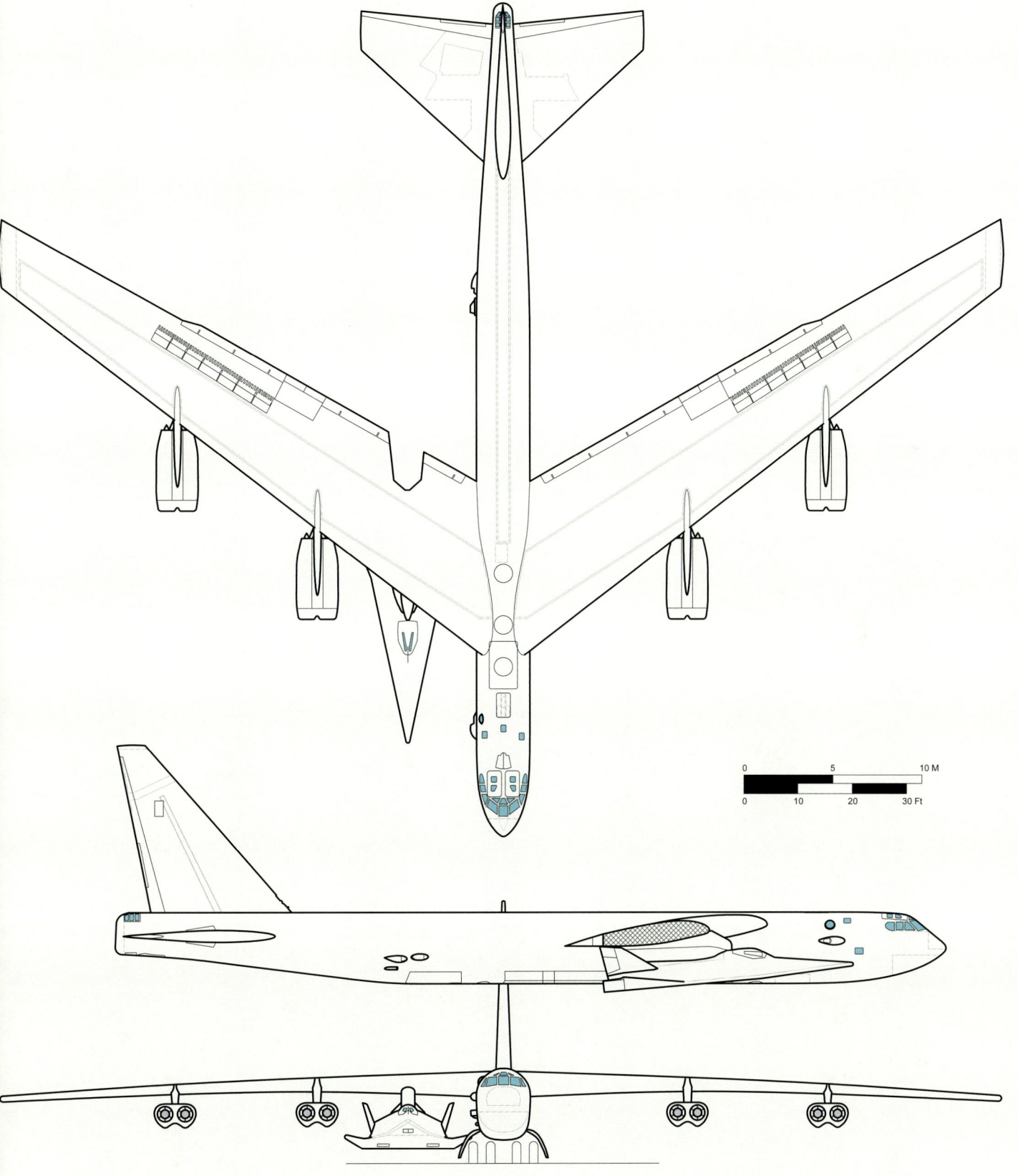

X-24C Baseline + NB-52B
SCALE 1/300

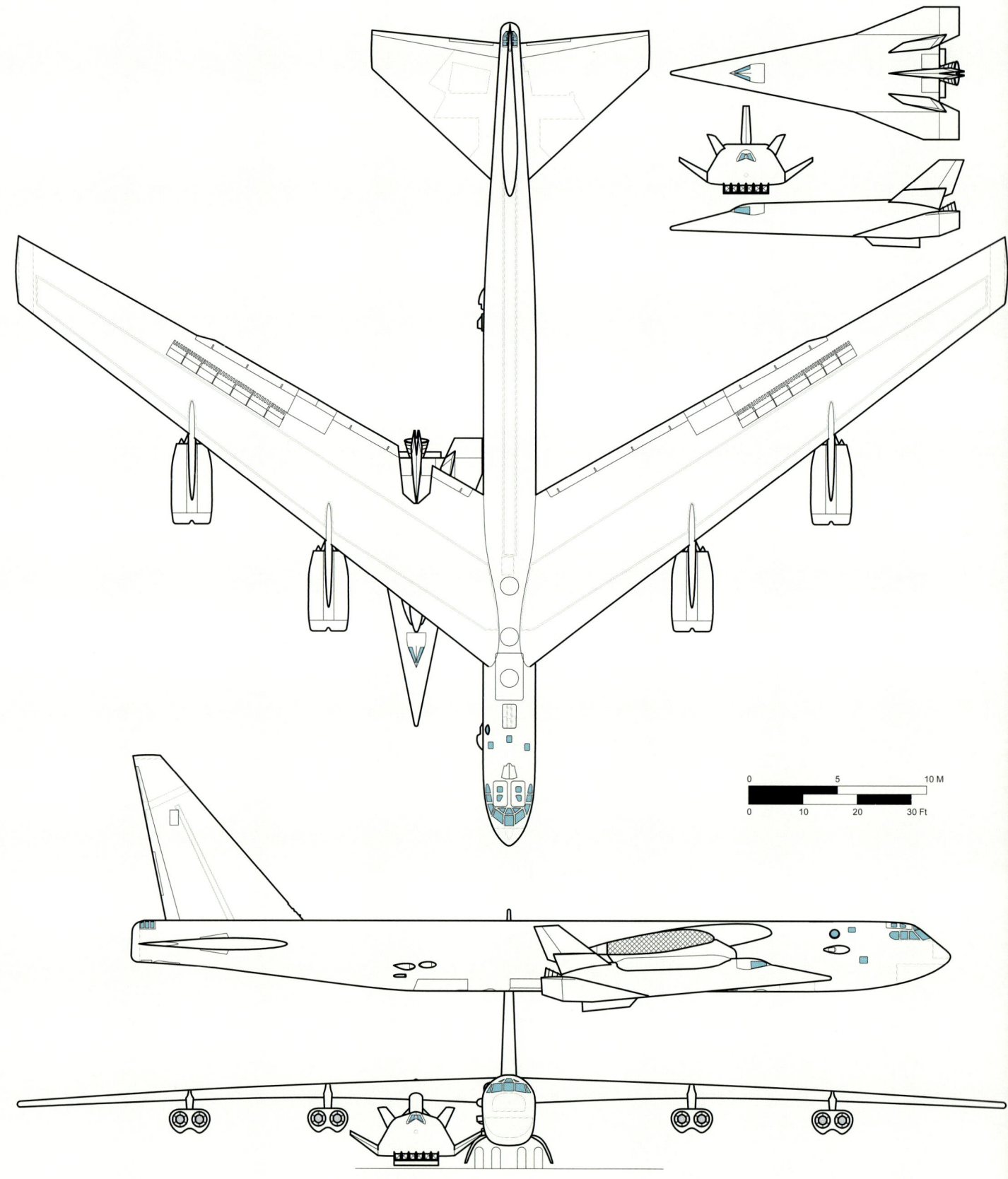

Rockwell Hypersonic Research Airplane
SCALE 1/90

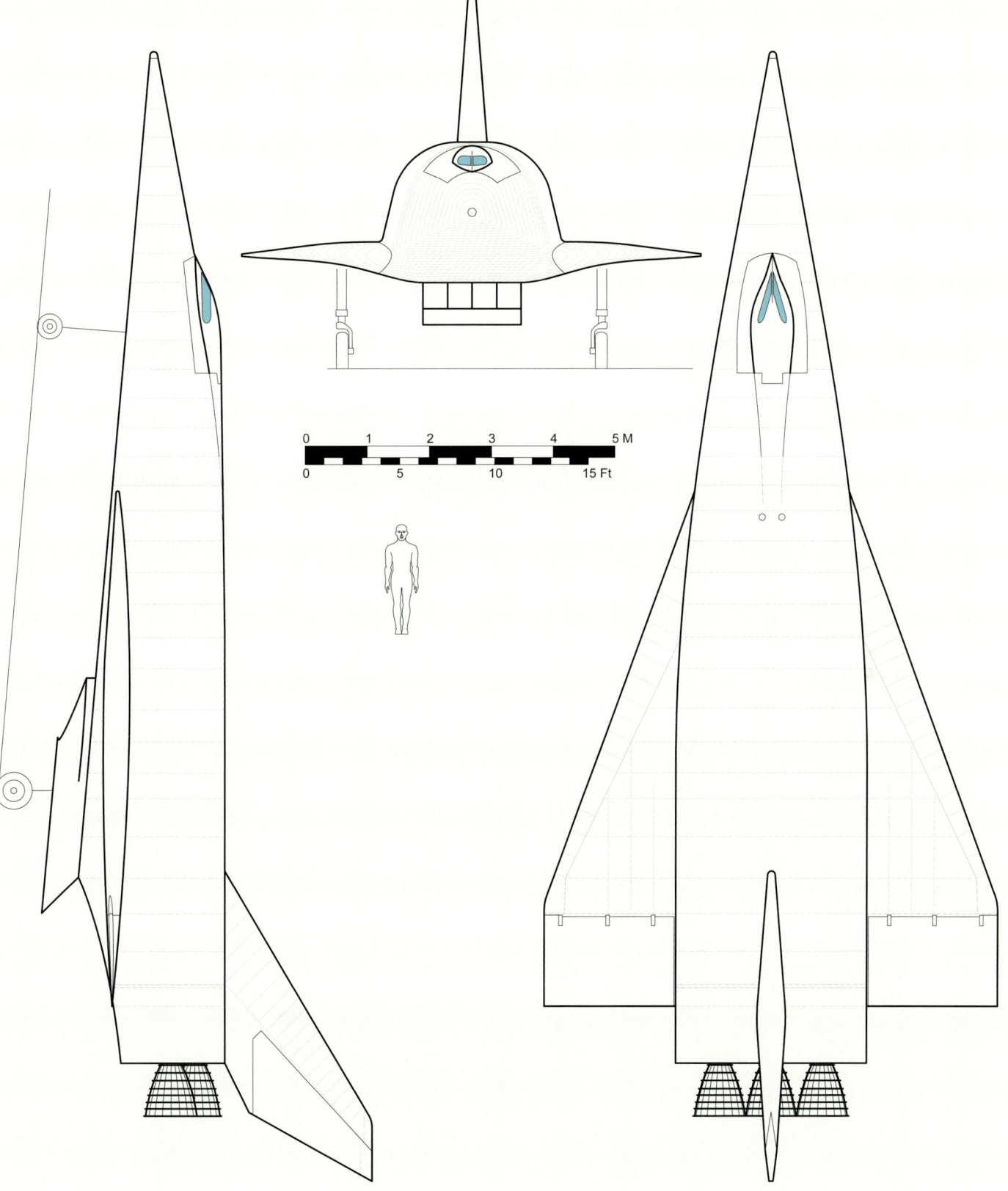

Rockwell HRA + NB-52B
SCALE 1/300

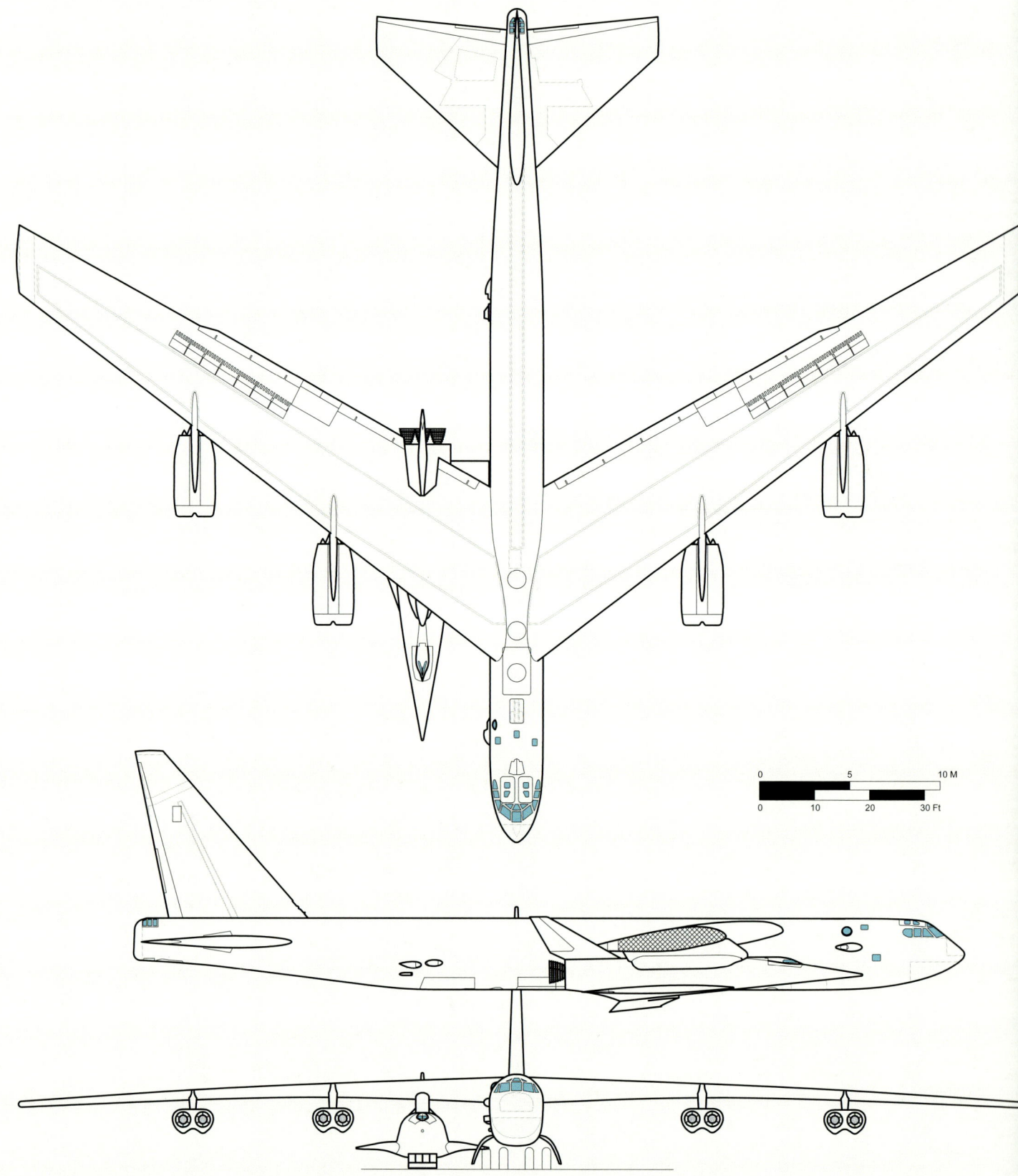

the Rockwell Hypersonic Research Aircraft from 1974. This bore some faint resemblance to the X-24B, but was a more conventional wing-body design and less of a lifting body. The HRA, developed at the Space Division with input from the Los Angeles Aircraft Division, was a modular design intended to be constantly changed and modified in order to test a wide array of features.

In its basic form it was a purely rocket-powered craft, using three Bell LR-81 'Agena' rocket engines. These burned the hypergolic mixture of UDMH and IRFNA; toxic and dangerous but easily storable and non-cryogenic.

The HRA had a flat-bottomed fuselage with sloping sides, something of an increased-fineness X-24C fuselage. But instead of large upward-tipped fins, the HRA had true and highly swept clipped-delta wings with no dihedral and almost flush with the flat underside of the craft. A conventional tail, looking much like the tail of the Space Shuttle Orbiter, completed the aerosurfaces. The tail would fit in the notch at the rear of the NB-52B's wing and the HRA's wings would just fit in the space permitted by the carrier plane's fuselage.

The underside of the fuselage formed a narrow triangular surface, sloping down from the nose to just ahead of the tail. The underside then swept back up. At the lowest point a scramjet package could be installed; the forward fuselage would serve as a compression inlet, the underside of the rear fuselage an exhaust nozzle. In order to feed the scramjet, volume was provided in the middle of the fuselage for a cylindrical tank capable of holding 1,440lb of liquid hydrogen.

The nose, wings and various skin panels were to be easily removable for replacement with experimental versions. Skins could be made of advanced high temperature metal, or covered in protective shielding, or given active cooling, depending upon the study. The airbreathing engines could be added or removed; external ramjets burning JP-5 could be added to the underside, the hydrogen tank serving to contain a much greater mass of hydrocarbon fuel.

More aggressively, the rockets could be removed and replaced with a pair of J79 turbojets and a pair of ramjets, fed from new inlets along the sides of the fuselage aft of the cockpit.

Just behind the cockpit was an instrumentation compartment. On many flights this would contain sensors and data recording equipment. But on other flights, the doors used to load the instruments through the underside of the craft would open in flight and eject military stores… missiles and bombs. How well such things would eject from a hypersonic aircraft was a subject of great interest. Would the weapon jettison safely and stably? Or would it be damaged passing through the aircraft's shockwave? Would the weapon disrupt air flow into the engine? Would it bounce back up and hit the aircraft? These were vital questions if the Air Force was to field hypersonic interceptors.

As it turns out, neither the Rockwell HRA nor any of the other kindred research aircraft of its time were built. So the Air Force never got to study hypersonic stores ejection, and never built a hypersonic interceptor.

B-52 Re-Engining Studies

The B-52H is the most recent major redesign of the B-52. Recent, of course, is a somewhat relative term; the design of the B-52H began in 1959, more than 60 years prior to the writing of this book. The 'H' variant was characterized by the inclusion of the Pratt & Whitney TF33-P-3, the first use of a turbofan engine on the B-52, and it has flown with this same model of engine for six decades.

This is not because engines have not improved since the early 1960s; indeed, the 1.4:1 bypass ratio of the TF33 is now shockingly low for modern subsonic turbofan engines. The Pratt & Whitney PW2000 that powers the C-17 Globemaster III, for example, has a bypass ratio of 6:1; the Pratt & Whitney PW1000G that powers the Airbus A220 has a bypass ratio of 12.5:1.

Many engines developed since the TF33 have provided improved fuel economy, increased thrust, or both. A new higher-thrust, reduced fuel consumption engine would not only make the B-52 more economical to operate by reducing the fuel used, it would allow the B-52 to fly further and stay aloft longer. It is therefore not surprising that both Boeing and the US Air Force have expressed interest over the years in re-engining the B-52. And yet so far that hasn't happened.

There are a number of reasons why – treaty compliance and ground clearance being two of the more important, not to mention the ever-present excuse of up-front cost. Replacing the old engines with new ones, especially engines of greatly different size and thrust, would not be a simple process. The existing nacelles and pylons would have to be dispensed with and all-new pylons would be needed, especially if the eight TF33s were replaced with four higher bypass, larger-diameter turbofans.

The wing structure itself would also doubtless require structurally reworking in order to withstand thrust of a higher level and from a somewhat different vector. Electrical, pneumatic and hydraulic controls would need to be re-routed, replaced or augmented. The flight controls for the aircraft could well require substantial modification too. If a standard B-52H were to lose thrust in a far outboard engine, compensation would be relatively straightforward. But if a four-engined version were to lose thrust in an outboard engine, countering the sudden loss of perhaps 40,000lb of offset thrust could be nightmarish.

Still, the siren call of improved fuel consumption has led to many studies over the years, most of which are poorly detailed publicly.

Boeing B-52X

One study for a major re-engining redesign is known to this author from a single Pratt & Whitney memo, including a letter from Boeing discussing the 'B-52X'. This mid-1975 study was, at least at the time of the memo, in the preliminary concept phase and included replacement of the standard eight Pratt & Whitney TF33 turbofans with options such as six or eight JT10D-2s or four JT9D-70Ds.

The letter provides three simple top view sketches of aircraft configurations with four of the 50,000lb thrust class JT9D-70D high bypass turbofans. The configurations differed from each other in that the first design replaced the existing B-52 engine pylons with new pylons at wing butt lines (WBL) of 410 and 720in; the second design kept the inboard pylon position and moved the existing outboard engine pylon to an extreme inboard position of WBL 218in; a third design showed a single pylon at WBL 410in that used a dual-engine pod.

Ground clearance was an issue for these designs, especially for Design 1. The outboard engine had a ground clearance of only 23in under fully loaded conditions.

Along with the change in engines, a few other changes were also made. Ride control canards, similar to those used on the B-1 bomber and demonstrated on the NB-52E Control Configured Vehicle demonstrator, were to be included. Their position was shown in planform view, but their anhedral/dihedral and vertical location are unknown and are shown in the reconstruction diagrams purely provisionally. It is unknown if a vertical fin such as the NB-52E had was also to be employed.

Curved extensions were also added to the main wing roots, continuing forward almost to the cockpit. These were similar to, though much larger than, the wing root extensions added to B-52s that carried the AGM-86B Air Launched Cruise Missile; in that case the root extensions were largely for Soviet reconnaissance satellite identification purposes, but the wing root extensions for the B-52X were almost

Boeing B-52X
Configuration 1
SCALE 1/300

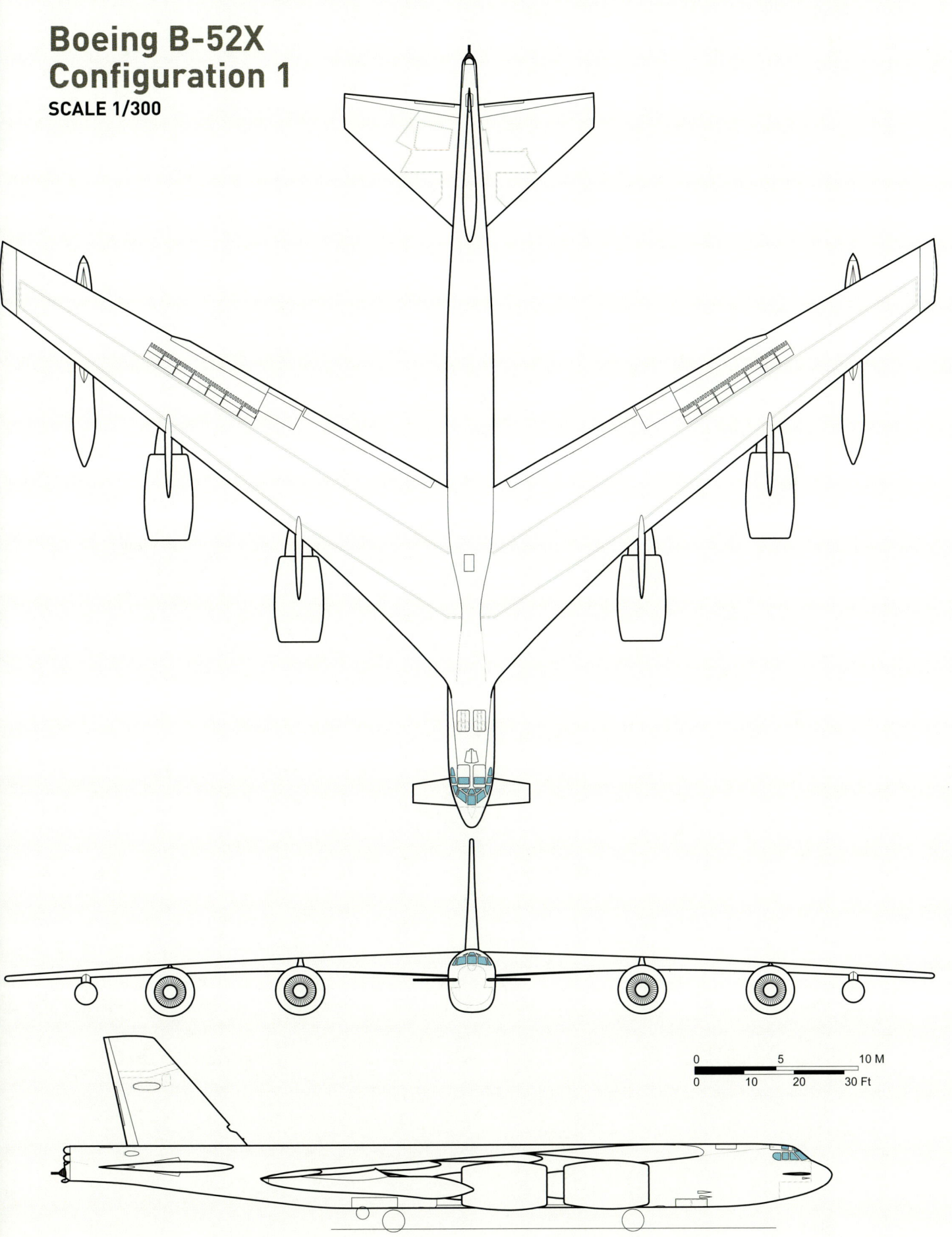

Boeing B-52X
Configuration 2
SCALE 1/300

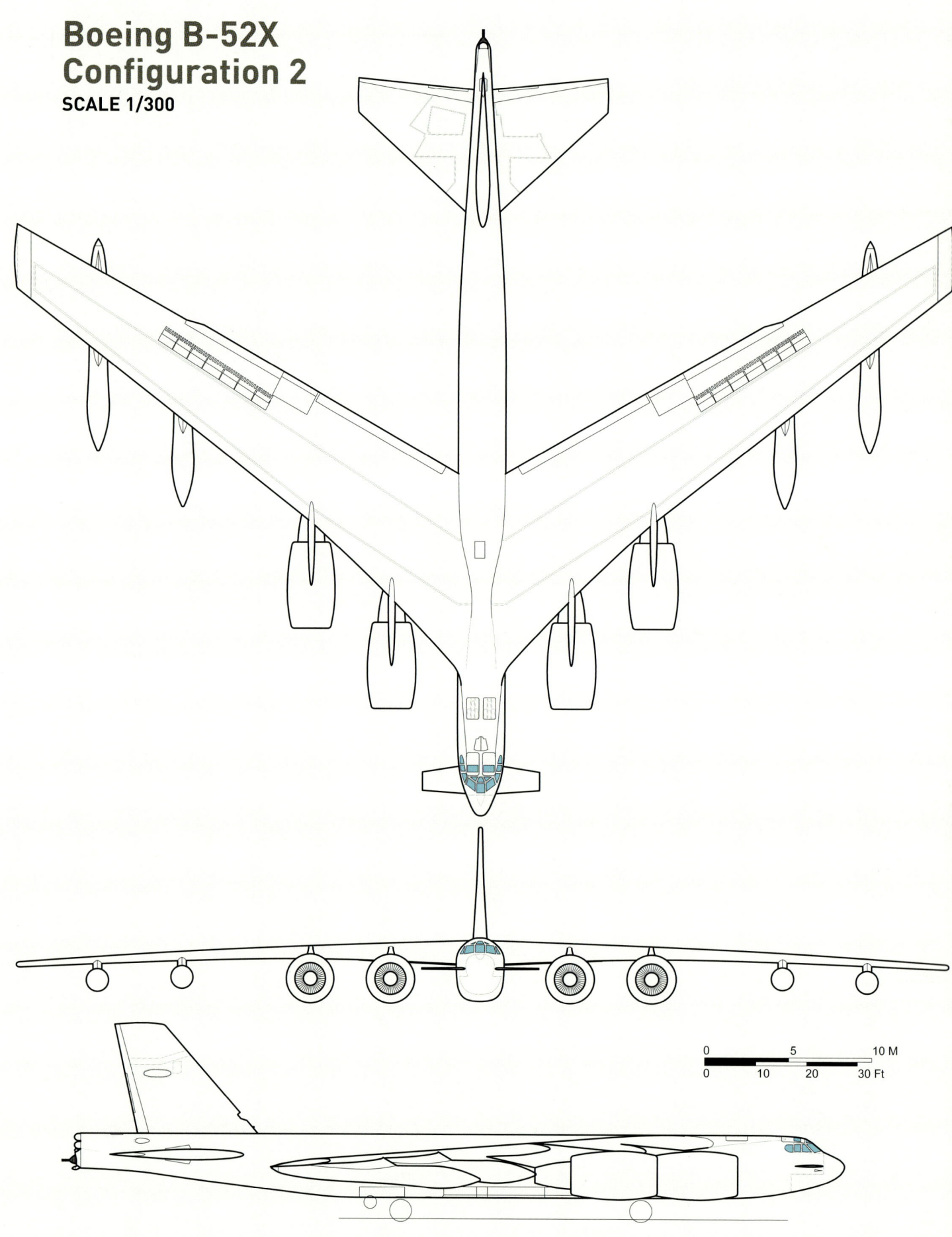

316

Boeing B-52X Configuration 3
SCALE 1/300

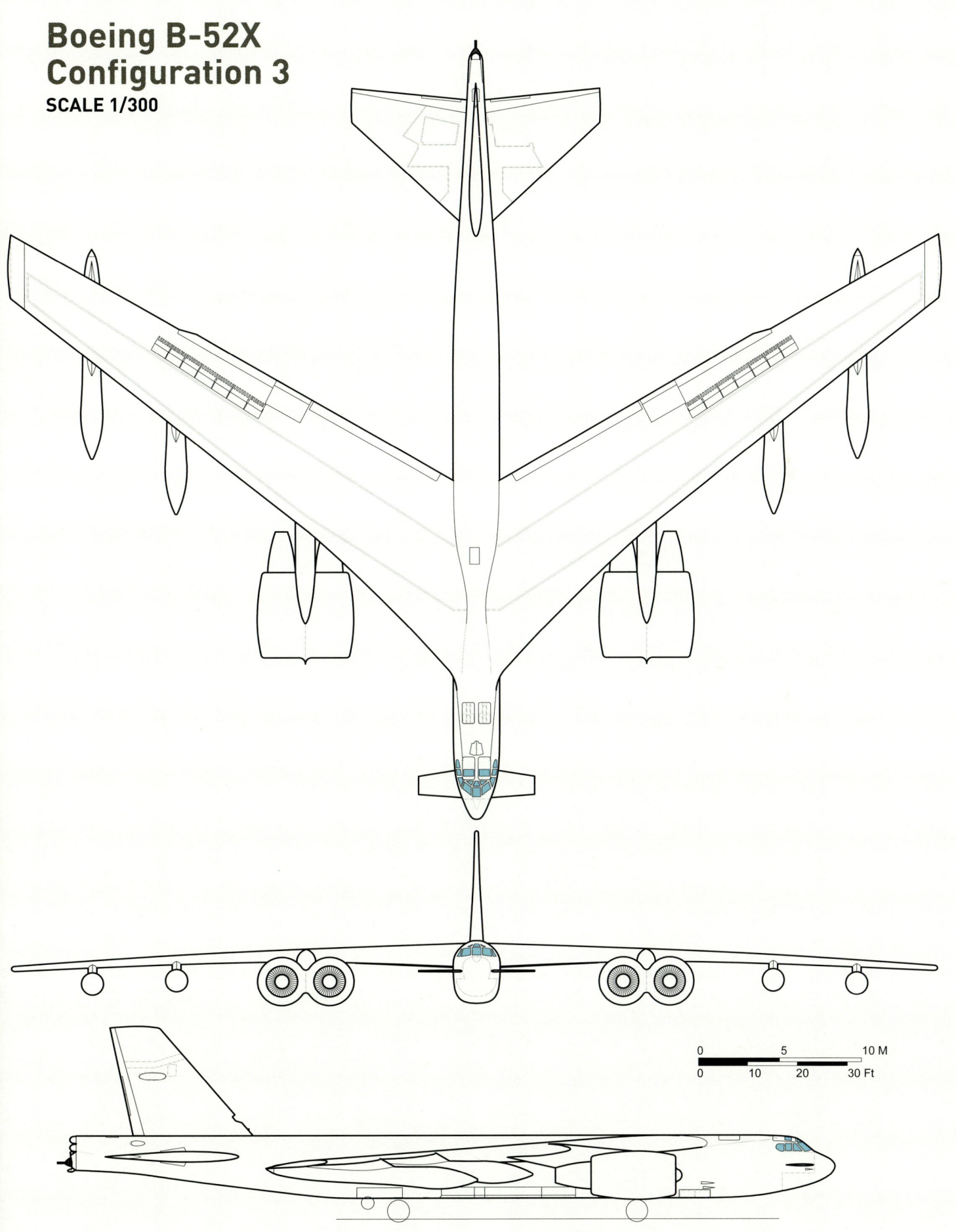

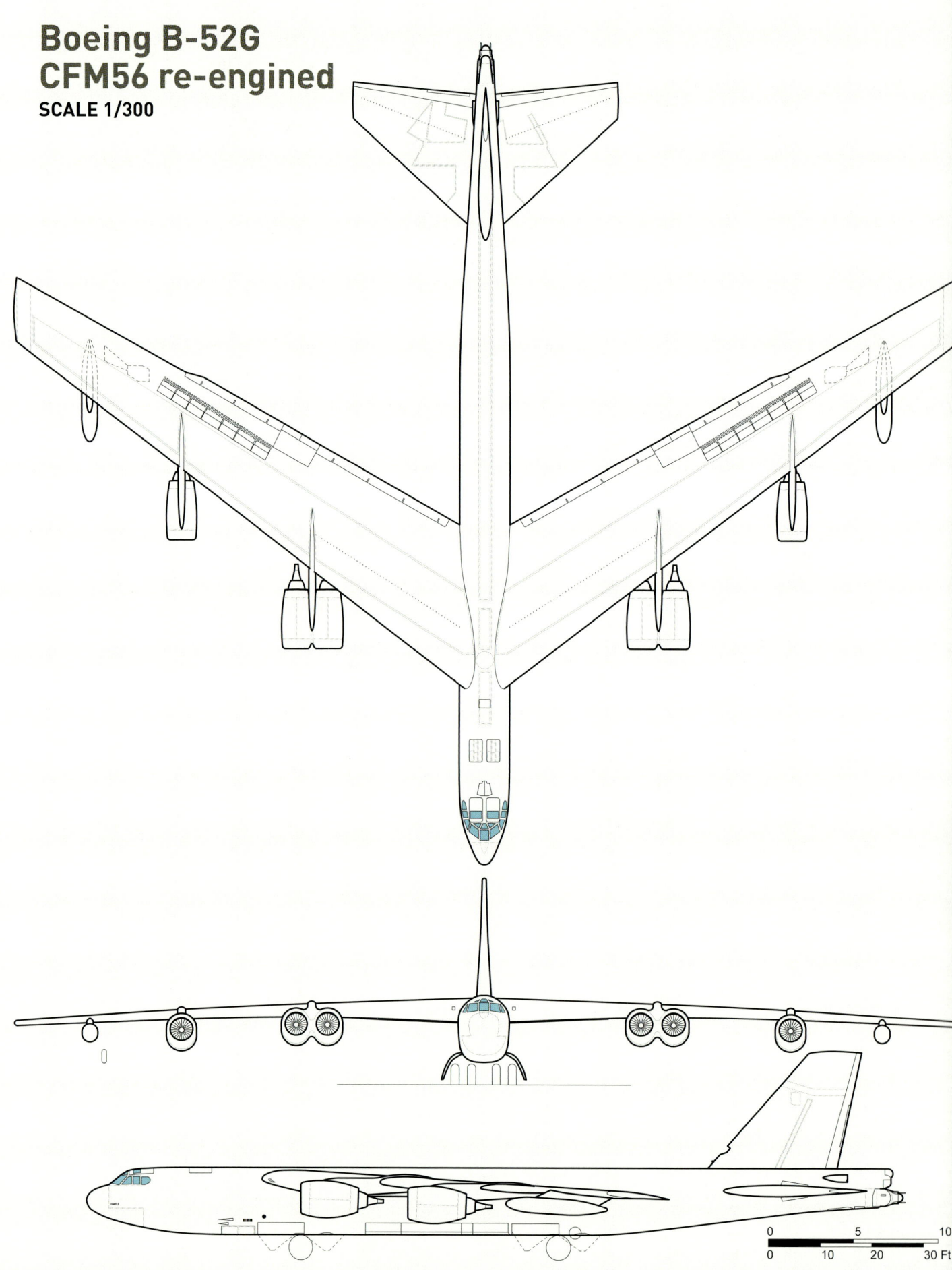

Boeing B-52G
CFM56 re-engined
SCALE 1/300

certainly meant for aerodynamic purposes. Some performance charts were provided for the engines, but not on the aircraft themselves.

Boeing B-52 with CFM56s

Known from a single undated blueprint, one concept for re-engining the B-52 saw a B-52G equipped with six jetliner-style high-bypass turbofans. The engine layout was much like that of the XB-52 when it tested the J75 engine… twin-engine nacelles on the inboard pylons and a single engine on the outboard pylons. The twin-engine nacelle was necessarily quite a bit larger than the two-engine nacelles on the standard B-52G.

While data on this concept is lacking, the engines are a close match for the CFM56-2 turbofan (manufactured by CFM International, a company jointly owned by Safran Aircraft Engines of France and GE Aviation of the USA). If this is accurate, it would date the concept to no earlier than 1973 or so, but probably much later, as studies of re-engining the B-52 with six CFM56s are known to have been performed in the mid to late 1980s.

The CFM56-2 engine was used to re-engine early jetliners as the 707 and DC-8; re-engining the B-52 with it would have certainly made sense. This engine could generate around 22,000lb of thrust; six of them would give the B-52 132,000lb of thrust compared to the 13,750lb of thrust possible from the J57-P-43WB (totalling 110,000lb).

Boeing B-52 with PW2037s

Pratt & Whitney described a concept for replacing the eight engines of the B-52G and B-52H with four of its PW207 high bypass turbofans in 1982. These engines were designed for and used on the first generation of Boeing 757 jetliners; Pratt & Whitney's suggestion was to use the same engines and the same nacelles on the B-52.

The pylon would be similar to that of the 737 and modified in design to fit the engine closely to the wing, with the top edge of the fan exhaust just ahead of and essentially level with the leading edge of the wing. This would give a ground clearance of about 33in compared to 56in for the J57 nacelles on the B-52G. Some relatively minor modifications to the engines would be needed for the higher altitudes the B-52s were expected to operate at.

Pratt & Whitney estimated that the PW2037 engines would burn fuel at a rate 40% less than the J57s and 30% less than the TF-33s. This would result in an increase in unrefuelled range of 65% and 45% compared to the B-52G and B-52H, respectively. This would also translate into considerable increasing in loiter time for maritime support missions (likely referring to anti-submarine missions).

For a mission radius of 1,000 nautical miles, the loiter time on station would be 18 or 19 hours; certainly impressive, but likely a bit of a nuisance for the crew who would have to stay awake through all that. On the other hand, for a mission radius of 4,000 nautical miles, loiter time on station would still be three or four hours.

Four PW2037 engines would provide 35% more takeoff thrust than eight J57s, shortening the takeoff roll by 27%, able to lift off in 6,400ft on a hot day. This would not only speed the launching of B-52s, it would increase the number of available fields that the B-52 could use, allowing greater dispersion. All of these performance improvements would also come with cost savings; Pratt & Whitney estimated that each B-52G with the new engines would save some 715,000 gallons of fuel per year.

Boeing B-52 with C-17 nacelles

Seven years later, LTV studied a very similar concept: using the nacelles from Boeing's own C-17 on the B-52G. The C-17 is equipped with PW2040 engines, very similar to the PW2037 engines of the 757, with similar nacelles. The available documentation indicates a very preliminary study, merely examining the possibilities of such an engine swap; the performance of the aircraft was not presented. However, it would most likely have been quite a bit like the earlier study with the PW2037. It was found that a high g turn on the ground with fully laden wing and tip tanks would reduce clearance from the runway to the bottom of the outboard nacelle to a mere 8in. Entirely new pylons would be needed.

Recent proposals

As this book was being written, the Air Force was expressing apparently serious interest in giving the B-52 modern engines. Several firms are proposing to use their turbofans to extend the lifespan of the already quite old B-52; with new engines, it might be expected that the B-52 could soldier on into the 2050s. This would put the B-52H airframes at well over 80 years of age and the B-52 design at well over a century. This would have been something like the US Navy going in to the Second World War with wooden-hulled sailing vessels or the Army riding into the Gulf War on horse-drawn wagons.

So it can be seen two ways: either as an amazing tale of a design that was made virtually perfect when first created… or as a condemnation of American military foresight, spending and priorities. If the United States gets into a shooting war with a near-peer adversary, the wisdom of reliance upon airframes pushing a century in age could be tested.

The firms actively pursuing the contract to re-engine the B-52H are General Electric, Pratt & Whitney and Rolls-

Boeing B-52H w/PW2037
SCALE 1/300

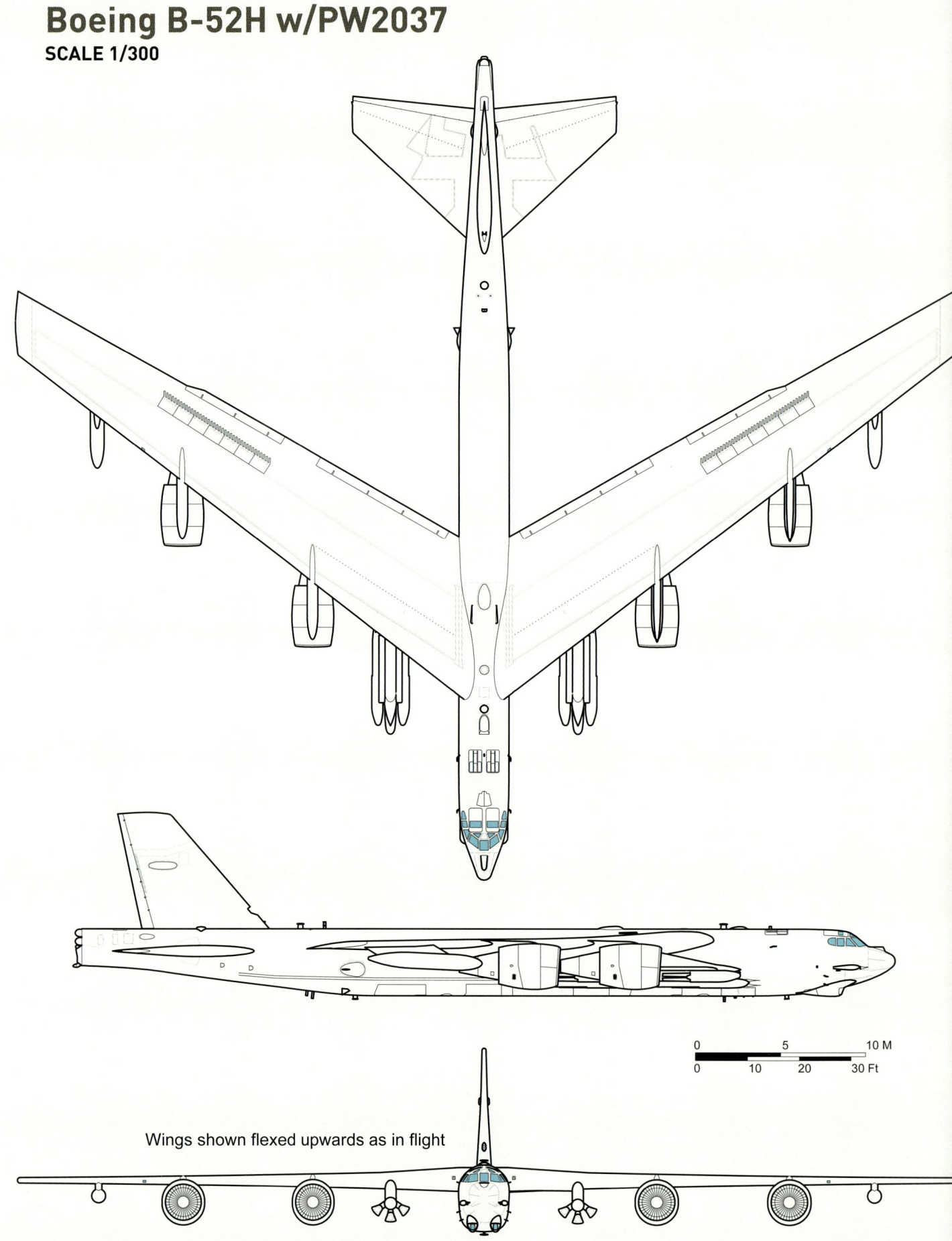

Wings shown flexed upwards as in flight

Boeing B-52G w/C-17 Nacelles
SCALE 1/300

Royce. All of the engines proposed have been off-the-shelf corporate jet engines. Roughly the same size as the TF33s they are meant to replace, they are more powerful and much more fuel efficient. The proposals are to keep the eight-engine layout, although the nacelles themselves would be completely replaced. The pylons would likely be largely or completely replaced as well. The configurations of the new pylons and nacelles have not, at the time of writing, been clearly shown.

General Electric has proposed two of its engines: the CF34-10 and the Passport. The former has a maximum thrust of 20,360lb to the latter's 18,900lb; the former has a bypass ratio of 5.4:1 to the latter's 5.8:1. The former has a fan diameter and length of 53 and 88.7in; the latter, 52 and 102.7in. The Passport first ran in 2013; the CF34 all the way back in 1982.

The Pratt & Whitney proposal is for PW800 engines. First run in 2012, maximum thrust is 16,000lb and bypass ratio is 5.5:1. Fan diameter is 50in and length is 130.4in.

Rolls-Royce has put forward the F130, based on the BR725 model. First run in 1995, maximum thrust is 16,900lb with a bypass ratio of 4.1:1. Fan diameter is 50in and length is 129.8in.

If the US Air Force elects to actually go through with a re-engining programme for the B-52H, chances are good that it will take ten years to accomplish and result in a designation change; both 'B-51I' and 'B-52J' have been put forward. The re-engining would no doubt be accomplished when the aircraft are already undergoing a complete overhaul of structure and avionics, resulting in aircraft that are as close to brand-new as possible.

It is also possible that some B-52H airframes currently mothballed at Davis-Monthan will be resurrected and given new life. Were this all to occur, it is very unlikely that the updated aircraft would have projected lifespans of less than a decade or two – putting their retirement well into the 2050s if not later. It is not inconceivable that some of these B-52s will be in service for more than a century. The last B-52 crews may include the great-great-grandchildren of those who first flew the aircraft.

CHAPTER 10

B-47 and B-52 Derived Transports

There had been a large number of cases where bombers were turned into transport planes during the Second World War – and where transport planes became bombers. The pre-war Germans were masters of this; famed wartime bombers such as the Heinkel He 111 and Dornier Do 17 started off as transports. They did not make very good airliners – but they were the best airliners that could be hammered out of aircraft intended from the beginning to violate the terms of the Treaty of Versailles.

Boeing was quite familiar with turning bombers into transports. Several B-17s were modified into various transports including the C-108 and CB-17; several B-17s impounded in Sweden during the war were similarly converted. However, these were modest modifications to the existing airframe, poorly optimized for passenger or cargo transport. More important was the Model 307 'Stratoliner', an airliner made by adding an entirely new and much more voluminous fuselage to the wings and tail of the B-17.

The fuselage was of circular cross section and was the first airliner to be pressurized for flight above most of the weather, providing a much smoother and more comfortable flight for the passengers. The advent of the Second World War curtailed mass production and only ten examples were built.

After the war, the B-29 was looked at for the transport role and eventually emerged as Model 367-1-1, better known as the C-97 'Stratofreighter'. As with Model 307, Model 367 used the wings, engines and tail of the B-29; unlike the Stratoliner, the fuselage of the B-29 was to some degree retained. A larger diameter cylindrical fuselage was built atop the existing fuselage, creating a 'double bubble' configuration. The C-97 was much more successful than the Stratoliner – nearly 1,000 of the military C-97 freighter (and subsequent KC-97 tanker) and more than 50 of the Model 377 'Stratocruiser' dedicated commercial passenger liner were built.

In the immediate post-war years, the Stratoliner (first flight in 1947) seemed quite advanced. But Boeing knew jet propulsion would eventually come to passenger transports. An obvious starting point for the design of the next generation of jet airliner would be to start with the jet bomber then under development: the B-47. It is perhaps interesting that while Boeing created a number of designs for jetliners based in part on the B-47 and B-52, it never seemed to design a simple conversion of the bombers into jetliners but took only general design ideas instead. And in the end, no cargo or passenger transport was ever created from either the B-47 or the B-52.

Boeing's initial efforts to develop a turbojet-powered airliner fell under the Model 473 designation. Initially Model 473 designs took inspiration from the contemporary B-47, but eventually switched over to the B-52. Most of the designs seem to have been studied only briefly. The early Model 473 designs were generally simpler and less elegant than the B-47 but the general configuration was retained. Even though designed all the way back in June of 1947, the basic layout for virtually all successful jetliners was already established. Apart from having a high shoulder-mounted wing, as opposed to the more usual low-mounted wing, the earliest Model 473 designs would fit on modern airfields.

The Model 473 design series did not directly result in an actual aircraft being constructed. Boeing's first jetliner, the 707, was derived from the Model 367-80 (better known as the Dash 80) prototype. The Model 367 series was a separate design programme from the Model 473 series, but for a good long while the two ran parallel to each other. In the end, the Model 473 studies diverged from the B-47 and B-52 and became wholly their own designs, trending towards and converging with the Model 367.

Boeing Model 473-10

The first Model 473 configurations known from diagrams are referred to simply as 'Model 473', though the data seems to indicate that this is perhaps the Model 473-10. The Model 473-10 was, based on data, similar in configuration but with smaller wings, and some references indicate that it dates from sometime in 1946. There were a few features that

Boeing Model 473
SCALE 1/200

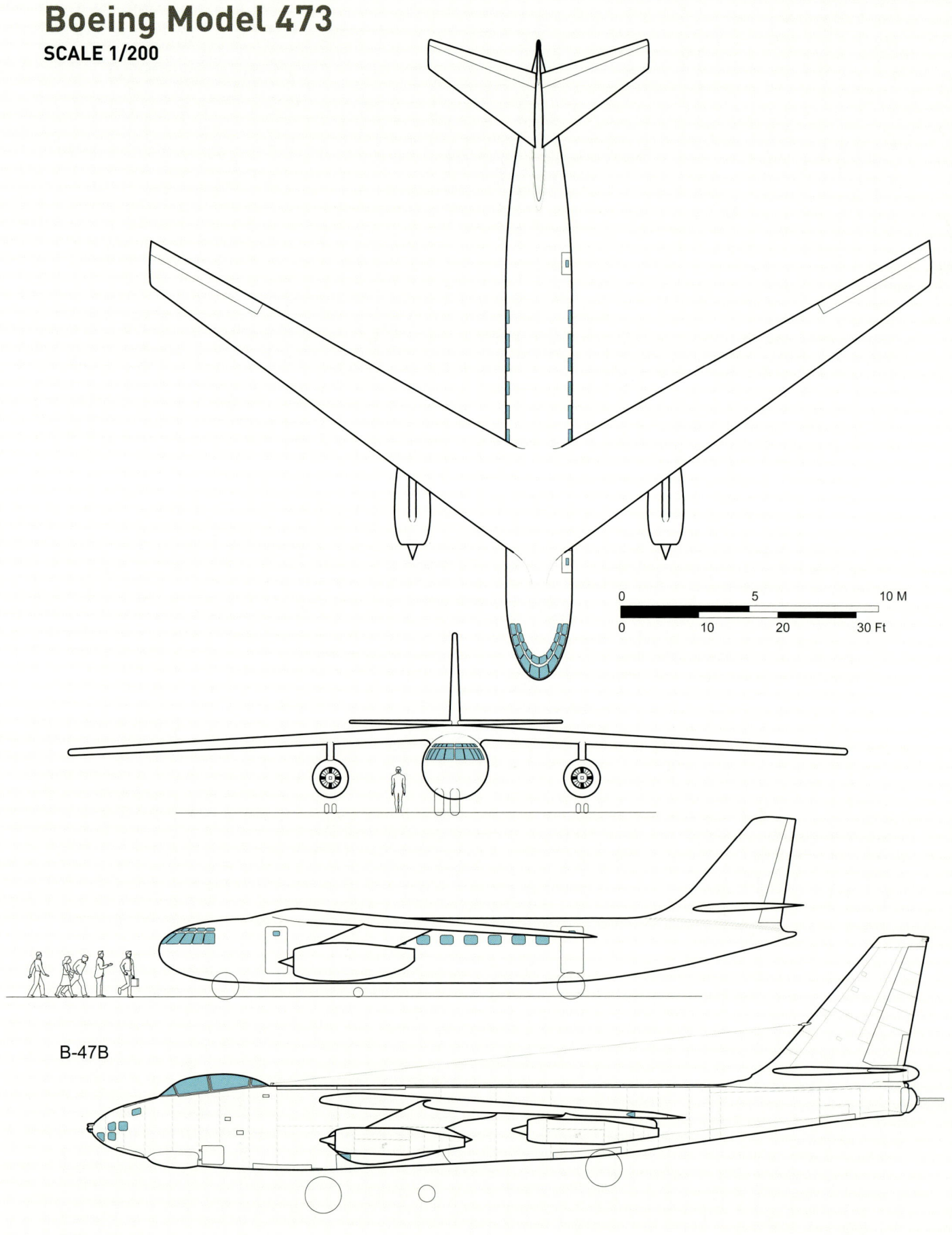

B-47B

Boeing Model 473
SCALE 1/144

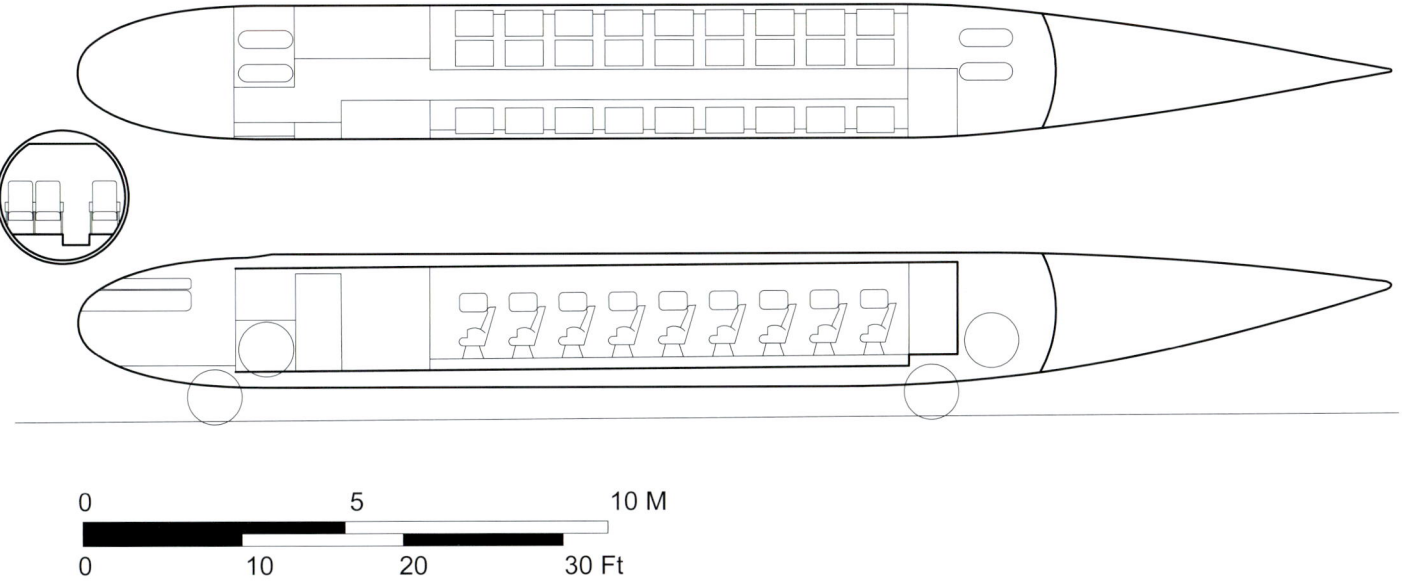

would set the Model 473-10 of June 1947 apart from modern jetliners visually. The basic layout common to airliners that would be designed for the next three quarters of a century – and likely much longer – was already established. The configuration was clearly not simply a modified B-47, but was noticeably smaller.

The fuselage was a circular cross-section cylinder with a fairly blunt nose and a gradually tapering tail. The two-seater cockpit canopy was quite expansive and flush with the blunt forward fuselage contour. The canopy itself was composed of a large number of individually framed windows providing a wide field of view. The landing gear used the B-47's 'bicycle' configuration, with two sizable twin-wheel bogeys, one fore, one aft, with small outrigger gear below the engine pods.

The main landing gear was stowed entirely within the cylindrical outer lines of the relatively narrow (100in diameter, though the data calls for 122in) fuselage, meaning that the landing gear bays intruded into the confines of the pressurized passenger compartment. Somewhat oddly the main gear was offset to starboard in the fuselage; this was done to provide space for crew/passenger passage in the otherwise rather cramped fuselage.

The Rolls-Royce Nene turbojets, each capable of producing 6,000lb of thrust, were individually suspended from the wings by pylons; small outrigger gear would deploy from the undersides of the nacelles. The wings were mounted high, the spars passing above the fuselage. While an unusual feature for jetliners today, shoulder mounted wings are not unknown, as the BAe 146 can attest. Wings are typically mounted low so that the landing gear can be stowed partially or completely within the wings without intruding into the fuselage, as was necessary on Model 473-10.

Seating consisted of nine rows in a 2+1 layout, though data shows 28 total passengers. Crew of three included one steward. The description called for 'high speed' and 'long range', but performance details are currently lacking.

Boeing Model 473-13

Model 473-13 (data from July 1948, but the design is possibly from some time earlier) was a four-engine configuration clearly derived from the 473-10. The layout was closer to that of the B-47 with two twin-engine B-47-style nacelles and a larger fuselage. The four engines were located in two dual-engine nacelles much like those of the inboard nacelles of the B-47.

This design could carry 36 passengers in nine rows of four with a central aisle. 473-13's wings were set somewhat lower on the fuselage than those of 473-10, while the cockpit canopy was raised and more conventional for the time, reasonably closely resembling the raised cockpit of Model 464-17.

As with 473-10, the landing gear was of bicycle arrangement (though located along the centreline, not shifted to the side) with small outriggers below the engine pods. Model 473-13 was, apart from the twin-jet nacelles, closer still to the modern conception of the jetliner.

Boeing Model 473-13
SCALE 1/200

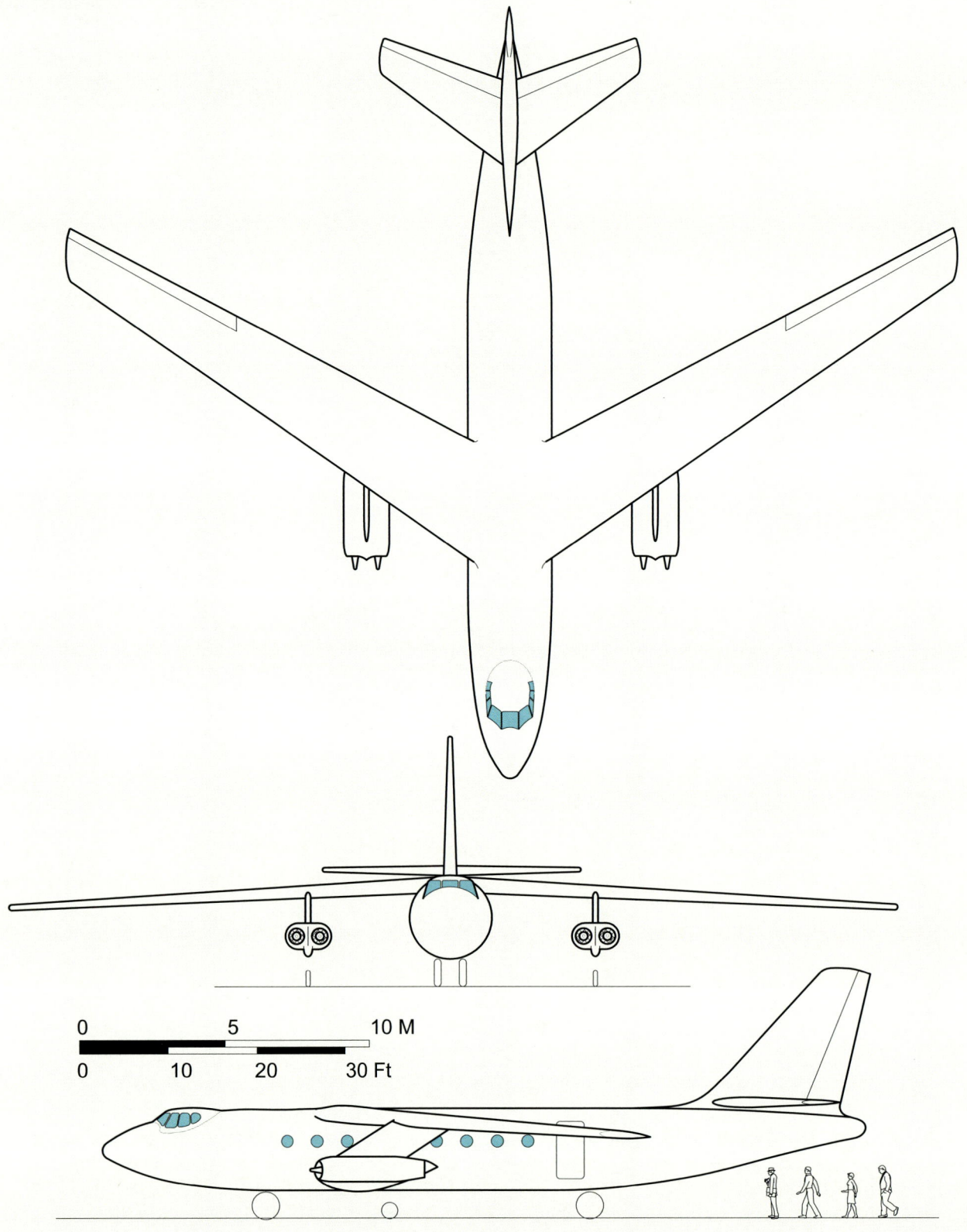

Boeing Model 473-14

Model 473-14 was quite similar to 473-13 in general configuration but with important changes. The wing was given less sweepback; sizable external fuel tanks were attached to the underside of the outer wings. The fuselage was slightly curvier and more rotund, yet had room for four additional passengers. The horizontal stabilizer was moved slightly upwards onto the vertical stabilizer and was reduced slightly in size, made possible by locating it slightly further aft. Passenger seating was in ten two-by-two rows.

Model 473-14 retained the landing gear arrangement of the prior models, but featured a somewhat more conventional nose gear that folded up under the

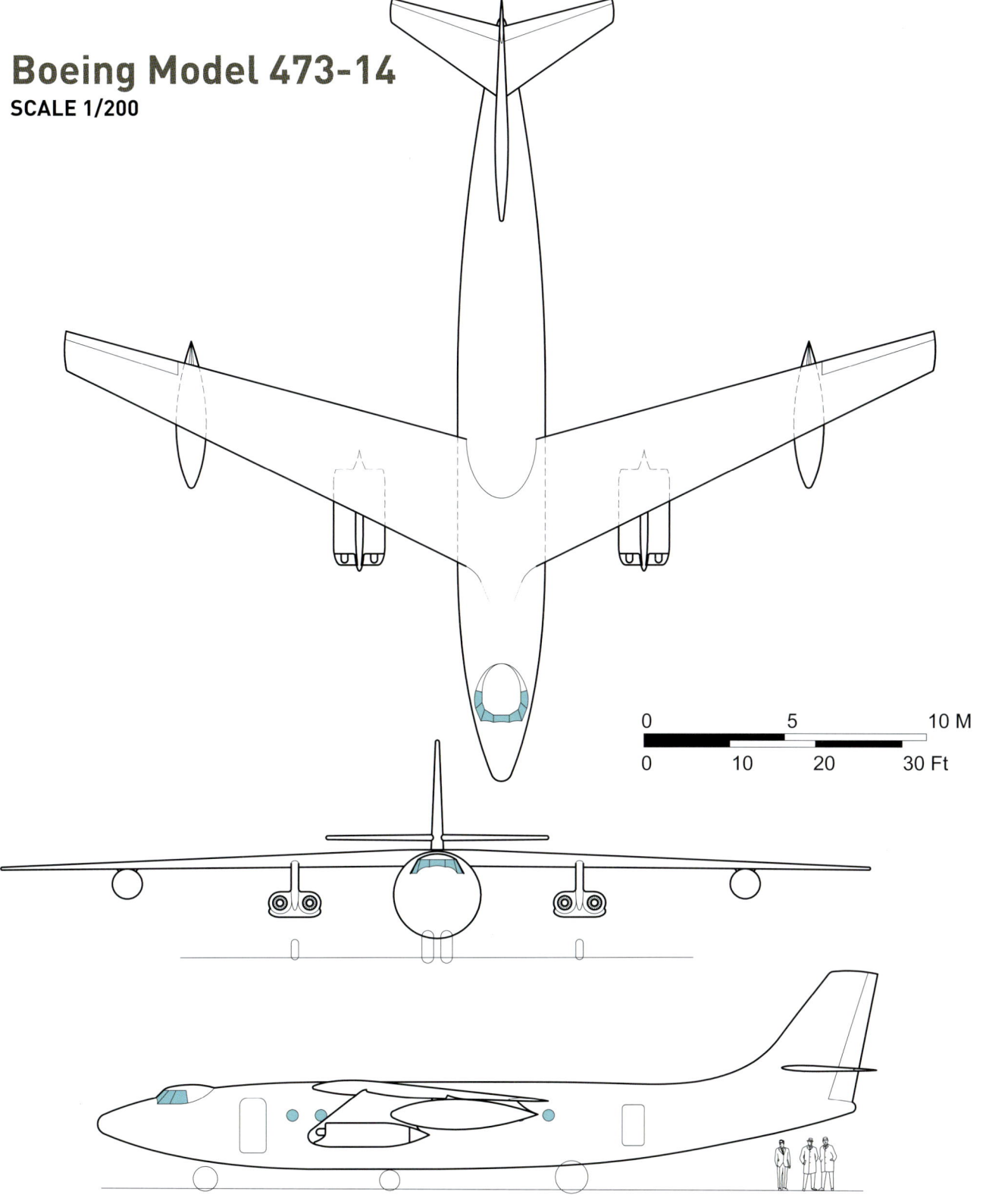

Boeing Model 473-14
SCALE 1/200

cockpit. The main aft gear retracted into a well within the rear of the pressurized cabin.

Boeing Model 473-14N

Model 473-14N was a minor modification of -14 into a cargo transport. This was done by the simple expedient of leaving out such passenger comforts as seats and windows and installing large cargo loading doors at fore and aft on the port side of the fuselage.

Boeing Model 473-23

Model 473-23 was the same basic configuration but scaled up slightly, with the same basic passenger load. This was to be a nonstop trans-Atlantic airliner, certainly an ambitious goal for a jet-powered aircraft designed in April 1949. In order to fly its 36 passengers more than 5,000 miles, Model 473-23 had fairly gigantic external fuel tank pods suspended beneath the inner wings. The dual-engine pods – now

Boeing Model 473-14
SCALE 1/144

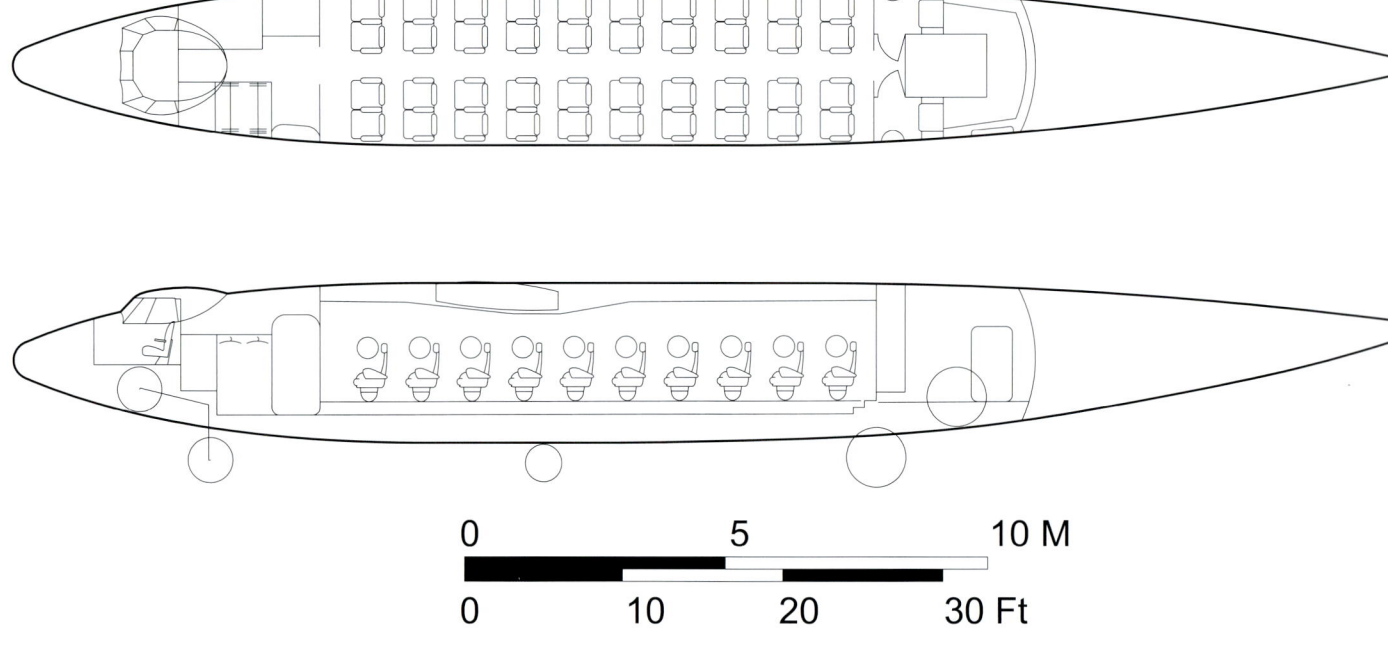

Boeing Model 473-14N
SCALE 1/144

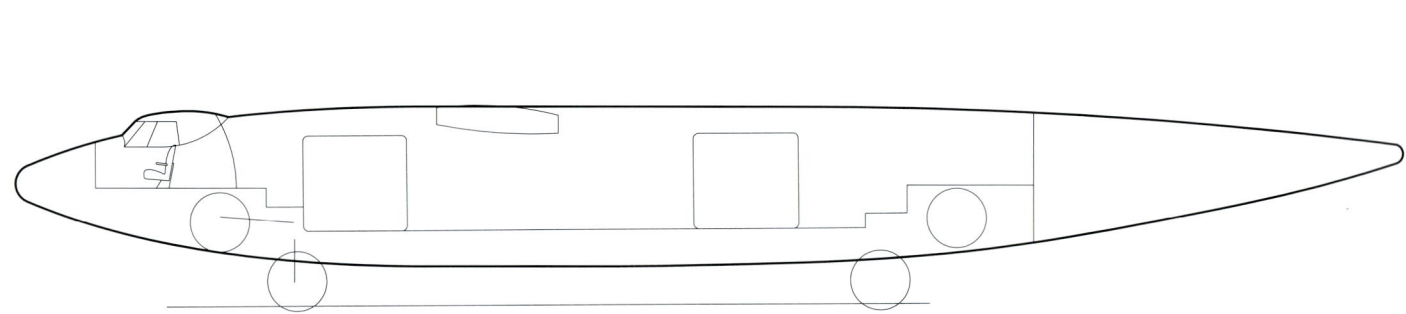

Boeing Model 473-14N
SCALE 1/200

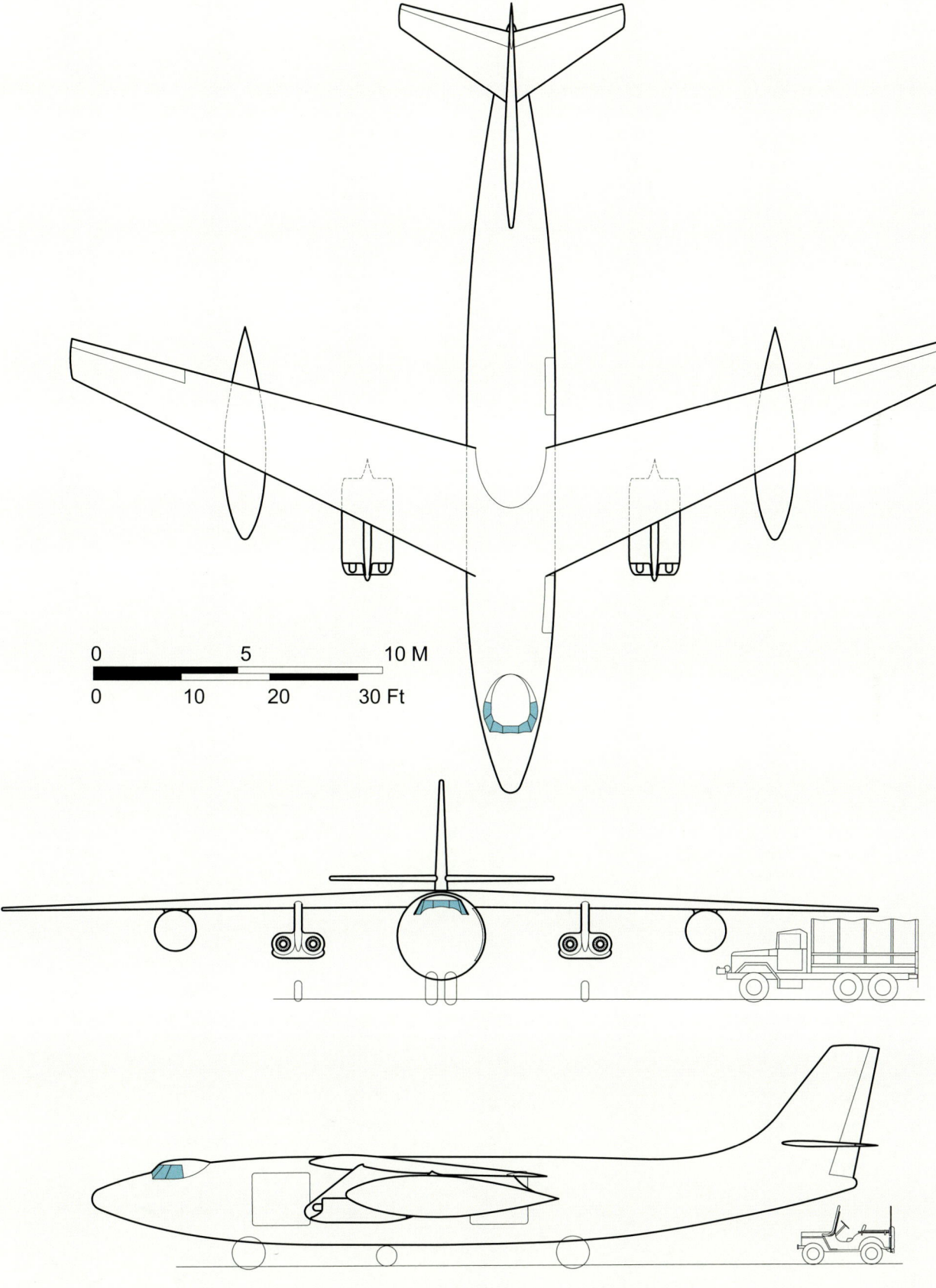

fitted with the same J57s that powered the B-47 – were moved well outboard, an arrangement certain to cause thrust imbalance headaches in the event of an engine out.

Boeing Model 473-24

This was the first of the series to take its cues from the B-52 rather than the B-47. Also dating from April 1949, the design was clearly based on the

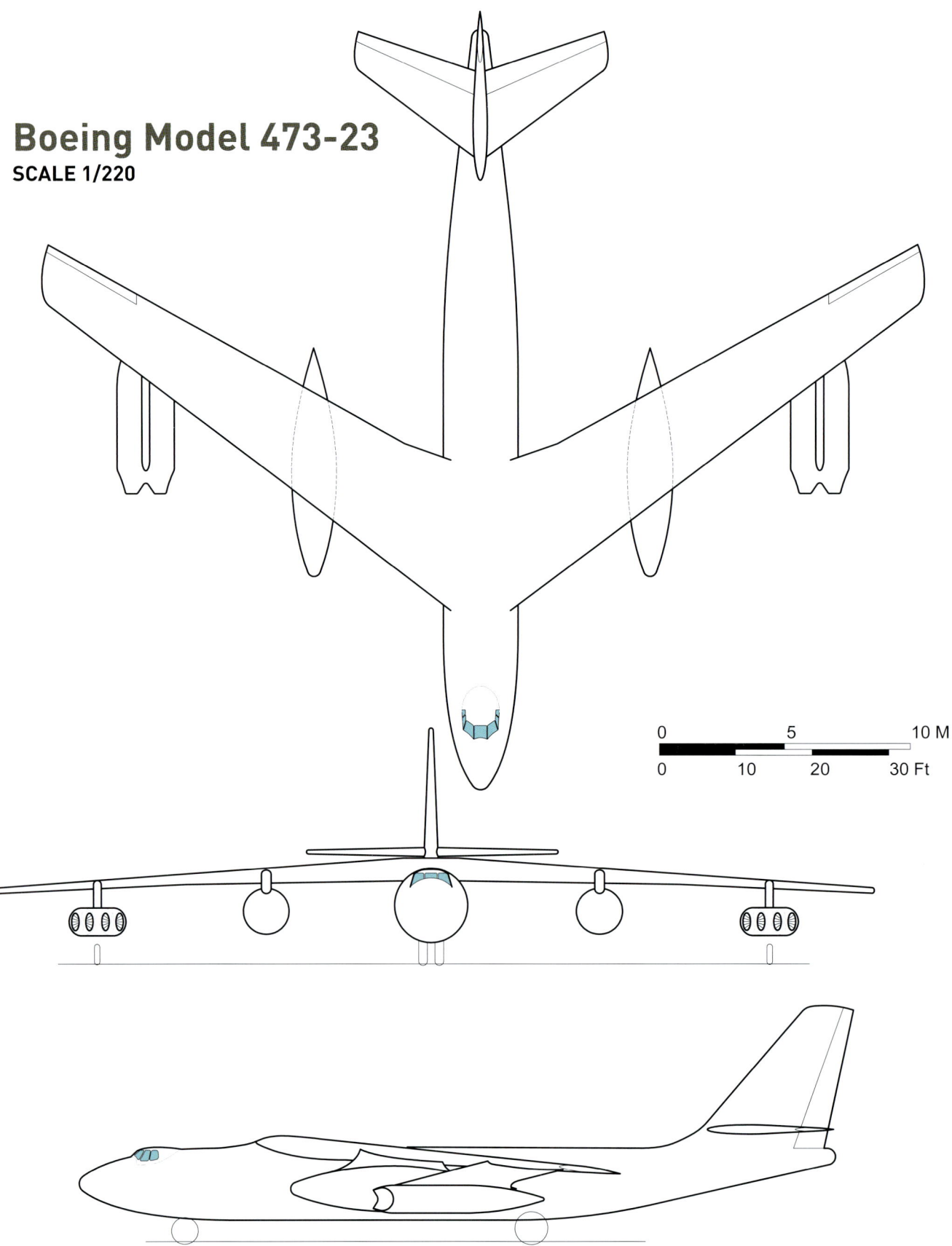

Boeing Model 473-23
SCALE 1/220

Boeing Model 473-24
SCALE 1/300

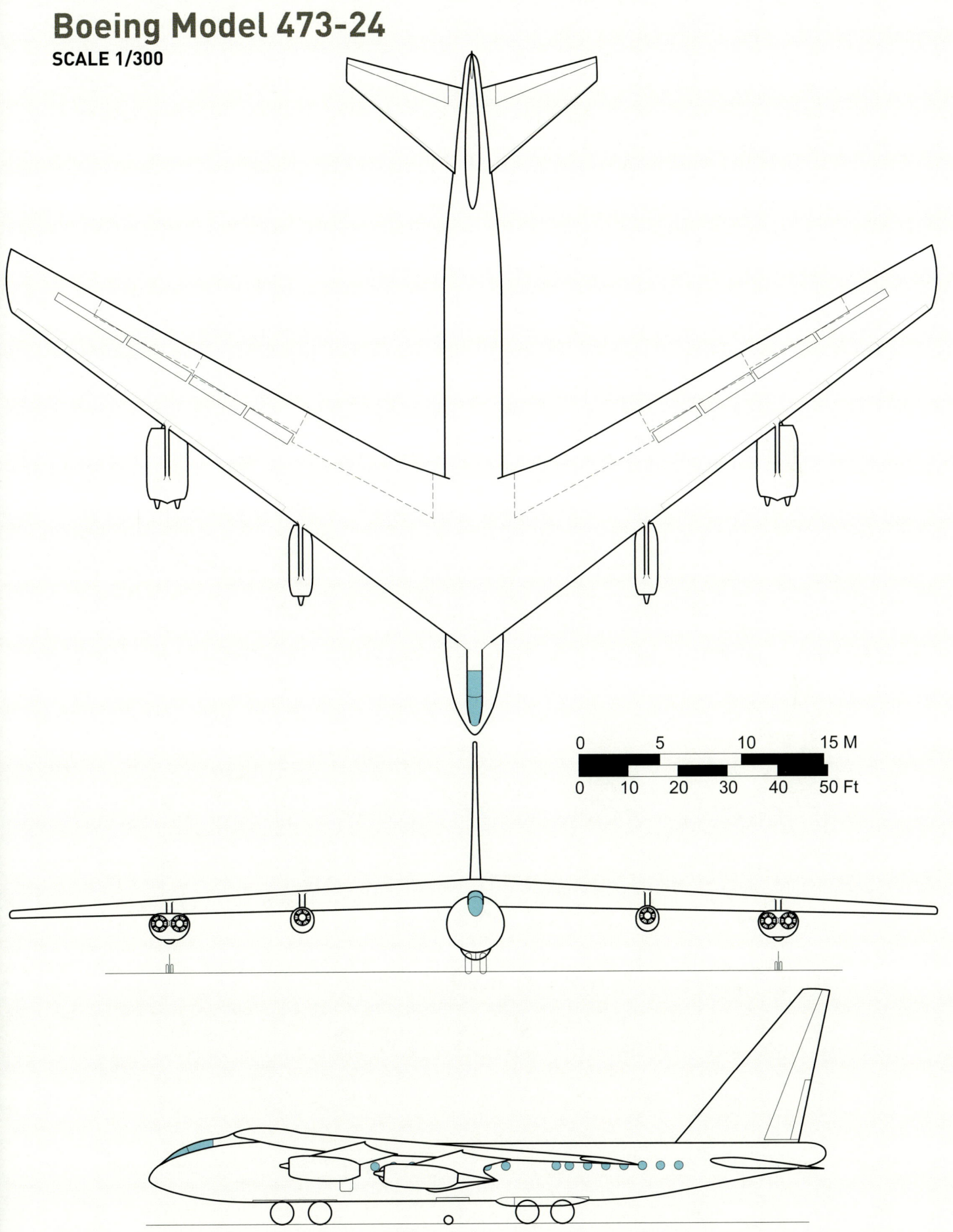

Model 464-49 B-52 configuration, using the same wings, tails and engine nacelles. The fuselage was new; instead of the roughly rectangular cross-section fuselage used on the bomber, the jetliner employed a cylindrical pressurized fuselage of 162in diameter. As with Model 473-23, 473-24 was meant to be a trans-Atlantic airliner, in this case with a passenger capacity of 78.

Model 473-24 had six engines rather than the bomber's eight. Curiously, the engine arrangement was the opposite of the B-47… each inboard pylon held a single engine while the outboard pylons held dual-engine nacelles. Compared to the bomber, the wings of the jetliner were raised above the fuselage to prevent the spars from having to pass through the pressurized compartment. The trailing edges of the wings were granted substantial flaps for lift at low speed. No ailerons were provided; control was by means of a long series of spoilers atop the wings.

The landing gear was similar to that proposed for the early B-52, but with some accommodations for the nature of the fuselage. The forward bogeys – offset slightly to starboard – retracted up into a wheel well beneath the wing and ahead of the passenger compartment. The rear bogeys were granted a slight bulge beneath the fuselage to provide a bit of added space beneath the passenger compartment. Seating was 13 rows of three by two, two rows of two by two and one final row of five seats side-by-side; cargo and baggage were loaded in a compartment below the passengers and ahead of the rear landing gear.

Boeing Model 473-25

The next design was similar, but intended for even greater passenger capacity. Model 473-25 used a longer but narrower (142in diameter) fuselage; similar to the Boeing Stratoliner. Volume in the fuselage was increased by using a 'double bubble' configuration. This resulted in a deeper fuselage that the landing gear fit within comfortably. Together with the fuselage extension forward of the wing, a new passenger compartment was squeezed in between the cockpit and the wings. The wings intruded within the top of the pressure cylinder, so the floor between the forward and aft passenger cabins was lowered by several steps to provide headroom.

This design had 16 rows of three by two seating, one lone pair of seats, two rows of two by two and an aft row of five seats. A cargo door, along with a pair of lavatories and the stewardess station, was located under the wings; a galley was further back in the aft cabin. Model 473-25 retained Model 473-24's 'reverse B-47' engine arrangement, with outrigger stabilizing gear deployed from the outboard nacelles. Further fuel was provided in the form of streamlined tanks under the outer wings.

Boeing Model 473-27/-28

The Boeing Model 473-27 and -28 of May 1949 returned to the B-47 scale of jetliner. The two were more or less the same aircraft, the only difference being that the -27 was a troop/cargo transport while the -28 was a jetliner. The overall configuration was much like that of Model 473-23, including the far-outboard dual-engine nacelles and the large inboard fuel tanks.

One noticeable advancement was incorporation of the thicker wing inboard section; as described earlier, this was one of the secrets of the B-52's success. This would also come in handy for jetliners; when the wings moved down to the bottom of the fuselage, rather than the top (as was still the case with Model 472-27/-28), the thicker wing inboard sections gave volume for the landing gear to partially or entirely fold into, so that the pressurized passenger cabin did not have to accommodate wheel wells.

Model 473-27 had seating for 52 troops, most in seats with their backs to the 120in diameter fuselage walls. This left a substantial free space down the centre. It's not clear if Model 473-27 was meant simply to fly troops form one airfield to another or if it was to open its door in flight and expel paratroopers, but it was equipped with two doors on the port side: a standard sized one at the rear of the passenger compartment and a larger one just aft of the cockpit. That door would be used for the loading of cargo as well as troops. Presumably the troop seats would fold up against the wall to maximize cargo volume.

The Model 473-28 airliner had rather more comfortable seating for 36 passengers. Model 473-27 had an additional 5,000lb of fuel capacity and some 15,000lb greater gross weight than the passenger liner.

According to concept art, both versions had the same rather unusual canopy arrangement over the cockpit. The available diagrams depict, rather simply, a canopy similar to earlier Model 473 designs. The odder, more interesting, canopy is reproduced here.

Boeing Model 473-29A

This June 1949 switched back to the B-52 again. As with Model 473-25, a double-bubble fuselage was chosen (150in diameter for the upper lobe, 122in for the lower). Passenger capacity was raised to 102, with the passenger cabin being on several slightly different levels in order to fit within the confines of the fuselage, given its tapering uplifted tail and wing spars passing through the forward section.

Model 473-29A was equipped with six J57 engines. However, a cargo variant of the same basic design

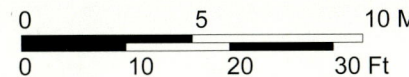

Boeing Model 473-24
SCALE 1/220

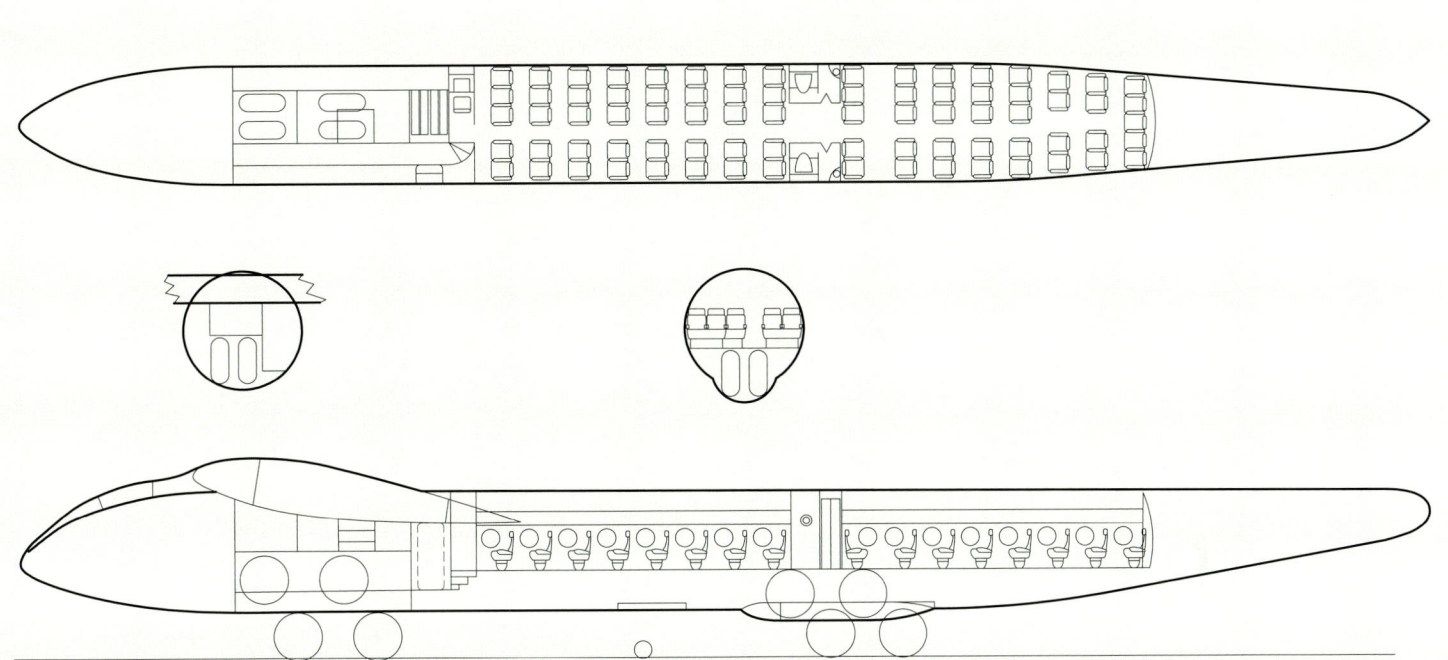

Boeing Model 473-24 compared to Model 464-49
SCALE 1/400

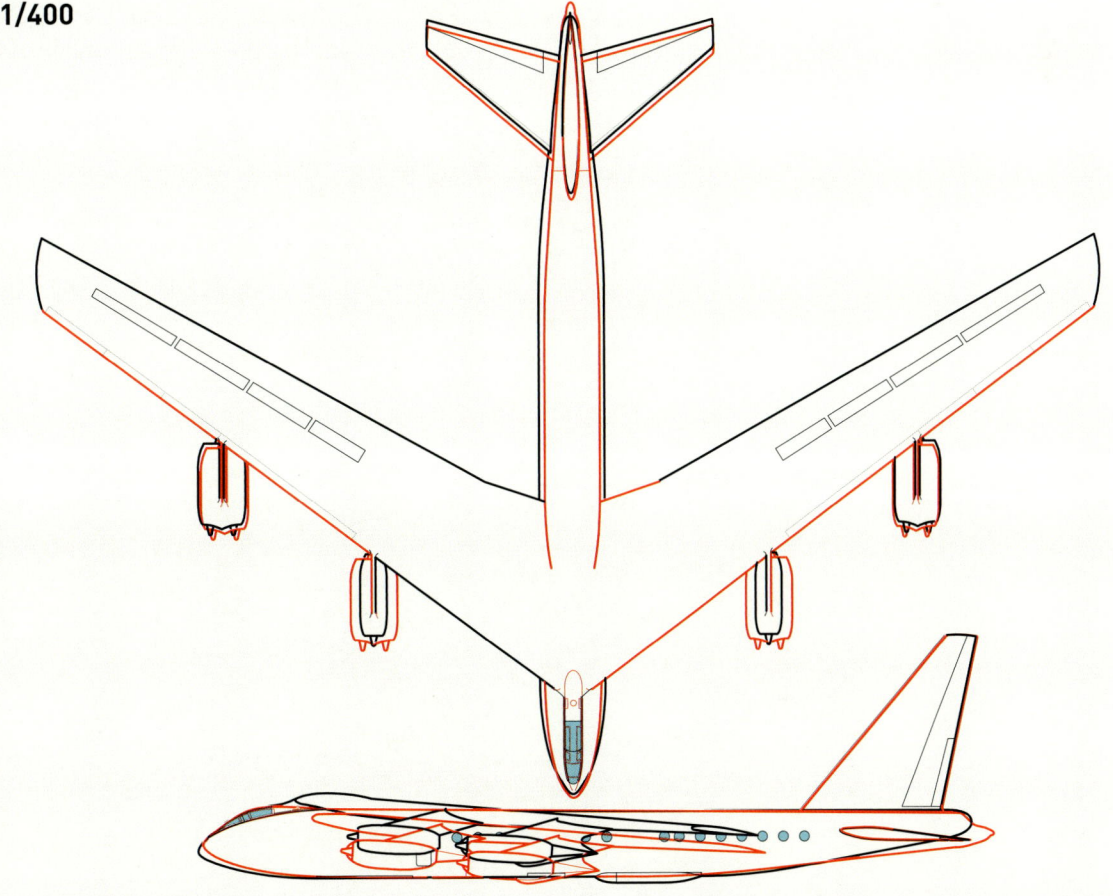

Boeing Model 473-25
SCALE 1/300

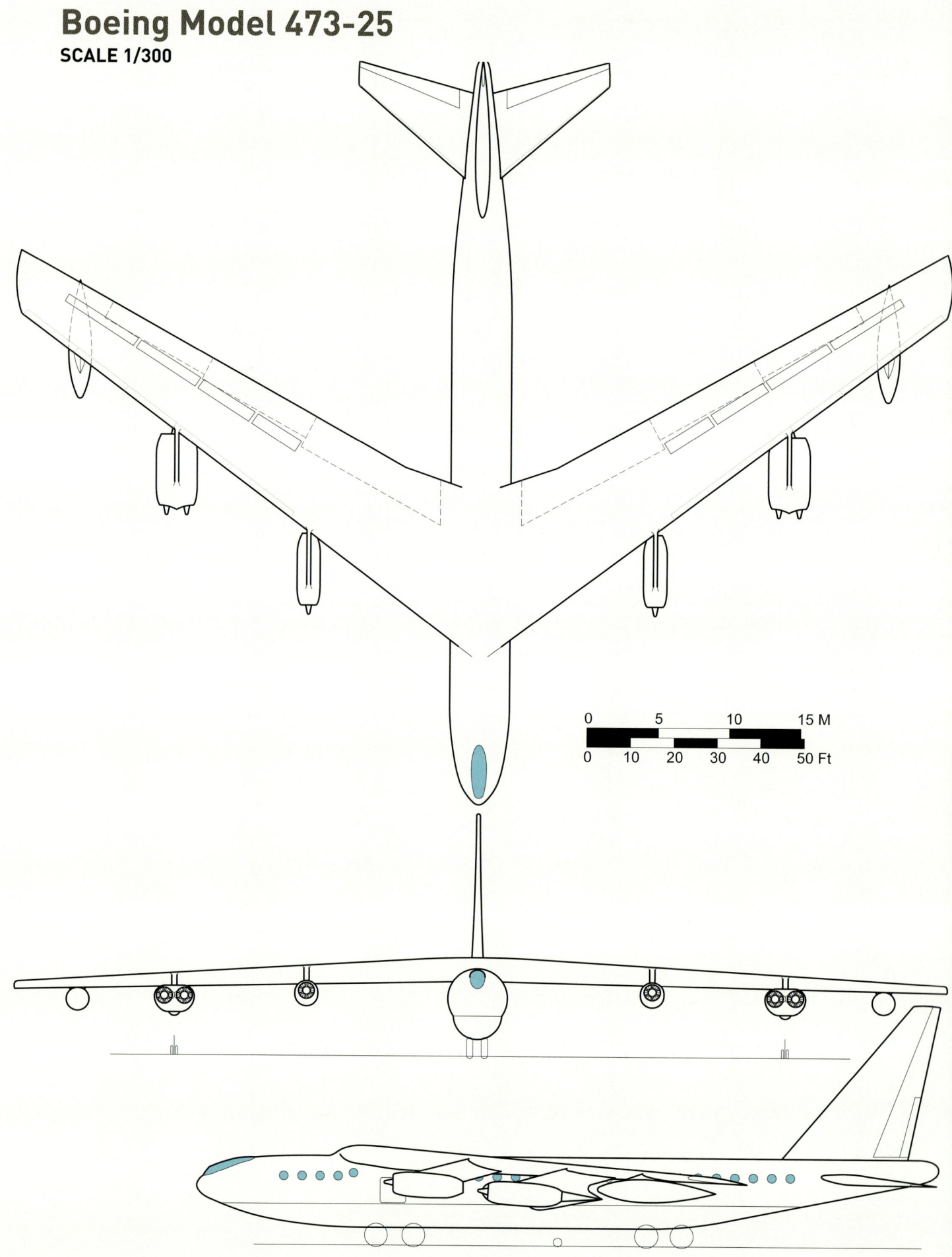

Boeing Model 473-25
SCALE 1/250

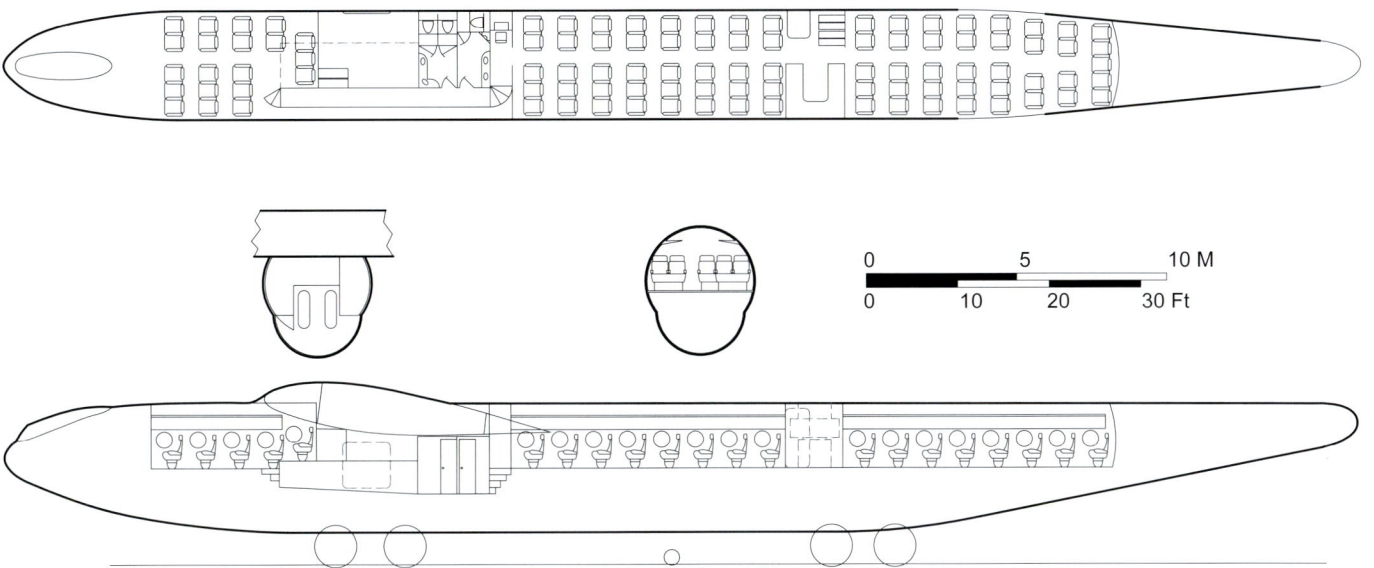

was also looked at, equipped instead with four J47 engines. That version could carry 52,940lb of cargo at the same speed and altitude as the passenger liner, for approximately the same range. Artwork depicts the same unusual canopy arrangement as for the Model 473-27/-28.

Boeing Model 473-29B

An outright cargo version was produced, Model 473-29B, by increasing the diameter of the upper lobe to 190in and the lower to 142in. Due to the increased drag and weight of the fuselage, it did not have quite the same cargo weight capacity as the Model 473-29A cargo variant, but it did have greater volume. It was also to be fitted with a rear cargo loading ramp – a standard feature on most large cargo jet aircraft. It would have been a substantially larger aircraft than the Lockheed C-130, though smaller than the proposed Douglas C-132 which would come along a few years later.

Boeing Model 473-30

And back to the B-47 for the Model 473-30 from June 1949. This configuration was similar to that of Model 473-28, though stretched in all dimensions and stripped of the external fuel tank pods. With a fuselage 11.8ft in diameter, it could carry 70 passengers a considerably greater distance, though perhaps not quite trans-Atlantic.

While the configuration was largely in line with Model 473s that had gone before, it had something of a blunt angularity to it. Here at last was a B-47-sized airliner where the top surface of the wing was actually below the top surface of the fuselage. Along with wingtip ailerons, the wings had an extensive series of spoilers. With the wings anhedral and the far outboard position of the nacelles, the deployable outrigger gear were quite short.

Boeing Model 473-31

Models 473-31, -32 and -33 of July 1949, were the last of the series that bore a distinct resemblance to the B-47, and the resemblance was fading fast. The requirements of passenger transport were drawing the design series further and further away from the B-47; the B-52 does not seem to have appeared again in the design series.

Model 473-31 was smaller than the preceding design, and correspondingly had both a smaller passenger load, 61, and a much shorter range. The fuselage diameter was the same but it was 11ft shorter. The configuration was much the same as that of Model 473-30, though the cockpit canopy – depicted only in the side view unfortunately – appears to have been different. The -32 and -33 had differences in engines and dimensions, but were visually distinct, largely in that -32 had a small external fuel tank under the wing and -33 had a large one.

Boeing Model 473-27/28
SCALE 1/200

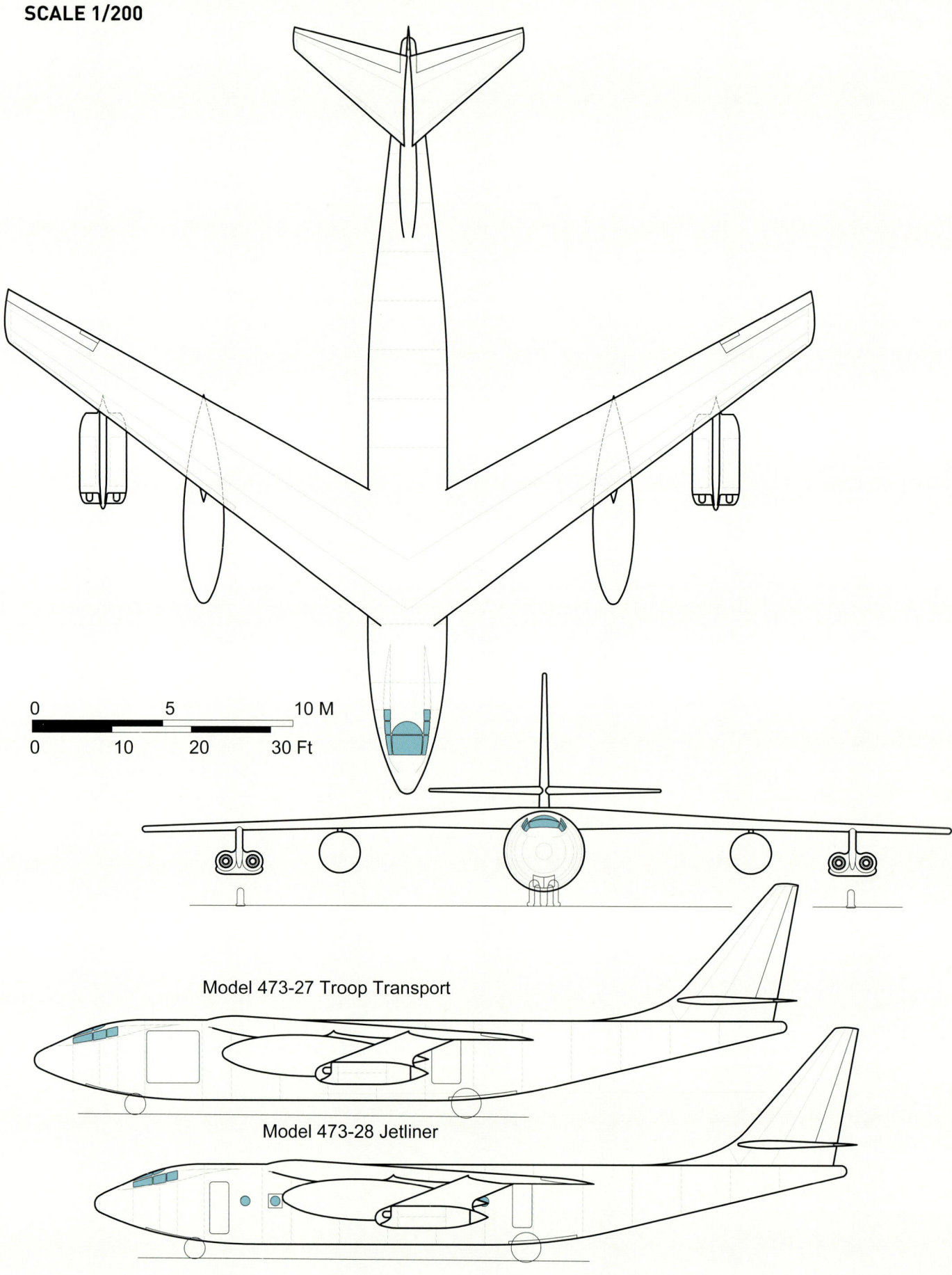

Model 473-27 Troop Transport

Model 473-28 Jetliner

Boeing Model 473-27/28

SCALE 1/160

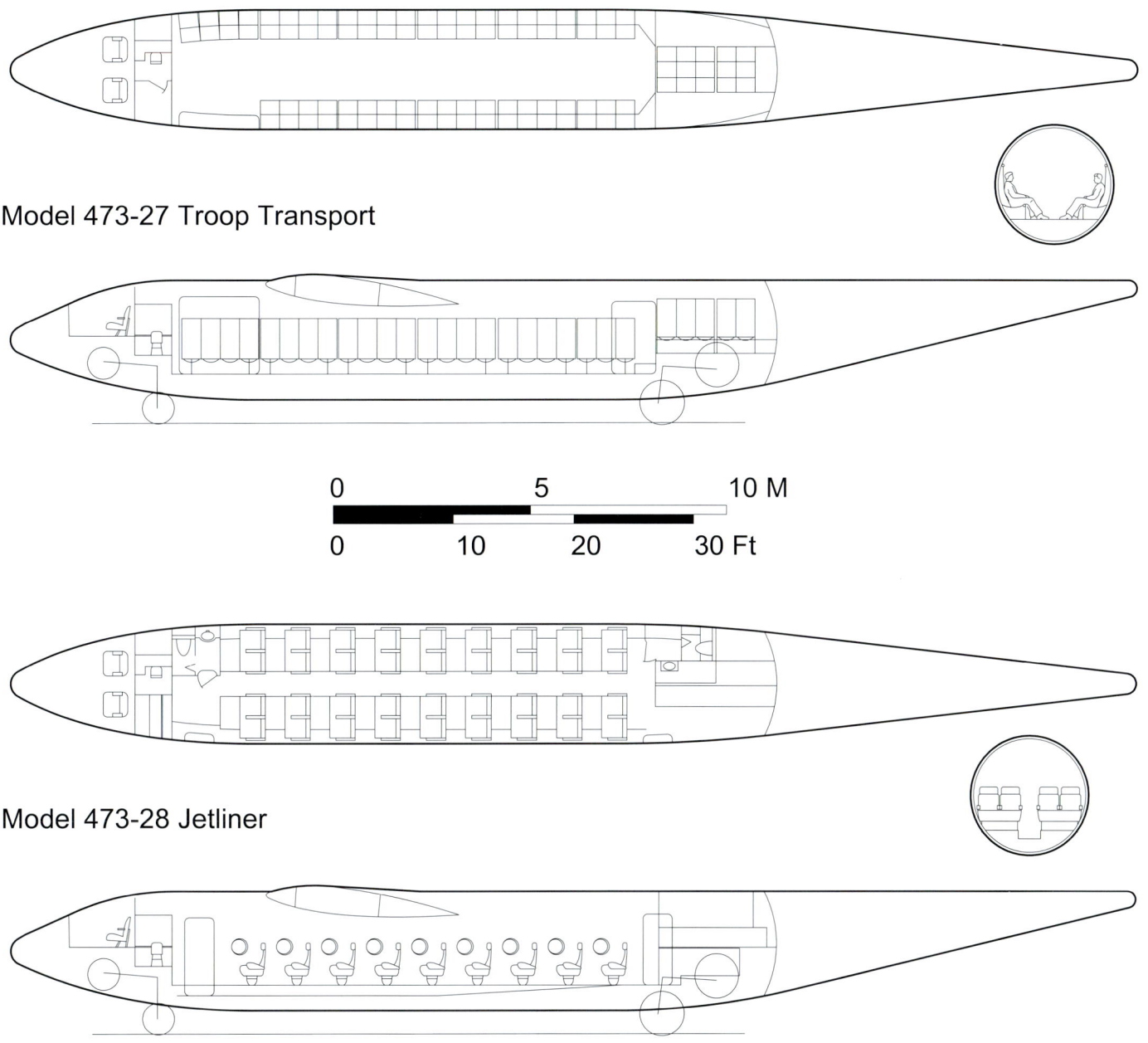

Model 473-27 Troop Transport

Model 473-28 Jetliner

The Model 473 design series would go on from here, continuing to at least Model 473-65 from December 1950. The designs would cover considerable ground, including turboprops as well as jets, but were no longer beholden to the B-47 or B-52 for their design features or inspirations.

Boeing Model 820

Boeing proposed a long range, high-speed air logistics system able to transport large and heavy military payloads to the Air Force in 1958. Included among these were battlefield missiles and ICBMs, previously unheard of as payloads. These are often finicky items to transport, large and low density (in particular liquid propellant ICBMs such as Atlas), often needing to be shipped while attached to a ground transporter.

At the same time, high density payloads such as armoured vehicles and construction equipment also need to be transported, so strategic airlifters need to be versatile and capacious... and fully capable of transporting large numbers of troops and other passengers.

The goal was evidently to transport everything needed to turn a barren patch of some distant land into a fully prepared launch site. Range and payload needed to be 4,500 nautical miles and 100,000lb.

While a variety of wildly different vehicles (including VTOL and supersonic configurations) were proposed under the Model 820 umbrella designation, the baseline design was a straightforward modification of the B-52. This was just a year after the cancellation of the Douglas C-132, a proposed large turboprop-powered cargo plane. The loss of the C-132 had left the USAF to struggle on with the C-124, C-130 and C-133 – none of which were particularly fast. The role of a fast strategic airlifter would eventually be filled by the Lockheed C-5 Galaxy, but the beginnings of that programme were still years away.

Boeing Model 820-100

Boeing proposed that the Model 820-100 could be made available relatively quickly, with initial operational capability by the mid 1960s. For the Model 820-100, the wings and engines of a B-52G and the tail surfaces of the B-52F would be mated to an all-new much larger fuselage with cargo loading doors both fore and aft. The new fuselage would be nearly the size of that of the later Lockheed C-5. Total volume within the pressurized cargo bay was 29,060cu ft, with 1,937.5sq ft of floor space. The forward pressure bulkhead would be behind the doors, as the doors were too large and complex to seal. An additional 5,250cu ft of unpressurized cargo space would be available within the closed nose doors.

The width of the fuselage would require the addition of 15ft to the wing centre section. The landing gear wheels and bogeys would be taken from the KC-135, giving the aircraft a true four-poster landing gear arrangement. The cockpit canopy bore only a meager resemblance to that of the B-52; the flight deck may have been taken from the B-52, but is not described in the available literature.

The cargo bay doors at the rear were fairly straightforward: two clamshells that hinged along the top edges. This arrangement would seem to preclude opening in flight. The nose doors, however, were much more complex – five separate doors, including one on the underside that would serve as a loading ramp, two at the nose that were hinged along the top and two more further along the fuselage that hinged at the rear.

Payloads considered included ICBMs, smaller ballistic missiles, surface-to-air missiles such as the BOMARC, troops and miscellaneous equipment. Data is vague, but it could carry 240 troops and a load of cargo; the total number of troops it could carry if dedicated to that task is not available. One potential mission described was the transportation of 7,000 tons of military equipment and personnel from a continental US base to Indo-China in 46 hours. This required a fleet size of 195 aircraft, each refuelling in Hawaii and the Marianas.

Boeing Model 820-101

The first alternate design replaced the B-52G wings and engines with a straight wing and turboprop engines. The four T57 turboprops would each have a four-bladed, 20ft diameter propeller and were mounted above the wings to provide adequate ground clearance for the props. And while the wing was much sturdier looking than the B-52 wing, much higher off the ground, and the four-poster landing gear had a much wider stance than the B-52, the wings were still to be granted small deployable outrigger gear to prevent the wingtips from striking the ground.

The wings were also to be reinforced and fitted with attachments to allow them to carry external stores. These stores are not described in the available literature, but might have included cargo pods and/or fuel tanks.

The fuselage was different in shape to that of the Model 820-100, with the cockpit – which bore no resemblance to that of the B-52 whatsoever – projecting well forward. All that remained of the B-52 were the tail surfaces. The landing gear no longer folded up into the wide underside of the fuselage, but into sponsons running much of the length of the fuselage.

The turboprops gave the aircraft a range of 4,580 nautical miles at a cruise speed of 403 knots.

Boeing Model 820-102

Model 820-102 returned to the configuration of the -100. The diagram indicates a change in engines to turbofans, though there is no text to define them. The design seems otherwise little changed from Model 820-100 except for a considerable redesign for the nose cargo loading door. Instead of a complex multi-piece clamshell, 820-102 had a simpler two-piece design, with left and right halves that hinged up to the sides.

Boeing Model 820-103

Model 820-103 used B-52G wings and surfaces married to a fuselage much like that of Model 820-101. Rather than the wide-bottomed fuselage used on the -101 and -102, the -103 fuselage was more circular in cross-section with external sponsons. The wings were reinforced for external stores in the same positions as the Hound Dog pylons. Range was 4,660 nautical miles at 454 knots.

Models 820-103A and -103B differed in having slightly reshaped sponsons.

Boeing Model 820-104

With Model 820-104, Boeing began to examine less conventional configurations. Similar to -101, -104 had

Boeing Model 473-29
SCALE 1/320

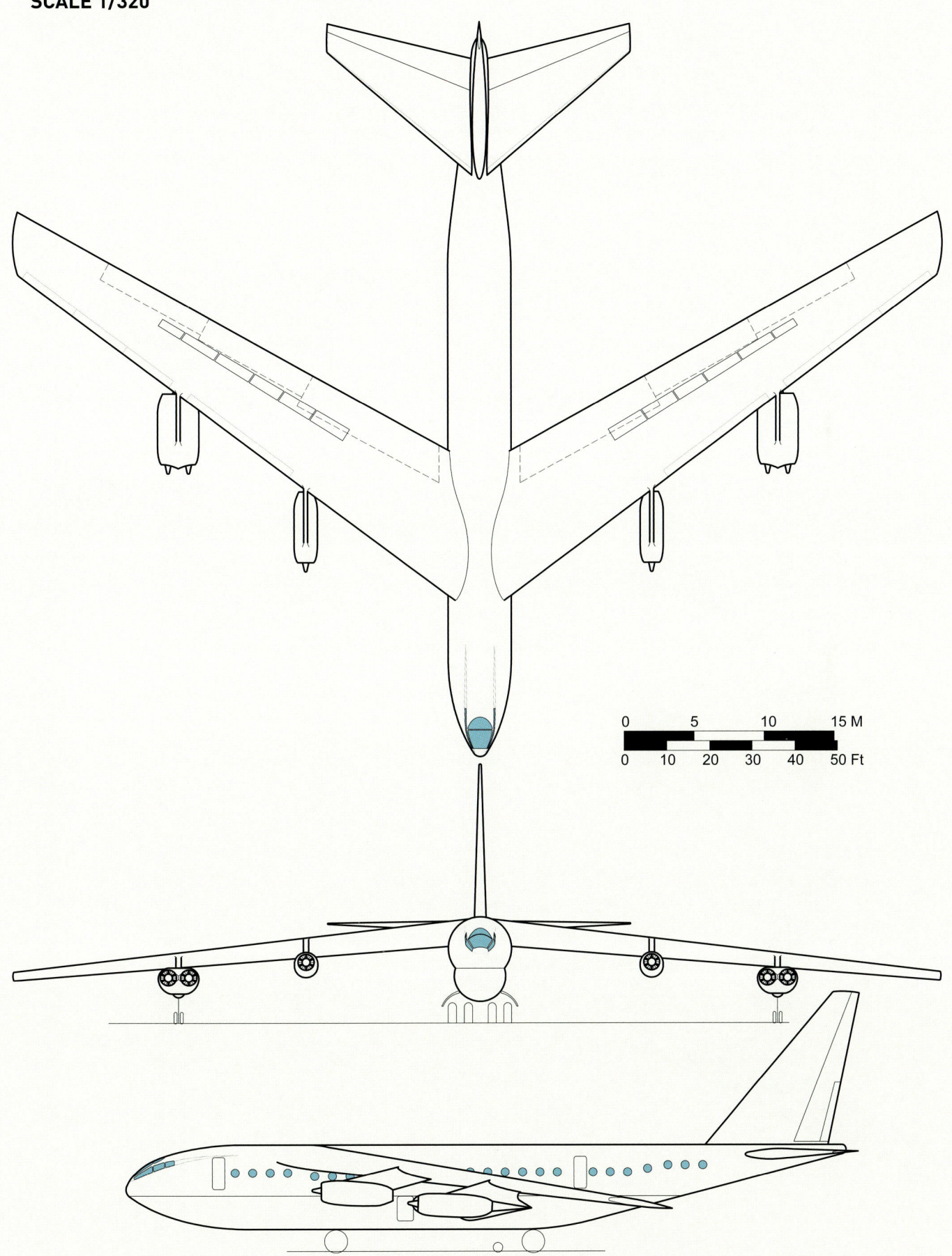

Boeing Model 473-29
SCALE 1/350

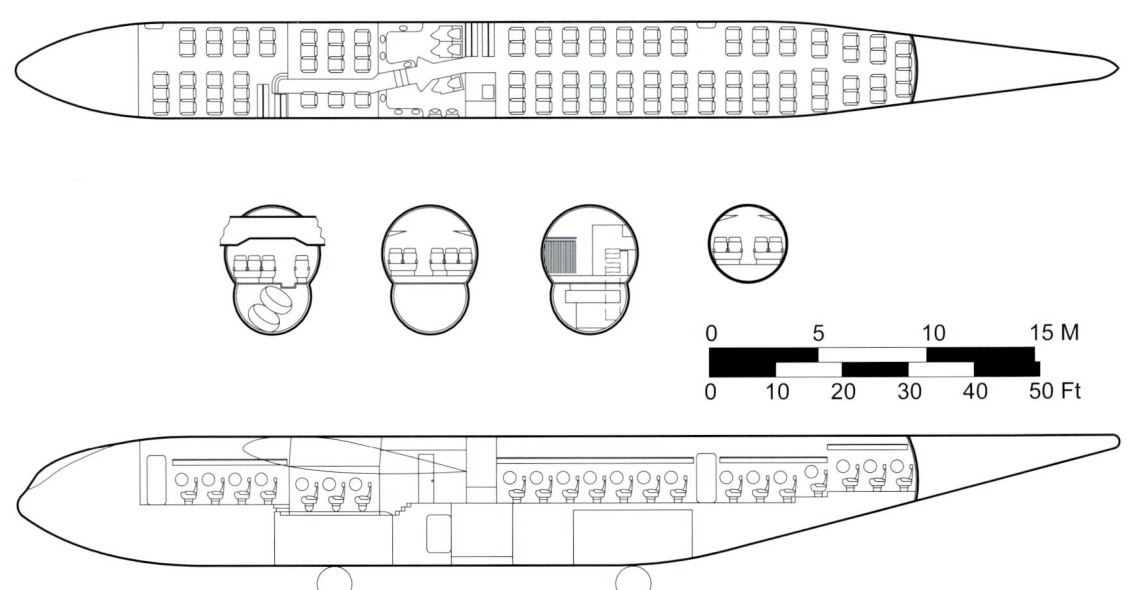

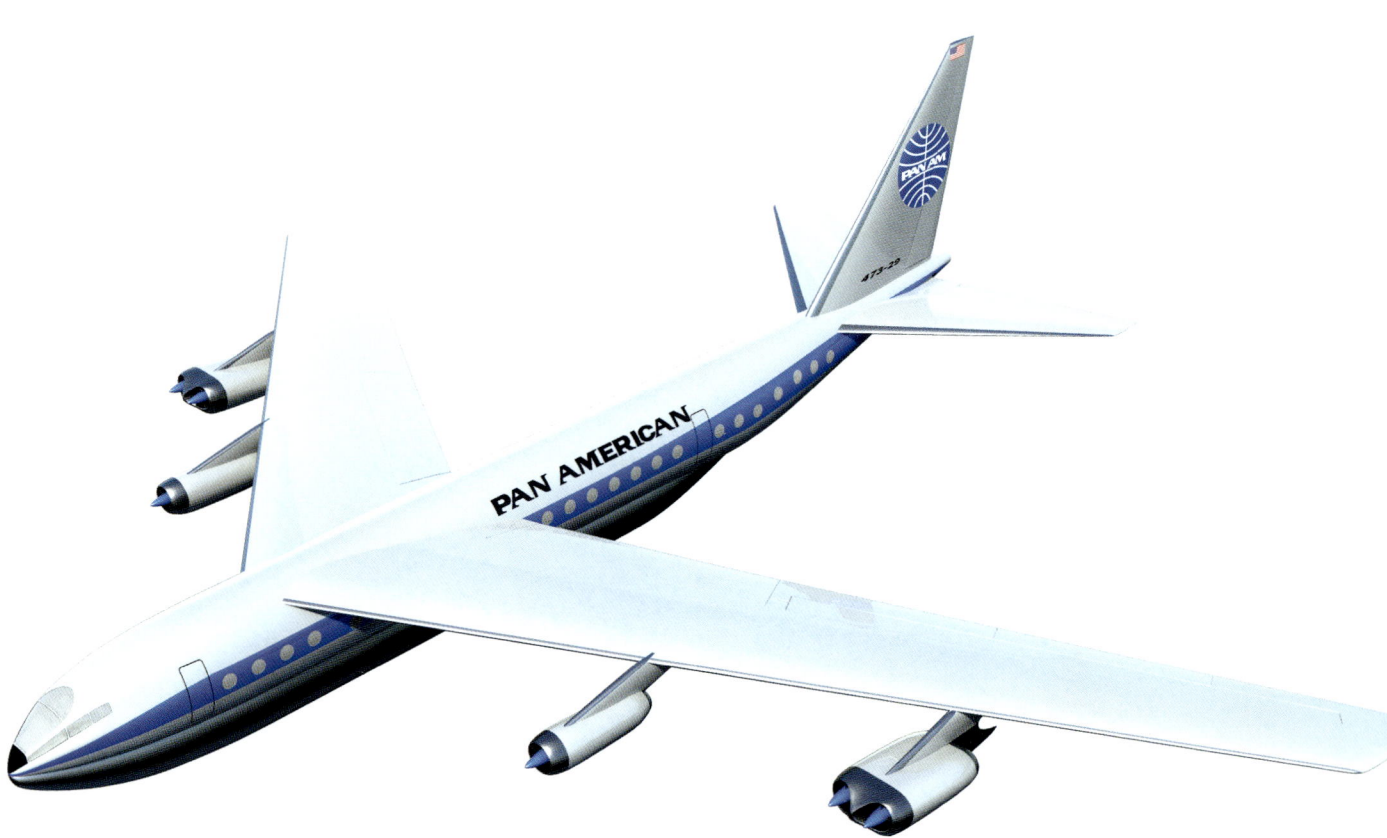

Boeing Model 473-29B
SCALE 1/350

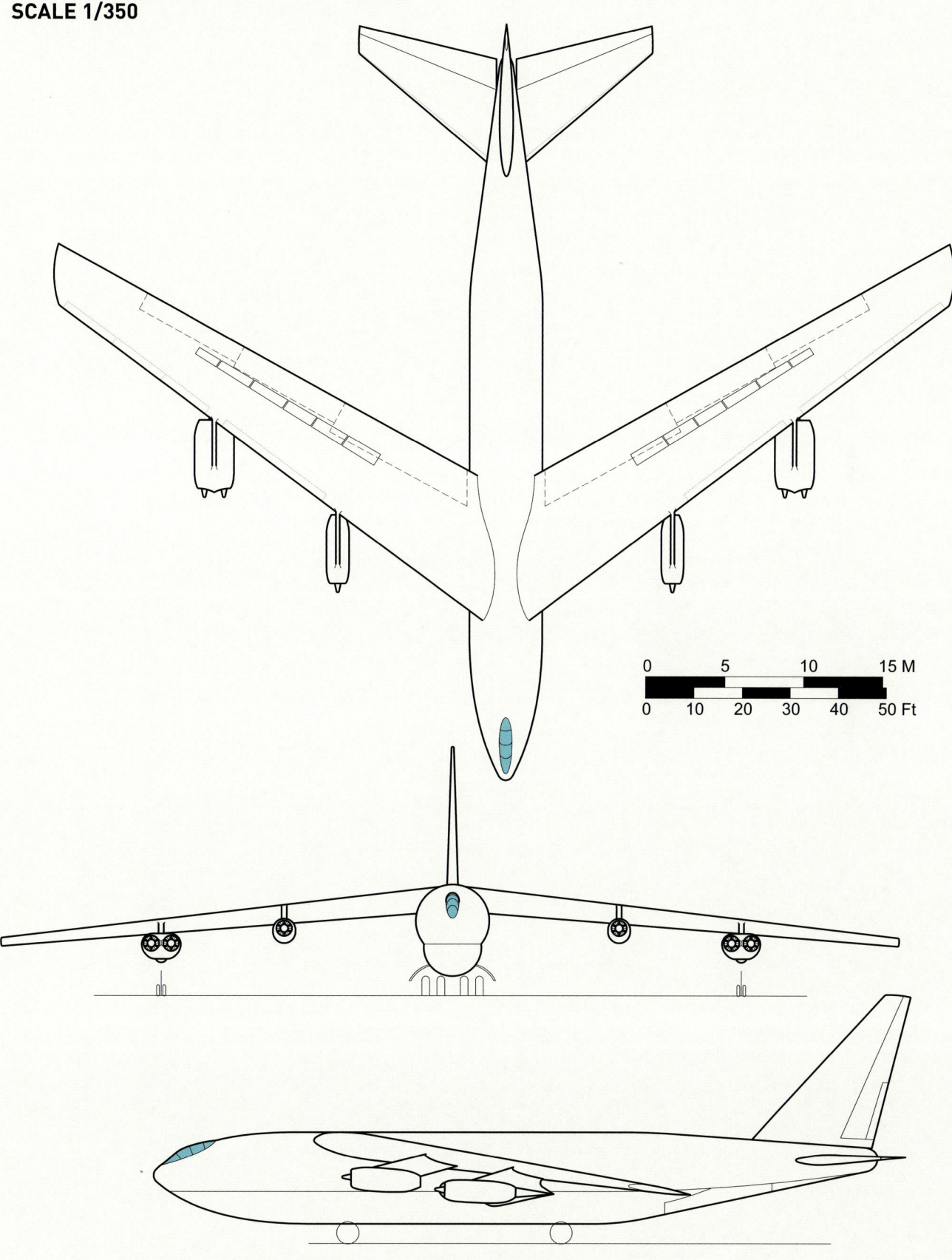

Boeing Model 473-29B
SCALE 1/350

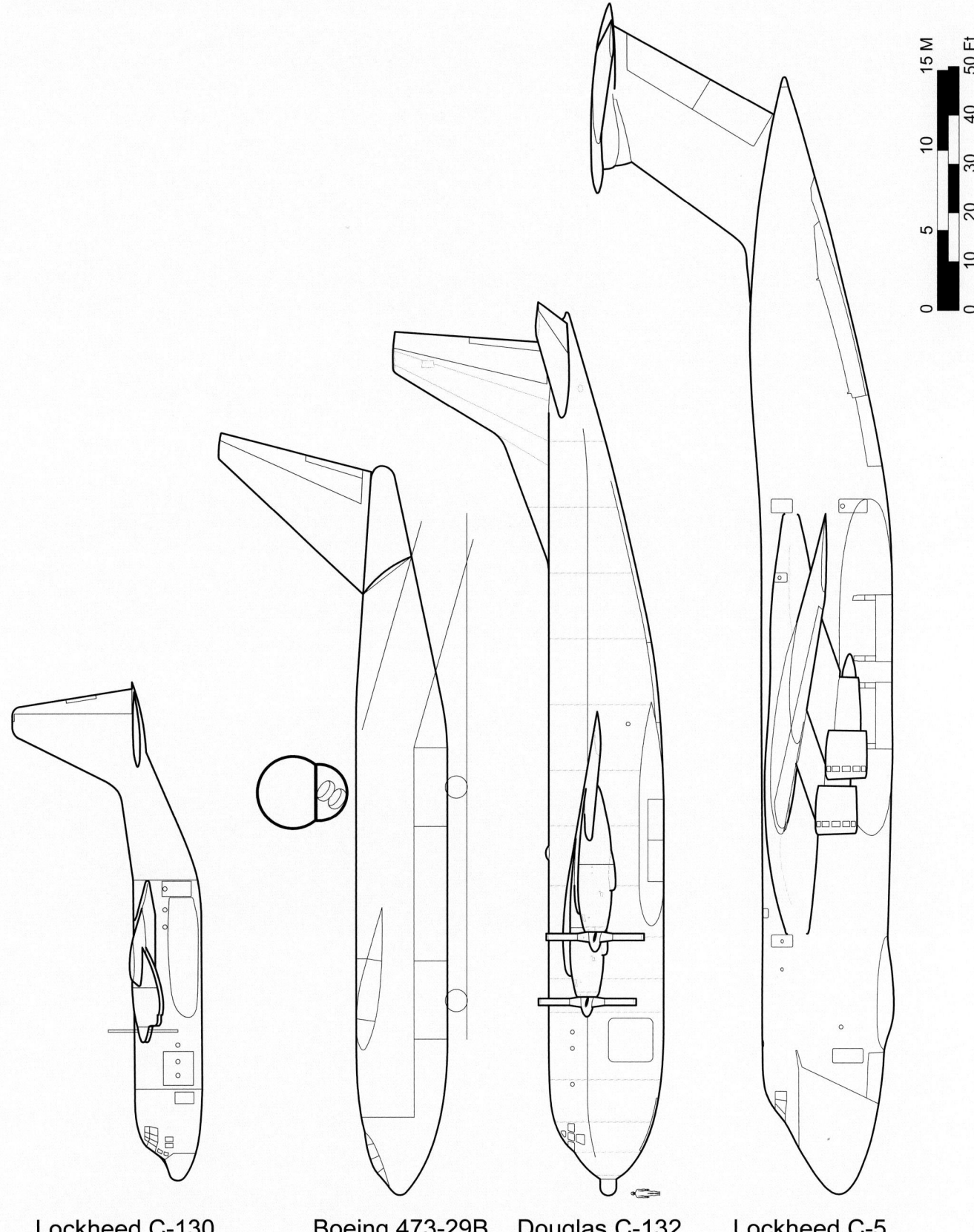

Lockheed C-130 Boeing 473-29B Douglas C-132 Lockheed C-5

Boeing Model 473-30
SCALE 1/250

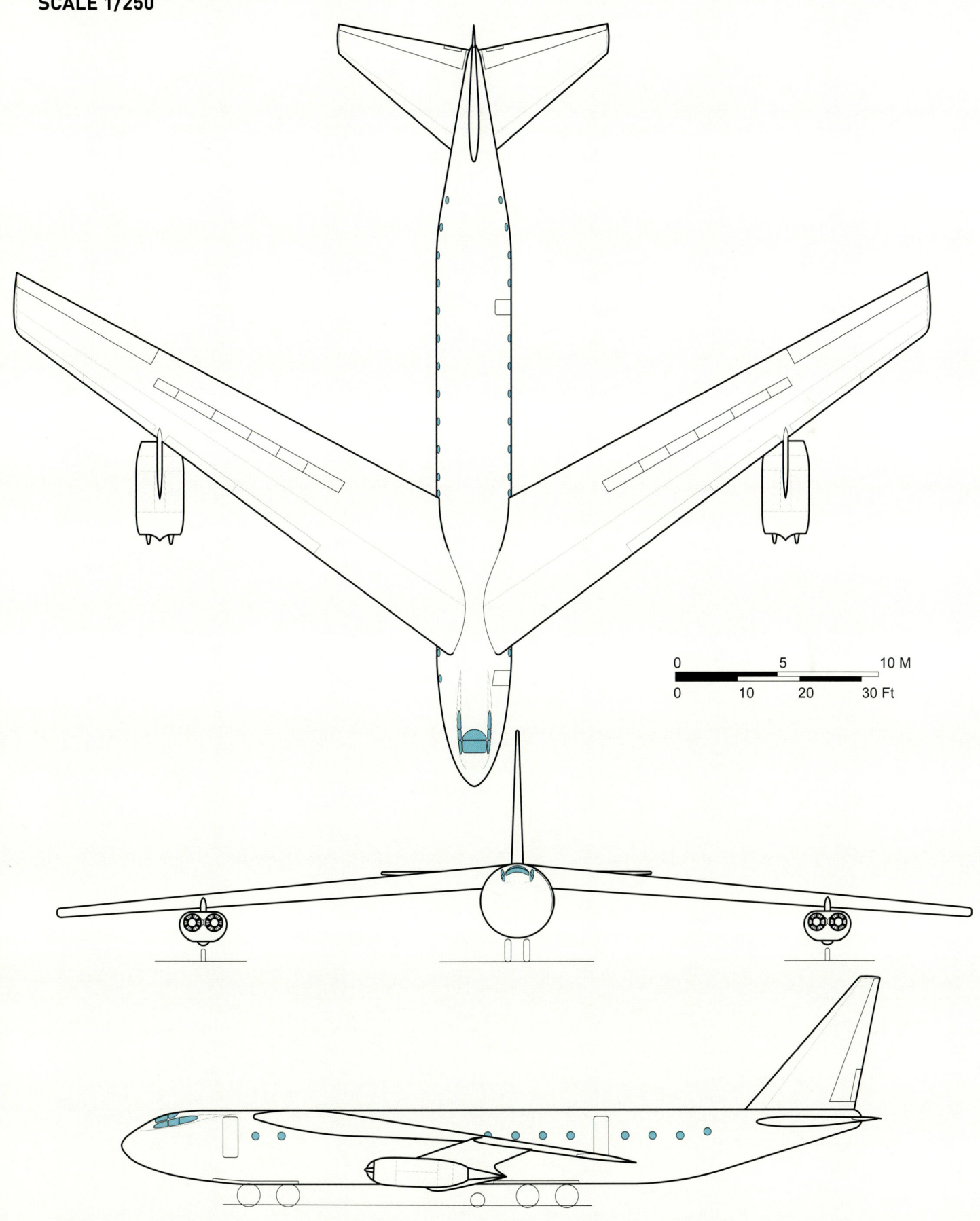

Boeing Model 473-30
SCALE 1/200

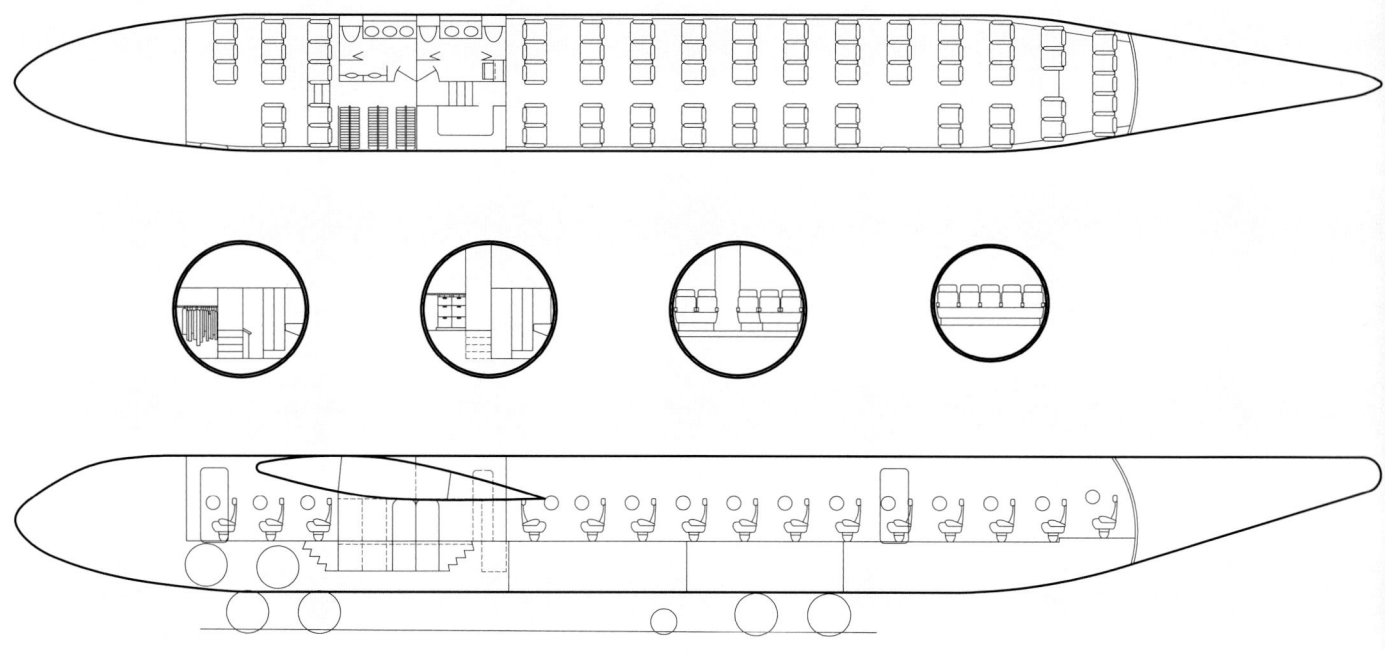

Boeing Model 473-31
SCALE 1/175

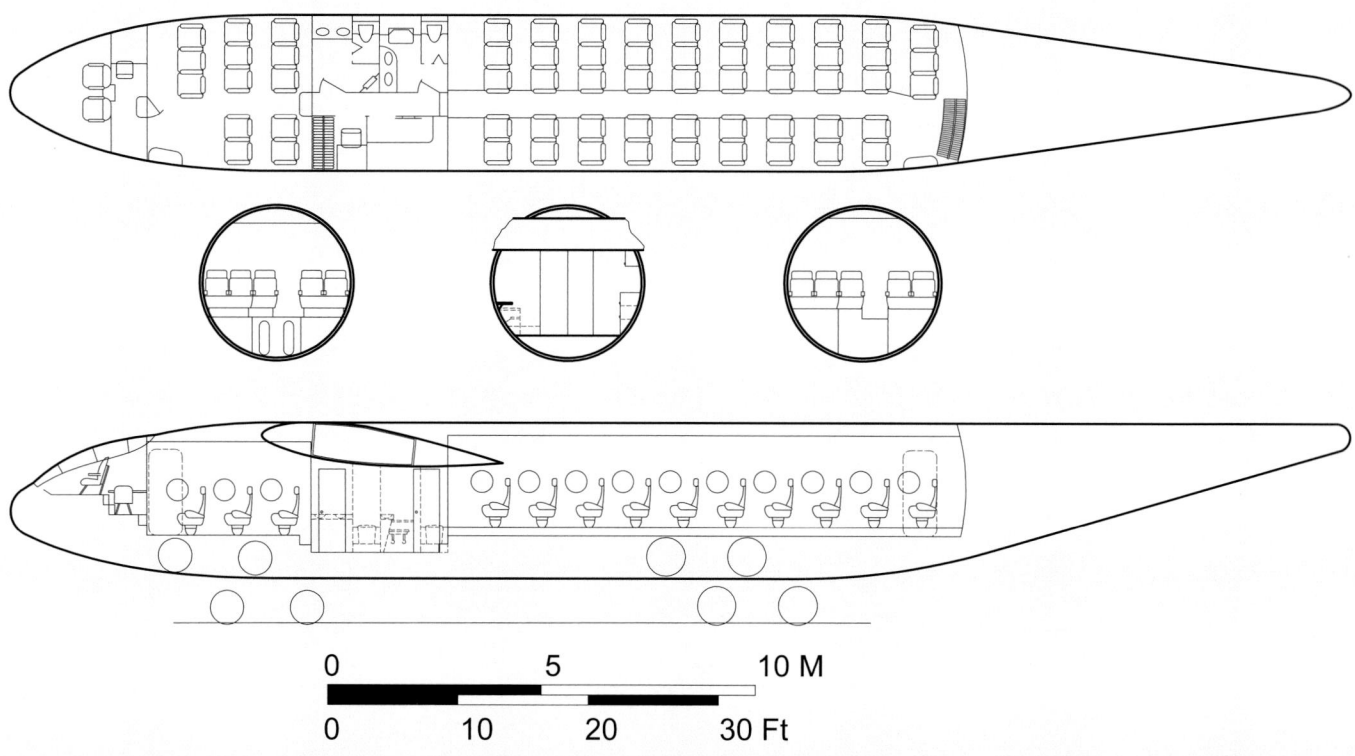

B-47 and B-52 Derived Transports

Boeing Model 473-31
SCALE 1/220

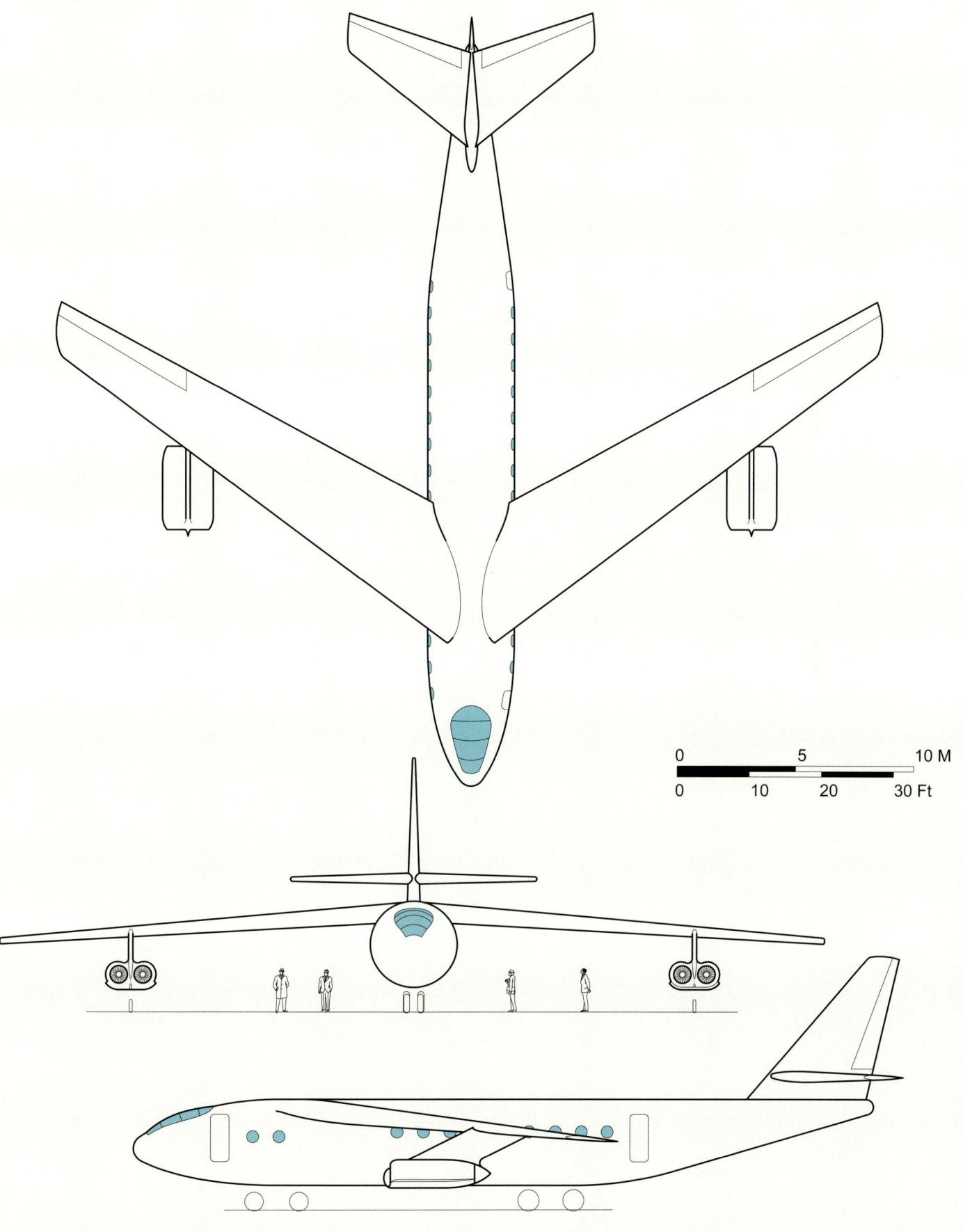

Boeing Model 820-100
SCALE 1/400

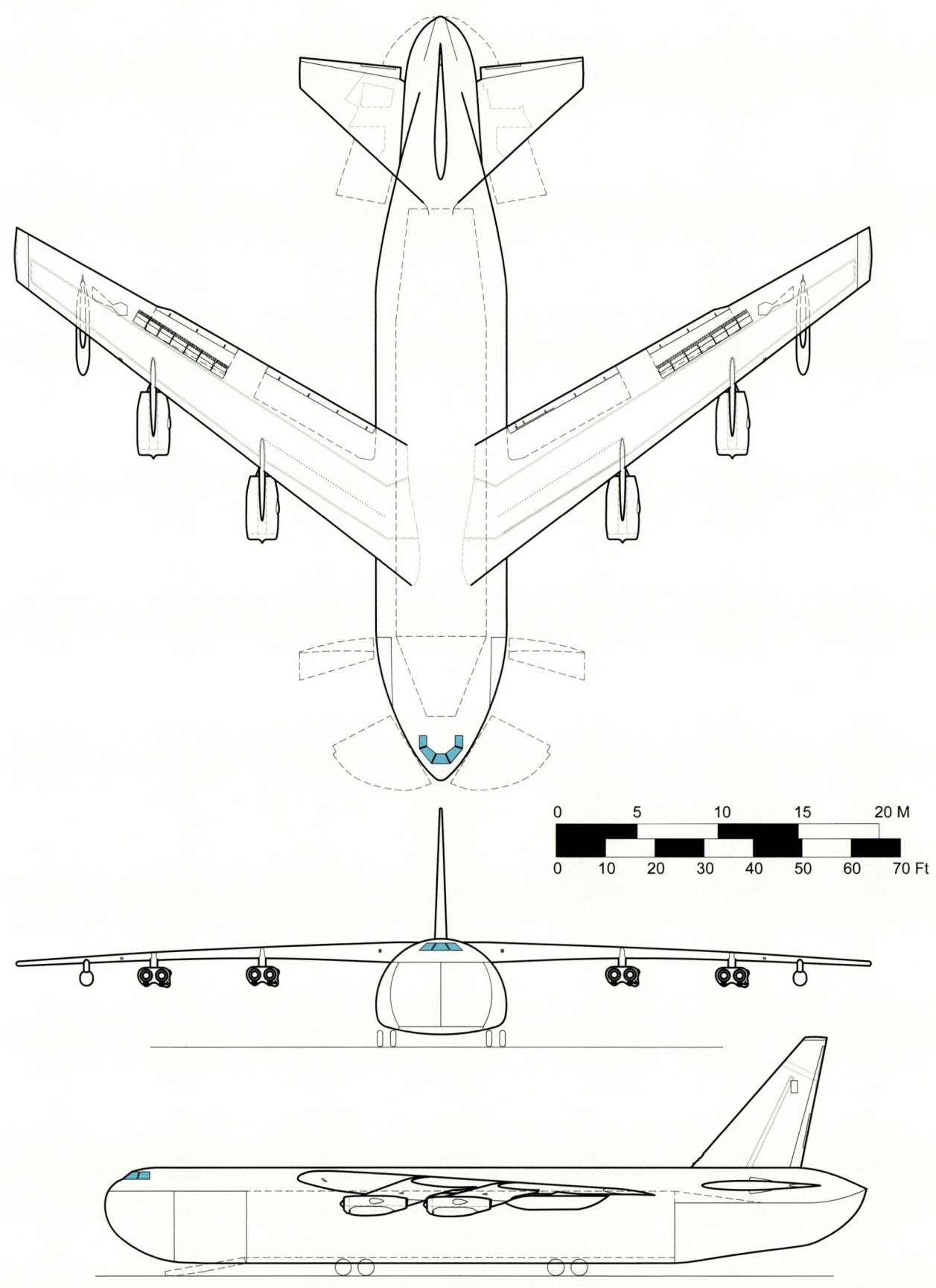

straight wings and turboprop engines. But those wings not only had a remarkable degree of dihedral, they were hinged to tilt upwards up to 45°. The turboprops were increased in number to six and were suspended below the wings under pylons, the propellers increased in diameter to 25ft. While performance data is currently lacking, this arrangement would have been optimized for short takeoff, very likely at the expense of range and speed.

The tail surfaces were still stock B-52F, but little else from the B-52 remained. The cockpit is depicted as being remarkably wide.

Boeing Model 820-105

Now came a major change – turning the cargo lifter into a flying boat. The wings were back to being those of the B-52G; the tail surfaces and the fuselage were all-new. Fixed floats were suspended beneath the wingtips to provide on-water stability. The eight engines, likely the same turbofans from prior models, would be on inverted pylons well above and ahead of the wings, away from sea spray.

Unlike the earlier Model 820 designs, 820-105 did not have a rear cargo door. Instead, the tail of the aircraft was tailored for hydrodynamics, with deployable water rudder and stubby upswept planes. Cargo loading was through the nose, which raised as a single piece much like the nose of the C-5 Galaxy, and through a rectangular door on the upper surface of the rear fuselage. This cargo loading door would be used for on-water loading, while the forward door would be used when the aircraft was beached.

A flying boat has certain definite advantages over a landplane for military cargo transport in that it can land wherever the breadth and depth and water permits. Oceans, lakes and large rivers all make adequate landing sites, with no need for preparation. Of course, actual airfields would be inaccessible for it; inland sites would only be possible if adjacent to sizable bodies of water. Flying boats, with their deeper bellies and increased cross-sectional area, also suffer from poorer aerodynamics.

Boeing Model 820-106

The Model 820-106 was another flying boat, but with a somewhat shallower fuselage and landing gear in sponsons on the side – turning it into an amphibian. The design also differed in that it had a sizable rectangular cargo loading door on the port side of the aft fuselage.

The shallower fuselage would be better for aerodynamics, but less advantageous for water takeoff. In order to improve water takeoff and landing performance, the underside would deploy downwards to form a planing surface. The fixed floats suspended below the outer wings would also deploy small stabilizing landing gear.

Boeing Model 820-100
SCALE 1/300

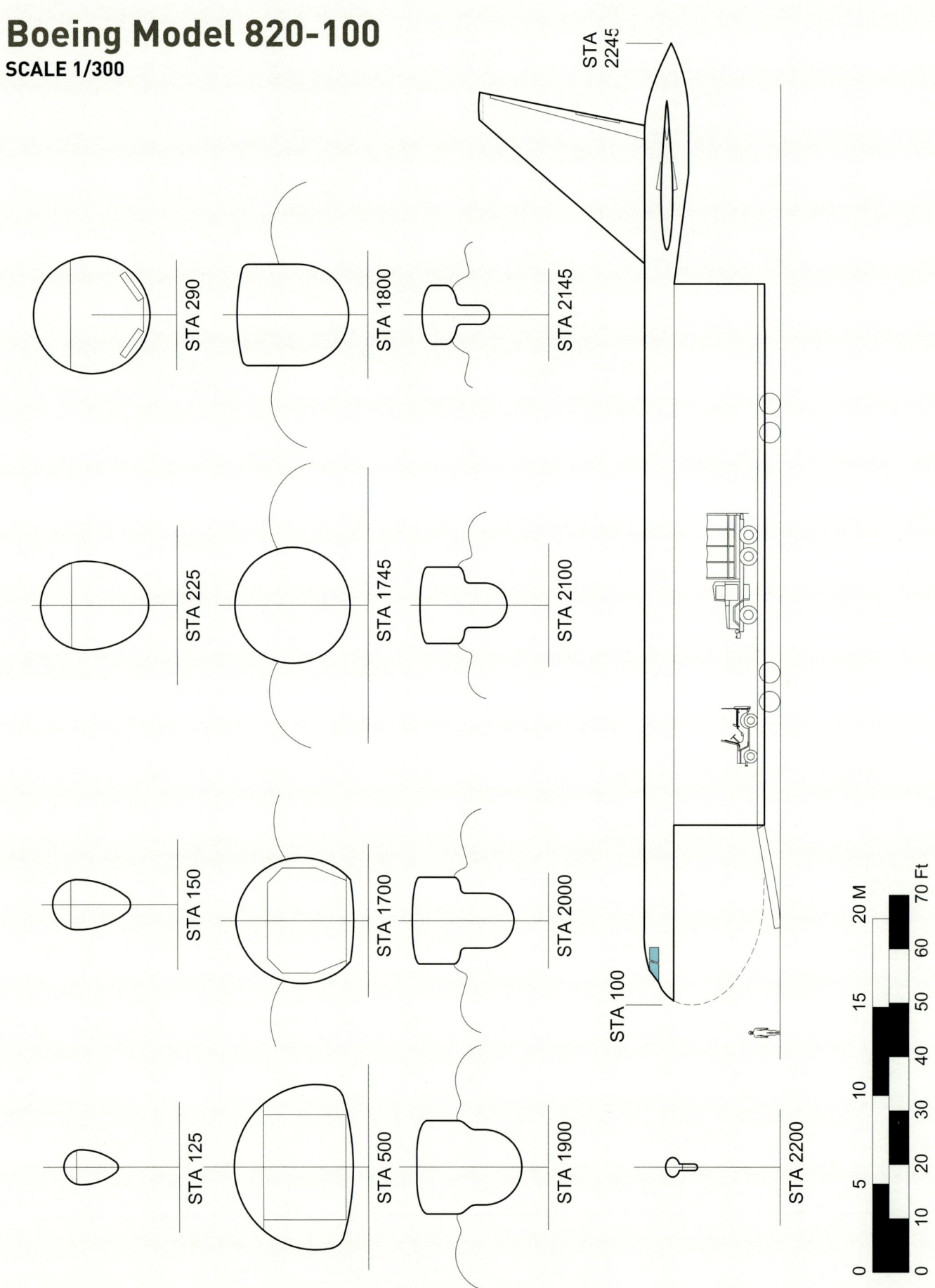

B-47 and B-52 Derived Transports

Boeing Model 820-101
SCALE 1/350

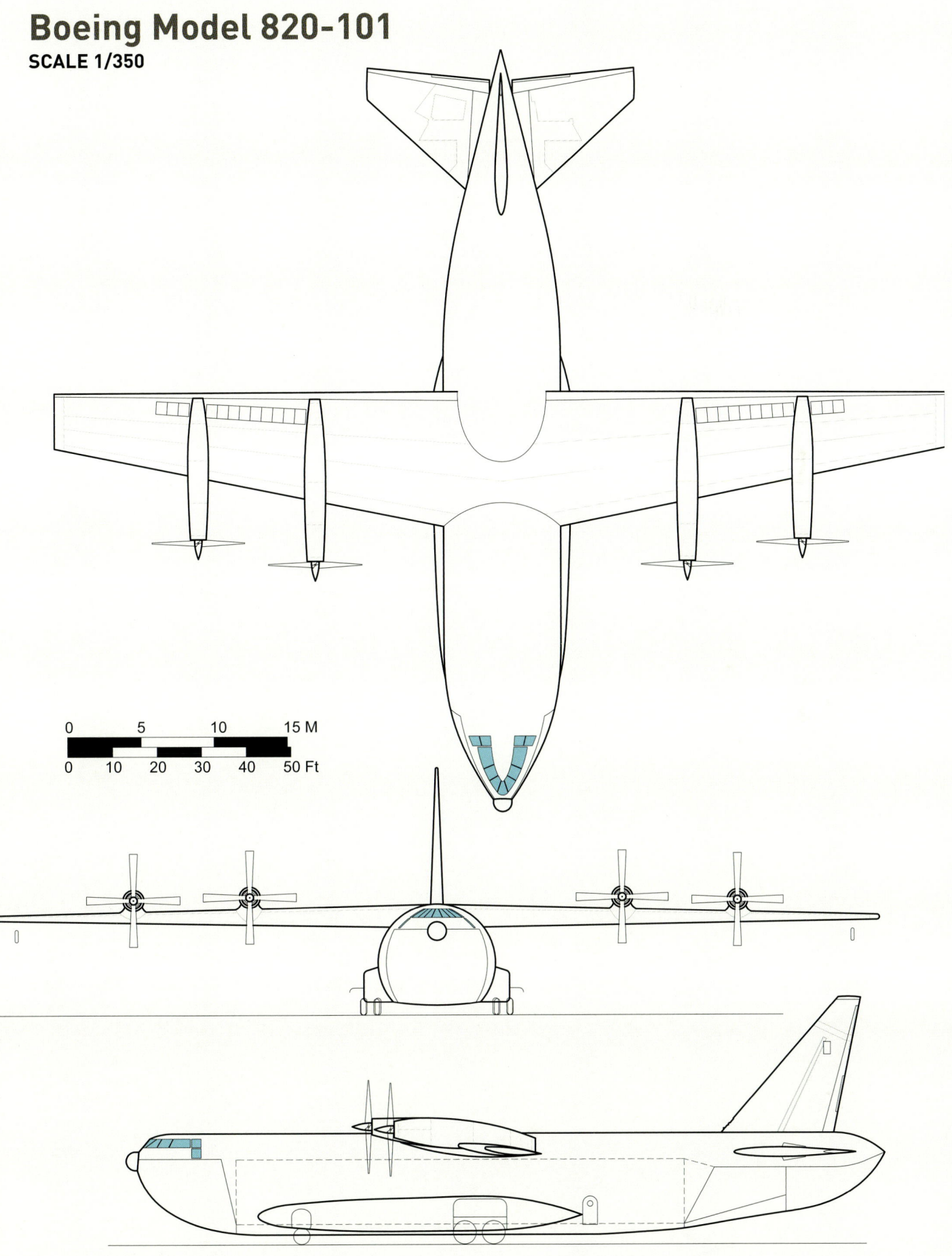

Boeing Model 820-102
SCALE 1/400

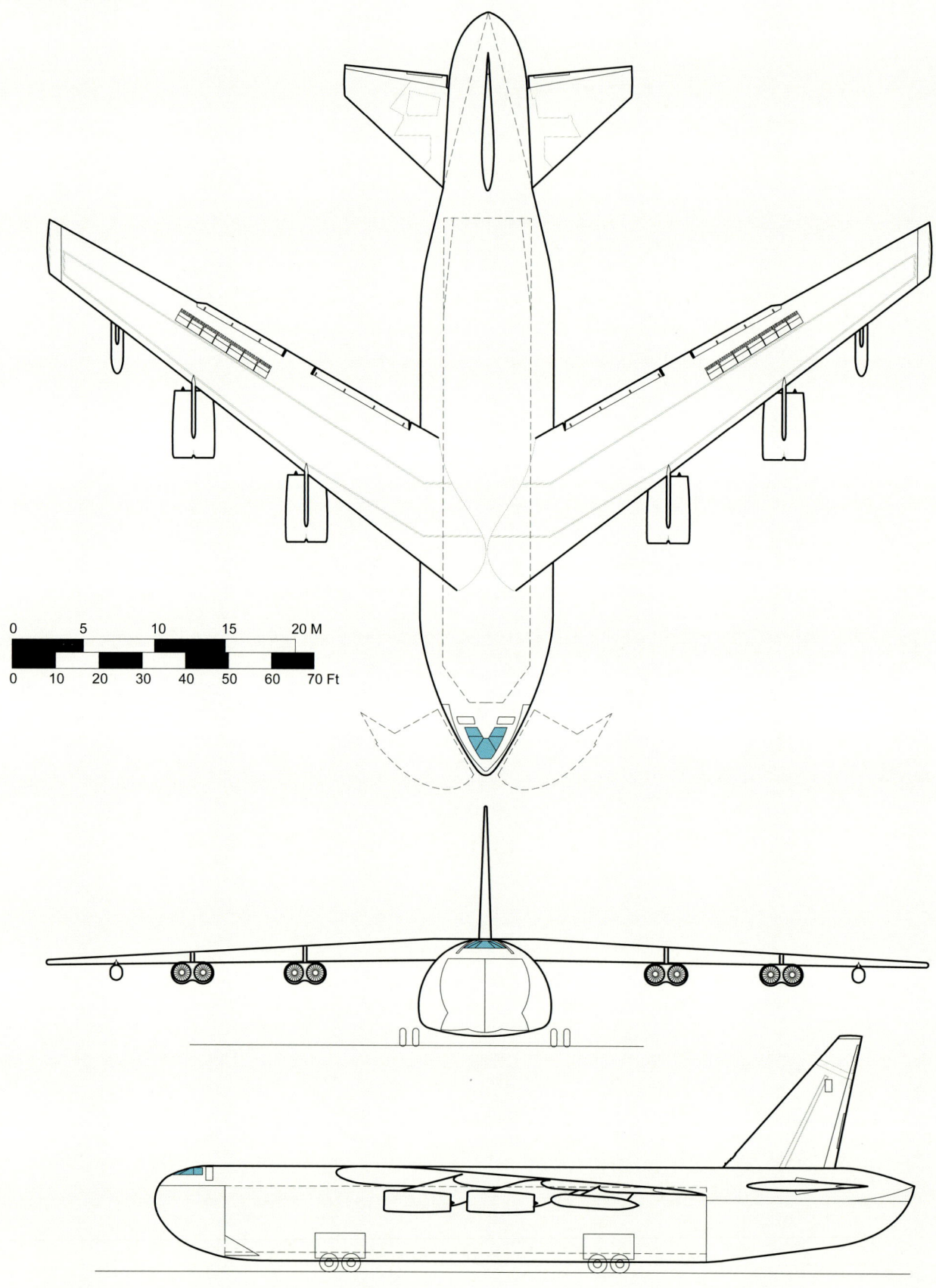

Boeing Model 820-103A
SCALE 1/400

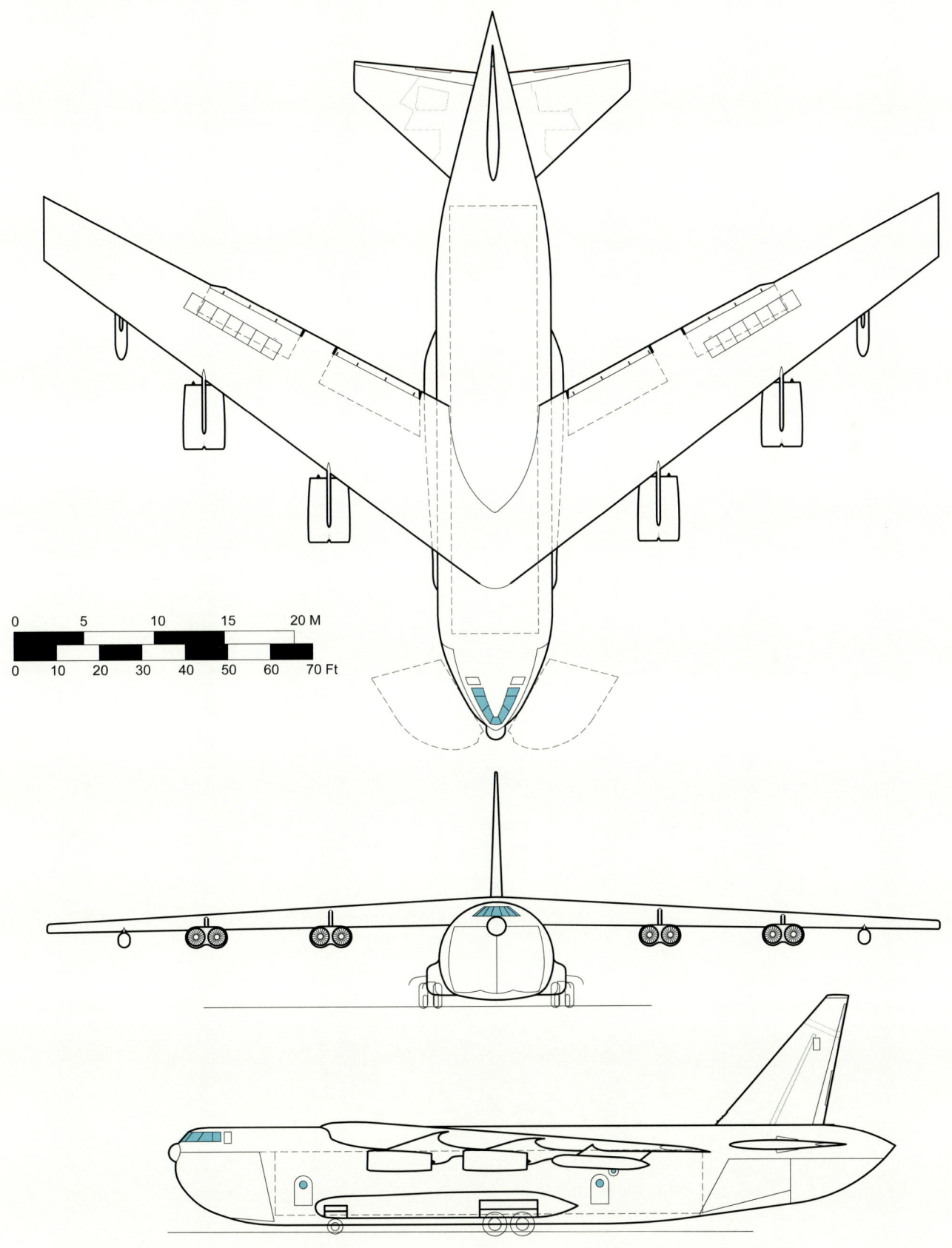

Boeing Model 820-104
SCALE 1/400

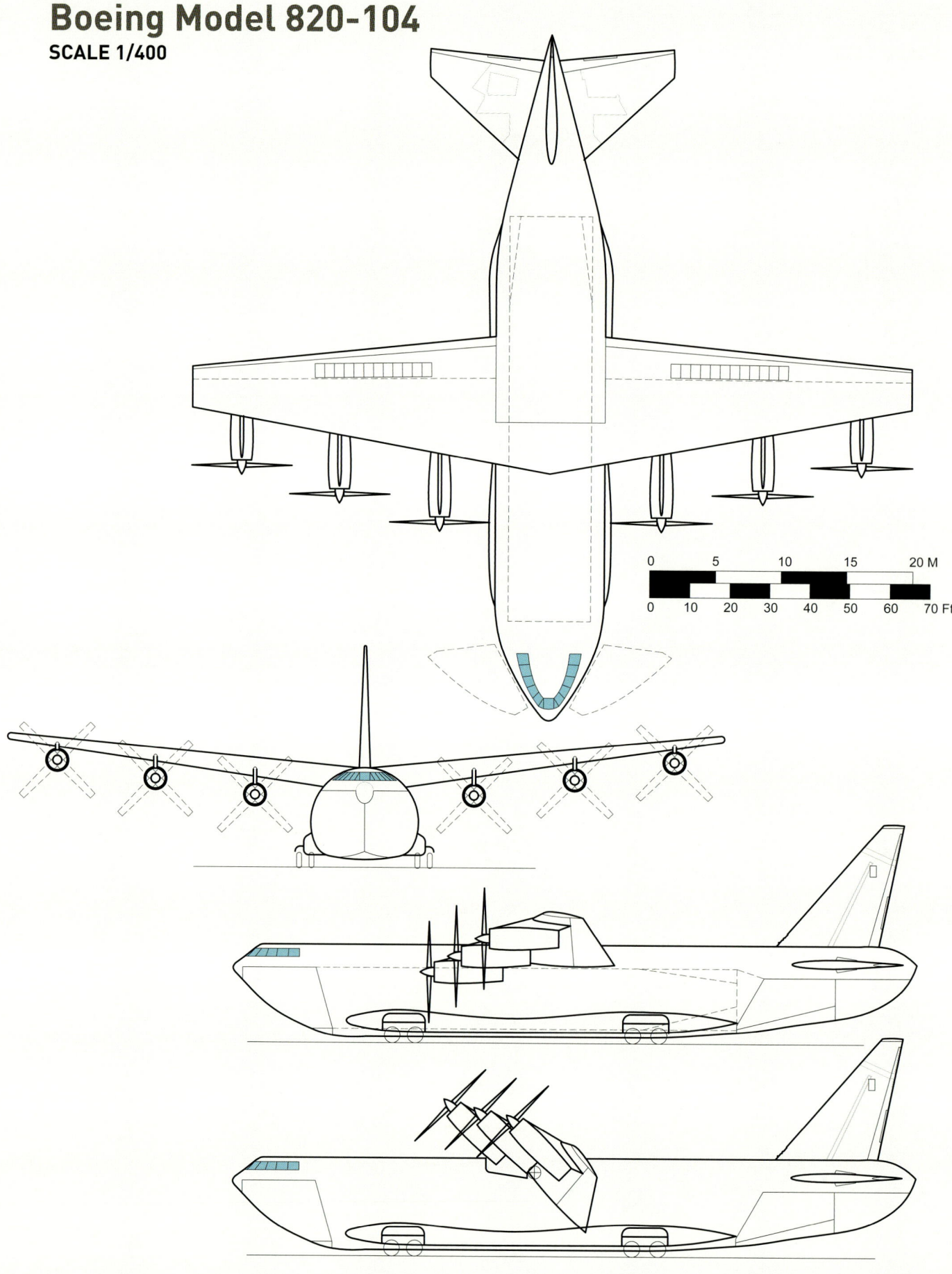

Boeing Model 820-105
SCALE 1/400

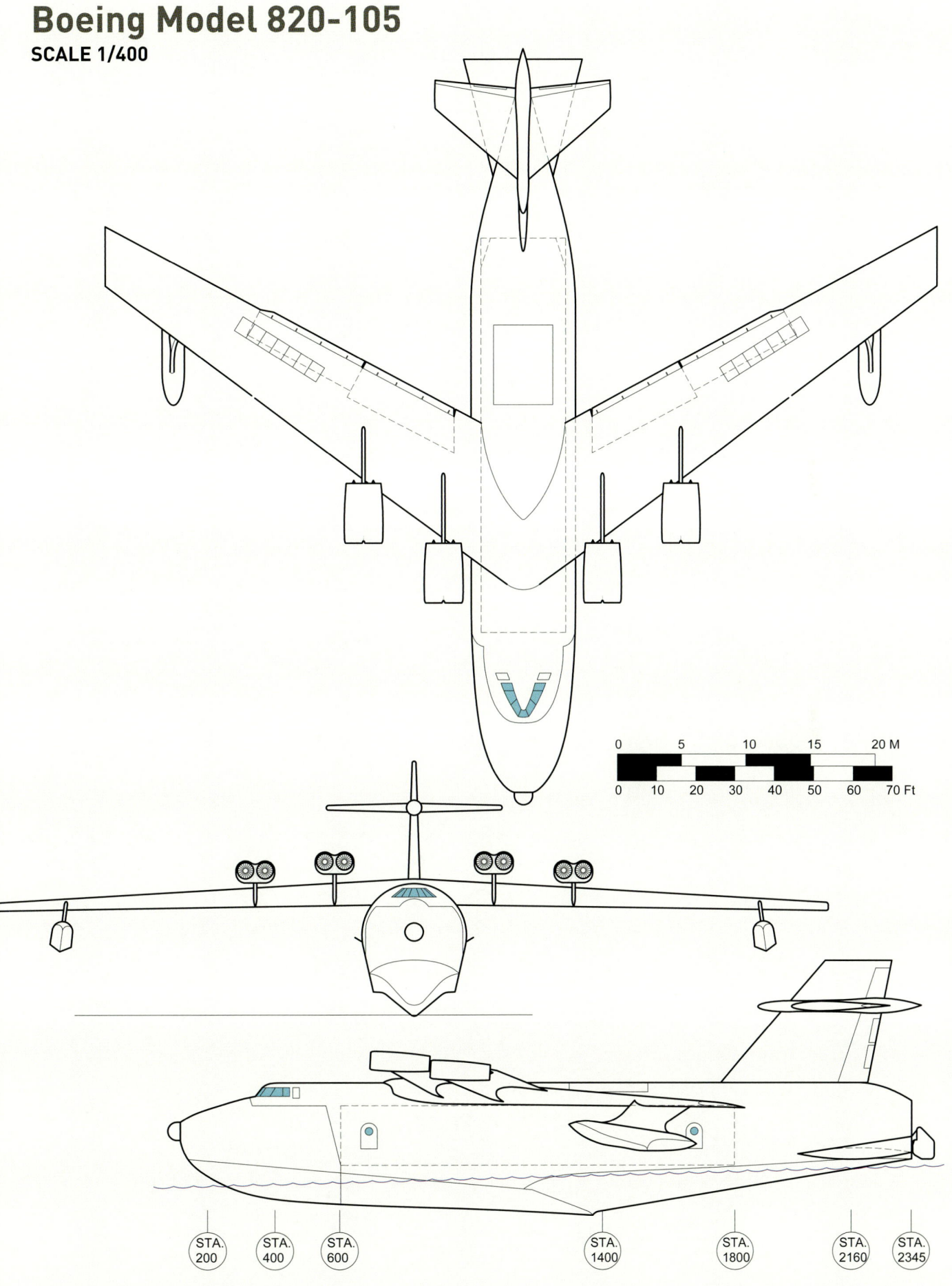

Boeing Model 820-105

SCALE 1/400

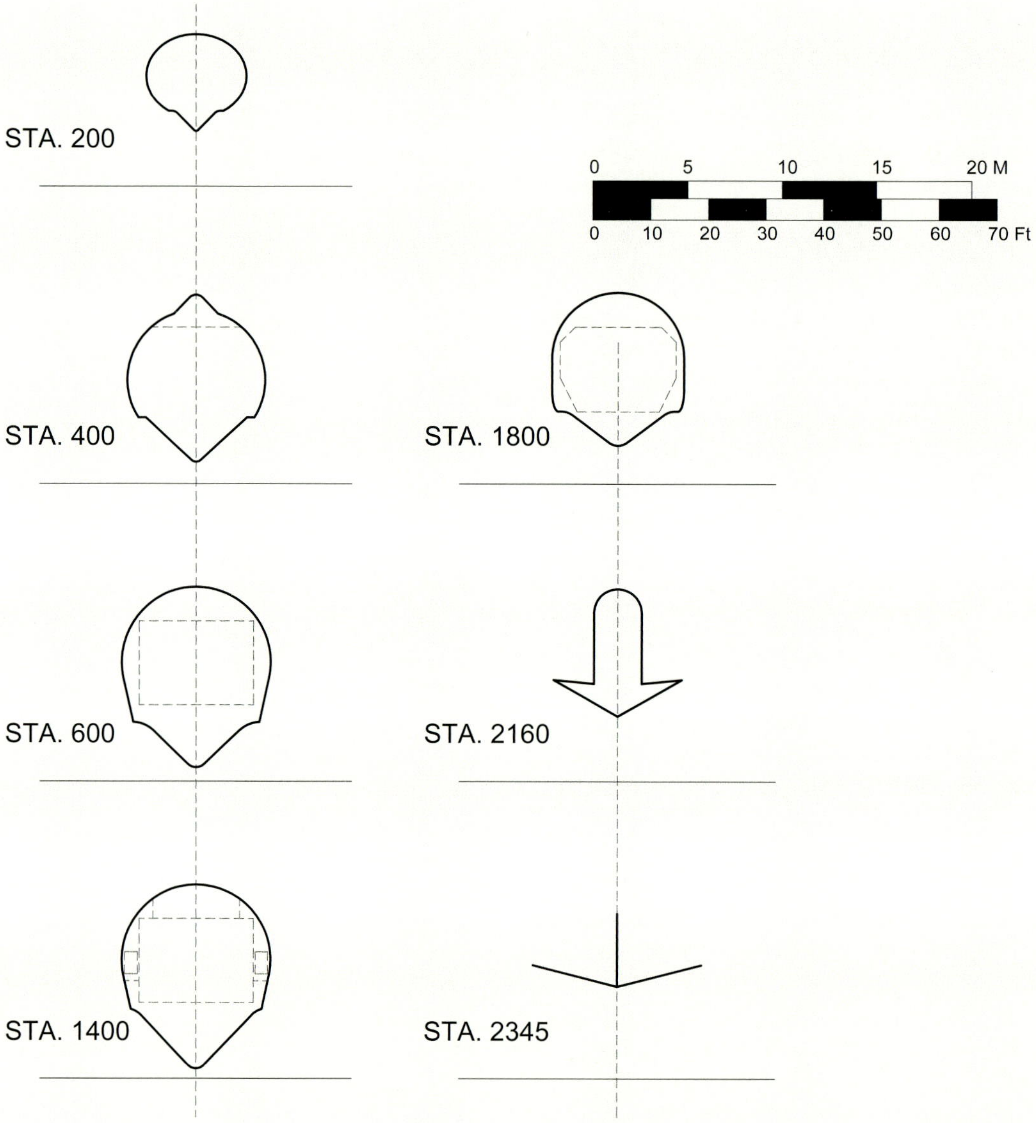

STA. 200
0 5 10 15 20 M
0 10 20 30 40 50 60 70 Ft
STA. 400
STA. 1800
STA. 600
STA. 2160
STA. 1400
STA. 2345

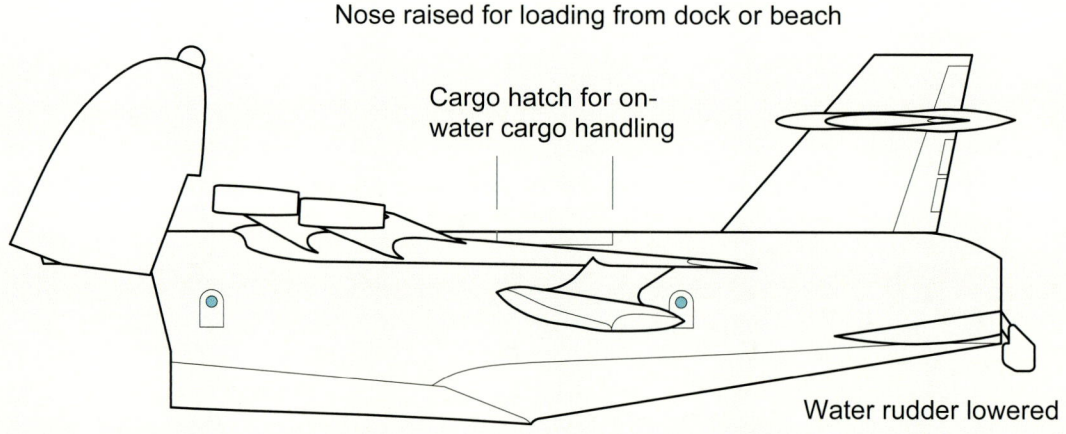

Nose raised for loading from dock or beach
Cargo hatch for on-water cargo handling
Water rudder lowered

Boeing Model 820-108

This design was similar to the original Model 820-100, but with four turbofans in two outboard nacelles and two inboard nacelles each with two nuclear-powered turbojets.

Span was greatly increased, with an increase in surface area. The added weight of the nuclear reactors and radiation shielding would mean the aircraft needed as much lift capacity as possible. The wings and horizontal stabilizers appear to be derived from those of the B-52, but the vertical stabilizer is new. Performance is lacking, but although the nuclear turbojets would have granted virtually unlimited range, the aircraft would also have required long runways and dedicated facilities to maintain the reactors.

The diagram presented here includes the Boeing Model 726-20 for scale. This was a contemporary design for a nuclear powered supersonic bomber using a similar nacelle configuration, through for only a single turbojet.

Boeing Model 820-108A

This design was a conventionally powered version of the baseline Model 820-108. The replacement of the inboard nuclear turbojet nacelles with a pair of conventional turbofan nacelles and a slight change in the cross-section of the landing gear sponsons seem to have been the only real changes.

Boeing Model 820-109

Model 820-109 was a nuclear powered version, relying fully on four nuclear turbojets. This would be practical only after nuclear propulsion had proven itself to be reliable enough as the sole powerplant for an aircraft; thus Model 820-109 would not be the first nuclear powered vehicle to be built.

Jet fuel would be stored in the wings outboard of the nacelles. This would be used in the nuclear turbojets during times when extra thrust was needed, such as during takeoff.

Boeing Model 820-110

Compared to some of the prior designs, Model 820-110 was positively staid. It was a conventional design with a relatively narrow (only 15ft diameter) cylindrical fuselage, B-52G wings without a span-increasing insert and with landing gear in long sponsons. The eight turbofans from the prior designs were replaced with four turbofans of much greater diameter, clearly early high-bypass turbofans.

Boeing Model 820-111

The last known Model 820 variant was also one of the most remarkable. It was similar to Model 820-104 in that it had a tiltable wing with six turboprops… but it differed in that it had two such wings, for a total of 12 turboprops. The wings could tilt upwards a full 90°, allowing VTOL performance.

As with most of the other Model 820 designs, performance data is lacking. It seems extremely likely that range for Model 820-111 would have been quite poor compared to most of the others – but it would not have needed expensive runways and airfields, just wide open spaces to set down. Where runways existed it would be able to tilt its wings less than 90° for a more fuel-economical short takeoff.

There was seemingly nothing left of the B-52 in this design, with even the vertical tail being new, and no horizontal stabilizers being used.

Boeing Model 877-1

Boeing submitted several responses to an Air Force System Study Directive for a 'Multipurpose Long Endurance Aircraft' in 1960. The MPLEA was to fulfill two roles: cargo carrier and carrier/launcher of the Skybolt air-launched ballistic missile. The Seattle division turned in designs based on then-current Boeing transport aircraft studies, but the Wichita division turned in a design based in part on the B-52. Model 877-1 used the wings and tail surfaces of the B-52G married to an entirely new fuselage, somewhat in the manner of the Model 820 designs.

The Model 877-1 fuselage was quite sleek and aerodynamic when compared to the Model 820 designs, with more than a passing resemblance to the Lockheed C-141. As with the C-141, the nose of Model 877-1 did not open up to provide drive-on access to the cargo bay. The rear featured a large loading ramp; a smaller cargo door was available on the port side of the forward fuselage for loading smaller items and troops.

Two further details set the Model 877-1 apart from the B-52 it was based on. The General Electric MF 288 engines the 877-1 used were a new type, the 'aft fan'. As the name suggests, this engine puts the fan at the rear of the engines, rather than the front, producing the appearance of the engines being mounted backwards. These engines were also proposed for use on retrofitted B-52s, though little information on that has so far come to light.

In addition, the 877-1 could be fitted with pylons for the AGM-87 Skybolt. A pylon carrying two missiles would be attached under the inboard wing; this appears to be the same as the pylons fitted to actual B-52s. But one more Skybolt would be suspended under each wing between the inboard and outboard engine nacelles, and a further Skybolt was suspended from pylons projecting from each side of the fuselage just aft of the wings, for a total of eight missiles. This

Boeing Model 820-106
SCALE 1/400

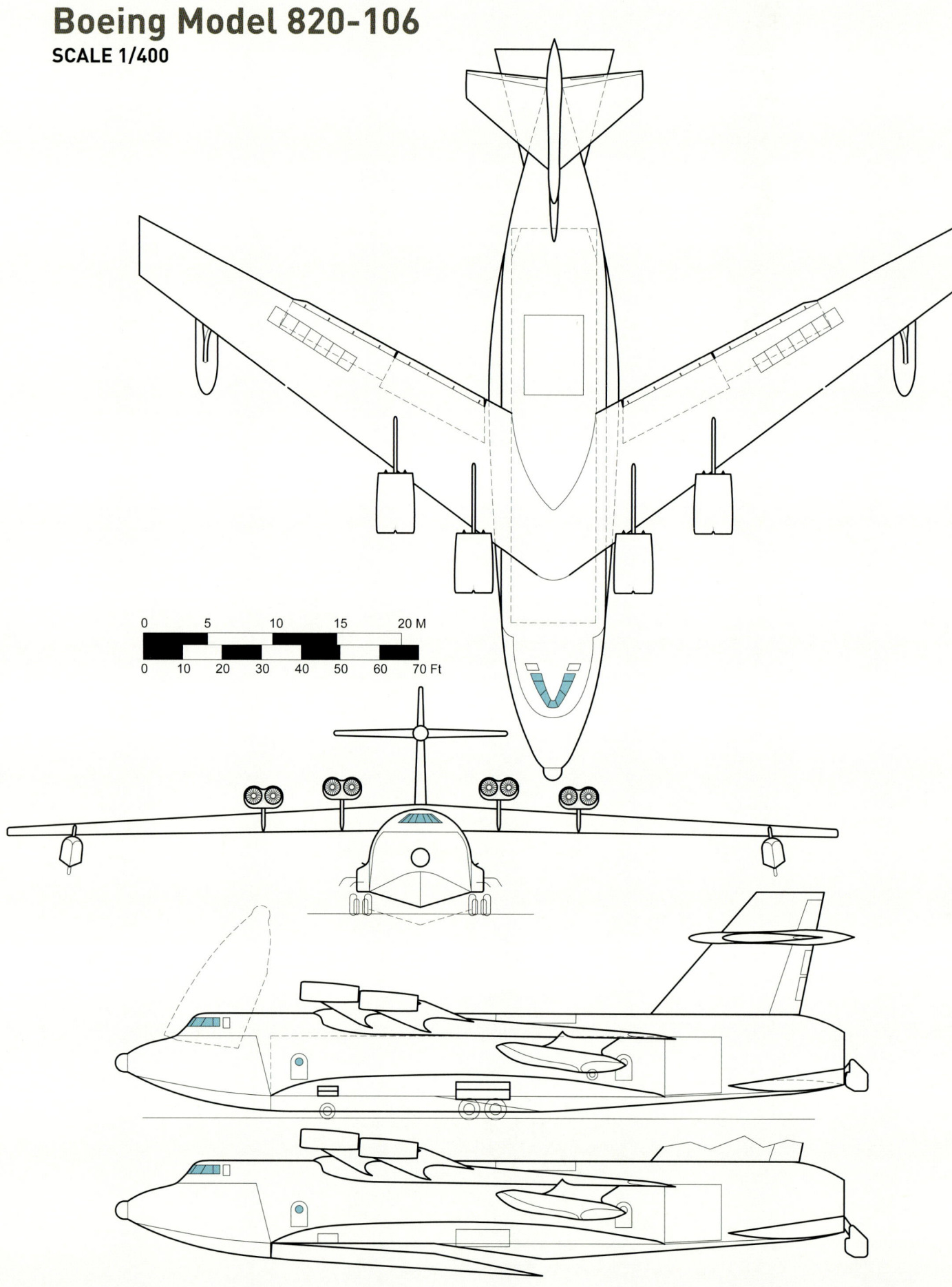

Boeing Model 820-108
SCALE 1/400

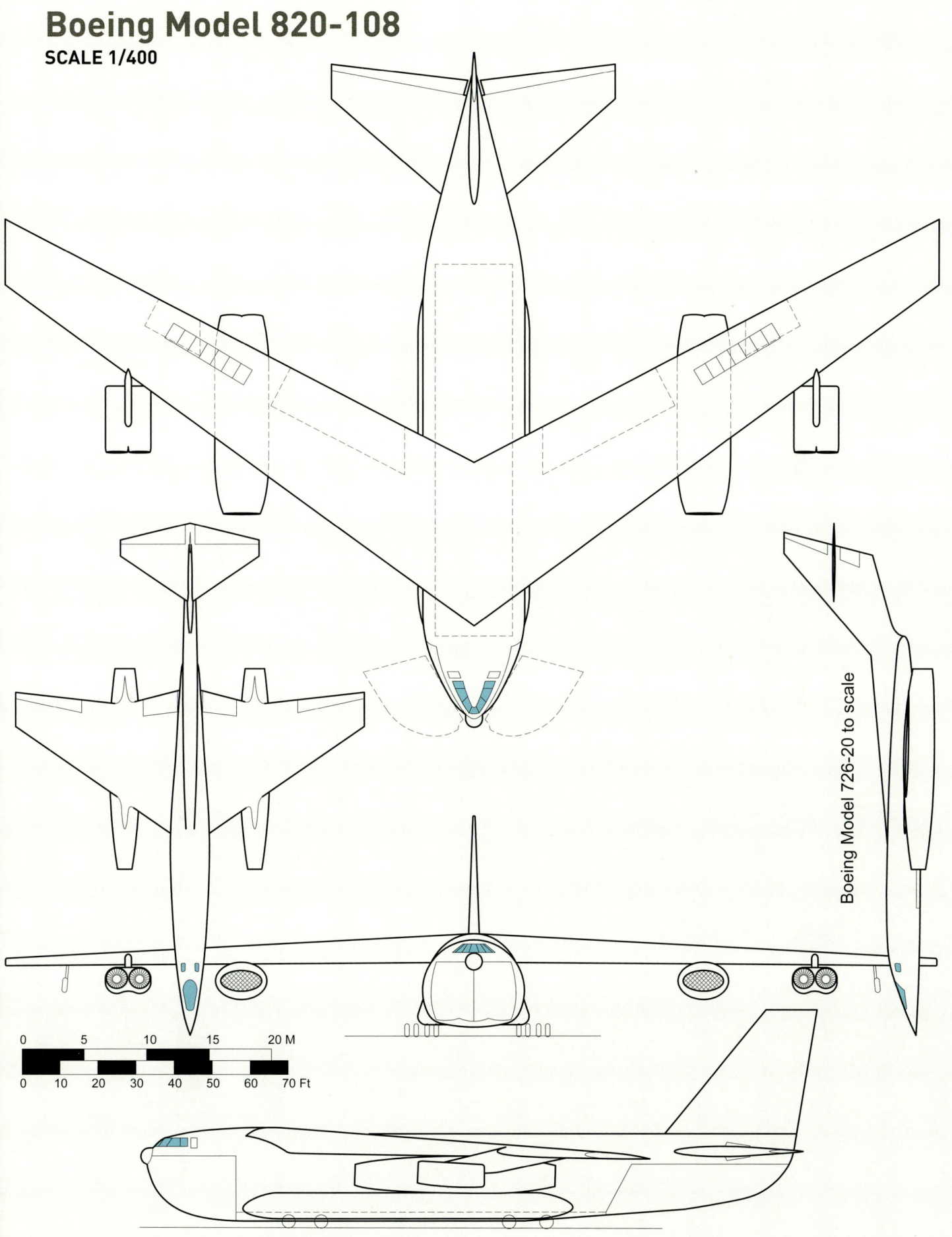

Boeing Model 820-108A
SCALE 1/400

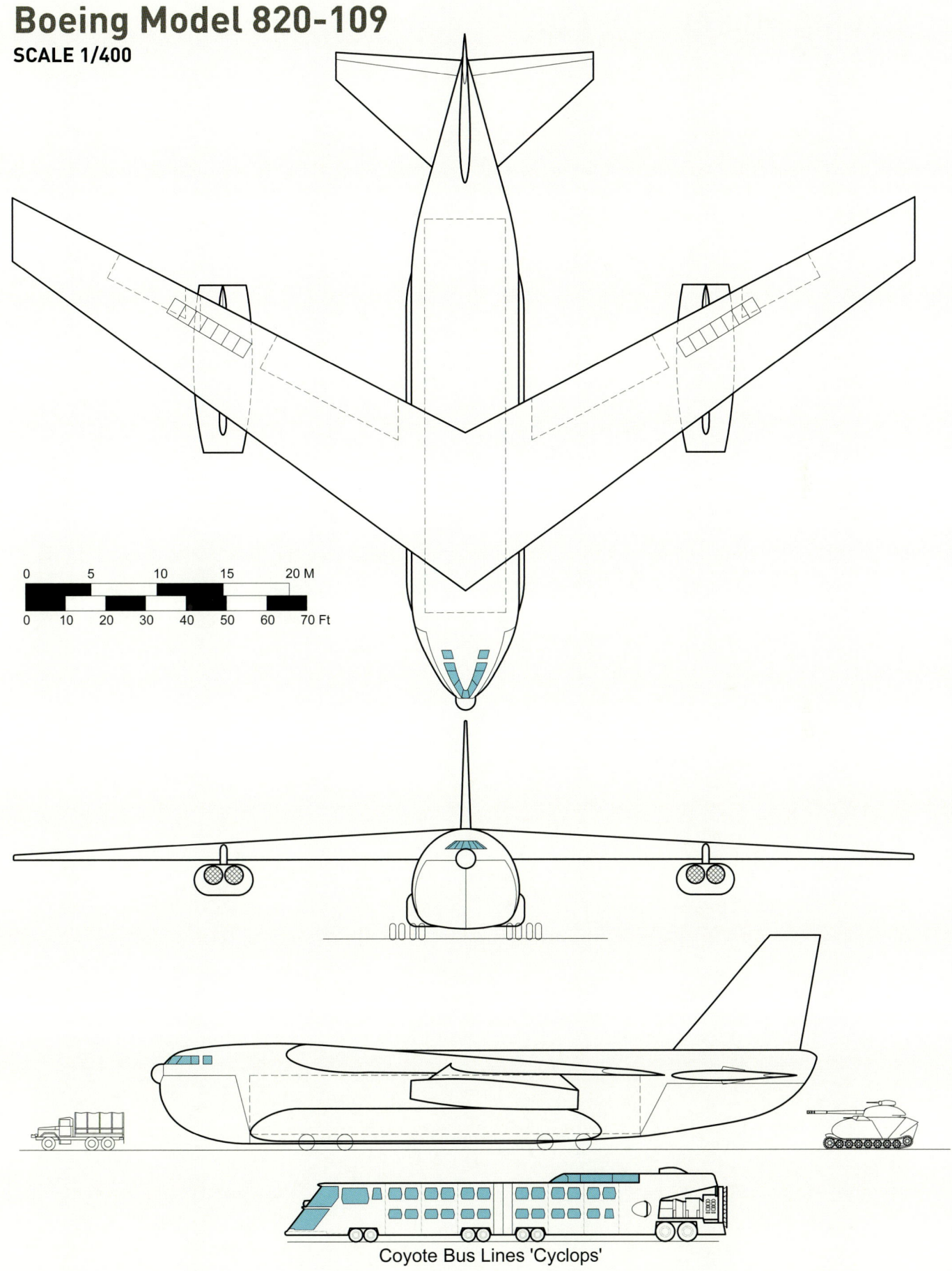

Boeing Model 820-110
SCALE 1/400

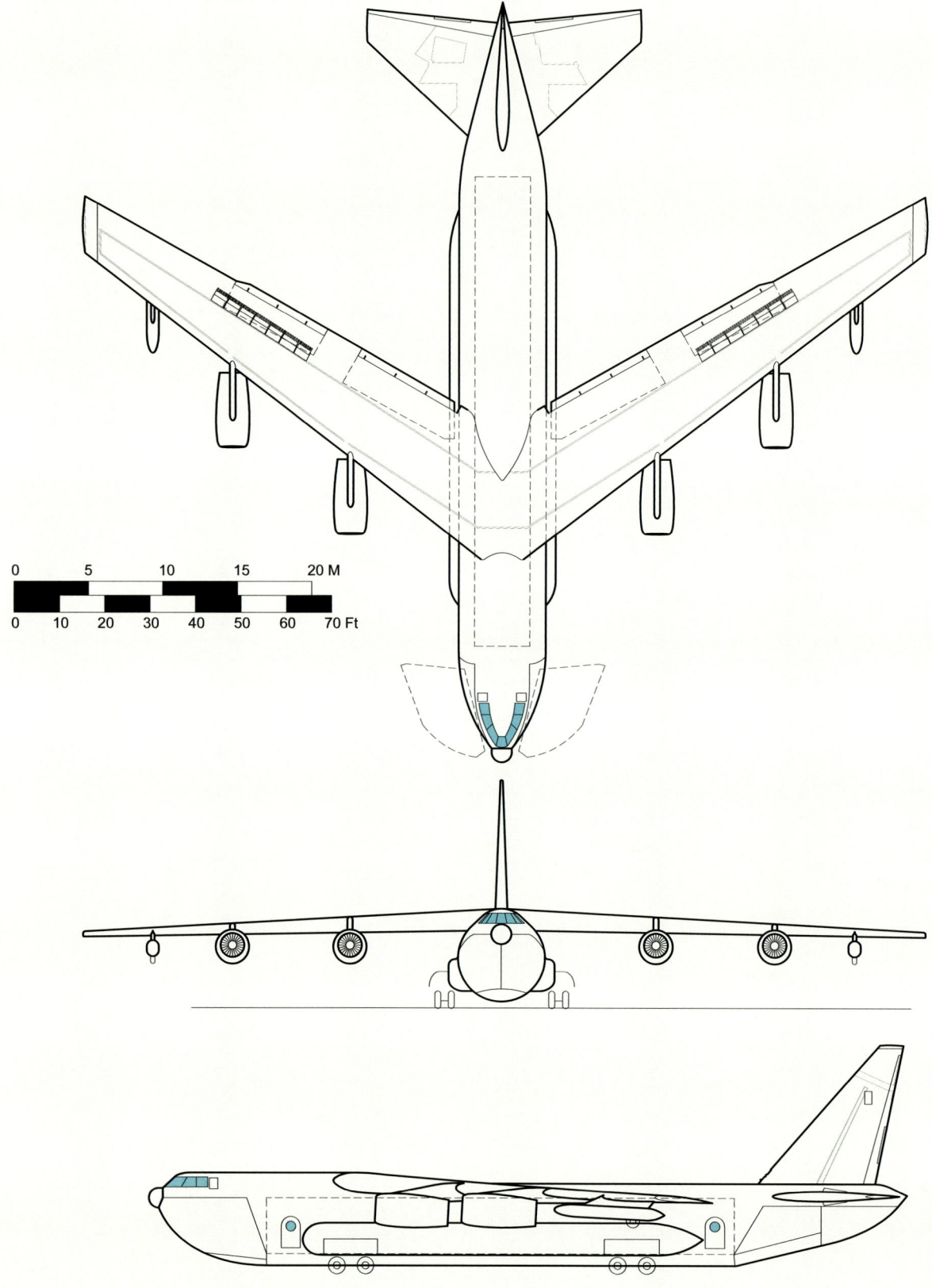

Boeing Model 820-111
SCALE 1/450

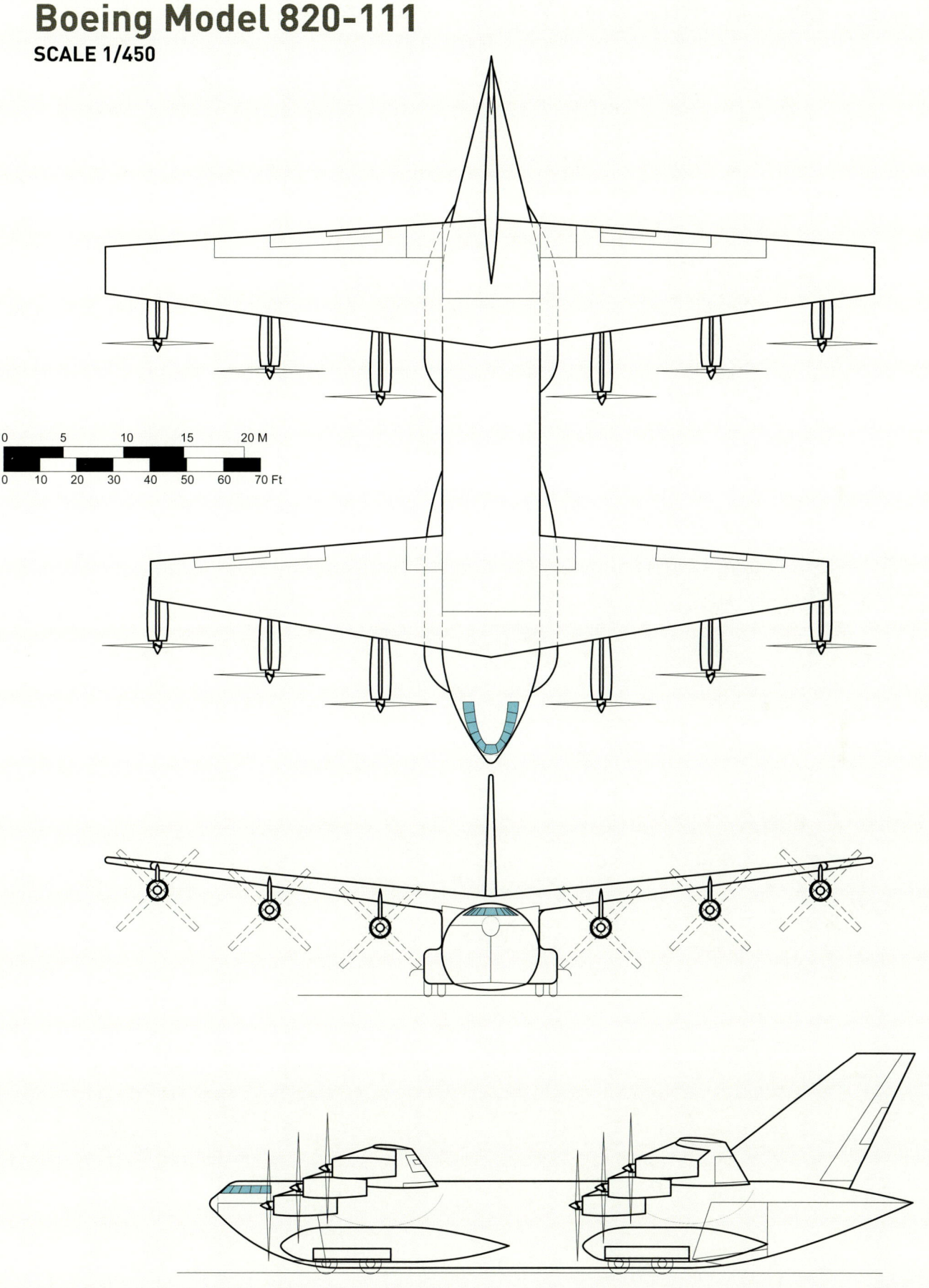

Boeing Model 820-111
SCALE 1/450

Bell V-44 to scale

Boeing Model 877-1
SCALE 1/350

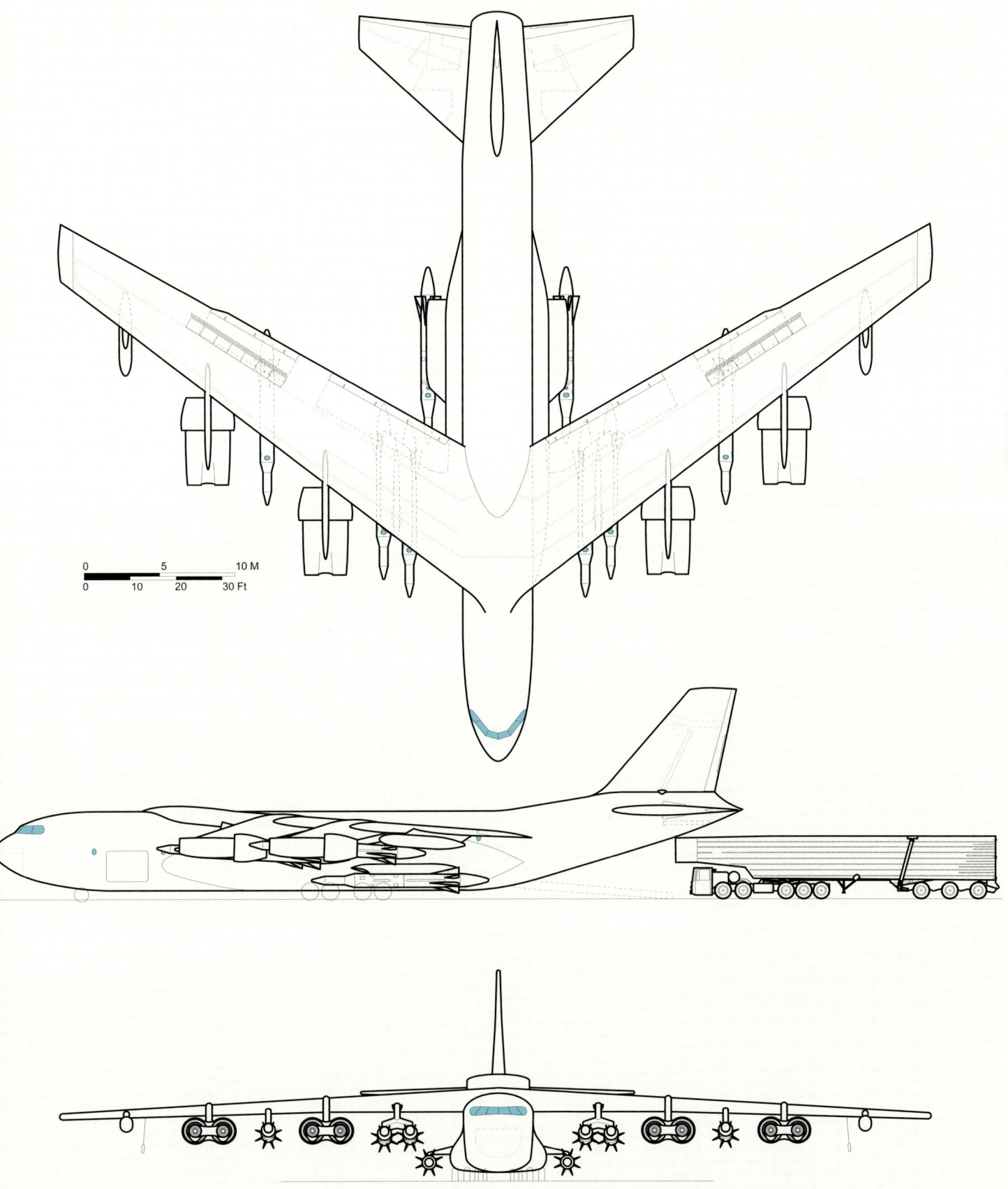

arrangement is similar to, though obviously distinct from, that of the eight-Skybolt B-52H AML II which had two missiles under each fuselage pylon.

Model 877-1 obviously would not fulfill both the logistics carrier and missile carrier role at the same time. But a cargo transport like this was something that the Air Force was quite interested in at the time. The same year the Air Force expressed interest in the MPLEA it also put out the requirement that led directly to the Lockheed C-141, which was also proposed as a Skybolt carrier (four missiles).

A cargo aircraft would seem a reasonable basis for a missile launcher; such aircraft are already capable of carrying heavy payloads, and usually over a long distance. Range is less of a necessity for ballistic missile launchers; rather, duration is more important. But there would be a down side, one unrelated to technology: who would be responsible for an aircraft like this? The Strategic Air Command would operate a nuclear missile launcher but the Material Air Transport Service/Military Airlift Command would be in charge of cargo transports. The jurisdictional fights between these commands would doubtless have proven protracted, with the possibility of the aircraft being passed around on a time share basis or the SAC ending up with its own transport aircraft.

Aero Spacelines 'Pregnant Princess'

As NASA began making the dream of rockets to the moon a reality, it needed a way to transport the very large launch vehicles from where they were built to where they would be launched. Often this meant transporting large cylindrical structures the size of mansions from California to Florida, a task readily describable as 'non trivial'.

Many companies put forward ideas for modifying aircraft to transport these relatively lightweight yet still vast payloads. One company actually went and did it: Aero Spacelines Corporation of Van Nuys, California, took several surplus Boeing 377 Stratocruiser turboprop transports and converted them into the 'Mini Guppy', 'Pregnant Guppy' and 'Super Guppy' transports by gutting the bodies and adding an all-new very large fuselage capable of holding large rocket stages.

The Guppys were used by NASA for transportation of Gemini, Saturn and Apollo hardware; the Super Guppy was capable of transporting the S-IVb stage from the Douglas manufacturing facility in southern California to Cape Canaveral in Florida. Formed about 1960, Aero Spacelines only lasted till 1967, when it was sold.

The larger S-II stage, also built in southern California, could not be carried by the Super Guppy, nor by any other aircraft. Aero Spacelines had an answer for that: an even larger 'Guppy', also to be built from a mothballed aircraft. In order to carry the S-II, the new Guppy would need to be substantially larger than the Boeing 377. Aero Spacelines found a proper foundation in the form of three Saunders-Roe 'Princess' flying boats.

The Princess was intended to be a crowning achievement of the British aviation industry in the years shortly after the Second World War, the largest all-metal flying boat ever built. The Princess would fly the long distance routes to connect the far-flung points of the Empire with the only kind of aircraft that could make it: large turboprop flying boats. But by the time the first Princess flew in 1952, it was becoming clear that the day of the flying boat wasn't just ending, it was over.

The de Havilland Comet and Boeing 707 soon drove the final nails into the coffin of water-based passenger transport aircraft, and the lone flying example of the Princess and two incomplete stablemates were mothballed in 1954. The three airframes were cocooned to protect them from vagaries of English weather, rumored to be somewhat damp from time to time.

In 1964, after several other failed attempts to revive the Princess airframes, Aero Spacelines contracted to purchase them for the purpose of Guppification. It seems that Aero Spacelines was looking at two somewhat different approaches: turboprops and turbojets. Neither concept currently has much data to back it up, at least as far as this author has seen.

The turboprop approach at least has a single sizable layout diagram to define it, found in the archive of the NASA-Marshall Space Flight Center in Huntsville, Alabama. The cockpit, wings and tail of the Princess would be mated to a gigantic, bulbous new fuselage. The unusual engines of the Princess – four Bristol Coupled-Proteus 610 turboprops (two engines hooked together through a single gearbox) and two conventional Bristol Proteus 620 turboprops – would be removed and replaced with six Rolls-Royce Tyne turboprops.

The wings would be extended by attaching them not to the fuselage directly but to new stubs, and the wingtip deployable floats would be replaced with conventional aerodynamic wingtip fairings. The new fuselage had a blunt round nose, split down the middle to open two doors. The cargo hold was 40ft in diameter, large enough to hold the 33ft-diameter S-II stage. Landing gear would be stored within four small faired pods on the lower extremities of the fuselage.

The result was an aircraft that looked as though it originated in a children's cartoon series, with a bulbous, oversized body fitted with undersized wings. But while the cargo it was meant to haul was voluminous, it was fairly light for its size; the aircraft

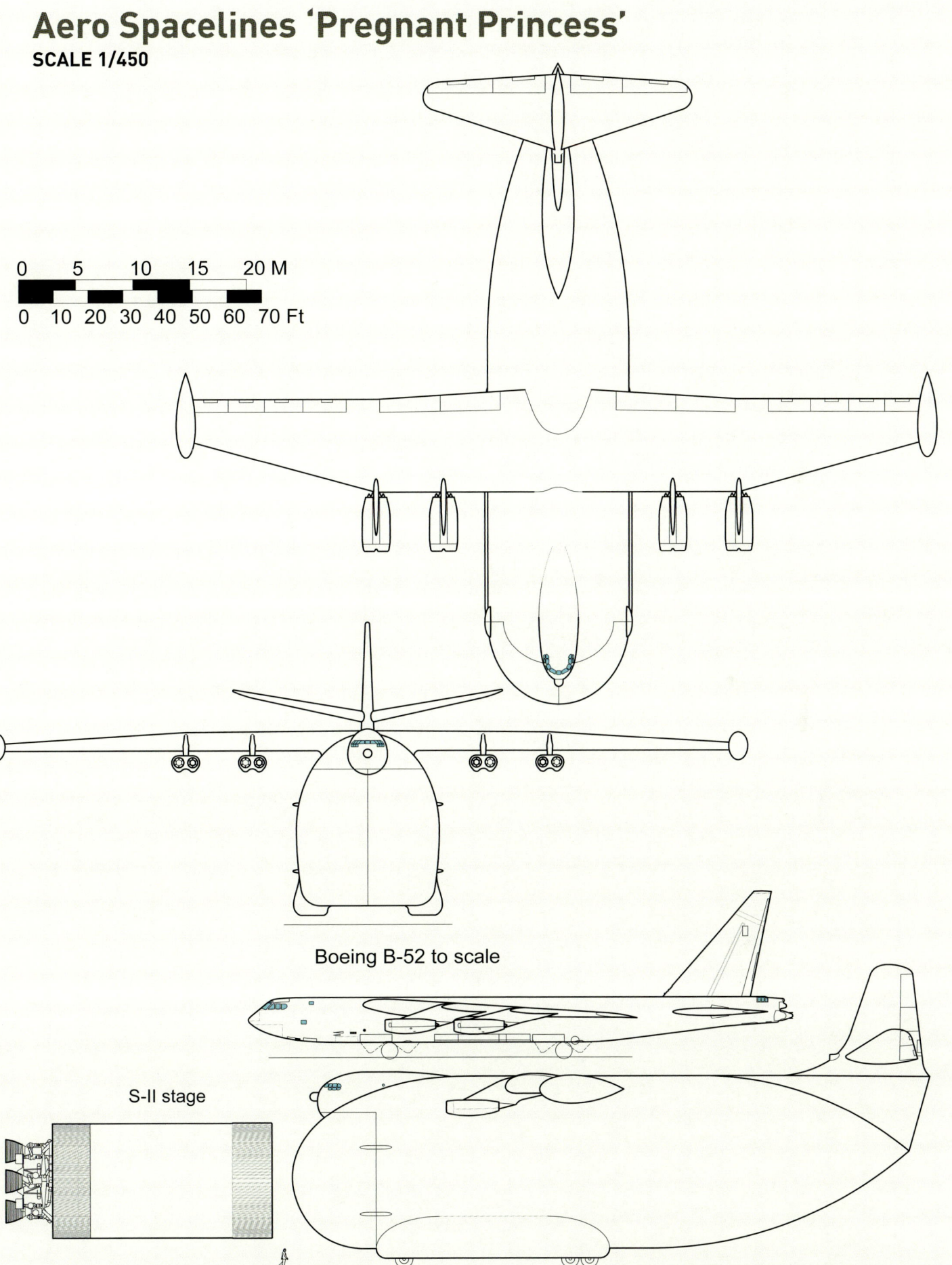

Aero Spacelines 'Colossal Guppy'
SCALE 1/500

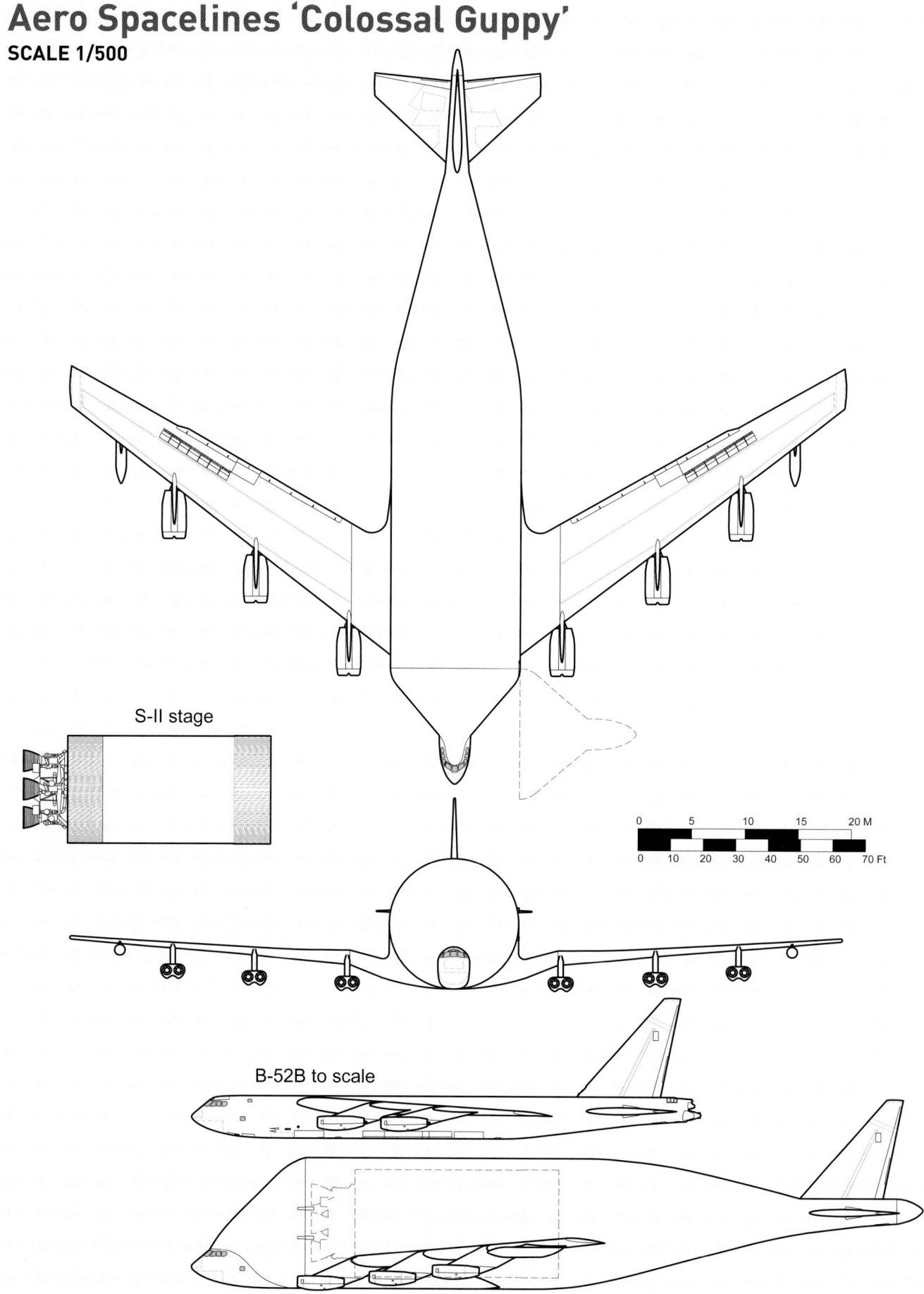

S-II stage

B-52B to scale

doubtless would have been able to fly, but it would have done so slowly and ponderously.

Aero Spacelines also examined a variation of the idea using turbojets rather than turboprops, known from a single artist's rendering and a few scraps of data. It was not a simple matter of swapping engines; instead, it was in many ways quite a bit different. The wings were not extended but retained the original span of the Princess and wingtip floats were replaced with teardrop shaped pods, presumably fuel tanks.

The contours of the aft fuselage were somewhat different, the cockpit appears to have been located slightly lower, the landing gear fairings would be converted to two longer fairings. The turbojets were not defined other than by their thrust ("18,000lb. thrust range") and their appearance. These were clearly to be engines and nacelles from the Boeing B-52 bomber; from the available artwork, the pre-B-52F configuration. It may have been something of a chore for Aero Spacelines to actually procure important portions of the B-52 bomber for the project, but by this time some early B-52s were already being mothballed.

It appears the jet-powered version preceded the turboprop version. The jet version was illustrated in the press as early as February 1964; the turboprop diagram is dated at the very end of June 1964. While performance data is lacking for both, it seems reasonable to surmise that analysis of the original jet version showed it to be incapable of flying from southern California to Florida. And thus more economical turboprops, as well as wing extensions, were added to improve range.

Unfortunately, when the cocooning material was removed from the stored Saunders-Roe Princesses, it was found that the British climate had done the airframes no favours; they were by that point essentially ruined. Conversion to anything other than scrap metal would be impractical and cost prohibitive. With no hope of ever being useful again the airframes were all broken up by 1967.

With the lack of an aircraft capable of carrying the S-II stages, they were instead shipped from California to Florida on barges, towed down the west coast of Mexico, east through the Panama Canal and then north through the Caribbean. Flying them would have been far faster, but it simply was not an option.

The official name of this aircraft does not seem to have been given. However, Aero Spacelines founder and President Jack Conroy referred to it as the 'Pregnant Princess', noting that the British might well have been slightly offended by that name had the project gone forward. Apparently princesses don't get pregnant.

The first version of the Aero Spacelines Pregnant Princess would have had a 220ft span, been 185ft long, had a gross take-off weight of 460,000lb and a maximum payload of 200,000lb. Propulsion was by eight J57s.

Aero Spacelines 'Colossal Guppy'

Aero Spacelines was not about to give up on the idea of aircraft conversions for transportation of the S-II stage. Another idea – dating to 1966 – took not just the engines, nacelles and pylons from the B-52, but virtually the whole aircraft. The 'Colossal Guppy' used everything from the B-52 except most of the fuselage. That was replaced with a far more voluminous fuselage, large enough (about 36ft in diameter) to fit the S-II stage.

The B-52's original engines were apparently not quite powerful enough to overcome the additional weight and drag, so the wings were attached to sizable wing stubs and two additional pairs of engines were added, bringing the total number of J-57 turbojets up to 12.

The 'Colossal Guppy' is sadly poorly described in available literature, with just a few artist concepts available (an early concept shows it with the extended wings, but without the additional engines). The reconstruction shown here is therefore somewhat provisional. It was still being shown in the press in the summer of 1967, but in the end was not built. Aero Spacelines Corporation was sold in 1967 after financial troubles; this doubtless played a part in the Colossal Guppy fading away.

Turbo-Three Virtus

From the beginning, the point of the Space Shuttle was that it could land like an airplane on a conventional runway. And it did not need to be a runway at the Kennedy Space Center; it could be a runway in California, or in Spain, or even on Easter Island if circumstances called for it. But to be launched again, the Shuttle would need to be returned to KSC.

For most of the development period of the Shuttle, the answer was clear: the Shuttle orbiter would use turbofan engines to fly itself back home. If these were not the jet engines that the orbiter would carry onboard throughout the mission for crossrange extension or go-around capability, then they would be engines that could be bolted onto the orbiter as required.

The problems with this plan were many and, in the end, insurmountable (or at any rate not worth the bother of surmounting). So an orbiter that could fly itself was off the table; no jet engines would be carried onboard and no mounting attachments for jet engines would be provided. Instead, the orbiter would be transported by a large carrier aircraft.

Conroy Virtus Turbo-Three Corporation
SCALE 1/750

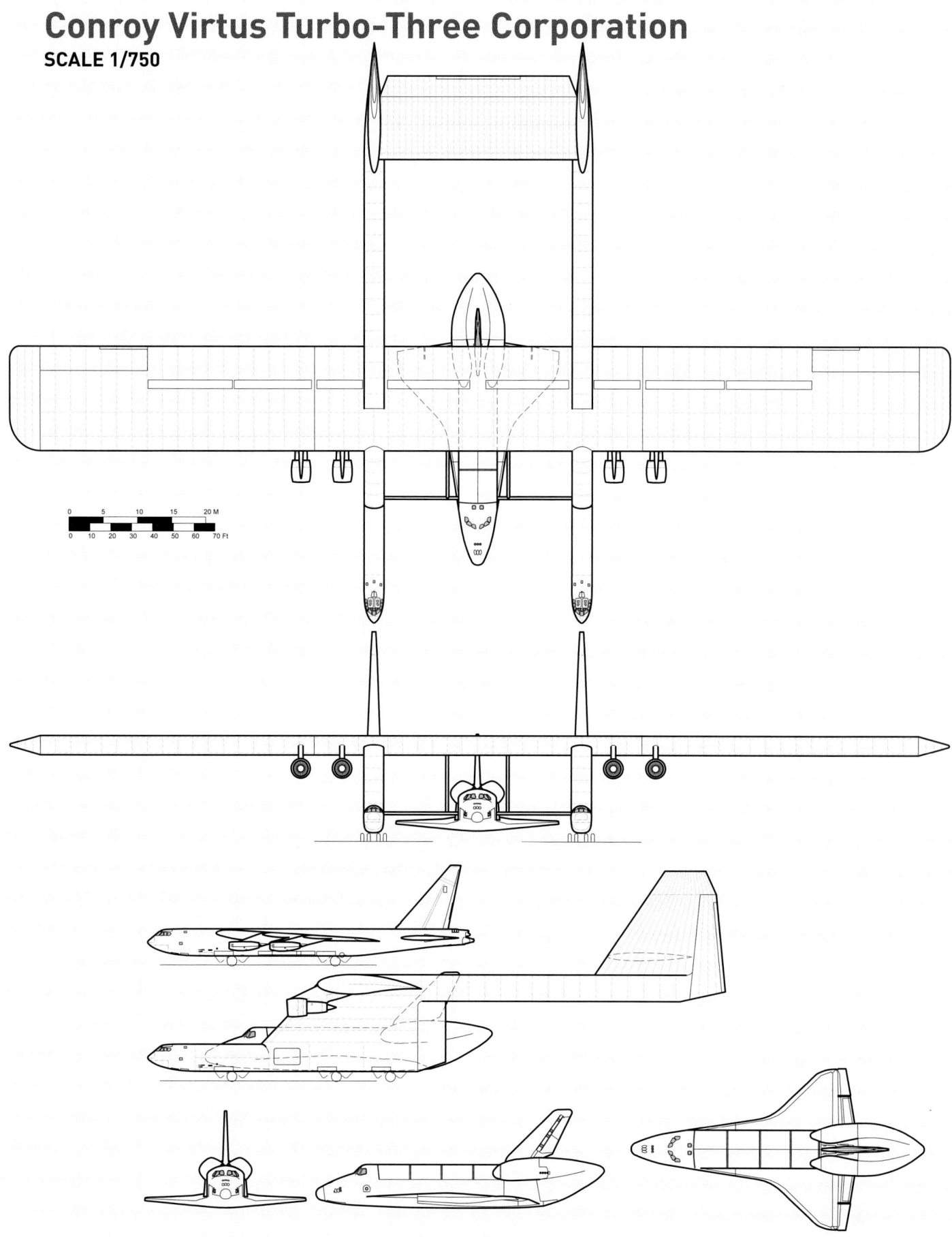

The Boeing 747 was selected as the Shuttle Carrier Aircraft and served in that role successfully for decades. Lockheed, unsurprisingly, proposed to use its C-5 Galaxy cargo jet for the role. Both Lockheed and Boeing also proposed highly modified twin fuselage variants of their C-5 and 747 carriers; the twin fuselage versions would have greater lift capability and would suspend the orbiter underneath the wing centre section.

This arrangement had the advantage of letting the carrier simply drop the orbiter during development tests. The sudden loss of weight would cause the carrier to spring upwards while gravity dragged the orbiter downwards; separation would be assured, and was presumed to be safer than releasing the orbiter off the top. Additionally, by suspending the orbiter, the need for additional lift equipment would be reduced. Instead of a massive crane to place the orbiter atop the aircraft, winches inside the carrier aircraft's wing centre section could simply haul the orbiter up to where it could be latched in place.

Lockheed and Boeing were not the only ones to realize the potential advantages of the multi-body approach for the Shuttle carrier aircraft. Another such design was put forward by John Conroy, designer and builder of the 'Guppy' series.

In 1968, Conroy formed another company following the sale of Aero Spacelines the previous year. This new firm was Conroy Aircraft and it gained success by converting a Canadair CL-44 into a large-capacity transport. This was done by adding a larger diameter fuselage, much as was done with the Guppy series, to create the 'Conroy Skymonster'. Conroy Aircraft also delved into the turboprop conversion of numerous piston engined aircraft, but went out of business in 1972.

One of the conversions was of the Douglas DC-3, forming the 'Turbo-Three'. This aircraft gave the name to yet another company founded by Conroy, the Turbo-Three Corporation. It was at this company in 1974 that Conroy proposed his most audacious aircraft conversion project: the Conroy Virtus.

While several versions of the Virtus were drawn up, the design that seems to have been studied in greatest detail used major fuselage elements from Boeing B-52s. Giants of the sky in their day, in the Virtus the B-52 elements would end up as little more than landing gear pods. These would be connected, by means of very stout pylons, to a large rectangular wing of 450ft span (aspect ratio of 9, surface area 22,166sq ft), and to a twin-boom tail. The Shuttle Orbiter, External Tank, two solid rocket motor cases or a dedicated cargo pod could be carried under the centre of the wing, between the pylons.

Ceiling for the Virtus was 35,000ft; cruise speed was 300mph, with a maximum range of 3,000 miles. Four Pratt & Whitney JT9D-3A turbofans provided 160,000lb of thrust. Maximum payload was 375,000lb; gross takeoff weight was 850,000lb.

The Virtus would have been easily capable of either ferrying the Shuttle Orbiter from place to place, or hauling it up to 35,000ft for drop tests. Unlike the 747 carrier aircraft, for those tests the Virtus would not need to dive out from underneath the orbiter; it would simply drop it. Additionally, the large volume between the pylons would permit the transport of all manner of outsize cargo that could not otherwise get proper transportation. A special cargo pod could be carried; shaped much like the Shuttle External Tank, the pod had an interior diameter of 35ft and an overall length of 184.8ft. It could comfortably fit the External Tank within. Empty weight of the pod was 60,000lb; it could contain 315,000lb of cargo.

At the time the Virtus received some publicity and even spent time in a NASA-Langley wind tunnel as a 0.0293 scale model, but nothing more than that came of it. Despite the advantages, the negatives were quite large… an essentially all-new, gigantic aircraft is going to be trickier than simply modifying an existing jetliner. Around 1990, the Myasishchev Design Bureau produced the M-90 series of outsize cargo lifters that followed much the same design practices, but these were never built either.

General data updates

Aircraft	Source Grade	Crew	Span	Wing area (sq ft)	Length	Engines	Dry weight (lbs)	Design Fuel (lb)	Design Payload (lbs)	Max Payload (lbs)	Gross weight (lb)	Range (n.mi.)	Cruise speed	Max speed	Ceiling (ft)
B-47 development															
Boeing Model 413	3	3	140 ft	1867	108.67 ft	4 GE TG-180	49,180	26,962			77,700				
Boeing Model 422	3	4	139 ft 6.8 in	1804.5	110 ft 9 in	2 turboprops									
Boeing Model 424	3	4	115	1100	91 ft	4 GE TG-180	42,155				80,000				
Boeing Model 425	3	4	139 ft 6.8 in	1867	108 ft 6 in	4 GE TG-180									
Boeing Model 426	3	4	114 ft. 7.74 in		91 ft	2 turboprops									
Boeing Model 432	3	4	113 ft 8 in	1300	101 ft 8 in	4 GE TG-180	56,295	33,047	8,432	18,000	100,000				41,700
Boeing Model 446	2	4	113 ft 8 in	1300	101 ft 8 in	4 GE TG-180 + 2 turbojets									
Focke Wulf mit 2 HeS 109-11	4	1	12.65 m	27.0 sq m	14.2 m	2 HeS 109-11	4225 kg	2775 kg	1000 kg		8100 kg		900 to 1000 kph		
Boeing Model 432-swept wing	2	4			101 ft 8 in	4 GE TG-180								550 mph	
Boeing Model 448	4	3	100 ft 0 in	1300	109 ft 3 in	6 GE TG-180	77,517	37,455	8,430	22,000	125,000		555 mph @ 25,000 ft	555 mph @ 35,000 ft	36,000
Boeing Model 450-1-1	4	3	100 ft 0 in	1300	108 ft 1 in	6 GE TG-180	75,186		8,585	22,000	125,000	4,100	560 mph @ 25,000 ft	640 mph @ 25,000 ft	36,000
Boeing Model 450-2-2	4		116' 0"		106' 9.2"	6 GE TG-180									
Douglas XB-43	4	3	116' 0"	1428	108' 0"	6 GE J35	76,000		22,000		125,000	4,000		578 mph	38,000
North American XB-45 (data B-45A)	4	4	89.0'	1175	75.3'	2 J47-GE-7+ 2 J47-GE-9 or -15	45,694	5746 gal	8,000	22,000	91,775	1036	377 @ 38-43 kft	492 kts @ 4,000 ft	46,000
Convair XB-46	3	3	113' 0"	1285	105' 9"	4 GE J35	43,436	35,360	8964		89,770	2870	439 mph	545 mph	40,000
Martin XB-48	4		98' 6"	1104	86' 7"	6 GE TG-180	51,884		10,000	22,000	99,500	1566	361 knots	454 knots	39,400
Lockheed L-173	3	8-10	173 ft 0 in	2000	128 ft 6 in	6 Westinghouse XJ40-6			10,000	40,000	240,000	4765	413 kts	530 kts at sea level	45,800
Douglas Model 1126	4	6	102.2 ft	1428	119.9 ft	6 Westinghouse XJ40-WE-6									
Northrop N-31	4		1540"	3070	889"	6 Westinghouse X40E2						2,781	520 mph		
Northrop N-31A (2 eng)	4		128.33 ft		74.67 ft	2 Turbodyne XT-37					161,540	2,400 radius	450 knots	510 mh @ 42,800 ft	
Northrop N-31A (4 eng)	4		128.33 ft		74.67 ft	4 Allison XT-40					175,400	2,400 radius	440 knots		
Republic AP-42	2		~136 ft 3 in		~150 ft	8 Turbojets									
450-3-3 XB-47	4	3	116 ft	1428	105.5"	6 General Electric J35-GE-7	74,623	9957 gal		22,000	162,500	2300	462 kts @ 35 kft	502 kts @ 15 kft	37,500
B-47A	4	3	116 ft	1428	108' 0"	6 General Electric J47-GE-11	73,149	9789 gal		22,000	150,500	2648	490 kts @ 35 kft	536 kts @ 10,500 ft	43,800
B-47B	4	3	116 ft	1428	106.7 ft	6 General Electric J47-GE-11	77,830	12,770 gal	10,000		181,440	3360	486 kts @ 35 kft	537 kts @ 1,800 ft	41,500
Boeing Model 450-16-17	4	3	116' 0"	1602	106' 10.2"	4 Westinghouse XJ40-WE-6	73,400	99,290	10,000		125,000	2,270 radius	432 kts	485 kts @ 35 kft	40,600
YB-56	3	3	116 ft	1428	106.7 ft	4 Allison J35-A-23	75,250	13583 gal	10,000	25,000	185,000	4225	426 knots		
Boeing Model 450-155-33	4	4	116' 0"	1570	109' 5"	4 Pratt & Whitney J57-P-1W	86,800				125,000			510 mh @ 42,800 ft	
B-47X	2	4	116' 0"	1570	109' 5"	4 Pratt & Whitney J57-P-1W									
Boeing Model 450-166-38	3	3	116' 0"	1920	107.1'	4 Pratt & Whitney J57-P-1W	81,910	18,011 gal	10,000		221,090	2900 radius	486 knots @ 45,050 ft	531 knots at 15kft	38,000
Boeing Model 450-9-3	3	3	116' 0"	1300	107' 5.6"	4 Allison Model 500 Turboprops	75,940	81,060	10,000		157,000		473 mph		
Boeing Model 450-30-10	4	3	116' 0"	1428	106' 10.2"	4 Allison T-40-A-8 turboprops	81,180	93,040			180,000	5690	460 mph		
Boeing Model 450-31-10	4	3	116' 0"	1428	106' 10.2"	4 Allison T-40-A-8 turboprops	76,190	91,464			180,000	6040	480 mph		
Boeing Model 450-32-10	4	3	116' 0"	1428	106' 10.2"	2 P&W J57 plus free turbine	85,350	81,160			180,000				
Boeing Model 450-33-10	4	3	116' 0"	1428	106' 10.2"	2 P&W JT-3A turboprops	84,840	74,724			180,000				
Boeing Model 450-59-10	4	3	116' 0"	1428	106' 10.2"	4 Allison T-40-A-8 turboprops	78,070	89,190			180,000	5986	465 mph	490 mh @ 41,300 ft	
Boeing Model 450-60-10	4	3	116' 0"	1428	106' 10.2"	4 P&W PT-2E Turboprops	74,730	92,530			180,000	6100	455 mph	483 mph @ 35,100	
B-47D	3	3	116' 0"	1428	106.8'	2 Wright YT49-W-1 2 GE J47-GE-23	79,800	13,979 gal	10,000		184,428	2717 radius	479 knots @35kft	519 knots @ 13,500 ft	41,500

Aircraft	Source Grade	Crew	Span	Wing area (sq ft)	Length	Engines	Dry weight (lbs)	Design Fuel (lb)	Design Payload (lbs)	Max Payload (lbs)	Gross weight (lb)	Range (n.mi.)	Cruise speed	Max speed	Ceiling (ft)
B-47E	4	3	116' 0"	1428	107.1'	6 General Electric J47-GE-25	79,074	18,000 gal	10,000		230,000	1780 radius	491 knots @ 35 kft	497 knots @ 20kft	31,500
Boeing Model 450-36-10	4	3	155' 0"	1650	106' 10.2"	4 Allison J35-A-23	75,350	91,104			208,000	5880	480 mph		
Boeing Model 450-57-10	4	3	145' 0"	1602	106' 10.2"	4 Allison J35-A-23	86,670	120,507			220,000	5840	480 mph	503 mph	
Boeing Model 701-333	3		81' 4"		123' 4"	4 General Electric J79					148,300				
Boeing Model 450-65-10A	4	3	87' 6"	2190	106' 10.2"	4 P&W XJ-57-P5 w/afterburners	85,680				180,000				
Boeing Model 450-65-10C	4	3	87' 6"	2190	113' 1.7"	4 P&W XJ-57-P5 w/afterburners					180,000				
Boeing Model 450-65-10F	4	3	116' 0"	2530	113' 1.7"	4 P&W XJ-57-P5 w/afterburners	91,370				200,000				
Boeing Model 450-151-31	4		87' 6"	2190	124' 7"	4 P&W J57-P1 w/afterburners	89,490		9,600		180,000	4500			
Boeing Model 450-152-27	4		87' 6"	2100	110' 11"	4 P&W J57-P1 w/afterburners	77,400		10,000		180,000	4026			
Boeing Model 450-154-32	4		87' 6"	2100	123' 8"	4 P&W J57-P1 w/afterburners	82,880		10,000		180,000	3940	490 mph	520 mph @ 42,350 ft	36,500
Boeing Model 450-139-10	4		116' 0"	1428	106' 10.2"	6 Allison Model 509 Fan Jets	73,400	82,730			170,000	2400		600 mph at sea level	36,500
Boeing Model 450-148-30	4	5	116' 0"	1428	106' 4.1"	2 T-49 turboprops 2 J67-W1 turbojets	88,280	61,870	16,000		168,200	2400		600 mph at sea level	
Boeing Model 450-150-30	4	5	116' 0"	1428	106' 4.1"	4 Westinghouse J-40-WE-6	73,400	60,270	16,000		220,000	2700	360 kts @ 23-40 kft		
LAMP LIGHT	3		116' 0"	1428		6 General Electric J47-GE-25			11,200						
All American Eng Pusher	3	1	~51.75'		~59.3'	4 Turbojets (J47?)									
B-36-B-47 Tip tow	2		462'												
499-1	3	5	170' 0"	9350	98' 6"	4 P&W J-57-P-5		15,954 gal			180,000			631 mph @ 22 kft	40,000
GAM-67 Crossbow	3	0	12' 7.8"		20' 0.2"	1 Continental J67-T-17 turbojet					2700	300	Mach 0.86		
DB-47E RASCAL	3	0	16.7'	114.65	32.0	1 Bell LR 67-BA-9	5857	287 gal (Fuel) 586 gal (Oxy)	2810		18,169	90		Mach 2.84 @ 70 kft	70,000
DB-47E Bold Orion	2	0	~80.5'		~31.9'	1 Thiokol TX-20 Sargeant 1 ABL X-248 Altair						1100			160 n.mi.
Canadair CL-52	3	3	116 ft	1428	106.7 ft	6 General Electric J47-GE-11 + 1 Orenda Iroquois									
Boeing Model 360	3		220 ft 0 in	5300	120 ft 0 in					40,000					
Boeing Model 361	4		200 ft 0 in	3970	130 ft 0 in					40,000					
Boeing Model 362	3	8	300 ft 0 in	9041,3	176 ft 0 in	eight undefined		223,400	10000	88,000	470000	10,000	350 mph @ 30k ft		40,000
Boeing Model 365	4	9	219 ft 0 in	4000	157 ft	four P&W X Wasp	124950	86410	10000	51,200	230000	N/A	189 mph @ 15 k ft	N/A	N/A
Boeing Model 384	4	9-12	204' 3.25"	3087	144 ft 4 in	4 P&W Wasp Major R-4360-5	103039	92030	10160	64,000	210500	7430		318 mph @ 30k ft	29,000
Boeing Model 385	3	9-15	251 ft 0 in	4500	161 ft 0 in	six P&W Wasp Major	160020	139180	10000	120,000	315000	8210		329 mph @ 30k ft	20,000
MCD 392	4	15	280'	7140	129' 4"	8 Allison V-3420	216,610		16,000	86,000	500,000	11,250 (16 klbs) 6,100 (120 klbs)		342 mph @ 20k ft	
B-36D	4	13	230' 0"	4772	162' 1"	6 P&W R-4360-41 radials 4 GE J47-GE-19 turbojets	160,974	170,148	10,000	24,000	357,500	6278	182 knt @ 10-25 kft	373 kts @ 34,500 ft	36,800
Consolidated 10-engine high performance patrol	3	5	148' 4"	2200	129' 9"	6 Westinghouse 25D turboprops 4 Westinghouse X24C-4A turbojets	93,481	68,000		41,274	167,000	1650 radius	325 mph	485 mph @ 25 kft	30,000
Consolidated heavy Bombardment	3	9	141' 8"	2000	145' 4"	4 Wright TGAA turboprops	90,430	55,000	22,275		175,000	5670	375 mph	477 mph @ 35 kft	41,000
Consolidated long range heavy bombardment	4	11	167'	4000	172' 6"	4 T-35 turboprops 4 TG-180 turbojets	116,443	98,200	10,000		235,000	2445 radius	364 mph	520 mph @ 35 kft	44,300
B-29 turboprop	2		141' 3"	1736	99' 0"	4 unspecified turboprops									
Boeing Model 461	3	5	164' 3.6"	1800	131' 6"	4 Wright T-35-1		39,400							
Boeing Model 462	4		221' 0"	3250	161' 2"	6 Wright T-35-1	87,960		82,040		360,000	3100 radius		382 knots	
Boeing Model 462-5	4		221' 0"		165' 4"	6 Wright T-35-1									
Boeing Model 464-17	4	11	205' 0"	3000	156' 0"	4 WAC 827 GTAI (T-35-5)	176,000			90,000	480,000				
Boeing Model 464-18	4	11	145'	1500	~128' 4"	2 Wright T-35-3	83,300		10,000		180,000	10,900		460 knots	

Aircraft	Source Grade	Crew	Span	Wing area (sq ft)	Length	Engines	Dry weight (lbs)	Design Fuel (lb)	Design Payload (lbs)	Max Payload (lbs)	Gross weight (lb)	Range (n.mi.)	Cruise speed	Max speed	Ceiling (ft)
Boeing Model 464-25	4	9	205' 0"	3300	147' 8"	4 XT35-W-3	149,340			12,000	400,000	12,665			
Boeing Model 464-27	4	9	205' 0"	3600	148' 8"	4 XT35-W-3	149,880			12,000	400,000				
Boeing Model 464-33	4	8	185' 0"	2600	124' 10"	4 XT35-W-3	149,880			12,000	260,000				
Boeing Model 464-34-3	4		185' 0"	2600	136' 4"	4 XT35-W-3					285,000				
Boeing Model 464-35	3	5	185'	2600	131.3'	4 Wright XT-35-W-5	132,410	127,500	15,000		280,000		435 knots	435 knots at 35,000 ft	42,100
Boeing Model 464-40-4	3		185'	2600	143' 4"	8 Westinghouse J40			15,000			6750		536 mph @ 35kft	45,200
Boeing Model 464-41	3	5	185' 0"	2600	130' 9"	4 XT35-W-3	133,560	21,400 gal	15,000		280,000				
Boeing Model 464-46	3		185' 0"	2600	130' 9"	6 Westinghouse XJ40-10	118,310				284,000				
Boeing Model 464-47	3		185' 0"	2600	130' 9"	6 P&W XJ-57	125,000				281,000				
"Hotel proposal"	3		185' 0"		138' 9"	8 P&W XJ-57-P-1		27,600 gal	10,000	25,000	330,000	6000		530 kts @ 35 kft	
Boeing Model 464-49	3	5	185' 0"	4000	137' 11"	8 P&W XJ-57	152,300	165,600	10,000	25,000	330,000	2660	453 knots	490 kts @ 49,400 ft	49,400
B-36G turboprop (B-60)	3		206' 4.79"	5239.11	168' 7.8"	4 Wright T35 turboprops									
YB-60	3	5	206.4'	5239	171.2'	8 P&W YJ57-P-3	136,292	42,106 gal	10,000	72,000	410,000	2910 radius	400 kts @ 37-53 kft	448 kts @35,332 ft	47,400
Douglas 1211-J	4	9	227' 6"	4700	160' 6"	4 turboprops	170,400				322,000	11,000		>450 knots	50,000
Douglas Model 1064	3		97'		83' 6"	4 turbojets									
Fairchild M-121	4	4	116' 6"	1900 3800 total	78' 10"	10 Westinghouse J46-WE-2	52,256	91,019 (AC) 85,550 (wing)	10,000		154,600 (AC) 251,600 (biplane)	5130 radius	487 knots	600 kts sea level	
RAND G-2	3	6	141'	4000	86.5'	4 turboprops		25,000 gal	10,000		300,000	8000	525 mph		
Northrop B-49	3	6	172.0'	4000	53.1'	8 Allison J35-A-15	88,442	14,542 gal		16,000	193,938	1483 radius	365 knots	428 kts @20,800 ft	45,200
Boeing Model 481-100	3	6	185' 0"	4500	73' 3"	8 Westinghouse XJ-40-WE6	118,780				320,000				
Boeing Model 474	3	8	135'	1500	118' 11"	4 Allison T-40-A2 turboprops	78,000		10,000		153,000	1878 radius	416 knots	440 knots	40,000
Boeing Model 479	3	7	150' 0"	2500	132' 1"	8 Westinghouse XJ-40-WE6	101,760		10,505		235,000	1737 radius	435 knots	550 knots	44,000
Boeing Model 483-11	3		130' 0"		114' 6"										
Boeing Model 483 delta	3		~111'		~76'										
Boeing Model 484-106	3	3	92.3'	2990	74.0'	10 Westinghouse XJ-40-WE-10	63,610	12,200 gal	10,250		150,000				
Boeing Model 484-230	3	4	110' 7"	8500	88.7'	6 turbojets		14,950 gal			175,000				
XB-52 Model 464-67	3	5	185' 0"	4000	152' 8"	8 P&W J-57-P-8	160,000	38,865 gal			405,000	3070 radius		483 kts @ 40 kft	39,000
B-52A	4	5	185' 0"	4000	156' 6.9"	8 P&W J57-P-1W	167,509	37,550 gal		34,000	420,000	3110 radius			
B-52 B	4	5	185' 0"	4000	~160' 1"	8 J57-P-1W/WA/WB/29W/29WA/18W	177,832	37,550 gal		63,000	420,000	3110 radius	453 kts @ 35-51 kft	546 kts @ 19,600 ft	47,300
B-52C/D	4	5	185' 0"	4000	156' 6"	8 J57-P-29WA/19W	170,126	41,500 gal		64,000	450,000	3305 radius	453 kts @ 33-51 kft	551 kts @ 20,200 ft	46,950
B-52E	4	5	185' 0"	4000	156' 6"	8 J57-P-29W/29WA/19W	172,720	41,500 gal		65,000	450,000	3320 radius	453 kts @ 33-51 kft	551 kts @ 20,200 ft	46,950
B-52F	4	5	185' 0"	4000	156' 6"	8 J57-P-43W/WA/WB	179,158	41,500 gal		65,000	450,000	3625 radius	453 kts @ 33-51 kft	553 kts @ 20,500 ft	47,600
B-52G	4	5	185' 0"	4000	157.6'	8 J57-P-43WB	158,737	47,975 gal		104,900	488,000	3915 radius	454 kts @ 33-52 kft	553 kts @ 20,500 ft	48,150
B-52H	4	5	185' 0"	4000	159' 4"	8 TF33-P-3	165,988	48,030 gal		105,200	488,000	4510 radius	456 kts @ 33-52 kft	555 kts @ 20,700 ft	47,800
Boeing Model 464-72	3	5	185' 0"	4000	152' 8"	4 P&W JT-3A turboprops	170,950				390,000				
Boeing Model 464-74	3	5	185' 0"	4000	152' 8"	4 P&W JT-3A turboprops	159,500				390,000				
Boeing Model 464-79-0	3	5	245' 0"	4560	152' 8"	8 P&W J57					445,000				
Boeing Model 464-149	3		230' 0"	4460	152' 8"	8 P&W J57P									
Boeing Model 464-EXT #1	3		230' 0"	4460	152' 8"	8 P&W J57									
Boeing Model 464-EXT #2	3		230' 0"	4460	152' 8"	8 P&W J57									
Boeing Model 724-15 (core)	4	4	93' 6"	2500	156' 8"	4 GE R56 AGT120	135,500		20,000		370,000			1725 knots	63,000
Boeing Model 724-15 full air vehicle	4	4	275' 2"	7500		4 GE R56 AGT120		365,700			580,000	3030 radius	489 knots		

Aircraft	Source Grade	Crew	Span	Wing area (sq ft)	Length	Engines	Dry weight (lbs)	Design Fuel (lb)	Design Payload (lbs)	Max Payload (lbs)	Gross weight (lb)	Range (n.mi.)	Cruise speed	Max speed	Ceiling (ft)
Boeing Model 464-197	4	3	122' 6"	5000	174' 4"	8 afterburning J67-W-1	172,140	38,868 gal	10,000	50,000	425,000				
Boeing Model 464-238	3	3	218' 0"	5600	156' 6.9"	8 P&W J67									
Boeing Model 464-239	3	3	185' 0"		156' 6.9"	8 P&W J67									
Boeing Model 464-245	3	5	185' 0"		156' 6.9"	4 J75		83,000 LH2 134,000 JP							
Boeing Model 464-246	3	5	197' 6"		180' 5"	6 J75		89,850 LH2 152,000 JP			375,000				
Boeing Model 491	3	6	205'	3000	109'	6 Allison T-40 turboprops		39,000 gal							
Fairchild XB-52 NEPA 1	3	3	185'	4000	~155' 9"	6 nuclear XJ-57					350,000		Mach 0.8 @ 35 kft		
Fairchild XB-52 NEPA 1	3	3	185'	4000	~151' 4"	6 nuclear XJ-53					340,000		Mach 0.8 @ 40 kft		
Fairchild NEPA N-14	3	7	185' 0"	4000	164' 8"	6 XJ-53 turbojets					340,000			480 kts @ 30 kft (nuc) 550 kts @ 21 kft (n+c)	41,500 (n) 46 kft (n+c)
Fairchild NEPA N-15	3	7	185' 0"	4000	149' 4"	4 XJ-53 turbojets					326,000			462 kts @ 27 kft (nuc) 520 kts @ 25 kft (n+c)	30,200 (n) 38 kft (n+c)
Fairchild NEPA N-16	3	7	185' 0"	4000	151'	6 XJ-53 turbojets					305,000			495 kts @ 30 kft (nuc) 555 kts @ 21 kft (n+c)	43 kft (n) 48 kft (n+c)
GE LF-2	2		185' 0"	4000	~153.5'	8 nuclear J73							Mach 0.75 @ 35 kft		
B-52 flying test bed	3		185' 0"	4000	~159.8'	8 J57-P-43WB 1 XNJ140E-1		157,000			450,000		Mach 0.6 @ 30 kft	Mach 0.84@ 44kft	
Boeing Model 702-138(1)-1 MX-2145	3		211'	5220	171'	8 Liquid Metal Cycle Turbojets			20,000		232,000			Mach 0.75	60,000
Boeing Model 703	4	5	140' 0"	2516	122' 1"	4 afterburning J40-WE-8	82,080		30,000		175,000				
Boeing Model 713-1-133	3	4	232.5'	9000	190.2'	14 P&W J75	260,400	329,600	10,000		600,000				
Boeing Model 713-1-133C	3	4	232.5'	9000	190.2'	16 GE X-84	252,650	337,350	10,000		600,000				
Boeing Model 713-1-138	3	4	245'	10,000	190.2'	16 GE X-84					600,000				
Boeing Model 809-1004	3		80' "	1270	152' 3"	34 J85 turbofans		320,000	10,000		463,000				
Standby Alert	3		200'	4475	156' 6.9"	4 P&W t57 turboprops	167,400				500,000				
B-52X 1	2		185' 0"	4000	157.6'	4 JT9D-70D turbofans									
B-52X 2	2		185' 0"	4000	157.6'	4 JT9D-70D turbofans									
B-52X 3	2		185' 0"	4000	157.6'	4 JT9D-70D turbofans									
CFM56 Re-engined B-52	4		185' 0"	4000	157.6'	6 CFM56 turbofans									
C-17 nacelles for B-52	3		185' 0"	4000	157.6'	4 PW2040 turbofans									
PW2037 re-engined B-52	4		185' 0"	4000		4 PW2037 turbofans									
Flashback	3	0	95.8' (dia)		297.3'										
XB-52/J75	2		185' 0"	4000	152' 8"	4 P&W J-57-P-8 2 J75									
JB-52E	2		185' 0"	4000	156' 6"	6 J57-P-29W 1 TF-39 or 1 JT-90 or 1 CF6									
NB-52E	2		185' 0"	4000	~173.3'	8 J57-P-29W		360,000							
NB-52	4		185' 0"	4000	~151.6	8 J57-P-1									
X-15	4	1	22' 4"	200	50.3'	XLR99-RM-1	11,374				31,275	275		Mach 5.92	314,750
X-15A-3	4	1	23.18'	603	62.43'	1 YLR-99	18,988				56,000			Mach 8.0	100,000+
HL-10	4	1	13.6'	160	21.17'	1 XLR-11 RM-13	5285	3536			10,009	45		Mach 1.86	90,303
M2-F3	4	1	9' 8"	160	22' 2"	1 XLR-11	5071				6,000	39		Mach 1.63	71,500

| 373

Aircraft	Source Grade	Crew	Span	Wing area (sq ft)	Length	Engines	Dry weight (lbs)	Design Fuel (lb)	Design Payload (lbs)	Max Payload (lbs)	Gross weight (lb)	Range (n.mi.)	Cruise speed	Max speed	Ceiling (ft)
X-24A	3	1	13' 8"	162	24' 6"	1 XLR11-RM-13	6300				11,450			1036	71,400
X-24B	3	1	19' 2"	330	24' 6"	1 XLR11-RM-13	7800				13,800			1164	74,130
X-38	3	0	12' 6"	162	24' 0"	N/A					14,900			500 mph	39,000
HiMAT		0	15.56'		22.5'	General electric J85-21		630			3370			Mach 1.6	
Shuttle Booster DTV	4	0	10.2' Dia.		51'	N/A					48,070				
F-15 RPRV	0	4.89 M		7.15 M	N/A					10,964 N			Mach 0.65	15,000 M
DAST "Blue Streak"	3	0	2.7 M	7.94 sq. m	7.15 M	Continental YJ69-T-406					853 kg			~Mach 1.2	
DAST-1	3	0	4.343 M	2.97 sq. m	8.61 M	Continental YJ69-T-406	730 g				1008 kg			~Mach 1.2	
DAST-2	3	0	227.84"	2.787 sq.m	8.61 M	Continental YJ69-T-406	870 kg	150			2350		M 0.8 @ 46,800 ft	~Mach 0.9	
OSC Pegasus	4	0	22' 0"	35 sq. ft	49' 2"	Orion 50S, Orion 50, Orion 38			700		42,000			Mach 25	108 n.mi.
X-43 Hyper-X	0	5' 0"		12' 0"	GASL scramjet							850+		Mach 9.736	109,440
X-24C-9	4	1	23' 0"	516.4	46.1'	YLR-99 RM 1	13,901	16,316			30,053			Mach 5.06	
X-24C baseline	3	1	24' 2.5"		52' 6"	XLR-99	16,137	39,300			55,437		Mach 6	Mach 7.4	
Rockwell HRA	4	1	~292'	539	~722'	3 Bell LR-81 Agena rockets		32,500			56,000			Mach 10+	~130,000
HYWARDS config A	3	1	37 '	1135	75.5'	XLR99-RM-1		40,000	60,800			500		7,000 ft/sec	
HYWARDS config B	3	1	32.5'	1075	80.5'	XLR99-RM-1		40,000	60,800			500		7,000 ft/sec	
Martin-Bell CTV	3	1	28.35'	395	57.5	F2+N2H4 rocket engine	13,620	55,200			16,690 (aircraft) 77,500 (AC & booster)			12,700 ft/sec	170,000
Republic CTV	3	1	23.8'	600	35'						16,000	50 to 400		supersonic	
Dyna Soar 814	4	1	18' 7.2"	330	34' 7.5"		6820				7800	Orbital	Mach 25	Mach 25	
Dyna Soar 814 Drop Test	4	1	18' 7.2"	330+50	56.67' 814 + booster	XLR-99		7500 (LOX) 6000 (NH3) 531 (H2O2)			30,135			Mach 6	
Dyna Soar Launcher	2	1		1020	64.5' (booster + DS) ~75.66' (overall)	3 150,000 lbf rocket engines		166,049			200,000	Orbital	Mach 25	Mach 25	
Dyna Soar 2050E glider	4	1	250"	345	424.14"						15,200				
Dyna Soar 2050E w/transition	4	1	250"	345	424.14"	1 Thiokol XM-92		2200+		1109	21,447.12			supersonic	
Convair "HAZEL" MC-10	4	1	67.71'	1985		1 pentaborane ramjet		6330			30,525	3200	Mach 3.0		131,400
McD ISINGLASS	3	1	27.5'		85.8'	1 P&W XLR-129	25,450	107,500			132,770	~Global		Mach 20	200,000
Junkers RT 8 Stage 1	3	1	11.8 M		30 M (fuselage)	1 rocket engine	21 tonnes				79 tonnes				
Junkers RT 8 Stage 2	3	1	8.8 M		20.5 M (fuselage)	3 rocket engines	13 tonnes				69 tonnes	Orbital	Mach 25	Mach 25	
Junkers Raumstransporter System B	2		15.4 M (s 1) 13.4 M (s 2)		25.9 M (stage 1) 14 M (stage 2)							Orbital	Mach 25	Mach 25	
ALCM AGM-86B		0	143.6"	11	249.0"	1 Williams F107-WR-101			280 lbs (W80)		2816	1550		500 mph	56,200
AGM-28B	4	0	12.2'	81.95	42.5'	1 Pratt & Whitney J52-P-3	6077	2195	1742		10,147	674		Mach 2.01	

Aircraft	Source Grade	Crew	Span	Wing area (sq ft)	Length	Engines	Dry weight (lbs)	Design Fuel (lb)	Design Payload (lbs)	Max Payload (lbs)	Gross weight (lb)	Range (n.mi.)	Cruise speed	Max speed	Ceiling (ft)
GAM-72A	4	0	5' 4.5"	28.01	12' 10.6"	1 General Electric J85-GE-7	923.89				1230				
Tagboard D-21B	5	0	19.08'	388.5	42.85'	1 Marquardt RJ-73	5300	5900	950		11,200	3000	Mach 3.3		80-95,000
D-21B w/booster	5	0	19.08'	388.5	46.27'	1 RJ-73 + 1 Avanti		9280 (solid)			24,100			9500 mph, 300+ miles	
Skybolt, Heavyweight warhead	4	0	73"		385.95'	1 Aerojet XM80, 1 XM81		8160	1950 (RV)		12,137	1150			
B-52H AML II	3		2220"		1875'	8 TF33-P-3			~90,824						
Longbow	3	0			~20 ft	2 solid rockets					4-5,000	2,000			
Model 473-10	3	3	100 ft	1000	82' 3"	2 Nene turbojets	34,160	2620 gal	28 pax		50,000				
Model 473-13	4	3+	100 ft	1000	87 ft	4 Westinghouse 24C-10 turbojets	35,960	18880	36 pax		63,000			535 mph at 35K ft	
Model 473-14	4	3	100 ft	1000	89' 8"	4 Westinghouse 24C-10 turbojets	36,540	17,880	40 pax/ 27,380 lbs		64,900				
Model 473-23	3		116 ft	1575	100 ft	4 P&W J-57 turbojets	74,450	59,620	36 pax/ 8700 lbs		145,000	5215	500 mph at 47,000 ft		
Model 473-24	3		185' 0"	4000	134' 6"	6 P&W J-57 turbojets	144,080	102,820	78 pax/ 15,660 lbs			5175	520 mph at 53,000 ft		
Model 473-25	3														
Model 473-27	3		101.5'	1200	94.5'	4 WE-24C-10 turbojets	43,910	25,070	52 troops	20,000	90,000	2090	500 mph at 40,000 ft		
Model 473-28B	3		101.5'	1200	94.5'	4 WE-24C-10 turbojets	45,310	20,500	36 pax	1,520	75,000	1985	500 mph at 40,000 ft		
Model 473-29	3		185 ft	4000	143.5'	6 J-57 turbojets	157,520	111,600	102 pax/ 20,400 lbs		300,000	5,087	500 mph at 50,000 ft		
Model 473-29B	2		185 ft	4000	149.17'	6 J-57 turbojets	154,620	128,600	44,160		330,000	5,100	500 mph at 50,000 ft		
Model 473-30	3		146.5'	2500	114'	4 Westinghouse XJ-40-WE6	87,530	54,000	70 pax/ 14,000 lbs		170,000	4,293	500 mph at 49,000 ft		
Model 473-31	3		113' 6"	1500	103.1' (fuselage)	4 5,2000 lbf turbojets	58,340	23,400	61 pax/ 12,200 lbs		94,000	2,250	500 mph at 46,000 ft		
Model 820-100	3		200'		179' 9"	8 Pratt & Whitney J57	219,000		100,000		550,000	4,500			
Model 820-101	3		200'	4475	170' 5"	4 P&W T57 turboprops	223,000		100,000		500,000	4,580	403 knots		
Model 820-102	3		200'		168' 4"	8 turbojets									
Model 820-103	3		210'		170' 5"	8 turbojets	219,000		100,000		550,000	4,660	454 knots		
Model 820-104	3		180'		170' 5" (fuselage)	6 turboprops									
Model 820-105	3		210'		194' 9"	8 turbojets					530,000				
Model 820-106	3		210'		191'	8 turbojets					575,000				
Model 820-108	3		245'		177' (fuselage)	4 chemical turbojets 4 nuclear turbojets									
Model 820-108A	3		245'		177' (fuselage)	8 turbojets									
Model 820-109	3		225'		170' 5" (fuselage)	4 nuclear turbojets									
Model 820-110	3		161' 10"		185'	4 turbofans									
Model 820-111	3		191' 3"		210' (aft) 185' (forward)	12 turboprops									
Model 877-1	3		190' 4"	4162.3	162' 7"	8 GE MF 288 Aft Fans									
Pregnant Princess	2		220		185	8 18klbf turbofans			>200,000						
Colossal Guppy	2		~240		226	12 P&W J57									
Conroy Virtus	3		450	22,300	275	4 P&W JT9D-3A	375,000			375,000	850,000	3,000	276 mph @ 35 kft	300 mph	35,000

References

Early bomber development
Ginter, Steve 'Consolidated Vultee XB-46,' Air Force Legends Number 221, 2016
'The XB-48 Jet Bomber,' Historical Office, Air Material Command, Wright-Patterson Air Force Base, July, 1950
'Off The drawing Board: Lockheed Model L-173,' Bill Slayton, *Crazed Plastic* #27
'Brochure No. 24,' Northrop Aircraft, Inc. May, 1950
'Mother Ships, Parasites and More: Selected USAF Strategic Bomber, XC Heavy Transport and FICON Studies, 1945-1954,' Jared Zichek, 2010.
'Standard Aircraft characteristics, B-45A,' July 9, 1951
'Characteristics Summary, B-45A,' June 6, 1949
'Characteristics Summary, XB-48,' April 7, 1949
'Model Specification for Bombardment Type Airplane Model B-43 for the U.S. Army Air Forces,' Douglas Aircraft Company, April 3, 1945

B-47 development
Lloyd, Alwyn 'Boeing's B-47 Stratojet,' Specialty Press, 2005
Lloyd, Alwyn 'B-47 Stratojet in Detail & Scale Vol. 18,' Detail & Scale, 1986
Natola, Mark 'Boeing B-47 Stratojet, A Photographic History,' Schiffer Publishing, 2010
Peacock, Lindsay 'Boeing B-47 Stratojet,' Osprey Publishing, 1987
Habermehl, C. Mike and Hopkins, Robert III 'Boeing B-47 Stratojet, Strategic Air Command's Transitional Bomber,' Crecy Publishing, 2018
'Characteristics Summary, XB-47,' November 30, 1949
'Characteristics Summary, B-47A,' October 19, 1949
'Standard Aircraft Characteristics, B-47A,' June 4, 1951
'Characteristics Summary, B-47B,' February 9, 1951
'Standard Aircraft Characteristics, B-47B,' February 9, 1951
'Characteristics Summary, B-47C,' January 5, 1951
'Standard Aircraft Characteristics, B-47C,' January 5, 1951
'Characteristics Summary, XB-47D,' November 30, 1949
'Standard Aircraft Characteristics, XB-47D,' November 6, 1952
'Characteristics Summary, B-47E,' April 5, 1956
'Standard Aircraft Characteristics, B-47E,' June 7, 1955
'Characteristics Summary, B-56,' March 1, 1950
'Characteristics Summary, Boeing Model 450-166-138,' January 5, 1951
'Standard Aircraft Characteristics, Boeing Model 450-166-138,' January 5, 1951
'Characteristics Summary, DB-47E,' February 1, 1956
'Standard Aircraft Characteristics, DB-47E,' February 1, 1956
'Characteristics Summary, GAM-63A,' January 10, 1958
'Standard Missile Characteristics, GAM-63A,' December 1, 1958
'Project MX-776 RASCAL Weapon System Quarterly Progress Report,' BMPR-38, Bell Aerospace Co, September 30, 1954
'Project MX-776 Quarterly Progress Report,' Report no. 56-981-021-046, Bell Aerospace Co, September 30, 1956
Termena, B., 'History of the RASCAL Weapon System, 1952-1958,' Historical Division, Office of Information services, Air Materiel Command, Historical Study Number 321, September 1959
'Defense of North America: Final Report of Project Lamp Light,' Massachusetts Institute of Technology, NR-ORI-078, four volumes, March 15, 1955
Brown, S., Holleman, E., 'Experimental and Predicted Lateral-Directional Dynamic Response Characteristics of a Large Flexible 35° Swept-Wing Airplane at an Altitude of 35,000 Feet,' Ames Aeronautical Laboratory, NACA TN 3874, December 1956
Mayo, A., 'Flight Investigation and theoretical Calculations of the Fuselage deformations of a Swept-Wing Bomber During Push-Pull Maneuvers,' Langley Aeronautical Laboratory, NACA RM L56L05, March 19, 1957
Diederich, F. and Zlotnick, M. 'Calculated Lift Distributions of a Consolidated Vultee B-36 and Two Boeing B-47 Airplanes Coupled at the Wing Tips,' Langley Aeronautical Laboratory, NACA RM L50I26, November 30, 1950
Anderson, C.E. 'Aircraft Wingtip Coupling Experiments,' *Aerophile*, Volume 2 Number 4, December 1980

US Patent 2,981,499, 'Aircraft with Auxiliary Launching Aircraft,' Raymond B. Janney II, assignor to All American Engineering Company, filed Dec. 11, 1956

US Patent 2,921,756, 'Composite Aircraft,' Darrel C. Borden, assignor to Goodyear Aircraft Corporation, filed April 20, 1956.

'Experimental Vehicle Study and Demonstration,' Bold Orion Termination Report, November 1, 1960

B-52 evolution

'Standard Aircraft Characteristics, B-29,' April 19, 1950

'The XB-52 Airplane,' Historical Office, Executive Secretariat, Air Materiel Command, Wright-Patterson Air Force Base, August 1949

'History – Boeing Model B-52 Airplane,' Boeing Airplane Company document D-13009, March 15, 1952

Zichek, Jared 'The B-52 Competition of 1946,' American Aerospace Archive #3

Zichek, Jared 'Mother Ships, Parasites & More: Selected USAF Strategic Bomber, XC Heavy Transport and FICON Studies, 1945-1954,' American Aerospace Archive #5

'Intercontinental High Speed Strategic Type Airplanes,' Fairchild Aircraft, Report No. R 121-106, March 31, 1951

Mandeles, M. 'The Development of the B-52 and Jet Propulsion: A Case Study In Organizational Innovation,' Air University Press, Maxwell Air Force Base, Alabama, 1998

Tagg, Lori 'Development of the B-52: the Wright Field Story,' History Office, Aeronautical Systems Center, Air Force Materiel Command, 2004

Bradley, Robert 'Convair Advanced Designs, Secret Projects from San Diego 1923-1962,' Specialty Press, 2010

Bradley, Robert 'Convair Advanced Designs II, Secret Fighters, Attack Aircraft and Unique Concepts 1929-1973,' Crecy Publishing, 2013

Jenkins, Dennis 'Convair B-36 'Peacemaker' Warbird tech Series Volume 24,' Specialty Press, 1999

Jenkins, Dennis 'Magnesium Overcast The Story of the Convair B-36,' Specialty Press, 2001

Jenkins, Dennis and Pyeatt, Don 'Cold War Peacemaker, The Story of Cowtown and the Convair B-36,' Specialty Press, 2010

Jacobsen, Meyers 'Convair B-36, A Comprehensive History of America's 'Big Stick',' Schiffer Publishing, 1997

'Standard Aircraft Characteristics, YB-60,' September 21, 1951

'Standard Aircraft Characteristics, YB-60,' July 11, 1952

Buttler, Tony 'American Secret Projects 4: Bombers, Attack and Anti-Submarine Aircraft 1945-1974,' Crecy Publishing, 2021

Lee, Ben 'Next Step in Bombers: B-36F or XB-52?' Aviation Week, November 20, 1950

'Consolidated Vultee's Big Eight-Jet Version of the B-36 Intercontinental Bomber,' Aviation Week, March 26, 1951

'Heavy Bombardment Airplane: Gas Turbine Propelled,' Consolidated Vultee Aircraft Corporation, ca. 1945

'Navy Search Landplane,' Consolidated Vultee Aircraft Corporation, ca. 1945

'Physical Data for Target Airplane Model G-2 High Subsonic All-Wing Bomber,' Project RAND, RM-62, June 30, 1949

'Standard Aircraft Characteristics, B-36D-III,' August 1, 1955

'Standard Aircraft Characteristics, YB-49,' December 20, 1949

'Characteristics Summary, YB-49,' May 4, 1949

'Long Range Bombardment Airplane M.C.D. 392,' Aircraft Laboratory, Engineering Division, AAF materiel Command, 1944

'Medium Bomber Proposals Project MX-948,' Boeing Airplane Company, January 1948

Convair's Swept Wing B36 Program,' Convair, FZP-36-006A, November 11, 1950

General B-52 references

Holder, W., Woodside, R. 'Boeing B-52 Stratofortess,' 2nd edition, Tab Books, 1988

Bowers, Peter 'Boeing Aircraft since 1916,' Putnam Aeronautical Books, 1989

Drendel, Lou 'B-52 Stratofortress Illustrated,' Aviation Art, 2018

Davies, Steve 'Boeing B-52 Stratofortress 1952 onwards (all marks),' Haynes Publishing, 2013

Wachsmuth, Wayne 'B-36 Peacemaker in detail & scale,' Kalmbach Books, 1995

Yenne, Bill 'B-52 Stratofortress, The Complete History of the World's Longest Serving and Best Known Bomber,' Zenith Press, 2012

Holder, William 'B-52 Boeing 'Stratofortress', Aero Publishers, 1975

Lloyd, Alwyn 'B-52 Stratofortress in detail & scale,' TAB Books, 1988

Doyle, David 'B-52 Stratofortress, Boeing's Iconic Bomber from 1952 to the Present,' Schiffer Publishing, 2018

Katz, Kenneth 'B-52G/H Stratofortress In Action,' Squadron/Signal Publications, 2012

Drendel, Lou 'Walk Around B-52 Stratofortress,' Squadron/Signal Publications, 1996

Drendel, Lou 'B-52 Stratofortress In Action,' Squadron/Signal Publications, 1975

Jenkins, Dennis and Rogers, Brian 'Boeing B-52G/H Stratofortress, Aerofax Datagraph 7,' Aerofax, 1990
Bowers, Peter 'Boeing B-52A/H Stratofortress, Aircraft Profile 245,' Profile Publications, 1972
Boyne, Walter 'Boeing B-52 A Documentary History,' Schiffer Publishing, 1994
Dorr, R., Peacock, L., 'B-52 Stratofortress Boeing's Cold War Warrior,' Oprey Publishing, 1995
Davies, S., 'Boeing B-52 Stratofortress Owners Workshop Manual,' Haynes Publishing, 2013
Montulli, L. 'Lessons Learned from the B-52 Program Evolution: Past, Present and Future,' Boeing Military Airplane company, AIAA-86-2639, October, 1986
Knaack, M., 'Encyclopedia of U.S. Air Force Aircraft and Missile Systems,' Office of Air Force History, 1988

B-52 development

'High Altitude Heavy Bomber Study, Model MX-2145,' Boeing Document D-14822, January 8, 1954
'Standard Aircraft Characteristics, XB-52,' October 6, 1950
'Characteristics Summary, XB-52,' October 6, 1950
'Standard Aircraft Characteristics, B-52B,' October 1, 1958
'Standard Aircraft Characteristics, B-52BC and D,' March 24, 1958
'Standard Aircraft Characteristics, B-52D,' October 5, 1954
'Characteristics Summary, B-52D,' October 5, 1954
'Standard Aircraft Characteristics, B-52E,' October 1, 1958
'Characteristics Summary, B-52E,' July 1964
'Standard Aircraft Characteristics, B-52F,' November 16, 1959
'Characteristics Summary, B-52F,' August 1964
'Standard Aircraft Characteristics, B-52G,' November 20, 1958
'Characteristics Summary, B-52G,' May 1961
Comassar, S. 'Comprehensive Technical Report General Electric Direct-Air-Cycle Aircraft Nuclear Propulsion Program: Aircraft Nuclear Propulsion Application Studies,' General Electric Flight Propulsion Laboratory Department, APEX-910, April 30, 1962
Hutton, J., McCulloch, J., Schmill, W., Ward, W., 'Studies of Fourteen Nuclear-Powered Airplanes,' Unites States Atomic Energy Commission, NEPA Division, NEPA-1639, September 1952
'NEPA Project, Quarterly Project Report for the Period April 1-June 30 1950,' NEPA Division, Fairchild Engine & Aircraft Corporation, Report No. NEPA 1484
'Preliminary Study Standby Alert Bomber,' Boeing Aircraft company, D2-3420, July 1958

Re-engining the B-52

Banach, H. 'Boeing-Wichita B-52X Studies,' Pratt & Whitney Interoffice Communication, July 25, 1975
Altman, R., Flynn, E. '727, B-52 Retrofit with PW 2037… Meeting Today's Requirements,' Pratt & Whitney Aircraft, SAE Paper 821443, October 1982
Santamaria, J., Sears, W., Boruff, W. 'Installation of C-17 Nacelle & Engine on B-52G,' LTV Aircraft Products Group, May 1989
Crandall, R. 'Flight Testing the C-5A/TF39 Power Plant Installation on the B-52 Flying Test Bed,' General Electric Company, AIAA paper 68-592, June 1968
Landis, Tony 'New Power for an Old Soldier—Re-engining the B-52 Stratofortress,' AFMC History & Museums Program

Boeing NB-52E

Hodges, Garold 'Active Flutter Suppression – B-52 Controls Configured Vehicle,' The Boeing Company, AIAA Paper 73-322, March 1973
Arnold, J. and Murphy, F. 'B-52 Control Configured Vehicles: Flight Test Results,' The Boeing Company, 1976
Hodges, G. and McKenzie, J. 'B-52 Control Configured Vehicles Maneuver Load Control System Analysis and Flight Test Results,' The Boeing Company, AIAA Paper 75-72
'B-52E CCV Flight Test Data Applicable to Parameter Estimation,' The Boeing Company, AFFDL-TR-75-131, December, 1975
Burris, P., Bender, M., 'Aircraft Load Alleviation and Mode Stabilization (LAMS), B-52 System Analysis, Synthesis, and Design,' The Boeing Company, AFFDL-TR-68-161, November 1969
Burris, P., Bender, M., 'Aircraft Load Alleviation and Mode Stabilization (LAMS) Flight Demonstration Test Analysis,' The Boeing Company, AFFDL-TR-68-164, December 1969
Arnold, J., Murphy, F. 'B-52 Control Configured Vehicles: Flight Test Results,' The Boeing Company
McKenzie, J., 'B-52 Control Configured Vehicles Ride Control Analysis and Flight Test,' The Boeing Company, AIAA Paper 73-782, August 1973

Proposed Payloads

HYWARDS

'Study of the Feasibility of a Hypersonic Research Airplane,' NACA report, Washington, D.C, September 3, 1957

Junkers Raumtransporters

Sharp, Dan 'British Secret Projects 5: Britain's Space Shuttle,' Crecy Publishing, 2016
Jahrbuch 1987 I, der Deutschen Gesellschaft fur Luft- und Raumfahrt e. V. (DGLR)

Jahrbuch 1990 II, der Deutschen Gesellschaft fur Luft- und Raumfahrt e. V. (DGLR)

ISINGLASS

'Flight Control of the Model 192,' CIA-RDP71B00265R000200130005-0

'Basic Concepts of Hypersonic Lifting Vehicles,' CIA-RDP71B00265R000200130015-9

'ISINGLASS,' CIA-RDP68B00724R000100070051-8

X-15A-3

'Technical Proposal for a Conceptual Design Study for the Modification of an X-15 Air Vehicle to a Hypersonic Delta Wing Configuration,' North American Aviation report NA-67-344 Volume 1, 17 May 1967.

Dyna Soar

Bellman, D., Washington, H. 'Preliminary Performance Analysis of Air Launching Manned Orbital Vehicles,' NASA Technical Memorandum X-636, 1962

'System 464L: Project Dyna-Soar, General Management Proposal' ER 10132P, March 24, 1958, Martin Co.

'General Management Proposal Project Dyna-Soar System 464L' Republic Aviation Corporation, MCXH-12317, March 1958

'DS-I Airborne Vehicle Design Summary,' Boeing Airplane Co, D5-4210, March 1959

Bellman, D., Washington, H., 'Preliminary Performance Analysis of Air Launching Manned Orbital Vehicles,' NASA-Langley, TM X-636, 1962

'Specification for Modification of the B-52 for the Dyna-Soar glider Air-Launch Program,' Boeing Airplane Company, D2-8135-1, June 1961

'Standard Aircraft Characteristics, X-20A,' March 1964

'Weight Analysis Report, Model X-20,' The Boeing Company, D2-81264-1, July 1964

Convair MC-10

'Project Hazel Propulsion, Structural Heating and Pressurization,' Report No. ZJ-026, Convair, October 31, 1958

'Project Hazel Aerodynamics,' Report No. ZA-282, Convair, October 1958

'Project Hazel Aircraft Design,' Report No. ZP-253, Convair, October 1958

'Project Hazel Aircraft Design Continued Studies,' Report ZP-266, Convair, March 1959

Flashback

'Subsonic Wind Tunnel test of the FBTV Configuration in Proximity to the B-52,' Cornell Aeronautical Laboratory, Report No. I36-023-1, December 1966

'Electromagnetic Radiation Test on the Flashback Test Vehicle,' Sandia Labs, April 19, 1965

NB-52B

Lockett, Brian 'Balls Eight, History of the NB-52B Stratofortress Mothership,' Lockett Books, 2009

X-15

Guenther, B., Miller, J., Panopalis, T., 'North American X-15/X-15A-2' Aerofax Inc, 1985

F-15 RPRV

Iliff, K., Maine, R., Shafer, M., 'Subsonic Stability and Control Derivatives for an Unpowered, Remotely Piloted 3/8-Scale F-15 Airplane Model Obtained from Flight Test,' Flight Research Center, NASA TN D-8136, 1976

Petersen, K., 'Evaluation of an Envelope-Limiting Device Using Simulation and Flight Test of a Remotely Piloted Research Vehicle,' Dryden Flight Research Center, NASA TN D-8216, April 1976

Thompson, Milton 'Flight Research: Problems Encountered and What They Should Teach Us,' NASA History Division, NASA SP-2000-4522, 2000

Pegasus

Isakowitz, Steven 'International Reference Guide to Space Launch Systems,' American Institute of Aeronautics and Astronautics, 1991

Always, Peter 'Rockets of the World,' Saturn Press, 1999

X-43

Marshall, L., Bahm, C., Corpening, G., Sherrill, R. 'Overview With Results and Lessons Learned of the X-43A Mach 10 Flight,' NASA-Dryen Flight Research Center/NASA-Langley Research Center, AIAA Preprint, 2005

Peebles, Curtis 'The X-43A Flight Research Program: Lessons Learned on the Road to Mach 10'

Harsha, Dr. P., Keel, L., Castrogiovanni, Dr. A., Sherrill, R. 'X-43A Vehicle Design and Manufacture,' AIAA, 2005

Peebles, Curtis 'Road to Mach 10: Lessons Learned from the X-43A Flight Research Program,' American Institute of Aeronautics and Astronautics, 2008

Joyce, P., Pomroy, J., Grindle, L., 'The Hyper-X Launch Vehicle: Challenges and Design Considerations for Hypersonic Flight Testing,' Orbital Sciences Corporation/NASA, AIAA 2005-3333

M2-F2/HL-10/X-38

McKinney, L., Boyden, R. 'Predicted Characteristics of M2-F2 Lifting Body Launched from B-52 Airplane with Comparison of Flight Results,'

Langley Research Center, NASA TM X-1514 April 1968

McKinney, L., Boyden, R. 'Predicted Launch Characteristics of the HL-10 Manned Lifting Entry Vehicle Launched From the B-52 Airplane,' Langley Research Center, NASA TM X-1668, November 1968

Anderson, Fred 'Northrop An Aeronautical History,' Northrop Corporation, 1976

Kempel, R., Painter, W., Thompson, M. 'Developing and Flight Testing the HL-10 Lifting Body: A Precursor to the Space Shuttle,' NASA Reference Publication 1332, April 1994

Reed, Dale 'Wingless Flight: The Lifting Body Story,' NASA SP-4220, 1997

Miller, Jay 'The X-Planes X-1 to X-45,' Midland Publishing, 2001

Kock, B., Painter, W., 'Investigation of the Controlablity of the M2-F2 Lifting-Body Launch from the B-52 Carrier Airplane,' Flight Research Center, NASA TM X-1713, December 1968

Petty, Chris 'Beyond Blue Skies: The Rocket Plane Programs That Led to the Space Age,' University of Nebraska Press, 2020

Reed, R., 'Wingless Flight: The Lifting Body Story,' University Press of Kentucky, 2002

DTV

Moog, R., Sheppard, J., Kross, D. 'Space Shuttle Solid Rocket Booster Decelerator Subsystem Drop Test Results,' Martin –Marietta, NASA-Marshall, AIAA paper 79-0463

'B-52B/DTV (Drop Test Vehicle) Flight Test Results – Drop Test Missions,' The Boeing Company, D500-10855-1, 5-14-1985

'Load and dynamic Assessment of B-52B-008 carrier Aircraft for Finned Configuration 1 Space Shuttle Solid rocket Booster Decelerator Subsystem Drop Test Vehicle,' Boeing Company, D3-11220-2, June 1978

DAST

McGehee, C. 'Design Verification and Fabrication of Active Control Systems for the DAST ARW-2 High Aspect Ratio Wing, Part 1,' The Boeing Co., NASA CR-177959, January 1986

Grose, D. 'The Development of the DAST I Remotely Piloted Research Vehicle for Flight Testing an active Flutter Suppression Control System,' University of Kansas, NASA CR-144881, February 1979

Eckstrom, C., 'Loads Calibrations of Stain Gage Bridges on the DAT Project Aeroelastic Research Wing (ARW-2),' NASA-Langley Research Center, NASA TM-87677, March 1986

Kotsabasis, A., 'The DAST-I Remotely Piloted research Vehicle Development and Initial Flight Testing,' University of Kansas, NASA CR-163105, February 1981

Eckstrom, C., 'Flight Measurements of Surface Pressures on a Flexible Supercritical Research Wing,' NASA-Langley Research Center, NASA TP-2501, December 1985

Hodges, G., McGehee, C. 'Final Design and Fabrication of an Active Control System for Flutter Suppression on a Supercritical Aeroelastic Research Wing,' The Boeing Co., NASA CR-165714, June 1981

HiMAT

Bellman, D., Kier, D. 'HiMAT – A New Approach to the Design of Highly Maneuverable Aircraft,' NASA, SAE paper 740859, October 1974

Brown, L. Jr., Roe, M., Quam, R. 'HiMAT Systems Development Results and Projections,' North American Aircraft Division, Rockwell International, SAE Paper 801175, October 1980

Matheny, N., Panageas, G. 'HiMAT Aerodynamic Design and Flight Test Experience,' NASA-Dryden, Rockwell International, AIAA Paper 81-2433, November 1981

Duke, E., Jones, F., Roncoli, R., 'Development and Flight Test of an Experimental Maneuver Autopilot for a Highly Maneuverable Aircraft,' Ames Research Center, NASA Technical Paper 2618, 1986

X-24C

'Experiments Impact on X-24C Flight Research Vehicle – Final Oral Review,' Martin-Marietta MCR-75-144, April 1975

Kirkham, F., Jones, R., Buck, M., Zima, W. 'Joint USAF/NASA Hypersonic Research Aircraft Study,' NASA-Langley, AFFDL, AIAA paper 75-1039, August, 1975

Jackson, L., Taylor, A. 'A Structural Design for the Hypersonic X-24C Research Aircraft,' NASA-Langley, AIAA paper 76-906, September 1976

Draper, A., Sieron, T., 'Evolution and Development of Hypersonic Configurations 1958-1990,' Flight Dynamics Directorate, Wright Laboratories, WL-TR-91-3067, September 1991

HRA

Kirkham, F., Jackson, L., Weidner, J., 'Study of a High-Speed Research Airplane,' NASA-Langley, Journal of Aircraft, November 1975

Van Camp, V., Williams, E., 'Hypersonic Research Airplane Propulsion for Boost and Test,' Rockwell International Corporation, AIAA Paper 74-990, August 1974

Van Camp, V., Williams, E., 'Propulsion Options for the Hypersonic Research Airplane,' Rockwell International Corporation, Journal of Aircraft, July 1975

D-21B

Bradley, A. 'Manufacturers Model Specification (D-21),' Lockheed Aircraft Corporation, Advanced Development Projects, SP-582, November 6, 1963
'Systems Descriptions (D-21),' Lockheed Aircraft Corporation, Advanced Development Projects, SP-790 Vol. 1, January 1, 1968

GAM-72A

'Contract Technical Compliance Inspection GAM-72A,' McDonnell Aircraft, September, 1960

GAM-77

'GAM-77 and GAM-77A Station Diagrams,' Space and Information Systems Division, North American Aviation, January 1961
'AGM-28A Characteristics Summary,' June 1972
'AGM-28A Standard Missile Characteristics,' August 1972
'AGM-28B Characteristics Summary,' January 1976
'AGM-28B Standard Missile Characteristics,' January 1977

GAM-87

'Status Report by Aircraft Installation Team Visiting USA from March 29, 1960, to April 12, 1960,' AH/600/02
'GAM-87A Pylon/Launcher/Thermo-Conditioning System Maintainability Design Review,' Douglas Aircraft Company, April 1962
Worman, C., 'History of the GAM-87A Skybolt Air-To-Surface Ballistic Missile, Volume 1' Historical Division, Information Office, Aeronautical Systems Division, March 1967
Cox, G., Kaston, C., 'American Secret Projects 3: US Airlifters since 1962,' Crecy Publishing, 2020

AGM 86

Werrell, Kenneth 'The Evolution of the Cruise Missile,' Air University Press, Maxwell Air Force Base, Alabama, 1985

Longbow

'Two New Ballistic Missiles Scrutinized,' Aviation Week, January 29, 1979

Transports

'Development Program for a Long-Range Military Air-Logistics System,' Boeing Airplane Company, D2-3022, 1958
'Super Guppy Modification of princess Flying Boat,' Aviation Week, February 10, 1964
'Modified B-52 proposed for Outsize Cargo,' Aviation Week, November 14, 1966
'Feasibility Study to Consider an Aircraft for the Air Launch and Air Transportation of the Space Shuttle Orbiter,' Turbo-Three Corporation, NASW-2627, February 28, 1974

Index

ADM-20 Quail 176, 187, 228, 261, 264, 271, 273
ADM-120 187
Aerojet
    15KS-1000 45
    YLR45-AJ-1 48
Aero Spacelines
    Colossal Guppy 367
    Mini Guppy 364
    Pregnant Guppy 364
    Pregnant Princess 364, 367
    Super Guppy 364
AGM-28 Hound Dog 174, 176, 187, 264, 267, 270
Airbus A220 314
Air Force Flight Dynamics Laboratory 234, 291, 307
All American Engineering 'Donkey' 92
Allegany Ballistics Laboratory Altair 69
Allison
    J35 7, 32, 43, 51, 95
    J71 51
    Model 500 aka T40 56, 60, 63, 208
    T56 56
    XT 40 33
Arnold, Hap 7
Atlas ICBM 236, 243, 276, 290, 337
Avro Canada Arrow 76
Avro Lancaster 33, 106

Bell
    BOMI 236, 237, 243
    Brass Bell 236, 237, 243
    HYWARDS 236, 237
    LR-81 Agena 313
    P-59 Airacomet 7
    ROBO 236, 237
    X-1 43
    X-2 236
    XLR67-BA-1 69
    XS-1 236
Betz, Albert 19
Boeing
    707 182, 231, 319, 323, 364
    737 319
    747 231, 369
    757 319
    B-17 Flying Fortress 7, 76, 106, 323
    B-29 Superfortress 7, 9, 15, 22, 33, 100, 102, 106, 109, 112, 114, 123, 130, 142, 208, 236, 280, 323
    B-29 Turboprop 114
    B-47 Stratojet 15, 19, 22, 30, 33, 36, 45, 48, 51, 56, 60, 63, 68, 69, 71, 73, 76, 77, 81, 83, 89, 92, 95, 99, 109, 118, 130, 131, 134, 135, 137, 139, 142, 144, 159, 163, 165, 167, 170, 174, 188, 192, 225, 259, 267, 276, 323, 325, 330, 332, 335, 337
    B-50 Superfortress 69, 142, 236, 280
    B-52A/General Electric LF-2 219
    B-52H AML II 278, 364
    B-52X 314
    B-55 142, 144
    B-59 81, 83, 199
    C-17 Globemaster III 314, 319
    C-97 Stratofreighter 323
    C-108 Flying Fortress 323
    CB-17 323
    CIM-10 / IM-99 BOMARC 268, 338
    DB-47A 76
    DB-47B 69
    DB-47E 53, 69
    EB-47E 68
    'Hotel' Design 137, 139
    JB-52E 231
    KC-97 Stratofreighter 63, 208, 323
    KC-135 Stratotanker 176, 208, 276, 338
    MB-47B 76
    Minuteman 276, 278, 292, 303
    Model 307 Stratoliner 323
    Model 334A 100
    Model 341 100
    Model 345 100
    Model 360 102
    Model 361 102
    Model 362 102
    Model 365 102, 106
    Model 367-1-1 323
    Model 367-80 208, 323
    Model 377 323, 364
    Model 384 106
    Model 385 106
    Model 413 9, 28, 30
    Model 422 9, 15
    Model 424 9, 15

Model 1126 39
Model 1155 157
Model 1211-J 157, 159
X-3 Stiletto 157
XB-42 Mixmaster 28, 157
XB-43 Jetmaster 25, 28
YB-43 28
Drones for Aerodynamic and Structural Testing (DAST) 293, 294

Edwards Air Force Base 243, 282, 288-291, 295
Eglin Field, Florida 109
Emerson A-2 45, 48
Evergreen Aviation Museum 300
Explorer VI 71

Fairchild
    M-121 150, 154-157
    NEPA N-14 216, 219
    NEPA N-15 216, 219
    NEPA N-16 219
    XB-52 NEPA 1 212, 213
    XB-52 NEPA 2 212, 213
Fairfield-Suisun Army Air Field, California 109
Flashback 231, 234
Focke-Wulf 1000 x 1000 x 1000 19, 25
Fort Worth airfield, Texas 109

GAM-63 RASCAL aka B-63 53, 56, 69, 73, 125, 127, 236
GAM-67 Crossbow aka B-67 73
GAM-72 (see ADM-20)
GAM-77 (see AGM-28)
GAM-87 Skybolt 182, 249, 267, 276, 278, 280, 355, 364
Gemini 284, 364
General Electric
    A-5 48
    CF34-10 322
    J31 7
    J33 7
    J47 32, 43, 48, 63, 76, 109, 335
    J73 81, 219
    J85 228, 261, 264, 294
    MF 288 355
    Passport 322
    TF39 231
    TG-180 7, 9, 15, 20, 28, 30, 43, 120
    X-84 199
    YJ-85 228
Gloster E.28/39 7
Gloster Meteor 7
Grumman X-29 120

Hamburger Flugzeugbau HFB 320 Hansa Jet 120
Handley-Page Victor 174
Heinkel He 111 323

Heinkel He 178 7
Hermann Göring Research Institute, Völkenrode, Germany 15
HL-20 Personnel Launch System 298
Hughes F-11 9
Hughes H-4 102
Hussain, Saddam 187

Ilyushin Il-28 30
International Space Station (ISS) 298
Johnson, Kelly 267, 270
Johnson, Lady Bird 167
Johnson, Lyndon B 174, 175
Johnson Space Center 298, 300
Junkers
    Raumtransporter 261
    RT 7 259, 261
    RT 8 257, 259, 261

Kennedy, John F 224
Kennedy Space Center 298, 367
Kirtland Air Force Base, New Mexico 231

LeMay, Curtis 51, 69, 144, 167
LGM-118 Peacekeeper ICBM 278
Lippisch, Alexander 150
Lockheed
    A-12 256, 267, 268
    C-5 Galaxy 231, 292, 338, 347, 369
    C-130 Hercules 56, 249, 270, 290, 335, 338
    C-141 355, 364
    CL-400 Suntan 205
    CL-839-28 304
    D-21 267, 268, 270, 271, 304
    F-117 182
    L-1011 Tri Star 298, 303
    L-173 36
    M-21 268, 270, 271
    P-38 Lightning 208
    SR-71 256, 257
    STAR Clipper 261
    TF-104G 294, 295
    U-2 99, 231
    UGM-27 Polaris 276, 278
    XFV 56
    YF-12 256
Lockheed Martin F-35 298
Lockheed Propulsion Company A-92 'Avanti' 270
Longbow 278, 280

M2-F1 284, 288-290
Marquardt RJ-73 Model MA20S-4 268
Martin
    Model 223 aka XB-48 33
    Model 236 120, 123

Model 247-1 33
P6M Sea Master 89
SV-5 290, 291
Titan I 234, 243, 249, 304
Titan II 231, 249
Titan III 246, 249, 253
Titan IV 298
X-23A PRIME 290
Martin-Bell 464L CTV 237
Martin Marietta Pershing II 278, 280
Martin Marietta X-24A 282, 290, 291, 298, 307
Martin Marietta X-24B 291, 307, 313
Martin Marietta X-24C 307, 313
MCD 392 112
McDonnell Model 192 ISINGLASS aka RHEINBERRY 257
McDonnell Douglas DC-10 231
McDonnell Douglas F-15 RPRV 291, 292, 294, 295
Mercury 284
Mikoyan-Gurevich MiG-21 175
MIT Project LAMP LIGHT 89
MX-776 69
MX-839 124
MX-948 33, 36, 39
MX-1018 99
MX-1712 81, 83, 144
MX-1964 81
MX-1965 81, 144, 199
MX-2145 219
MX-2276 236
Myasishchev M-90 369

National Advisory Committee for Aeronautics (NACA)
  NACA Ames M1 284
  NACA Langley 45
National Aeronautics and Space Administration (NASA)
  NASA Ames 288-290
  NASA Dryden 284, 288, 290, 292, 293, 303
  NASA Langley 284, 288, 289, 291, 293, 294, 298, 303, 369
  NASA Marshall Space Flight Center 364
NASA X-30 National AeroSpace Plane (NASP) 257, 300, 303
NASA X-38 298, 300
NASA X-43A (aka Hyper-X) 300, 303
National Air and Space Museum 282, 289, 295
National Museum of the United States Air Force 282, 290
Nixon, Richard 175
North American
  A-5 Vigilante 150
  B-45 Tornado 30, 33
  D435-1-4 304
  DF-100F 76
  P-51 Mustang 30, 95
  RB-45C 30
  SM-64 Navaho 89, 157, 264
  T-39 Sabreliner 288, 289
  X-15 236, 237, 243, 246, 249, 257, 280, 282, 288, 290-295, 300, 303, 307
  X-15A2 304
  X-15A3 257, 270, 304
  XA2J Super Savage 56
  XB-45 30
  XB-70 174, 176, 237, 246, 261
  XSSM-A-2 157
Northrop
  B-2 Spirit 187
  B-35 36, 102
  B-49 136, 63
  F-5 228, 288, 289
  HL-10 288-291
  M2-F2 282, 288-290
  M2-F3 288-290, 295
  N-31 36
  N-31A 36
  T-38 Talon 228, 288, 289
  Turbodyne V 36
  X-21 288
  X-216H Flying Wing 102
  XSSM-A-3 Snark 157
  YB-49 130
Northrop-Hendy T-37 159

OCS Pegasus 295, 298, 303
Operation Buster-Jangle 30
Operation Desert Storm 182, 273
Operation Enduring Freedom 187
Operation Iraqi Freedom 187
Operation Paddlewheel 231
Operation Paperclip 95
Orenda Iroquois 73, 76

Peterson, Bruce 288
Piasecki CH-21 288
PILOT 290
Pima Air Museum 282
Point Mugu Sea Test Range 243
Pontiac Catalina 284
Power Jets W.1 7
Power Jets W.2B/23 (see Rolls-Royce Welland I)
Pratt & Whitney
  J52 264, 267
  J57 51, 53, 55, 60, 89, 92, 142, 159, 165, 170, 175, 182, 205, 224, 225, 228, 231, 267, 319, 330, 332, 367
  J75 176, 199, 205, 208, 228, 231, 319
  JT9D-70D 314, 369
  JT10D-2 314

JT90 231
PT-2E aka T34 63
PW207 319
PW800 322
PW1000G 314
PW2000 314
PW2037 319
PW2040 319
R-4360 Wasp 106
TF33 182, 231, 314, 322
XJ-57 135, 139, 213
XLR-129 257
X Wasp 102
Project Aphrodite 76
Project BRASS RING 76
Project WEARY WILLIE II 76

RAND Corporation G-2 159
Reaction Motors
    XLR11 282, 288
    XLR99 237, 288
Redwing Cherokee 170
Republic
    464L CTV 243
    AP-42 36
    EF-84D 99
    F-12 Rainbow 9
    F-105 Thunderchief 76
    XGF-84H Thunderscreech 60
Rocketdyne AR-2-1 243
Rockwell
    B-1 Lancer 187, 271, 273, 278, 314
    HiMAT 294, 295
    HRA 313
Rolls-Royce
    Nene 325
    Olympus 199
    Tyne 364
    Welland I 7
Ryan
    BQM-34E/F Firebee II 293-295

Fireball 7

SA-2 Guideline 175
SALT II 273
Sänger, Eugen 236, 257
Saturn 284, 298, 364
Saunders-Roe Princess 364, 367
Silverbird 236, 237, 257, 261
Soyuz 284, 298
Space Shuttle Solid Rocket Booster Drop Test Vehicle 292
Sputnik 228, 237
Stalin, Joseph 112
START 290
Strategic Air and Space Museum 300
Stratolaunch Roc 102, 298
Symington, Stuart 130, 163

Thiokol
    TX-20 Sergeant 69
    XM-92 249, 253
Tirpitz 114
Turbo-Three Virtus 367, 369

University of Göttingen 19

Vickers Valiant 174
Vogt, Richard 95
Vought TF-8A Supercritical Wing 293

Westinghouse
    19XB-2A 28
    J40 48, 51, 95, 225
    XJ40-WE-6 134, 135
World Trade Center 187
Wright Air Development Center 219
Wright Field Aircraft Laboratory 9, 95, 112, 118, 124, 137, 142, 155, 188, 192, 276
Wright-Patterson Air Force Base 231

Yeager, Chuck 43